Recent Innovations in Biotechnology

Recent Innovations in Biotechnology

Edited by Wendell Carter

SYRAWOOD
PUBLISHING HOUSE

New York

Published by Syrawood Publishing House,
750 Third Avenue, 9th Floor,
New York, NY 10017, USA
www.syrawoodpublishinghouse.com

Recent Innovations in Biotechnology
Edited by Wendell Carter

© 2019 Syrawood Publishing House

International Standard Book Number: 978-1-68286-675-7 (Hardback)

Cataloging-in-Publication Data

Recent innovations in biotechnology / edited by Wendell Carter.
    p. cm.
Includes bibliographical references and index.
ISBN 978-1-68286-675-7
1. Biotechnology. 2. Biotechnology industries--Technological innovations.
3. Genetic engineering. I. Carter, Wendell.
TP248.2 .R43 2019
660.6--dc23

TABLE OF CONTENTS

Permissions

List of Contributors

Index

PREFACE

I am honored to present to you this unique book which encompasses the most up-to-date data in the field. I was extremely pleased to get this opportunity of editing the work of experts from across the globe. I have also written papers in this field and researched the various aspects revolving around the progress of the discipline. I have tried to unify my knowledge along with that of stalwarts from every corner of the world, to produce a text which not only benefits the readers but also facilitates the growth of the field.

Biotechnology integrates the fields of biology and engineering. It uses biological systems and models to develop various products or processes for industrial purposes. Modern developments in the field of biotechnology have also facilitated advancements in the field of medicine. Principles of biotechnology are also applied in areas like genetics, immunology, etc. This book contains some path-breaking studies in the field of biotechnology. It traces the progress of this field and highlights some of its key concepts and applications. The various studies that are constantly contributing towards advancing technologies and evolution of this field are examined in detail. With state-of-the-art inputs by acclaimed experts of this field, this book targets students and professionals.

Finally, I would like to thank all the contributing authors for their valuable time and contributions. This book would not have been possible without their efforts. I would also like to thank my friends and family for their constant support.

Editor

Metabolic engineering of *Escherichia coli* for the production of hydroxy fatty acids from glucose

Yujin Cao[1], Tao Cheng[1], Guang Zhao[1], Wei Niu[2], Jiantao Guo[2], Mo Xian[1*] and Huizhou Liu[1*]

Abstract

Background: Hydroxy fatty acids (HFAs) are valuable chemicals for a broad variety of applications. However, commercial production of HFAs has not been established so far due to the lack of low cost routes for their synthesis. Although the microbial transformation pathway of HFAs was extensively studied decades ago, these attempts mainly focused on converting fatty acids or vegetable oils to their hydroxyl counterparts. The use of a wider range of feedstocks to produce HFAs would reduce the dependence on oil crops and be expected to cut down the manufacturing cost.

Results: In this study, the industrially important microorganism *Escherichia coli* was engineered to produce HFAs directly from glucose. Through the coexpression of the acetyl-CoA carboxylase (ACCase) and the leadless acyl-CoA thioesterase ('TesA), and knockout of the endogenous acyl-CoA synthetase (FadD), an engineered *E. coli* strain was constructed to efficiently synthesize free fatty acids (FFAs). Under shake-flask conditions, 244.8 mg/L of FFAs were obtained by a 12 h induced culture. Then the fatty acid hydroxylase (CYP102A1) from *Bacillus megaterium* was introduced into this strain and high-level production of HFAs was achieved. The finally engineered strain BL21ΔfadD/pE-A1'tesA&pA-acc accumulated up to 58.7 mg/L of HFAs in the culture broth. About 24 % of the FFAs generated by the thioesterase were converted to HFAs. Fatty acid composition analysis showed that the HFAs mainly consisted of 9-hydroxydecanoic acid (9-OH-C10), 11-hydroxydodecanoic acid (11-OH-C12), 10-hydroxyhexadecanoic acid (10-OH-C16) and 12-hydroxyoctadecanoic acid (12-OH-C18). Fed-batch fermentation of this strain further increased the final titer of HFAs to 548 mg/L.

Conclusions: A robust HFA-producing strain was successfully constructed using glucose as the feedstock, which demonstrated a novel strategy for bioproduction of HFAs. The results of this work suggest that metabolically engineered *E. coli* has the potential to be a microbial cell factory for large-scale production of HFAs.

Keywords: Hydroxy fatty acid, *Escherichia coli*, Fatty acid hydroxylase, Acetyl-CoA carboxylase, Acyl-CoA thioesterase, Acyl-CoA synthetase

Background

The depletion of the earth's fossil energy resources and global climate change have stimulated us to develop environmentally friendly processes to produce fuels and chemicals. Hydroxy fatty acids (HFAs) are important fine chemicals which have a hydroxyl group in the carbon chain of fatty acids. Due to their unique attributes, HFAs have wide applications in different fields such as surfactants, lubricants, cosmetics or antimicrobials [1, 2]. They are also used as the intermediates for the production of a variety of value-added products [3]. More importantly, HFAs could serve as the precursors for the preparation of the next generation plastics, polyhydroxyalkanoates (PHAs) [4]. PHAs are completely biodegradable and possess good thermoplastic or elastomeric properties. Therefore, PHA bioplastics offer an alternative to conventional petrochemical-derived plastics [5].

Now, HFAs are commercially unavailable due to the lack of low cost routes for their synthesis. Chemical

* Correspondence: xianmo@qibebt.ac.cn; liuhuizhou@qibebt.ac.cn
[1]CAS Key Laboratory of Biobased Materials, Qingdao Institute of Bioenergy and Bioprocess Technology, Chinese Academy of Sciences, Qingdao 266101, China
Full list of author information is available at the end of the article

catalysts for specific hydroxylation reactions on the selective carbon atom of the fatty acyl chain are limited [6]. On the other hand, HFAs make up an interesting group of natural compounds among plants, bacteria, yeasts and fungi. A number of microorganisms capable of producing HFAs from fatty acids or vegetable oils have been isolated. For example, *Bacillus pumilus* could hydroxylate oleic acid on the 1, 2, and 3 carbon atoms to produce hydroxy oleic acids [7]. *Candida tropicalis* also excretes HFAs as by-products when cultured on n-alkanes or fatty acids as the carbon source [8]. Enzymes catalyzing the bioconversion of fatty acids to HFAs have been identified as the cytochrome P450 monooxygenases (CYPs). CYPs responsible for the hydroxylation of fatty acids have been cloned from several *Bacillus* species including *B. megaterium* [9], *B. subtilis* [10], *B. anthracis* [11] and *B. cereus* [12]. The CYP102A1 from *B. megaterium* is the most thoroughly studied member of these enzymes. Heterologous expression of this enzyme in *E. coli* indicated that the whole-cell biocatalyst showed the maximum activity to pentadecanoic acid and the resulting products were only 1, 2 and 3 HFAs [13]. This bioconversion has been demonstrated at the 2 L scale fermentor level under oxygen limitation, showing that 12-, 13-, and 14-hydroxypentadecanoic acids can be produced in the g/L range [14]. Recombinant *E. coli* cells harboring another fatty acid hydroxylase P450foxy from the fungus *Fusarium oxysporum* [15] could also convert saturated fatty acids with a chain length of 7–16 carbon atoms to their 1, 2 and 3 hydroxyl derivatives [16].

The above studies used fatty acids or their derivatives as the feedstocks for production of HFAs. Compared with the plant oil resources, renewable sugars from biomass are more easily available. In our previous study, we constructed an engineered *E. coli* strain for the direct production of HFAs from glucose through producing free fatty acids (FFAs) by a thioesterase and further converting FFAs to HFAs using a fatty acid hydroxylase [17]. However, production of HFAs of this strain was still too low. Here, the *E. coli* strain was further improved to enhance production of HFAs. The native *E. coli* acetyl-CoA carboxylase (ACCase) and a leadless thioesterase 'TesA were overexpressed to boost the host cell to produce FFAs. The fatty acid degradation pathway was blocked by disrupting the endogenous acyl-CoA synthetase (FadD). And the FFAs were then converted to HFAs by the fatty acid hydroxylase CYP102A1 (Fig. 1). The finally engineered strain was evaluated under fed-batch conditions and showed a promising perspective for large-scale production of HFAs.

Results and discussion
Expression of the recombinant enzymes in E. coli
With the aim to express the ACCase, 'TesA and CYP102A1 enzymes, we cloned the coding regions of the corresponding genes into plasmids pACYCDuet-1 pET28a, or pET30a under the control of T7 promoter. To verify the expression levels of the recombinant proteins, *E. coli* BL21(DE3) was transformed by the expression vectors pE-'tesA, pA-acc, pE-A1, pE-A1'tesA or a combination of these vectors. The resulting recombinant strains were grown in liquid LB medium followed by IPTG induction. The bacterial cells were collected and subjected to ultrasonication, and the lysates were then analyzed by SDS-PAGE. Figure 2 showed the gel electrophoresis patterns of samples from different recombinant strains. We noted distinct bands of the expected sizes from protein extracts of the recombinant strains compared with the control strain harboring pET28a. SDS-PAGE analysis of the recombinant strain carrying pE-'tesA revealed the band of the molecular weight 20.5 kDa (lane 2), which corresponded to the size of the leadless 'TesA [18]. Strain BL21/pA-acc gave all the bands of the four subunits of ACCase (lane 3). The recombinant strain expressing both 'tesA and ACCase displayed the protein bands for the two genes (lane 4). Strain BL21/pE-A1 showed a band corresponding to the molecular weight of CYP102A1 (119 kDa, lane 5) [19].

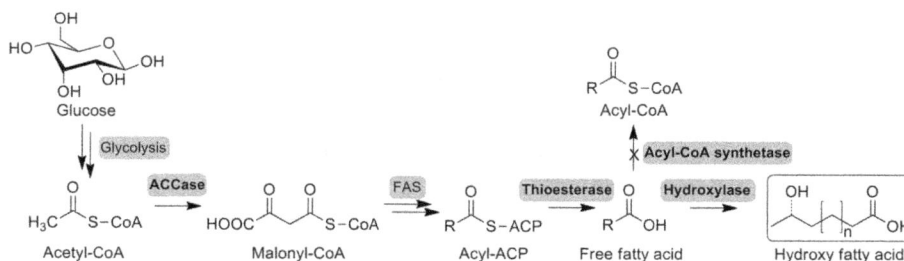

Fig. 1 Metabolic pathway from glucose to HFAs in engineered *E. coli*. Glucose is degraded into acetyl-CoA through glycolysis. Acetyl-CoA carboxylase (ACCase) catalyzes the irreversible carboxylation of acetyl-CoA to produce malonyl-CoA. The discrete, monofunctional type II fatty acid synthases (FAS) act on malonyl-CoA to synthesize fatty acyl-ACPs. Then thioesterase hydrolyzes the acyl-ACPs bond to form FFAs. At last, fatty acid hydroxylase transforms FFAs to HFAs. Acyl-CoA synthetase responsible for fatty acid degradation is knocked out to block fatty acids and HFAs degradation

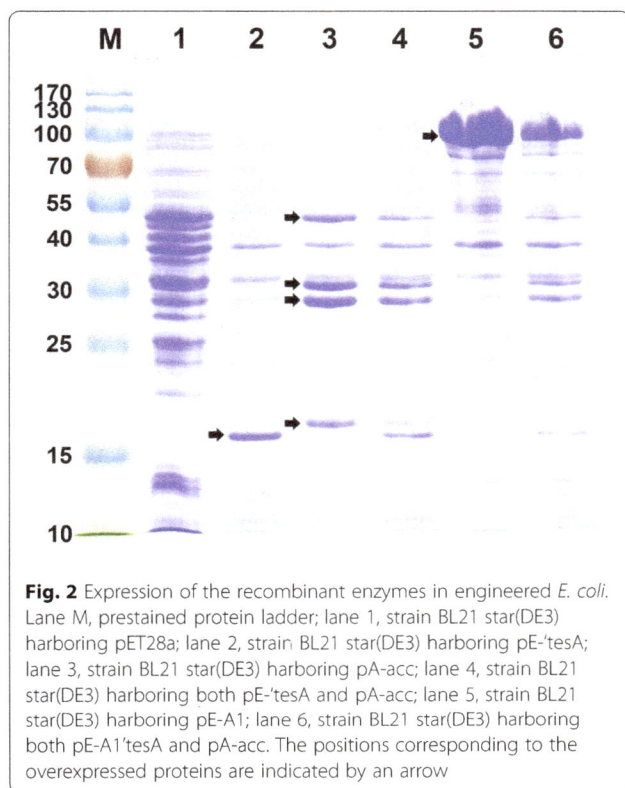

Fig. 2 Expression of the recombinant enzymes in engineered *E. coli*. Lane M, prestained protein ladder; lane 1, strain BL21 star(DE3) harboring pET28a; lane 2, strain BL21 star(DE3) harboring pE-'tesA; lane 3, strain BL21 star(DE3) harboring pA-acc; lane 4, strain BL21 star(DE3) harboring both pE-'tesA and pA-acc; lane 5, strain BL21 star(DE3) harboring pE-A1; lane 6, strain BL21 star(DE3) harboring both pE-A1'tesA and pA-acc. The positions corresponding to the overexpressed proteins are indicated by an arrow

Unlike the membrane-bound CYPs in eukaryotic systems, the bacterial CYPs usually exist in a soluble form [20]. Therefore, the CYP102A1 enzyme could function properly in the cytoplasm of the recombinant cells. The finally engineered strain BL21/pE-A1'tesA&pA-acc gave all the bands of the recombinant proteins (lane 6).

Production of FFAs by engineered E. coli

CYP102A1 catalyzes the hydroxylation of FFAs to form HFAs. Therefore, the first step to produce HFAs from glucose is to create an intracellular FFAs pool. Many studies have been performed to synthesize FFAs using *E. coli* [21]. In this research, we constructed a recombinant *E. coli* strain efficiently producing FFAs mainly from three aspects. To increase the cellular FFAs level of *E. coli*, the leadless native thioesterase 'TesA was overexpressed in strain BL21 star(DE3). About 108.5 mg/L of FFAs were produced after being induced for 12 h, which is similar to previous studies [22]. To enhance the precursor supply for fatty acids biosynthesis, the ACCase, which catalyzes the irreversible carboxylation of acetyl-CoA [23], was further coexpressed with 'TesA. Shake-flask fermentation of strain BL21/pE-'tesA&pA-acc accumulated up to 188.6 mg/L of FFAs in the culture. To eliminate fatty acid degradation, the acyl-CoA synthetase (FadD) participating in the β-oxidation pathway [24] was knocked out, resulting in strain BL21ΔfadD. Then the recombinant plasmids pE-'tesA and pA-acc were cotransformed into this strain. The engineered strain BL21ΔfadD/pE-'tesA&pA-acc was evaluated for production of FFAs and 244.8 mg/L of FFAs were synthesized, about 2.3-fold increase when compared with strain BL21/pE-'tesA. Gas chromatography - mass spectrometry (GC-MS) analysis (Additional file 1) showed that FFAs in these strains mainly consisted of C8:0, C10:0, C12:1, C12:0, C14:1, C14:0, C16:1, C16:0, and C18:1 fatty acids, with C14, C16 and C18 fatty acids as the dominant constitutes (Table 1).

Identification of HFAs from the CYP102A1 expressing strain

As shown above, FFAs of different chain length and saturation were efficiently produced by the recombinant strains. In order to convert these FFAs to their hydroxyl counterparts, the fatty acid hydroxylase CYP102A1 was further coexpressed in these FFA overproducing strains. To identify the HFAs produced by CYP102A1, the extracts from the culture broth of strain BL21/pE-A1'tesA coexpressing 'tesA and CYP102A1 were derivatized to their methyl esters and then analyzed by GC-MS. The mass spectrums of the hydroxy fatty acid methyl esters (HFAMEs) prepared from a 12 h - induced culture were shown in Additional file 1. Qualitative analysis was performed using a National Institute of Standards and Technology (NIST) - library search program. Four types of HFAs, 9-hydroxydecanoic acid methyl ester (9-OH-C10), 11-hydroxydodecanoic acid (11-OH-C12), 10-hydroxyhexadecanoic acid (10-OH-C16) and 12-hydroxyoctadecanoic acid (12-OH-C18), were detected in this strain. It has been reported that the fatty acid hydroxylase CYP102A1 has a broad substrate specificity

Table 1 FFAs composition produced by different engineered strains

Strains	C8:0	C10:0	C12:0	C12:1	C14:0	C14:1	C16:0	C16:1	C18:1	Total
BL21/pE-'tesA	1.36 (1.3 %)	1.65 (1.5 %)	8.44 (7.8 %)	5.47 (5.0 %)	10.64 (9.8 %)	5.88 (5.4 %)	42.0 (38.7 %)	12.4 (11.4 %)	20.7 (19.1 %)	108.5
BL21/pE-'tesA&pA-acc	2.15 (1.2 %)	2.86 (1.5 %)	15.1 (8.1 %)	9.6 (5.1 %)	18.9 (10.1 %)	10.2 (5.5 %)	71.3 (38.2 %)	21.3 (11.4 %)	35.2 (18.9 %)	186.6
BL21ΔfadD/pE-'tesA&pA-acc	3.64 (1.5 %)	3.89 (1.6 %)	20.4 (8.3 %)	13.1 (5.3 %)	25.2 (10.3 %)	13.2 (5.4 %)	90.3 (36.9 %)	28.3 (11.6 %)	46.8 (19.1 %)	244.8

The unit value for the fatty acids was mg/L

[20]. This enzyme could catalyze the hydroxylation of saturated or unsaturated fatty acids with a chain length of 12–22 carbons [25]. The hydroxylation always occurred in the subterminal position while the terminal methyl group of these substrates was never hydroxylated. The hydroxyl position could also be altered by rational mutagenesis of specific amino acid sites [26]. Here we further identify 10-OH-C16 and 12-OH-C18 from the mixture of hydroxylated products in addition to the subterminal ω-HFAs. It seems that CYP102A1 could oxidate the double bonds of the two kinds of unsaturated fatty acid, palmitoleic acid (C16:1Δ9) and cis-vaccenic acid (C18:1Δ11), and generate 10-OH-C16 and 12-OH-C18 HFAs. C16 or C18 HFAs at the subterminal positions were not identified in our engineered strain. Although these FFAs made up the major portion in the total fatty acid profiles, the catalytic activity of CYP102A1 towards them was much lower. This result was in accordance with many previous studies. CYP102A1 was more efficient toward medium-chain fatty acids and the catalytic activity of this enzyme decreased when the fatty acid chain length was greater than 15 [13, 17]. Therefore, we cannot detect any C16 or C18 HFAs at the subterminal positions from the CYP102A1 overexpressing strain.

Unlike the fungal fatty acid hydroxylases [27], the bacterial CYP102A1 does not act on the terminal position of the fatty acid chain. Therefore, the hydroxyl products of this enzyme could not be degraded by the host's endogenous enzymes, such as the fatty alcohol oxidase [28]. These HFAs were stable in the fermentation broth and could accumulate without deleting the ω-HFAs degradation enzymes. In addition, CYP102A1 is a self-sufficient fatty acid hydroxylase [29]. It consists of a heme-binding domain and a FMN/FAD-containing domain, and catalyzes the electron transfer from NADPH, via FAD, FMN, and heme, to O_2, resulting in the formation of a hydroxyl group on carbon atoms without the help of other enzymes [30]. This is different from many fatty acid hydroxylases which require the ferredoxin (Fdx) reductase domain to obtain reducing equivalents from NADPH [31]. Thus, the CYP102A1-based HFA-producing system would be much easier to be operated.

Evaluation of HFAs producing ability of different strains under shake-flasks conditions

To investigate the supply of FFAs on HFAs production, different E. coli strains including BL21/pE-A1'tesA, BL21/pE-A1'tesA&pA-acc and BL21ΔfadD/pE-A1'tesA&pA-acc were cultivated under shake-flask conditions. Fatty acids produced by these strains were extracted, derivatized and analyzed by GC-MS. The quantities of FFAs and HFAs were determined by comparison of the chromatographic peak areas with the internal standard (C20 or 12-OH-

C12). Cell density, FFAs and HFAs accumulated in the fermentation broth of different recombinant strains were calculated and shown in Fig. 3. It could be seen that all the three recombinant strains grew to a similar OD_{600} after 12 h induction (about 3.0-3.5). Strain BL21/pE-A1'tesA produced 77.5 mg/L of FFAs and 36.5 mg/L of HFAs. When ACCase was overexpressed, the final titer of FFAs and HFAs reached 143.4 mg/L and 40.3 mg/L, respectively. Production of FFAs was greatly improved in this strain, but there was only a slight increase in production of HFAs. The FFAs accumulated seemed not to be efficiently converted to HFAs by the fatty acid hydroxylase. When E. coli native fadD gene was knocked out, the finally engineered strain BL21ΔfadD/pE-A1'tesA&pA-acc showed an enhanced ability to produce HFAs. The titer of HFAs reached 58.7 mg/L, which is 1.6-fold to the original strain. Compositions of HFAs in these strains were shown in Table 2. 11-Hydroxydodecanoic acid and 12-hydroxyoctadecanoic acid made up the major HFAs constituents. The deletion of fadD could block both fatty acids and degradation of HFAs [13]. Thus, production of FFAs and HFAs was increased in this strain. The productivity of HFAs per cell dry weight (CDW) of strain BL21ΔfadD/pE-A1'tesA&pA-acc reached 44.3 mg/gCDW (1 OD_{600} = 0.43 gCDW). The enhancement of production of FFAs was much greater than production of HFAs along with the introduction of ACCase and knockout of fadD. Only 24 % of the FFAs were converted to HFAs in this finally engineered strain. These results indicated that the rate-limiting step for HFAs production was the fatty acid hydroxylase CYP102A1 [32].

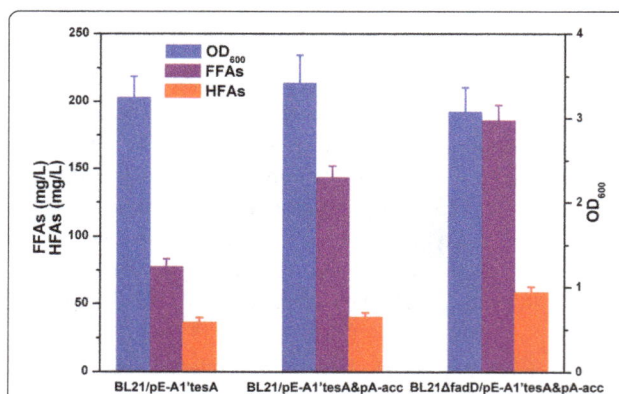

Fig. 3 Comparison of HFAs production of several different strains under shake-flask conditions. Data were obtained after each strain was induced for 12 h in liquid LB medium supplemented with 20 g/L glucose. BL21/pE-A1'tesA, strain BL21 star(DE3) expressing B. megaterium CYP102A1 and native E. coli 'tesA; BL21/pE-A1'tesA&pA-acc, further overexpressing native E. coli ACCase; BL21ΔfadD/pE-A1'tesA&pA-acc, knockout of native fadD gene while coexpressing the three enzymes

Table 2 HFAs composition produced by different engineered strains

Strains	9-OH-C10	10-OH-C16	11-OH-C12	12-OH-C18	Total
BL21/pE-A1'tesA	4.17 (11.4 %)	3.31 (9.1 %)	18.7 (51.1 %)	10.4 (28.4 %)	36.5
BL21/pE-A1'tesA&pA-acc	4.80 (11.9 %)	3.91 (9.7 %)	20.2 (50.1 %)	11.4 (28.3 %)	40.3
BL21ΔfadD/pE-A1'tesA&pA-acc	7.16 (12.2 %)	5.87 (10.0 %)	29.3 (49.8 %)	16.4 (28.0 %)	58.7

The unit value for the HFAs was mg/L

We can expect to achieve a higher titer of HFAs by improving the efficiency of this enzyme.

HFAs production at the fermentor scale

Microbial production of HFAs is currently achieved using the bacteria *Pseudomonas* sp. [33] or nonconventional yeasts *Candida* sp. [34] that produce selective CYPs as the hosts. Compared with these strains, *E. coli* has many advantages such as a clear genetic background, high convenience to be genetically modified, and good growth properties with low nutrient requirements [35]. Here, we tested our recombinant *E. coli* strain using high-density fermentation strategy. Based on the results obtained by the shake-flask cultivations, the finally engineered *E. coli* strain BL21ΔfadD/pE-A1'tesA&pA-acc was cultured in a 5 L-scale laboratory fermentor. Cell density, residual glucose concentration and products accumulation were monitored over the course of fed-batch fermentation. Figure 4 shows the time profiles of cell density and production of HFAs during 24 h fed-batch fermentation. The bacteria grew very fast at the first 12 h post-induction to an OD_{600} of approximate 70. FFAs and HFAs also accumulated rapidly in the culture broth. The highest production of FFAs and HFAs were obtained after 12 h induction, that is, 2.81 g/L and 548 mg/L. The volumetric productivities of FFAs and HFAs were 234 mg/(L·h) and 45.7 mg/(L·h), respectively. Both of the titers of FFAs and HFAs decreased to some extent in the following fermentation processes.

Compared with the HFA-producing process using fatty acids as the feedstock [1, 36–38], the current production and yield obtained by this engineered *E. coli* strain is still

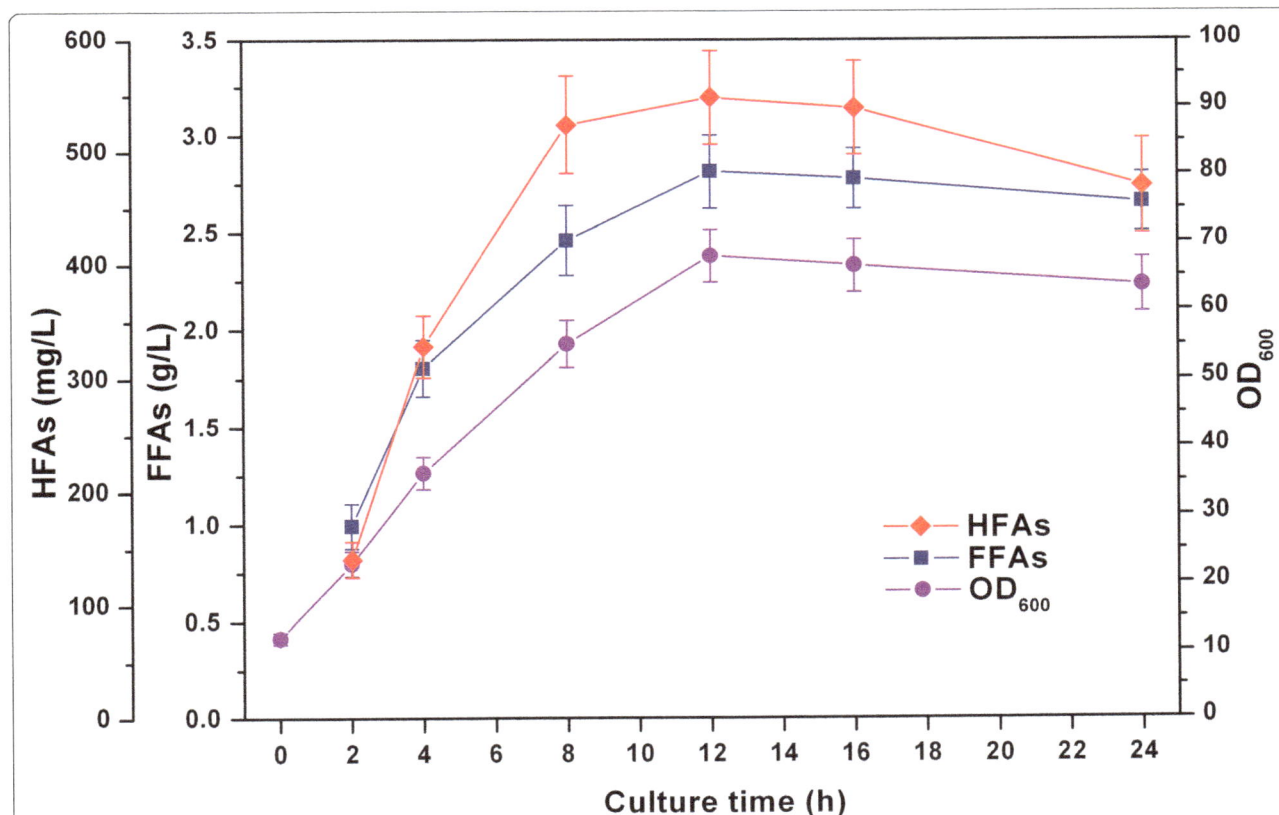

Fig. 4 Time courses of cell density (OD_{600}), FFAs and HFAs production during fed-batch culture of the finally engineered strain BL21ΔfadD/pE-A1'tesA&pA-acc. Cultivation was conducted in a 5 L fermentor with an initial volume of 2 L of rich growth medium

too low. This might be due to that these processes used quite different mechanisms to synthesize HFAs. The double bond hydratases were employed in previous work directly acting on unsaturated fatty acids to generate HFAs. The catalytic activity of the hydratases was more efficient than the P450 monooxygenase used in this study, leading to much higher productivity and yield. However, the use of fatty acids or plant oils increases the raw material cost since they are more expensive than glucose and other sugars. These carbohydrates have the potential to be manufactured from the easily available lignocellulosic biomass resources. The biotransformation of fatty acids also needed to first grow the cells with glucose or other carbon sources. The yield was overestimated for neglecting the consumption of the carbon sources. In addition, the yeast-based HFAs-producing strains always take several days to reach the maximum titer, while the whole fermentation process only requires less than 24 h for this engineered strain.

The production and yield of HFAs in the present study could be enhanced from several aspects. Biosynthesis of FFAs is the first rate-limiting step in our HFA-producing system. Fatty acid biosynthesis from glucose requires carbon fluxes through glycolysis to generate pyruvate which is further dehydrogenated to acetyl-CoA. Acetyl-CoA is then carboxylated to form malonyl-CoA, the precursor of the bacterial type II fatty acid synthases [39]. Fatty acyl-ACPs of different chain-length were finally cleaved by thioesterases into FFAs. Numerous effects have been conducted to improve the ability of *E. coli* to synthesize FFAs, but the highest titer of FFAs achieved up to now was roughly 9 g/L [40]. To obtain an even higher production of HFAs, the FFAs pool must be further increased. Fatty acid hydroxylase is another key enzyme for production of HFAs. Although the CYP102A1 enzyme has many excellent attributes, its catalytic activity is much lower than many other hydroxylases, e.g., lipoxygenase, hydratase and diol synthase [41]. Therefore, the use of more efficient fatty acid hydroxylases in our producing system would be helpful to improve production of HFAs.

Conclusions

In this study, a robust HFA-producing *E. coli* strain was successfully constructed. Four distinct genetic alterations targeted at the HFA metabolic pathways were introduced into the host strain BL21 star(DE3), including knockout of the endogenous *fadD* gene, which encodes the acyl-CoA synthetase, to block fatty acid β-oxidation; overexpression of native *E. coli* ACCase to enhance the supply of malonyl-CoA, the precursor for fatty acid biosynthesis; overexpression of a leadless thioesterase 'TesA to render the host releasing FFAs; and further introducing a hydroxylase CYP102A1 to hydroxylate the FFAs into HFAs. Under fed-batch conditions, up to 548 mg/L of HFAs were produced by the finally engineered strain BL21ΔfadD/pE-A1'tesA&pA-acc. The volumetric productivity of HFAs reached 45.7 mg/(L · h). Although the current production of this work is far from industrial application, it opens the door to employing the enormous power of metabolic engineering in this experimentally friendly organism for HFAs biosynthesis. This engineered *E. coli* would give some implication to industrial-scale production of HFAs in the future.

Methods

Bacterial strains and plasmids construction

A list of bacterial strains and recombinant plasmids was presented in Table 3. *E. coli* DH5α was used for gene cloning and *E. coli* BL21 star(DE3) was used as the host for the expression of the recombinant proteins. The chromosomal *fadD* gene of strain BL21 star(DE3) responsible for fatty acid degradation was knocked out using the Red recombination strategy in a previous study, resulting strain BL21ΔfadD [42].

Table 3 Strains and plasmids used in this study

Strains or plasmids	Genotype/Description	Sources
Strains		
E. coli BL21 star(DE3)	F^- *ompT hsdS*$_B$ (r_B^- m_B^-) *gal dcm rne131* (DE3)	Invitrogen
E. coli BL21 star(DE3) ΔfadD	Knockout of *fadD* encoding acyl-CoA synthetase	[42]
Plasmids		
pET28a(+)	Kan^r *oripBR322 lacI*^{q} *T7p*	Novagen
pET30a(+)	Kan^r *oripBR322 lacI*^{q} *T7p*	Novagen
pACYCDuet-1	*Cm*^{r} *oriP15A lacI*^{q} *T7p*	Novagen
pE-'tesA	pET30a(+) harboring *E. coli 'tesA*	This study
pE-A1	pET28a(+) harboring *B. megaterium CYP102A1*	This study
pE-A1'tesA	pET28a(+) harboring both *E. coli 'tesA and B. megaterium CYP102A1*	This study
pA-acc	pACYCDuet-1 harboring *E. coli ACCase*	[43]

Primers used for plasmids construction was provided in Table 4. The four subunits of native *E. coli* acetyl-CoA carboxylase were cloned into a single expression vector pACYCDuet-1, resulting pA-acc in another study [43]. The *CYP102A1* gene [GenBank: J04832] was PCR amplified from *B. megaterium* ATCC 14581 genomic DNA and cloned into the restriction sites *Nco*I/*Eco*RI of vector pET28a, resulting pE-A1. The *'tesA* gene [GenBank: EG11542] (encoding a leadless version of native *E. coli* thioesterase I without the N-terminal 26 amino acids) was amplified from *E. coli* K12 genome and cloned into the restriction sites *Nde*I/*Bgl*II of vector pET30a, resulting pE-'tesA. Then PCR reaction was performed using pE-'tesA as the template and a primer pair that allowed the amplification of the T7 promoter sequence along with the *'tesA* structural gene. The PCR product T7'tesA was then cloned into pE-A1 between *Eoc*RI and *Xho*I sites to create pE-A1'tesA, which was used for the coexpression of the two genes. Successful gene cloning was verified by colony PCR, restriction mapping and direct nucleotide sequencing.

Media and culture conditions

Luria-Bertani (LB) medium (10 g/L tryptone, 5 g/L yeast extract, 10 g/L NaCl) was used for DNA manipulation, protein expression and shake-flasks cultivation. Rich growth medium (20 g/L tryptone, 10 g/L yeast extract, 5 g/L NaCl and 5 g/L $K_2HPO_4 \cdot 3H_2O$) was used for fermentor-scale cultivation. $MgSO_4$ (0.12 g/L) and trace elements (1 ml per liter, 3.7 g/L $(NH_4)_6Mo_7O_{24} \cdot 4H_2O$, 2.9 g/L $ZnSO_4 \cdot 7H_2O$, 24.7 g/L H_3BO_3, 2.5 g/L $CuSO_4 \cdot 5H_2O$, 15.8 g/L $MnCl_2 \cdot 4H_2O$) were autoclaved or filter-sterilized separately and added prior to initiation of the fermentation. 50 mg/L of kanamycin or 34 mg/L of chloramphenicol were supplemented when necessary. Under shake-flask conditions, the bacterial cultures were first grown at 37 °C and 180 rpm. 0.5 mM of isopropyl-β-D-thiogalactopyranoside (IPTG) was added at an OD_{600} of about 0.6 to induce the expression of recombinant proteins and production of HFAs. Then the culture temperature was shifted to 30 °C after adding the inducer.

Table 4 PCR primers designed for plasmids construction

Oligonucleotide primers	Sequences
'tesA_ F_NdeI	GGAATTC**CATATG**GCGGACACGTTATTGATTCTGGG
'tesA_R_BglII	GA**AGATCT**TATGAGTCATGATTTACTAAAGGC
A1_ F_NcoI	CATG**CCATGG**GCATGACAATTAAAGAAATGCCTCAG
A1_ R_EcoRI	CCG**GAATTC**TTACCCAGCCCACACGTCTTTTG
T7'tesA_F_EcoRI	CCG**GAATTC**TAATACGACTCACTATAGGGG
T7'tesA_R_XhoI	CCG**CTCGAG**TTATGAGTCATGATTTACTAAAGGC

Protein expression and SDS-PAGE analysis

Recombinant *E. coli* strains harboring pE-A1, pE-'tesA, pE-A1'tesA, pA-acc or the combination of these plasmids were induced for 4 h to express the recombinant proteins. Then cells were collected from 1.5 ml of bacterial cultures by centrifugation and resuspended in 50 mM of Tris–HCl buffer (pH 8.0). Cell pellets were disrupted using a probe-type sonicator (VCX130, Sonics, USA) at 4 °C. The resulting crude extracts were centrifuged and the supernatants with the soluble proteins were recovered, mixed with equal volume of 2× sodium dodecyl sulfate (SDS) sample buffer, heated at 100 °C for 10 min and then analyzed by SDS-polyacrylamide gel electrophoresis (PAGE) according to a standard procedure. Protein bands were visualized with Coomassie Brilliant Blue staining.

Fed-batch fermentation

For large-scale production of HFAs, fed-batch cultures were carried out in a Biostat B plus MO5L fermentor (Sartorius Stedim Biotech GmbH, Germany) containing 2 L of rich growth medium. 50 ml of inoculum was prepared by incubating the culture in shake flasks at 37 °C overnight. After inoculation, the fermentation was first operated in a batch mode and the control settings were: 37 °C, pH 7.0, airflow at 2 L/min and stirring speed at 400 rpm. The dissolved oxygen (DO) was kept above 20 % by associating with the stirring speed. After the initial glucose was nearly exhausted, fed-batch mode was commenced by feeding a concentrated glucose solution (65 %) at appropriate rates to maintain the residual glucose at a low level. When OD_{600} reached about 12, 0.5 mM of IPTG was used to induce recombinant proteins expression and production of HFAs. Then the culture temperature was switched to 30 °C. Samples of fermentation broth were taken at appropriate intervals to determine cell density, residual glucose, production of FFAs and HFAs.

Fatty acids and HFAs extraction and the corresponding methyl esters preparation

To extract the FFAs and HFAs from the fermentation broth, the culture broth was acidified with 6 M hydrochloric acid to pH < 2. Eicosanoic acid (C20), 10-hydroxydecanoic acid methyl ester (10-OH-C10) or 12-hydroxydodecanoic acid (12-OH-C12) from a 50 mg/mL stock solution in ethyl acetate were added to the culture broth before extraction to serve as the internal standards. The acidic materials were extracted with equal volume of ethyl acetate. The collected organic layer was evaporated with nitrogen and then the extracts were exposed to sulfuric acid/methanol (1:99, by volume) at 70 °C for 1 h to generate fatty acids or HFAs methyl esters

(FAMEs or HFAMEs). The FAMEs and HFAMEs were then extracted with n-hexane.

Analytical methods

Cell growth of the *E. coli* culture in shake-flasks or fermentors was monitored by determining the optical density at 600 nm (OD_{600}) of appropriate dilutions using an UV–vis Spectrophotometer (Cary 50, Varian, USA).

The concentration of residual glucose was quantified by a glucose oxidase-peroxidase assay using an SBA-40D Biological Sensing Analyzer (Biology Institute of Shandong Academy of Sciences, China).

The resulting FAMEs and HFAMEs were analyzed an Agilent Trace GC 7890A system coupled to a quadrupole detector (5975C). The GC was equipped with a 30 m HP-5 ms column (internal diameter 0.25 mm, film thickness 0.25 μm), an ion source temperature of 220 °C and EI ionization at 70 eV. The method used a 10:1 split ratio and nitrogen as carrier gas with a linear velocity of 1 ml/min. The temperature program was an initial hold at 100 °C for 2 min, ramping at 10 °C per min to 200 °C followed by a temperature gradient of 5 °C per min to 280 °C and a final hold at 280 °C for 5 min. Since authentic standards for the HFAMEs were not available, these compounds were identified by searching the NIST Mass Spectral Library [44]. Quantification of FFAs and HFAs were performed by comparison to the internal standard.

Competing interests
The authors declare that they have no competing interests.

Authors' contributions
YC, MX and HL design of the experiment; YC, TC and GZ performed the experiments; YC, GZ, WN, JG and MX analyzed the primary data; YC and MX drafted the manuscript; TC, GZ, WN, JG and HL revised the manuscript. All authors read and approved the final manuscript.

Acknowledgements
This work was sponsored by National Natural Science Foundation of China (No. 21202179, 21376255 and 31200030), Key Program of the Chinese Academy of Sciences (KGZD-EW-606-1-3), Taishan Scholars Climbing Program of Shandong (No. tspd20150210).

Author details
[1]CAS Key Laboratory of Biobased Materials, Qingdao Institute of Bioenergy and Bioprocess Technology, Chinese Academy of Sciences, Qingdao 266101, China. [2]Department of Chemistry, University of Nebraska-Lincoln, Lincoln, NE 68588, USA.

References
1. Lu W, Ness JE, Xie W, Zhang X, Minshull J, Gross RA. Biosynthesis of monomers for plastics from renewable oils. J Am Chem Soc. 2010;132:15451–5.
2. Cao Y, Zhang X. Production of long-chain hydroxy fatty acids by microbial conversion. Appl Microbiol Biotechnol. 2013;97:3323–31.
3. Song JW, Jeon EY, Song DH, Jang HY, Bornscheuer UT, Oh DK, et al. Multistep enzymatic synthesis of long-chain α, ω-dicarboxylic and ω-hydroxycarboxylic acids from renewable fatty acids and plant oils. Angew Chem Int Ed Engl. 2013;52:2534–7.
4. Gross RA, Lu W, Ness J, Minshull J: Production of an α-carboxyl-ω-hydroxy fatty acid using a genetically modified *Candida* strain. 2012, US8158391. http://www.freepatentsonline.com/8158391.html.
5. Choi JI, Lee SY. High-level production of poly(3-hydroxybutyrate-co-3-hydroxyvalerate) by fed-batch culture of recombinant *Escherichia coli*. Appl Environ Microbiol. 1999;65:4363–8.
6. Liu C, Liu F, Cai J, Xie W, Long TE, Turner SR, et al. Polymers from fatty acids: Poly(ω-hydroxyl tetradecanoic acid) synthesis and physico-mechanical studies. Biomacromolecules. 2011;12:3291–8.
7. Lanser A, Plattner R, Bagby M. Production of 15-, 16- and 17-hydroxy-9-octadecenoic acids by bioconversion of oleic acid with *Bacillus pumilus*. J Am Oil Chem Soc. 1992;69:363–6.
8. Craft DL, Madduri KM, Eshoo M, Wilson CR. Identification and characterization of the CYP52 family of Candida tropicalis ATCC 20336, important for the conversion of fatty acids and alkanes to α, ω-dicarboxylic acids. Appl Environ Microbiol. 2003;69:5983–91.
9. Ruettinger RT, Wen LP, Fulco AJ. Coding nucleotide, 5′ regulatory, and deduced amino acid sequences of P-450BM-3, a single peptide cytochrome P-450:NADPH-P-450 reductase from *Bacillus megaterium*. J Biol Chem. 1989;264:10987–95.
10. Gustafsson MCU, Roitel O, Marshall KR, Noble MA, Chapman SK, Pessegueiro A, et al. Expression, purification, and characterization of Bacillus subtilis cytochromes P450 CYP102A2 and CYP102A3: Flavocytochrome homologues of P450 BM3 from Bacillus megaterium. Biochemistry. 2004;43:5474–87.
11. Furuya T, Shibata D, Kino K. Phylogenetic analysis of *Bacillus* P450 monooxygenases and evaluation of their activity towards steroids. Steroids. 2009;74:906–12.
12. Chowdhary PK, Alemseghed M, Haines DC. Cloning, expression and characterization of a fast self-sufficient P450: CYP102A5 from *Bacillus cereus*. Arch Biochem Biophys. 2007;468:32–43.
13. Schneider S, Wubbolts MG, Sanglard D, Witholt B. Biocatalyst engineering by assembly of fatty acid transport and oxidation activities for *in vivo* application of cytochrome P-450(BM-3) monooxygenase. Appl Environ Microbiol. 1998;64:3784–90.
14. Schneider S, Wubbolts MG, Oesterhelt G, Sanglard D, Witholt B. Controlled regioselectivity of fatty acid oxidation by whole cells producing cytochrome P450BM-3 monooxygenase under varied dissolved oxygen concentrations. Biotechnol Bioeng. 1999;64:333–41.
15. Kitazume T, Tanaka A, Takaya N, Nakamura A, Matsuyama S, Suzuki T, et al. Kinetic analysis of hydroxylation of saturated fatty acids by recombinant P450foxy produced by an *Escherichia coli* expression system. Eur J Biochem. 2002;269:2075–82.
16. Kitazume T, Yamazaki Y, Matsuyama S, Shoun H, Takaya N. Production of hydroxy-fatty acid derivatives from waste oil by *Escherichia coli* cells producing fungal cytochrome P450foxy. Appl Microbiol Biotechnol. 2008;79:981–8.
17. Wang X, Li L, Zheng Y, Zou H, Cao Y, Liu H, et al. Biosynthesis of long chain hydroxyfatty acids from glucose by engineered *Escherichia coli*. Bioresour Technol. 2012;114:561–6.
18. Cho H, Cronan JE. *Escherichia coli* thioesterase I, molecular cloning and sequencing of the structural gene and identification as a periplasmic enzyme. J Biol Chem. 1993;268:9238–45.
19. Narhi LO, Fulco AJ. Characterization of a catalytically self-sufficient 119,000-dalton cytochrome P-450 monooxygenase induced by barbiturates in *Bacillus megaterium*. J Biol Chem. 1986;261:7160–9.
20. Munro AW, Leys DG, McLean KJ, Marshall KR, Ost TWB, Daff S, et al. P450 BM3: the very model of a modern flavocytochrome. Trends Biochem Sci. 2002;27:250–7.

21. Lennen RM, Pfleger BF. Engineering *Escherichia coli* to synthesize free fatty acids. Trends Biotechnol. 2012;30:659–67.

22. Steen EJ, Kang Y, Bokinsky G, Hu Z, Schirmer A, McClure A, et al. Microbial production of fatty-acid-derived fuels and chemicals from plant biomass. Nature. 2010;463:559–62.

23. Cronan Jr JE, Waldrop GL. Multi-subunit acetyl-CoA carboxylases. Prog Lipid Res. 2002;41:407–35.

24. Yoo JH, Cheng OH, Gerber GE. Determination of the native form of FadD, the *Escherichia coli* fatty acyl-CoA synthetase, and characterization of limited proteolysis by outer membrane protease OmpT. Biochem J. 2001;360:699–706.

25. Capdevila JH, Wei S, Helvig C, Falck JR, Belosludtsev Y, Truan G, et al. The highly stereoselective oxidation of polyunsaturated fatty acids by cytochrome P450BM-3. J Biol Chem. 1996;271:22663–71.

26. Dietrich M, Do TA, Schmid RD, Pleiss J, Urlacher VB. Altering the regioselectivity of the subterminal fatty acid hydroxylase P450 BM-3 towards γ- and δ-positions. J Biotechnol. 2009;139:115–7.

27. Bae J, Park B, Jung E, Lee PG, Kim BG. *fadD* deletion and *fadL* overexpression in *Escherichia coli* increase hydroxy long-chain fatty acid productivity. Appl Microbiol Biotechnol. 2014;98:8917–25.

28. Gatter M, Förster A, Bär K, Winter M, Otto C, Petzsch P, et al. A newly identified fatty alcohol oxidase gene is mainly responsible for the oxidation of long-chain ω-hydroxy fatty acids in *Yarrowia lipolytica*. FEMS Yeast Res. 2014;14:858–72.

29. Boddupalli SS, Estabrook RW, Peterson JA. Fatty acid monooxygenation by cytochrome P-450BM-3. J Biol Chem. 1990;265:4233–9.

30. Hilker BL, Fukushige H, Hou C, Hildebrand D. Comparison of *Bacillus* monooxygenase genes for unique fatty acid production. Prog Lipid Res. 2008;47:1–14.

31. Scheps D, Honda Malca S, Richter SM, Marisch K, Nestl BM, Hauer B. Synthesis of ω-hydroxy dodecanoic acid based on an engineered CYP153A fusion construct. Microb Biotechnol. 2013;6:694–707.

32. Sung C, Jung E, Choi KY, Bae JH, Kim M, Kim J, et al. The production of ω-hydroxy palmitic acid using fatty acid metabolism and cofactor optimization in *Escherichia coli*. Appl Microbiol Biotechnol. 2015;1–10.

33. Martin-Arjol I, Llacuna J, Manresa Á. Yield and kinetic constants estimation in the production of hydroxy fatty acids from oleic acid in a bioreactor by Pseudomonas aeruginosa 42A2. Appl Microbiol Biotechnol. 2014;98:9609–21.

34. Girhard M, Tieves F, Weber E, Smit M, Urlacher V. Cytochrome P450 reductase from *Candida apicoia*: versatile redox partner for bacterial P450s. Appl Microbiol Biotechnol. 2013;97:1625–35.

35. Cao Y, Zhang R, Sun C, Cheng T, Liu Y, Xian M. Fermentative succinate production: An emerging technology to replace the traditional petrochemical processes. Biomed Res Int. 2013;2013:12.

36. Joo YC, Seo ES, Kim YS, Kim KR, Park JB, Oh DK. Production of 10-hydroxystearic acid from oleic acid by whole cells of recombinant *Escherichia coli* containing oleate hydratase from *Stenotrophomonas maltophilia*. J Biotechnol. 2012;158:17–23.

37. Kim BN, Joo YC, Kim YS, Kim KR, Oh DK. Production of 10-hydroxystearic acid from oleic acid and olive oil hydrolyzate by an oleate hydratase from *Lysinibacillus fusiformis*. Appl Microbiol Biotechnol. 2012;95:929–37.

38. Oh HJ, Kim SU, Song JW, Lee JH, Kang WR, Jo YS, et al. Biotransformation of linoleic acid into hydroxy fatty acids and carboxylic acids using a linoleate double bond hydratase as key enzyme. Adv Synth Catal. 2015;357:408–16.

39. Magnuson K, Jackowski S, Rock CO, Cronan JE. Regulation of fatty acid biosynthesis in *Escherichia coli*. Microbiol Rev. 1993;57:522–42.

40. Janßen HJ, Steinbüchel A. Fatty acid synthesis in *Escherichia coli* and its applications towards the production of fatty acid based biofuels. Biotechnol Biofuels. 2014;7:1–26.

41. Kim KR, Oh DK. Production of hydroxy fatty acids by microbial fatty acid-hydroxylation enzymes. Biotechnol Adv. 2013;31:1473–85.

42. Cao Y, Liu W, Xu X, Zhang H, Wang J, Xian M. Production of free monounsaturated fatty acids by metabolically engineered *Escherichia coli*. Biotechnol Biofuels. 2014;7:59.

43. Cao Y, Jiang X, Zhang R, Xian M. Improved phloroglucinol production by metabolically engineered *Escherichia coli*. Appl Microbiol Biotechnol. 2011;91:1545–52.

44. Kim S, Koo I, Jeong J, Wu S, Shi X, Zhang X. Compound identification using partial and semipartial correlations for gas chromatography–mass spectrometry data. Anal Chem. 2012;84:6477–87.

2

Enhancing the thermostability of α-L-rhamnosidase from *Aspergillus terreus* and the enzymatic conversion of rutin to isoquercitrin by adding sorbitol

Lin Ge[1,2†], Anna Chen[1,2†], Jianjun Pei[1,2], Linguo Zhao[1,2*], Xianying Fang[1,2], Gang Ding[3], Zhenzhong Wang[3], Wei Xiao[3*] and Feng Tang[4]

Abstract

Background: Thermally stable α-L-rhamnosidase with cleaving terminal α-L-rhamnose activity has great potential in industrial application. Therefore, it is necessary to find a proper method to improve the thermal stability of α-L-rhamnosidase.

Results: In this study, addition of sorbitol has been found to increase the thermostability of α-L-rhamnosidase from *Aspergillus terreus* at temperatures ranging from 65 °C to 80 °C. Half-life and activation free energy with addition of 2.0 M sorbitol at 70 °C were increased by 17.2-fold, 8.2 kJ/mol, respectively. The analyses of the results of fluorescence spectroscopy and CD have indicated that sorbitol helped to protect the tertiary and secondary structure of α-L-rhamnosidase. Moreover, the isoquercitrin yield increased from 60.01 to 96.43% with the addition of 1.5 M of sorbitol at 70 °C.

Conclusion: Our findings provide an effective approach for enhancing the thermostability of α-L-rhamnosidase from *Aspergillus terreus*, which makes it a good candidate for industrial processes of isoquercitrin preparation.

Keywords: Thermostability, α-L-Rhamnosidase, Sorbitol, Enzymatic conversion

Background

Isoquercitrin, a derhamnosylation product of rutin, has been widely acknowledged with several biological activities, including anti-mutagenesis, anti-virus, anti-hypertensive, anti-proliferative effects, lipid peroxidation, oxidative-stress protection as well as other pharmacological effects [1–4]. Isoquercitrin recently has been proved to exhibit better bioactivity than rutin [5]. However, isoquercitrin is rarely found in nature, and therefore it has hardly been isolated. Isoquercitrin is structurally similar to rutin that is abundantly present in plant [6]. Therefore, rutin can be an ideal precursor for preparing isoquercitrin. So far, several methods for the transformation of rutin to isoquercitrin

have been investigated, including acid hydrolysis, heating, microbial transformation, and enzymatic transformation techniques [7–9] the last of which has become the favored method due to its economic merits, eco-friendly status and applicability to the food industry.

The enzyme α-L-rhamnosidase (RASE, E.C. 3.2.1.40), which is able to release terminal α-L-rhamnose exclusively from glycosides, glycolipids, and other natural products [10–14], is widely distributed in nature and has been purified from animal tissues, plants, yeasts, fungi and bacteria [15–22]. This enzyme as catalyst has been generally used in a large number of industrial applications. For instance, removing naringin that is the bitter component of grapefruit juices and other citrus juices [23], enhancing the aroma of wines and grape juice [11], studying of the structure of plants and bacterial polysaccharides [24], producing of rhamnose from various natural products [25]. Furthermore, α-L-rhamnosidase is used not only for the derhamnosylation of natural

* Correspondence: lg.zhao@163.com; xw@kanion.com
†Equal contributors
[1]Co-Innovation Center for Sustainable Forestry in Southern China, Nanjing Forestry University, 159 Long Pan Road, Nanjing 210037, China
[3]Jiangsu Kanion Pharmaceutical Co., Ltd, 58 Haichang South Road, Lianyungang 222001, Jiangsu Province, China
Full list of author information is available at the end of the article

compounds but also for the reverse rhamnosylation of various small organic compounds [26].

However, the majority of the reported α-L-rhamnosidases are not thermally stable at higher temperatures [21, 27–29], which limits their industrial use. Therefore, it is necessary to find a proper method to improve the thermal stability of α-L-rhamnosidase with potential industrial roles. The thermostability of an enzyme hinges on multifarious causes, including amino acid composition, metal ions, pH as well as others [30–32]. Currently the thermostability of enzymes can be improved by chemical modification, immobilization, treatment with additives and protein engineering [33–36]. Among them, the effects of additives on the thermostability of the enzyme has received increasing attention from researchers because addition of additives to protein solution and changing its microenvironment provides a simple but practical means of increasing the stability of the enzyme [33]. Polyhydroxy compound sorbitol is widely used as the enzyme stabilizer [33, 37, 38]. Different mechanisms have been attributed to the stabilizing effect of sorbitol, which include preferential hydration, preferential exclusion from the denatured protein, and coating effect [39–41]. However, the effect of sorbitol at the molecular level is not known in detail. It is generally believed that the improvement of the thermostability of the enzyme by sorbitol is due to solvophobic interactions and hydrogen bonds [33].

Enhancement of thermal stability is beneficial for most of the biotechnological applications of proteins. However, the recombinant α-L-rhamnosidase of *Aspergillus terreus* has relatively low thermostability (half-life < 30 min at 70 °C) [42], which severely limits its industrial application. Naturally occurring osmolytes such as amino acids, polyols and salts are known to protect proteins against thermal inactivation by stabilizing the thermally unfolded proteins. In the present investigation, the effect of sorbitol on the thermostability of α-L-rhamnosidase was studied, and stability properties of the enzyme in the absence and the presence of sorbitol was determined, and the mechanisms responsible for sorbitol improving the thermostability of the enzyme were analyzed. Moreover, the effect of sorbitol on the enzymatic conversion of rutin to isoquercitrin was investigated. To the best of our knowledge, no report has been published concerning the application of polyhydroxy compound for improving the thermostability of the enzyme so as to enhance the enzymatic conversion of rutin to isoquercitrin.

Methods
Bacterial strains and plasmids
The gene encoding α-L-rhamnosidase (NCBI accession number JN899401.1) from *A. terreus* (Rha) was optimized based on the preferred codon usage of *P. pastoris* and synthesized by Shanghai Generay Biotech Co. Ltd.

(Shanghai, China). The synthetic gene (MRha) was inserted into plasmid pPICZαA (Invitrogen, USA) to generate the expression vector pPICZαA/MRha, which were linearized with SacI and transformed into yeast strain *P. pastoris* KM71H (Invitrogen, USA) by electroporation. The sequence alignment of Rha and MRha was shown in Additional file 1: Figure S1.

Production and purification of α-L-rhamnosidase
Yeast strain *P. pastoris* KM71H was cryopreserved at −80 °C in 15% (v/v) glycerol ($OD_{600 \ nm}$ = 50–100), added into YPD (1% (w/v) yeast extract; 2% (w/v) peptone; 2% (w/v) glucose) medium, or kept on YPD plates (YPD medium with 2% (w/v) agar). For the *P. pastoris* cultivation the following media were used: YPD, BMGY (1% (w/v) yeast extract; 2% (w/v) peptone; 100 mM potassium phosphate, pH 6.0; 1.34% (w/v) YNB (Invitrogen, USA); 4×10-5% (w/v) biotin; 1% (v/v) glycerol). *P. pastoris* KM71H was cultivated on a rotary shaker at 28 °C and 180 rpm.

The mycelium was removed from the cultivation medium by filtration. The filtrate was precipitated with ammonium sulphate to 80% saturation. The enzyme precipitate was dissolved in 50 mM Tris-HCl buffer (pH 7.5) and the enzyme solution was dialyzed against 50 mM Tris-HCl buffer (pH 7.5) at 4 °C for 48 h. The dialysate was loaded onto a column of DEAE Sepharose Fast Flow (Amersham Bioscience, USA) equilibrated with 50 mM Tris-HCl buffer (pH 7.5). The enzyme was eluted from the column using linear gradient (20–300 mM NaCl). Fractions containing α-L-rhamnosidase were pooled and dialyzed against distilled water at 4 °C for 4 h and then against 10 mM phosphate buffer (pH 6.5) at 4 °C for 20 h. The purified enzyme was aliquotted and stored at −80 °C.

Protein concentration determination
The protein concentration was quantified by the Bradford method [43], using the TaKaRa Bradford protein assay kit with bovine serum albumin (Dalian, TaKaRa Biotechnology, China) as the standard.

Assay of α-L-rhamnosidase
α-L-Rhamnosidase activity was assayed using *p*NPR (Sigma Aldrich, USA) as substrate. One unit of enzymatic activity was defined as the amount of enzyme releasing 1 μmol of *p*-nitrophenol per minute in 100 mM citrate-phosphate buffer at pH 6.5 and 65 °C. After incubation of the reaction mixture at 65 °C for 10 min, the liberated *p*-nitrophenol was determined spectrophotometrically at 405 nm under alkaline conditions (100 μL of the reaction mixture was added to 300 μL of 1 M Na_2CO_3).

Thermostability determination

The thermostability of α-L-rhamnosidase was determined by incubating the purified enzyme (1.13 mg/mL) in the presence and absence of sorbitol in 50 mM phosphate buffer (pH 6.5) at 70 °C for 1 h. And the samples were then centrifuged at 10000 rpm for 10 min. The residual α-L-rhamnosidase activity was determined as previously described. Measurements were performed in triplicate.

Thermal stability profile

The purified enzyme (1.13 mg/mL) in the presence and absence of sorbitol was incubated at temperatures ranging from 60 to 85 °C in 50 mM phosphate buffer (pH 6.5). Aliquots were removed after 10 min of incubation, and the residual α-L-rhamnosidase activity was determined as previously described. The values of T_{50} for the presence and absence of sorbitol, defined as the temperature at which 50% of the initial activity was retained, were determined from the plots of residual activity (%) versus temperature. Measurements were performed in triplicate.

Kinetics of thermal inactivation

The purified enzyme (1.13 mg/mL) in the presence and absence of sorbitol was incubated at temperatures ranging from 65 to 75 °C in 50 mM phosphate buffer (pH 6.5). Aliquots were removed at scheduled times, and the residual α-L-rhamnosidase activity was determined as previously described. Half-life was calculated from the first-order rate constants of inactivation, which were obtained from linear regression in logarithmic coordinates. The activation free energy ($\Delta G^{\neq}$) was calculated as previously described [44]. Measurements were performed in triplicate.

Kinetic parameters

The Michaelis-Menten parameters, K_M and $Vmax$ were determined from Michaelis–Menten plots by measuring the initial reaction rates with different substrate concentrations at pH 6.5 and 65 °C. Measurements were performed in triplicate.

Surface hydrophobicity analysis

The surface hydrophobicity (H_0) of the purified enzyme was determined using a fluorescence probe called ANS (Sangon Biotech, Shanghai, China), as previously described [45]. According to the method previously described [46], protein concentrations were diluted (0.02, 0.04, 0.06, 0.08 and 0.1 mg/ml) in 10 mM phosphate buffer solution (pH 6.5). Then, aliquots (5 µL) of ANS (8.0 mM in the same buffer) were added to 1 ml of sample. The fluorescence intensity at 495 nm was measured using a Cell imaging multifunctional test system (BioTek, Cytation3, America) with excitation at 370 nm.

H_0 was calculated by linear regression analysis using the initial slope of the fluorescence intensity versus protein concentration plot. Measurements were performed in triplicate.

Intrinsic fluorescence emission spectroscopy

Intrinsic fluorescence emission spectroscopy of α-L-rhamnosidase samples (0.1 mg/ml) in the presence and absence of sorbitol were measured in 10 mM phosphate buffer (pH 6.5) at room temperature using a fluorescence spectrophotometer (LS55, PE, America). To minimise the contribution of tyrosine residues to the emission spectra, the protein solutions were excited at 295 nm, and emission spectra were recorded from 320 to 400 nm at a scanning speed of 300 nm/min. The excitation and emission slit widths were 15 nm and 5 nm, respectively. All spectra were recorded in triplicate and corrected for the fluorescence of a protein-free sample.

Analysis of CD spectrum

A MOS-450 CD spectrometer (Biologic, Claix, Charente, France) was used for CD analysis. The CD spectra with UV (190–240 nm) region were recorded with a 2 mm path-length cell at room temperature. The spectra were obtained as the average of four scans with a bandwidth of 0.1 nm, a step resolution of 1 nm and a scan rate of 1 nm/s. The CD spectra of α-L-rhamnosidase (0.1 mg/ml) in the presence and absence of sorbitol were recorded in 10 mM phosphate buffer (pH 6.5), and corrected by subtracting control spectra of protein-free buffer solutions. Analysis of the protein secondary structure was performed using Dichroweb [47]. Four secondary structures, α-helix, β-sheet, turn, and random coil, were calculated.

Enzymatic hydrolysis of rutin and analysis of enzymatic hydrolysate

Rutin (Sangon Biotech, Shanghai, China) was treated with purified α-L-rhamnosidase, and the enzymatic hydrolysate was analyzed using an HPLC 1200 system (Agilent, USA) and a C18 column (4.6 × 250 mm; i.d. 5 µm; S.No. USAG008115, USA) with distilled water (A) and methanol (B) at A/B ratios 60:40 and run times of 30 min. The flow rate was 1.0 mL/min and the column was maintained at 30 °C, and detection was performed by monitoring the absorbance at 368 nm.

All enzymatic reactions were carried out in a temperature-controlled heating water bath. In this study, disodium hydrogen phosphate-citrate buffer (pH 4.5–6.5) were used. The typical reaction mixture (400 µL) contained disodium hydrogen phosphate-citrate buffer (pH 6.5), 8 mM of rutin and sorbitol. The reaction was started by adding the buffered solution of α-L-rhamnosidase from *Aspergillus terreus*, and the mixtures were incubated at 70 °C for different amounts of time at various sorbitol concentrations, pH values, substrate concentrations, enzyme concentrations while the other conditions were fixed in a temperature-controlled heating water bath. The reaction was stopped by adding 1 mL methanol. The crude hydrolysis products of rutin were then centrifuged at 10000 rpm for 10 min, and the supernatant solutions were filtered through a 0.45 μm filter before injection into the HPLC. Each value represents the mean of three independent measurements. The isoquercitrin yield and isoquercitrin concentration was calculated as follows.

$$\text{Isoquercitrin yield } (\%)$$

$$= \frac{\text{molar amount of isoquercitrin (mM)}}{\text{initial molar amount of rutin (mM)}} \times 100$$

$$\text{Isoquercitrin concentration (mM)}$$

$$= \text{isoquercitrin yield } (\%)$$

$$\times \text{ initial rutin concentration (mM)}$$

Results
Sorbitol enhanced the thermostability of α-L-rhamnosidase

From the practical point of view, thermostability is one of the most important characteristics to be considered for applying enzymes in industrial processes [48]. Therefore, the effects of different concentrations of sorbitol on the thermostability of α-L-rhamnosidase were determined. As shown in Fig. 1, the thermostability of α-L-rhamnosidase was increased in the presence of sorbitol ranging in molar concentration from 0.2 M to 2.5 M after incubation at 70 °C for 1 h. When the sorbitol concentration was less than 2.0 M, the thermostability

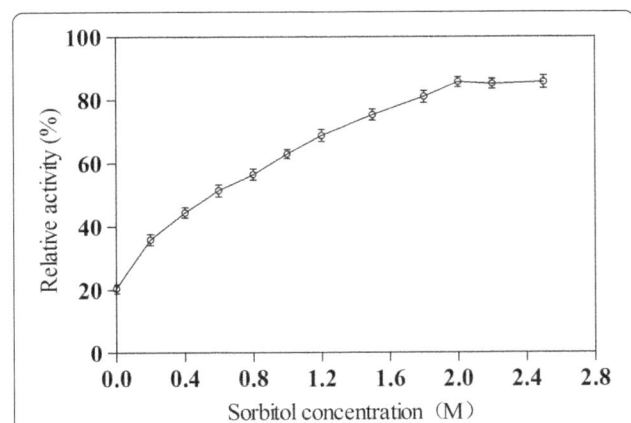

Fig. 1 Effects of sorbitol, at different concentrations, on the thermostability of α-L-rhamnosidase at 70 °C. The level of thermostability before each incubation was defined as 100%. Each value represents the mean of three independent measurements

enhancement of α-L-rhamnosidase was significantly increased following the increase of the sorbitol concentration. However, when the sorbitol concentration was more than 2.0 M, the thermostability enhancement of α-L-rhamnosidase was almost constant, which increased 65.3% by comparison without sorbitol. It clearly illustrated that 2.0 M sorbitol concentration is a point of inflection in thermostability enhancement of α-L-rhamnosidase resulted from adding sorbitol, and also it is the maximum peak for the rate of rise based on the increase curve of the α-L-rhamnosidase thermostability, For this reason, our further studies were focused on the reaction system with adding 2.0 M sorbitol.

Stability properties
The thermal stability profile of the presence and absence of sorbitol was determined by incubation during 10 min at temperatures ranging from 60 to 85 °C. As illustrated in Fig. 2a, after addition of 2.0 M sorbitol, the thermostability of enzyme was increased at temperatures ranging from 60 to 85 °C. Consequently, the value of T_{50} was increased by about 7 °C for α-L-rhamnosidase after addition of 2.0 M sorbitol.

The kinetics of thermal inactivation of the absence of sorbitol was determined by incubating the enzyme at several temperatures ranging from 65 to 75 °C during 1 h (Fig. 2b). The kinetics of thermal inactivation of the presence of sorbitol was also determined by incubating the enzyme at several temperatures ranging from 65 to 75 °C during 1 h (Fig. 2c). The results showed that addition of 2.0 M sorbitol decreased thermal inactivation rates. The half-life of α-L-rhamnosidase was increased at each temperature after addition of 2.0 M sorbitol (Table 1), which prolonged the half-life at 65 °C, 70 °C and 75 °C by 4.5-fold, 17.2-fold and 30.3-fold, respectively. The activation free energy ($\Delta G^{\neq}$) for thermal denaturation of α-L-rhamnosidase was increased at each temperature after addition of 2.0 M sorbitol (Table 1), which increased 4.2 kJ/mol, 8.2 kJ/mol and 10.0 kJ/mol of the activation free energy ($\Delta G^{\neq}$) at 65 °C, 70 °C and 75 °C, respectively. Taking into account the enzyme in the practical application, 70 °C was chose in the following experiment.

The kinetic parameters of the enzymes were analyzed using pNPR as a substrate at pH 6.5 and 65 °C. As shown Table 2, addition of sorbitol had a little effect on the values of K_M and $kcat/K_M$.

Surface hydrophobicity analysis
The surface hydrophobicity of α-L-rhamnosidase was measured after incubation at 70 °C for 4 h. As shown Table 3, the addition of different concentrations of sorbitol noticeably affected the surface hydrophobicity of α-L-rhamnosidase. Compared to the absence of sorbitol,

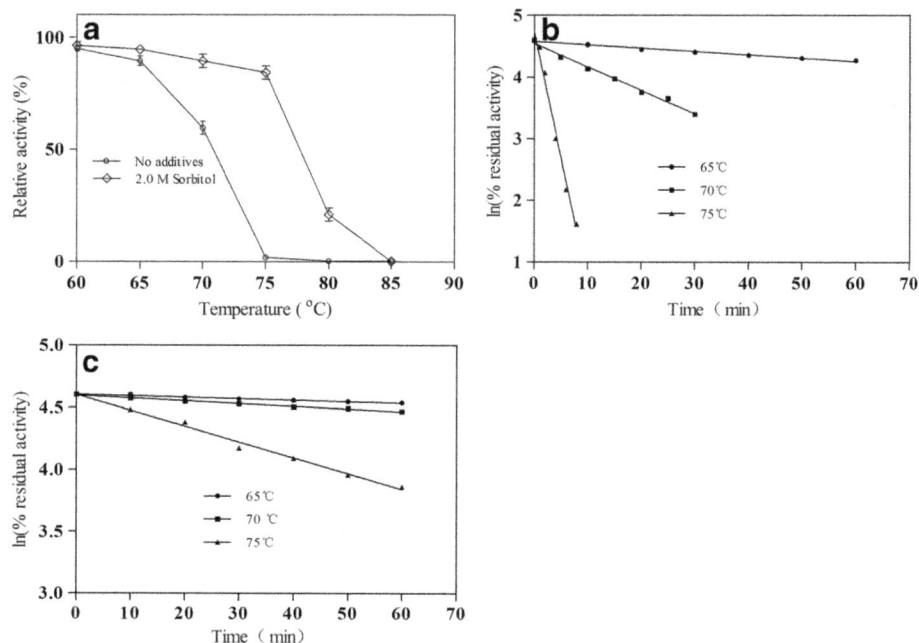

Fig. 2 a Thermal stability profile of α-L-rhamnosidase with and without 2.0 M of sorbitol incubated at different temperatures ranging from 60 °C to 85 °C for 10 min. **b** The kinetics of thermal inactivation of α-L-rhamnosidase without sorbitol incubated at different temperatures ranging from 65 to 75 °C for several time intervals. **c** The kinetics of thermal inactivation of α-L-rhamnosidase with sorbitol incubated at different temperatures ranging from 65 to 75 °C for several time intervals. The level of thermostability before each incubation was defined as 100%. Each value represents the mean of three independent measurements

the addition of 0.6 M and 1.0 M of sorbitol increased the surface hydrophobicity of α-L-rhamnosidase by 23.5 and 27%, respectively.

Emission fluorescence spectroscopic analysis

The fluorescence emission spectra of native α-L-rhamnosidase was examined. As shown in Fig. 3a, when excited at 295 nm, native α-L-rhamnosidase exhibited a maximum fluorescence emission at 342 nm. However, the maximum fluorescence intensity decreased rapidly, which was from 619.5 to 356.6 fluorescence intensity, and the fluorescence maximum exhibited a slight red shift, which was from 342 nm to 347 nm after incubation at 70 °C for 4 h. When sorbitol was added to the enzyme solution, the fluorescence intensity increased (Fig. 3b). Moreover, the

fluorescence maximum showed a blue shift with the addition of sorbitol, which was from 348.5 nm to 346 nm.

CD spectroscopy

It is known that the CD spectra in far-UV region reflects the secondary structure of protein [49]. To investigate the effect of sorbitol on the structure of α-L-rhamnosidase, CD measurements in the far UV (190–240 nm) were performed to reveal the changes that occurred in α-L-rhamnosidase secondary structure, both in the absence and presence of sorbitol, during incubation at 70 °C for 4 h (Fig. 4). Dichroweb was used to convert these CD spectra into the relative contributions of the secondary structural elements (α-helix, β-sheet, turn, and random coil) to the overall structure of α-L-rhamnosidase (Table 4). The α-helix content of the enzyme decreased

Table 1 Half-life and activation free energy ($\Delta G^{\neq}$) of α-L-rhamnosidase with and without 2.0 M of sorbitol

Temperature (°C)	Half-life (min)		$\Delta G^{\neq}$ (kJ/mol)	
	Control[a, b]	2.0 M Sorbitol[a, b]	Control[a, b]	2.0 M Sorbitol[a, b]
65	127.9 ± 1	580.3 ± 4	86.3 ± 0.4	90.5 ± 0.6
70	18.2 ± 0.2	313.7 ± 2	82.0 ± 0.3	90.2 ± 0.8
75	1.8 ± 0.1	54.6 ± 0.4	76.5 ± 0.5	86.5 ± 0.6

[a]Control, α-L-rhamnosidase without 2.0 M sorbitol
[b]Values are the means ± SD ($n = 3$)

Table 2 Kinetic parameters of α-L-rhamnosidase with and without sorbitol

Sample	Specific activity[A] (U/mg)	K_M^A (mM)	k_{cat}^A (s^{-1})	k_{cat}/K_M (s^{-1}M^{-1})
No additives	451.5 ± 12.5	0.481 ± 0.01	407.6 ± 11.6	8.5 × 10^3
0.6 M Sorbitol	414.5 ± 11.6	0.493 ± 0.02	379.2 ± 10.6	7.7 × 10^3
1.0 M Sorbitol	354.4 ± 10.4	0.542 ± 0.02	362.5 ± 10.7	6.7 × 10^3

[A]Values are the means ± SD (n = 3)

substantially after incubation at 70 °C for 4 h, and correspondingly, the random coil content increased substantially (Fig. 4a; Table 4).

Effect of sorbitol on the enzymatic conversion of rutin to isoquercitrin

The enzymatic conversion of rutin to isoquercitrin with sorbitol concentrations of 0.6 M to 1.8 M was tested to investigate the effect of different concentrations of sorbitol. As shown in Fig. 5a, the addition of 0.6 M to 1.8 M of sorbitol did enhance the enzymatic conversion of rutin to isoquercitrin. However, isoquercitrin yield did not increase with an increase in sorbitol concentration. The maximum increase in isoquercitrin yield (41.21%) was observed at 1.5 M of sorbitol because a high sorbitol concentration may decrease the mass transfer coefficient so as to reduce the isoquercitrin yield. Hence, the optimal sorbitol concentration to add to an enzymatic reaction system was calculated to be 1.5 M.

The effect of a buffer pH range (4.5–6.5) on the enzymatic conversion of rutin to isoquercitrin was investigated. As shown in Fig. 5b, the optimal pH of the enzymatic conversion of rutin to isoquercitrin by α-L-rhamnosidase in the absence or presence of sorbitol was 5.5. The addition of sorbitol enhanced the enzymatic conversion of rutin to isoquercitrin in the buffer pH range of 4.5–6.5. Furthermore, the enhancements of sorbitol on enzymatic conversion of rutin to isoquercitrin did not increase with the buffer pH. The addition of sorbitol at a buffer pH of 5.5 increased isoquercitrin yields by 11.5%. Based on this result, the optimal buffer pH for this enzymatic reaction system was calculated to be 5.5.

Table 3 Surface hydrophobicity values of α-L-rhamnosidase with and without sorbitol before and after incubation

Samples	H$_0$ values[c]
Control[a]	41483 ± 179.0
No additives[b]	289507 ± 112.1
0.6 M Sorbitol[b]	368005 ± 110.5
1.0 M Sorbitol[b]	357237 ± 238.1

[a]Control, α-L-Rhamnosidase before heating
[b]Samples after heating at 70 °C for 4 h
[c]Values are the means ± SD (n = 3)

To investigate the effect of substrate concentrations on the enzymatic conversion of rutin to isoquercitrin, a substrate concentration range from 8 mM to 20 mM was tested. As shown in Fig. 5c, as the concentration of rutin increased, the isoquercitrin concentration increased continuously. However, when the concentration of rutin was higher than 16 mM, the isoquercitrin concentration began to decrease. This result indicates that the substrate concentration plays an important part in the biosynthesis of isoquercitrin. As the substrate concentration increased, substrate inhibition of the reaction gradually became obvious. Therefore, the optimal concentration of rutin for isoquercitrin production by α-L-rhamnosidase was calculated to be 16 mM.

To investigate the effects of enzyme concentration and reaction time on the enzymatic conversion of rutin to isoquercitrin in the absence and presence of sorbitol, a range of reaction times, from 2 h to 10 h, was tested. As shown in Fig. 5, the isoquercitrin yield in the absence and presence of sorbitol almost always reached the maximum after 10 h of reaction. Therefore, the optimal reaction time for the enzymatic conversion of rutin to isoquercitrin was calculated to be 10 h. Compared to the absence of additive, the addition of 1.5 M of sorbitol increased the isoquercitrin yield from 60.01 to 96.43%. Moreover, the enhancements of isoquercitrin yield increased with the increase of reaction time.

Discussion

The aim of this study was to investigate the effect of sorbitol on the thermostability of α-L-rhamnosidase from *Aspergillus terreus* and enzymatic conversion rutin to isoquercitrin, and to analyze the mechanisms through which the stabilization was achieved. In this study, the increases in thermostability increased as the increase of the sorbitol concentration in the range of 0.2 to 2.0 M at 70 °C. However, the enhancement in thermostability was almost constant in the range of 2.0 to 2.5 M. The greatest increase in stability, a 17.2-fold increase in half-life, was seen at 2.0 M of sorbitol. The results of stability properties indicated addition of 2.0 M sorbitol at temperatures ranging from 60 to 85 °C enhance the thermostability of α-L-rhamnosidase. Furthermore, the results of the kinetics of thermal inactivation can be inferred that addition of 2.0 M of sorbitol serves to maintain the active conformation of enzyme after thermal treatment of high temperature ranging from 65 to 80 °C [50]. The effect of sorbitol on the thermostability of enzymes is not generally predictable. The addition of sorbitol has been reported to be able to increase the thermostability of several enzymes, for example, xylanase from *Thermomonospora sp.* [33], glucose dehydrogenase from *Haloferax mediterranei.* [37] and fungal α-amylase [51]. The results place α-L-rhamnosidase from *Aspergillus*

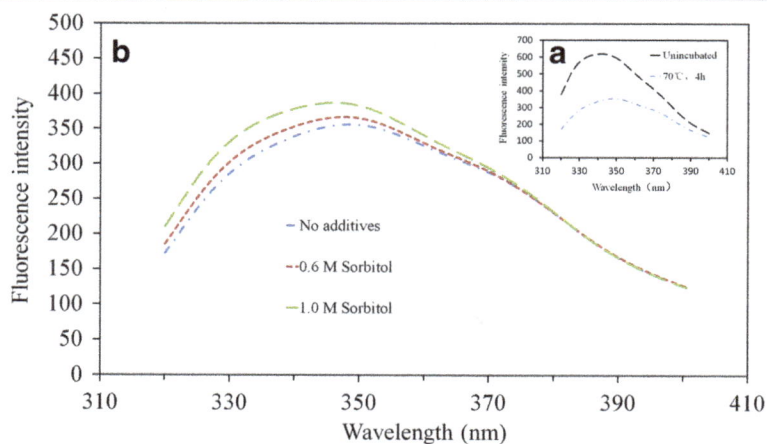

Fig. 3 The intrinsic fluorescence spectra of α-L-rhamnosidase. **a** Without sorbitol before and after incubation at 70°C for 4 h. **b** Incubated at 70°C for 4 h with different concentrations of sorbitol in 10 mM phosphate buffer at pH 6.5. All spectra were corrected for the fluorescence of a protein-free sample. Each value represents the mean of three independent measurements

terreus among those enzymes that are stabilized by the addition of sorbitol, and suggest that the addition of sorbitol may make α-L-rhamnosidase from *Aspergillus terreus* suitable for use in industrial processes.

In this study, the addition of sorbitol enhanced the surface hydrophobicity of α-L-rhamnosidase at 70 °C. The result is consistent with the reports of other researchers [46, 52–54], which shows that the sorbitol decreases the direct interaction of water and protein due to the preferential hydration, thereby increasing the thermostability of α-L-rhamnosidase.

In addition, the fluorescence and CD data suggest that the effects of sorbitol on α-L-rhamnosidase are definite. The measurement of intrinsic protein fluorescence is widely used to investigate the changes of tertiary conformation of proteins, because the fluorescence of internal tryptophan residues is particularly sensitive to various perturbations. The α-L-rhamnosidase from *Aspergillus terreus* contains 31 tryptophan residues. The sharp decreased in fluorescence intensity and red-shift of the fluorescence maximum that accompanied heating of the protein to 70 °C, indicating a shift of these residues to a more hydrophilic environment, was reversed by the addition of sorbitol. The extent of the reversal was in proportion to their ability to increase its thermostability. Since these data were relatively easy to obtain and correlated with thermostability, intrinsic fluorescence is considered the simplest predictor of thermostabilizing

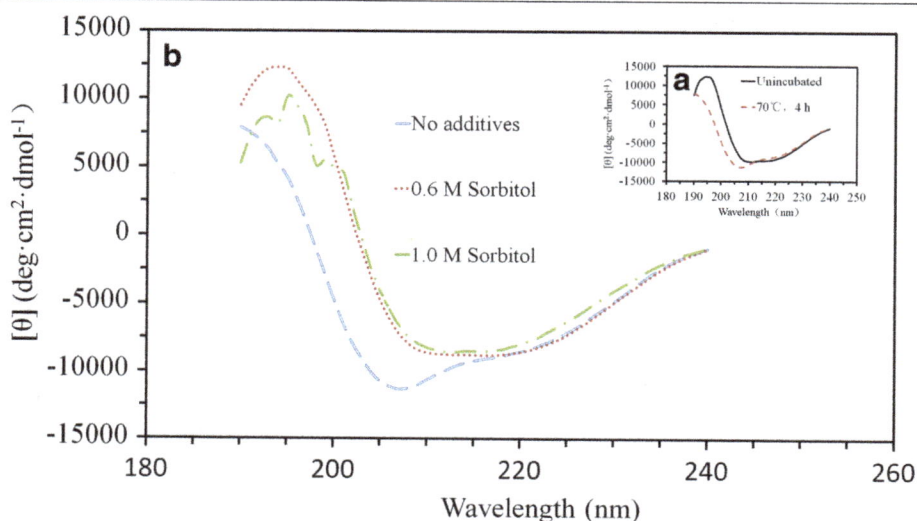

Fig. 4 Far-UV CD spectra of α-L-rhamnosidase. **a** Without sorbitol before and after incubation at 70 °C for 4 h. **b** Incubated at 70 °C for 4 h with different concentrations of sorbitol in 10 mM phosphate buffer at pH 6.5. All spectra were corrected for the signal generated by a protein-free sample

Table 4 Contents of estimated secondary structures of α-L-rhamnosidase with and without sorbitol before and after incubation

Samples	α-Helix (%)	β-Sheet (%)	Turn (%)	Random coil (%)
Control[a]	30	28.88	20.1	21.12
No additives[b]	22	22.48	23.26	32.27
0.6 M Sorbitol[b]	27.17	29.6	21.21	22.12
1.0 M Sorbitol[b]	27.11	31.28	19.56	22.05

[a]Control, α-L-Rhamnosidase before heating
[b]Samples after heating at 70 °C for 4 h

ability [46]. According to CD data, the addition of different concentrations of sorbitol increased the contents of α-helix and β-sheet, but they decreased the content of random coil (Fig. 4b and Table 4), which indicated the addition of sorbitol to mixture preserved secondary structures in the protein. The result was consistent with the report of other researcher [46], which could explain why addition of sorbitol could improve the thermostability of α-L-rhamnosidase from *Aspergillus terreus*.

Although there have been many reports about improving thermal stabilization of enzymes by polyols, there is no reports about adding a polyol to improve the yield of enzymatic hydrolysis product. In this study, this is the first time to apply sorbitol to improve the

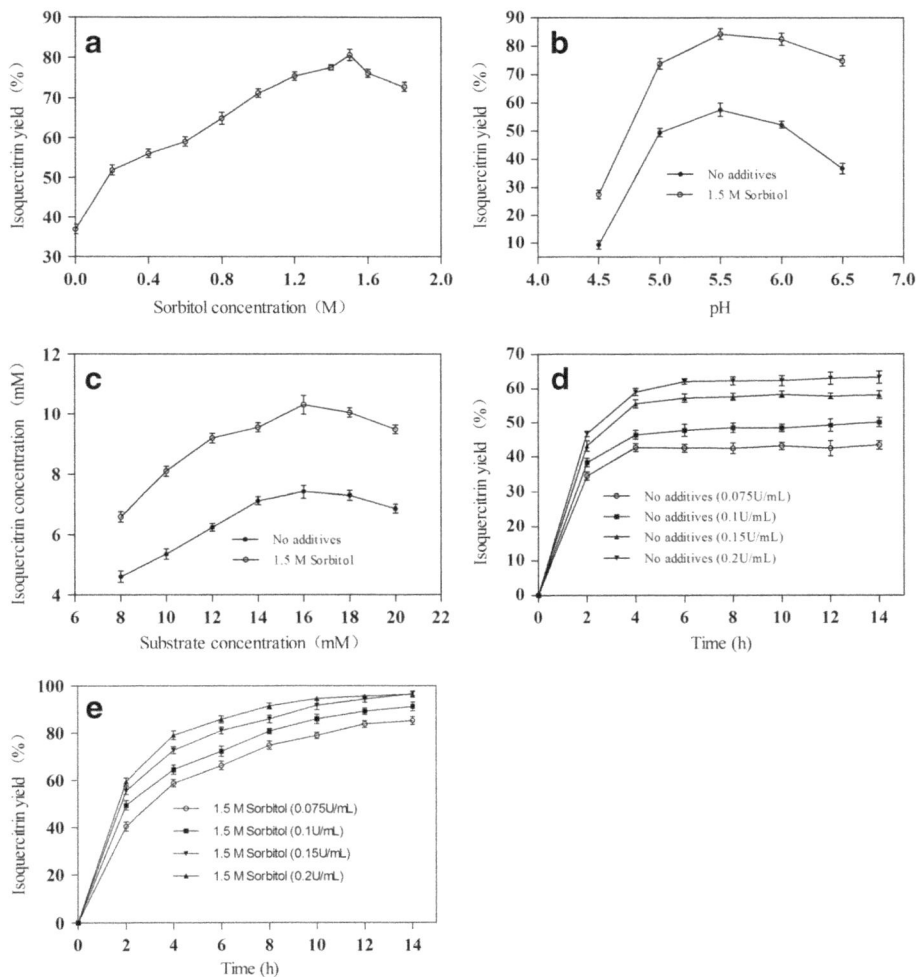

Fig. 5 Effect of sorbitol on the enzymatic conversion of rutin to isoquercitrin. **a** Incubated at 70 °C for 4 h with different concentration of sorbitol in disodium hydrogen phosphate-citrate buffer at pH 6.5, 8 mM of rutin, and 0.1 U/mL of α-L-rhamnosidase; (**b**) Incubated at 70 °C for 4 h with 1.5 M of sorbitol in disodium hydrogen phosphate-citrate buffer at pH range from 4.5 to 6.5, 8 mM of rutin, and 0.1 U/mL of α-L-rhamnosidase; (**c**) Incubated at 70 °C for 4 h with different concentration of rutin in disodium hydrogen phosphate-citrate buffer at pH 5.5, 1.5 M of sorbitol, and 0.1 U/mL of α-L-rhamnosidase; (**d**) Incubated at 70 °C for up to 12 h with different enzyme concentration without 1.5 M sorbitol in disodium hydrogen phosphate-citrate buffer at pH 5.5, 16 mM of rutin. **e** Incubated at 70 °C for up to 12 h with different enzyme concentration with 1.5 M sorbitol in disodium hydrogen phosphate-citrate buffer at pH 5.5, 16 mM of rutin. Each value represents the mean of three independent measurements

thermostability of the enzyme so as to enhance the enzymatic conversion of rutin to isoquercitrin. Addition of sorbitol slightly decreased the value of K_M and $kcat/K_M$, however, compared to the absence of additive, the isoquercitrin yield increased by 1.6 times after addition of 1.5 M of sorbitol, which was attributed to the addition of sorbitol to improve the thermostability of α-L-rhamnosidase. The result indicates that the addition of sorbitol to the reaction mixture makes α-L-rhamnosidase from *Aspergillus terreus* more suitable for use in industrial processes.

Conclusions

Addition of sorbitol enhanced the thermostability of α-L-rhamnosidase from *Aspergillus terreus* at temperatures ranging from 65 °C to 80 °C. Half-life and activation free energy with addition of 2.0 M sorbitol at 70 °C were increased by 17.2-fold, 8.2 kJ/mol, respectively. Moreover, the isoquercitrin yield increased by 1.6-fold with the addition of 1.5 M of sorbitol at 70 °C. The results suggest that the reaction system by adding sorbitol has great potential to promote enzymatic conversion of rutin to isoquercitrin production.

Abbreviations
ANS: 8-Anilino1-naphthalenesulfonic acid ammonium salt; BMGY: Buffered glycerol-complex medium; CD: Circular dichroism; HPLC: High performance liquid chromatography; pNPR: *p*-Nitrophenyl α-L-rhamnopyranoside; YNB: Yeast nitrogen base; YPD: Yeast extract peptone dextrose medium

Acknowledgements
Not applicable.

Funding
This work was supported by the National Natural Science Foundation of China (Grant No. 31570565), Special Fund for Forest Scientific Research in the Public Welfare (Grant No. 201404601), the Jiangsu "333" project of cultivation of high-level talents (Grant No. BRA2015317), the 11th Six Talents Peak Project of Jiangsu Province (Grant No. 2014-JY-011), the Jiangsu Key Lab of Biomass-based Green Fuels and Chemicals (Grant No. JSBGFC 14013), and A Project Funded by the Priority Academic Program Development of Jiangsu Higher Education Institutions (PAPD) as well as the Doctorate Fellowship Foundation of Nanjing Forestry University.

Authors' contributions
LG and ANC designed and performed the experiments and drafted the manuscript. JJP and XYF performed purification and characterization. GD contributed to analyze the data of the circular dichroism and surface hydrobicity. ZZW and FT helped to analyze the intrinsic fluorescence emission spectroscopy and revise the manuscript. LGZ and WX directed the over-all study and had given final approval of the version to be published. All authors had read and approved the final manuscript.

Competing interests
The authors declare that they have no competing interests.

Author details
[1]Co-Innovation Center for Sustainable Forestry in Southern China, Nanjing Forestry University, 159 Long Pan Road, Nanjing 210037, China. [2]College of Chemical Engineering, Nanjing Forestry University, 159 Long Pan Road, Nanjing 210037, China. [3]Jiangsu Kanion Pharmaceutical Co., Ltd, 58 Haichang South Road, Lianyungang 222001, Jiangsu Province, China. [4]International centre for bamboo and rattan, 8 FuTong East Street, Beijing 100714, China.

References
1. You HJ, Ahn HJ, Ji GE. Transformation of Rutin to Antiproliferative Quercetin-3-glucoside by *Aspergillus niger*. J Agr Food Chem. 2010;58: 10886–92.
2. Walle T. Absorption and metabolism of flavonoids. Free Radic Biol Med. 2004;36:829–37.
3. Amado NG, Cerqueira DM, Menezes FS, da Silva JFM, Neto VM, Abreu JG. Isoquercitrin isolated from Hyptis fasciculata reduces glioblastoma cell proliferation and changes beta-catenin cellular localization. Anti-Cancer Drug. 2009;20:543–52.
4. Gasparotto A, Gasparotto FM, Lourenco ELB, Crestani S, Stefanello MEA, Salvador MJ, da Silva-Santos JE, Marques MCA, Kassuya CAL. Antihypertensive effects of isoquercitrin and extracts from *Tropaeolum majus L.* evidence for the inhibition of angiotensin converting enzyme. J Ethnopharmacol. 2011;134:363–72.
5. Valentova K, Vrba J, Bancirova M, Ulrichova J, Kren V. Isoquercitrin: pharmacology, toxicology, and metabolism. Food Chem Toxicol. 2014;68:267–82.
6. Jiang P, Burczynski F, Campbell C, Pierce G, Austria JA, Briggs CJ. Rutin and flavonoid contents in three buckwheat species *Fagopyrum esculentum, F-tataricum, and F-homotropicum* and their protective effects against lipid peroxidation. Food Res Int. 2007;40:356–64.
7. Weignerova L, Marhol P, Gerstorferova D, Kren V. Preparatory production of quercetin-3-beta-D-glucopyranoside using alkali-tolerant thermostable alpha-L-rhamnosidase from *Aspergillus terreus*. Bioresour Technol. 2012;115:222–7.
8. Wang J, Sun GX, Yu L, Wu FA, Guo XJ. Enhancement of the selective enzymatic biotransformation of rutin to isoquercitrin using an ionic liquid as a co-solvent. Bioresour Technol. 2013;128:156–63.
9. Rajal VB, Cid AG, Ellenrieder G, Cuevas CM. Production, partial purification and characterization of alpha-l-rhamnosidase from *Penicillium ulaiense*. World J Microbiol Biotechnol. 2009;25:1025–33.
10. Bokkenheuser VD, Shackleton CH, Winter J. Hydrolysis of dietary flavonoid glycosides by strains of intestinal *Bacteroides* from humans. Biochem J. 1987;248:953–6.
11. Caldini CBF, Pifferi PG, Lanzarini G, Galante YM. Kinetic and immobilization studies on the fungal glycosidases for the aroma enhancement in wine. Enzyme Microb Technol. 1994;16:286–91.
12. Habelt K, Pittner F. A rapid method for the determination of naringin, prunin, and naringenin applied to the assay of naringinase. Anal Biochem. 1983;134:393–7.
13. Roitner MST, Pittner F. Characterization of naringinase from *Aspergillus niger*. Monatshefte Fur Chemie. 1984;115:11758–63.
14. Yoshinobu TTK, Takahisa N, Hiroshi T, Shigetaka O. Prevention of hesperidin crystal formation in canned mandarin orange syrup and clarified orange juice by hesperidin glycosides. Sci Technol Int. 1995;1:29–33.
15. Qian SL, Yu HS, Zhang CZ, Lu MC, Wang HY, Jin FX. Purification and characterization of dioscin-alpha-L-rhamnosidase from *pig liver*. Chem Pharm Bull. 2005;53:911–4.
16. Puri M, Kalra S. Purification and characterization of naringinase from a newly isolated strain of *Aspergillus niger* 1344 for the transformation of flavonoids. World J Microbiol Biotechnol. 2005;21:753–8.

17. Park SY, Kim JH, Kim DH. Purification and characterization of quercitrin-hydrolyzing alpha-L-rhamnosidase from *Fusobacterium K-60*, a human intestinal bacterium. J Microbiol Biotechnol. 2005;15:519–24.

18. Bourbouze RP-SF, Percheron F. Rhamnodiastase et α-L-rhamnosidase de *Fagopyrum esculentum*. Phytochemistry. 1975;14:1279–82.

19. McMahon HZB, Fugelsang K, Jasinski Y. Quantification of glycosidase activities in selected yeasts and lactic acid bacteria. J Ind Microbiol Biotechnol. 1999;23:198–203.

20. De Lise F, Mensitieri F, Tarallo V, Ventimiglia N, Vinciguerra R, Tramice A, Marchetti R, Pizzo E, Notomista E, Cafaro V, et al. RHA-P: isolation, expression and characterization of a bacterial α-l-rhamnosidase from *Novosphingobium sp.* PP1Y. J Mol Catal B Enzym. 2016;134:136–47.

21. Li L, Yu Y, Zhang X, Jiang Z, Zhu Y, Xiao A, Ni H, Chen F. Expression and biochemical characterization of recombinant alpha-l-rhamnosidase r-Rha1 from *Aspergillus niger* JMU-TS528. Int J Biol Macromol. 2016;85:391–9.

22. Qian SRGL, Wang HY, Zhang CZ, Yu HS. Isolation and characterization of dioscin-alpha-L-rhamnosidase from *bovine liver*. J Mol Catal B-Enzym. 2013;97:31–5.

23. Soares NFF, Hothkiss JH. Naringinase immobilization in packaging films for reducing naringin concentration in grapefruit juice. J Food Sci. 1998;63:61–5.

24. Michon F, Katzenellenbogen E, Kasper DL, Jennings HJ. Structure of the complex group-specific polysaccharide of group B *Streptococcus*. Biochemistry. 1987;26:476–86.

25. Ellenrieder GBS, Daz M. Hydrolysis of supersaturated naringin solutions by free and immobilized naringinase. Biotechnol Tech. 1998;12:63–5.

26. De Winter K, Simcikova D, Schalck B, Weignerova L, Pelantova H, Soetaert W, Desmet T, Kren V. Chemoenzymatic synthesis of alpha-L-rhamnosides using recombinant alpha-L-rhamnosidase from *Aspergillus terreus*. Bioresour Technol. 2013;147:640–4.

27. Zhang R, Zhang BL, Xie T, Li GC, Tuo Y, Xiang YT. Biotransformation of rutin to isoquercitrin using recombinant alpha-L-rhamnosidase from *Bifidobacterium breve*. Biotechnol Lett. 2015;37:1257–64.

28. Yadav V, Yadav PK, Yadav S, Yadav KDS. Alpha-L-Rhamnosidase: a review. Process Biochem. 2010;45:1226–35.

29. Rojas NL, Voget CE, Hours RA, Cavalitto SF. Purification and characterization of a novel alkaline alpha-L-rhamnosidase produced by *Acrostalagmus luteo albus*. J Ind Microbiol Biotechnol. 2011;38:1515–22.

30. Xie JC, Zhao DX, Zhao LG, Pei JJ, Xiao W, Ding G, Wang ZZ. Overexpression and characterization of a Ca2+ activated thermostable beta-glucosidase with high ginsenoside Rb1 to ginsenoside 20(S)-Rg3 bioconversion productivity. J Ind Microbiol Biotechnol. 2015;42:839–50.

31. Plaza L, Duvetter T, Van der Plancken I, Meersman F, Van Loey A, Hendrickx M. Influence of environmental conditions on thermal stability of recombinant *Aspergillus aculeatus* pectinmethylesterase. Food Chem. 2008;111:912–20.

32. Koschorreck K, Schmid RD, Urlacher VB. Improving the functional expression of a *Bacillus licheniformis* laccase by random and site-directed mutagenesis. BMC Biotechnol. 2009;9:12.

33. George SP, Ahmad A, Rao MB. A novel thermostable xylanase from *Thermomonospora sp.* influence of additives on thermostability. Bioresour Technol. 2001;78:221–4.

34. Kusano M, Yasukawa K, Inouye K. Effects of the mutational combinations on the activity and stability of thermolysin. J Biotechnol. 2010;147:7–16.

35. Jiang H, Zhang SW, Gao HF, Hu N. Characterization of a cold-active esterase from *Serratia sp* and improvement of thermostability by directed evolution. BMC Biotechnol. 2016;16:7.

36. Liu JZ, Wang M. Improvement of activity and stability of chloroperoxidase by chemical modification. BMC Biotechnol. 2007;7:23.

37. Jose Maria Obon AM, Iborra JL. Comparative thermostability of glucose dehydrogenase from *Haloferax mediterranei*. Effects of salts and polyols. Enzyme Microb Technol. 1995;19:352–60.

38. Pazhang M, Mehrnejad F, Pazhang Y, Falahati H, Chaparzadeh N. Effect of sorbitol and glycerol on the stability of trypsin and difference between their stabilization effects in the various solvents. Biotechnol Appl Biochem. 2016;63:206–13.

39. Timasheff SN. Control of protein stability and reactions by weakly interacting cosolvents: the simplicity of the complicated. Adv Protein Chem. 1998;51:355–432.

40. Mehrnejad F, Ghahremanpour MM, Khadem-Maaref M, Doustdar F. Effects of osmolytes on the helical conformation of model peptide: molecular dynamics simulation. J Chem Phys. 2011;134:035104.

41. McClements DJ. Modulation of globular protein functionality by weakly interacting cosolvents. Crit Rev Food Sci Nutr. 2002;42:417–71.

42. Gerstorferova D, Fliedrova B, Halada P, Marhol P, Kren V, Weignerova L. Recombinant alpha-L-rhamnosidase from *Aspergillus terreus* in selective trimming of rutin. Process Biochem. 2012;47:828–35.

43. Bradford MM. A rapid and sensitive method for the quantitation of microgram quantities of protein utilizing the principle of protein-dye binding. Anal Biochem. 1976;72:248–54.

44. Darias R, Villalonga R. Functional stabilization of cellulase by covalent modification with chitosan. J Chem Technol Biotechnol. 2001;76:489–93.

45. Hou DH, Chang SKC. Structural characteristics of purified glycinin from *soybeans* stored under various conditions. J Agr Food Chem. 2004;52:3792–800.

46. Li CM, Li WW, Holler TP, Gu ZB, Li ZF. Polyethylene glycols enhance the thermostability of beta-cyclodextrin glycosyltransferase from *Bacillus circulans*. Food Chem. 2014;164:17–22.

47. Whitmore L, Wallace BA. Protein secondary structure analyses from circular dichroism spectroscopy: Methods and reference databases. Biopolymers. 2008;89:392–400.

48. Klibanov AM. Stabilization of enzymes against thermal inactivation. Adv Appl Microbiol. 1983;29:1–28.

49. Ranjbar B, Gill P. Circular dichroism techniques: biomolecular and nanostructural analyses- a review. Chem Biol Drug Des. 2009;74:101–20.

50. Fernandez M, Villalonga ML, Fragoso A, Cao R, Banos M, Villalonga R. Alpha-Chymotrypsin stabilization by chemical conjugation with O-carboxymethyl-poly-beta-cyclodextrin. Process Biochem. 2004;39:535–9.

51. Combes MGD. Effect of polyols on fungal alpha-amylase thermostability. Enzyme Microb Technol. 1989;10:673–7.

52. Liu Y, Zhao GL, Zhao MM, Ren JY, Yang B. Improvement of functional properties of peanut protein isolate by conjugation with dextran through Maillard reaction. Food Chem. 2012;131:901–6.

53. Liu FF, Ji L, Zhang L, Dong XY, Sun Y. Molecular basis for polyol-induced protein stability revealed by molecular dynamics simulations. J Chem Physics. 2010;132:225103.

54. Murakami S, Kinoshita M. Effects of monohydric alcohols and polyols on the thermal stability of a protein. J Chem Phys. 2016;144:125105.

3

Protoplast transformation as a potential platform for exploring gene function in *Verticillium dahliae*

Latifur Rehman[†], Xiaofeng Su[†], Huiming Guo, Xiliang Qi and Hongmei Cheng[*]

Abstract

Background: Large efforts have focused on screening for genes involved in the virulence and pathogenicity of *Verticillium dahliae*, a destructive fungal pathogen of numerous plant species that is difficult to control once the plant is infected. Although *Agrobacterium tumefaciens*-mediated transformation (ATMT) has been widely used for gene screening, a quick and easy method has been needed to facilitate transformation.

Results: High-quality protoplasts, with excellent regeneration efficiency (65 %) in TB3 broth (yeast extract 30 g, casamino acids 30 g and 200g sucrose in 1L H_2O), were generated using driselase (Sigma D-9515) and transformed with the GFP plasmid or linear GFP cassette using PEG or electroporation. PEG-mediated transformation yielded 600 transformants per microgram DNA for the linear GFP cassette and 250 for the GFP plasmid; electroporation resulted in 29 transformants per microgram DNA for the linear GFP cassette and 24 for the GFP plasmid. To determine whether short interfering RNAs (siRNAs) can be delivered to the protoplasts and used for silencing genes, we targeted the *GFP* gene of Vd-GFP (*V. dahliae* GFP strain obtained in this study) by delivering one of four different siRNAs—19-nt duplex with 2-nt 3' overhangs (siRNA-gfp1, siRNA-gfp2, siRNA-gfp3 and siRNA-gfp4)—into the Vd-GFP protoplasts using PEG-mediated transformation. Up to 100 % silencing of *GFP* was obtained with siRNA-gfp4; the other siRNAs were less effective (up to 10 % silencing). *Verticillium* transcription activator of adhesion (*Vta2*) gene of *V. dahliae* was also silenced with four siRNAs (siRNA-vta1, siRNA-vta2, siRNA-vta3 and siRNA-vta4) independently and together using the same approach; siRNA-vta1 had the highest silencing efficiency as assessed by colony diameter and quantitative real time PCR (qRT-PCR) analysis.

Conclusion: Our quick, easy transformation method can be used to investigate the function of genes involved in growth, virulence and pathogenicity of *V. dahliae*.

Keywords: *Verticillium dahliae*, Driselase, Transformation, siRNAs

Background

V. dahliae, the causal agent of *Verticillium* wilt, is one of the most destructive plant pathogens, affecting over 400 plant species, including important ornamental, horticultural, agronomical and woody plants [1, 2]. Its control is difficult because it produces microsclerotia, which can survive in soil for several years [3]. Moreover, no effective fungicides or other chemicals are available to overcome this pathogen once the plant is infected.

Despite a great deal of research, the molecular mechanisms behind the pathogenicity of this fungus have remained unclear [1].

An efficient transformation system is necessary for genetic manipulation and functional genomics studies of fungi [4]. A number of methods, *Agrobacterium tumefaciens*-mediated transformation (ATMT), PEG-mediated transformation, electroporation, and particle bombardment, have been used to transform different fungal species, including *V. dahliae*, with variable efficiencies [5–7]. ATMT is widely used for transforming various materials such as protoplasts, spores or hyphae of several fungal species [8, 9]. In *V. dahliae*, ATMT has

* Correspondence: chenghongmei@caas.cn
[†]Equal contributors
Biotechnology Research Institute, Chinese Academy of Agricultural Sciences, Beijing 100081, China

been used for targeted gene disruption [7, 10–13] or deletion [14–16], but it is laborious and time-consuming. PEG-mediated transformation of *V. dahliae* was first reported in 1995 and has yielded a high efficiency of transformation [17–19]; however, obtaining the large amounts of high-quality protoplasts crucial to the success of the method is difficult for many fungal species including *V. dahliae*. Moreover PEG-mediated transformation results in a high percentage of transient transformation. Yet for filamentous fungi, this method has been ideal because it is simple and fast [20].

RNA interference (RNAi) is an effective tool to investigate gene function in various organisms [21–24]. In filamentous fungi, plasmid constructs expressing dsRNA have been applied for this purpose with a silencing efficiency between 70–90 % [25–28]. Although using this procedure for silencing gene is efficient, designing an RNAi plasmid is laborious. Synthetic siRNA can be used for the same purpose, e.g., incubation of synthetic siRNA with germinating spores of *A. nidulans* successfully silenced the target gene, leading to reduced mycelial growth [29]. In another study, synthetic dsRNA had in vitro antifungal activity against *adenylate cyclase*, *DNA polymerase alpha subunit* and *DNA polymerase delta subunit* in *F. oxysporum* and hindered spore germination [30]. However, transforming spores of certain fungal species like *V. dahliae* with siRNA is difficult as we have tried different ways to transform but obtained no satisfactory results (unpublished data). Thus, methods to directly transform protoplasts with synthetic siRNA to elucidate gene function have been sought. As reported for *Fusarium* sp. HKF15, siRNAs designed against hydroxymethyl glutaryl coenzyme A reductase (*hmgR*) and farnesyl pyrophosphate synthase (*fpps*) were used to transform protoplasts and knockdown these genes, however the silencing was effective for only 48 h [31].

Our main objective was to develop an easy and quick method to transform *V. dahliae* and facilitate screening of essential genes. First, we developed a protocol using one enzyme to obtain high quality protoplasts from *V. dahliae* with excellent regeneration efficiency in TB3 broth. We then compared variations in PEG-mediated transformation and electroporation methods to develop the most efficient protocol to transform the protoplasts with the GFP plasmid (circular and linear). We also used synthetic siRNAs (19-nt duplex with 2-nt 3′ overhangs) targeting the *GFP* gene in the GFP-transformed strain (Vd-GFP) and the *Vta2* gene, a regulatory gene that is essential for growth and conidiation of *V. dahliae* [16], in the wild-type strain (Vd-wt) using PEG-mediated transformation to test whether the siRNAs can enter the protoplasts and inhibit the expression of these genes. Our results indicated that PEG-mediated transformation is more effective than electroporation. Moreover the

transformation efficiency for siRNAs and the linear GFP cassette was significantly higher than with the circular GFP plasmid.

Our method of protoplast isolation, regeneration and transformation has advantages over other available methods in its rapidity and ease for generating protoplasts using a single enzyme and transforming the protoplasts with high efficiency. These techniques are conducive for the study of gene function using siRNA silencing or gene deletion in a short period of time.

Methods

Fungal growth and spore harvesting

Strain V991 of *V. dahliae*, a highly toxic and defoliating wild-type pathogenic strain, provided by Prof. Guiliang Jian of the Institute of Plant Protection, Chinese Academy of Agricultural Sciences (CAAS), was cultured on PDA plates at 25 °C for 7-10 days. Sterile distilled water was added to the plates to harvest spores by gently scraping the agar with a sterile loop. The resulting suspension was filtered through a sterile 40 µm nylon filter (Falcon, REF352340) and centrifuged at 4000 rpm for 5 min. The final spore concentration was adjusted to 1.5×10^7/mL.

Protoplast isolation

Driselase (Sigma D-9515), selected after comparing with a variety of enzymes (cellulase: Sigma C1184; snailase: BBI SB0870; lysozyme: Sigma 62970), was prepared by dissolving 500 mg *driselase* in 25 ml NaCl (0.7 M) and centrifuged at 12,000 rpm for 10 min. The supernatant was taken and purified using 0.22 µm filters (MILLEX®GP). Two milliliters of *V. dahliae* spores (1.5×10^7/ml) were cultured in 100 mL Complete Medium (CM: yeast extract 6 g, casein acid hydrolysate 6 g and 10 g sucrose in 1L H_2O) for 16–24 h at 28 °C and 150 rpm. Mycelia were then separated from the culture and medium using a sterile 40 µm nylon filter, then washed 2–3 times with 0.7 M NaCl. The harvested mycelia were aseptically transferred to 10 ml of the driselase solution and incubated at 33 °C for 0.5–3.5 h at 60 rpm. The preparation was then checked every 30 min for protoplast release. After the incubation time, the mixture was then filtered using a sterile 40 µm nylon filter to remove any hyphal fragments, and the protoplasts were centrifuged at 2800 rpm for 5 min. The supernatant was discarded, and the pellet was washed 2–3 times either with 1 M sorbitol, in case the protoplast has to be used for electroporation, or with STC buffer (20 % sucrose, 10 mM Tris-HCl pH 8.0, and 50 mM $CaCl_2$), if used for PEG-mediated transformation. The concentration of protoplasts was adjusted with either 1 M sorbitol or STC to 10^6/ml.

Regeneration of protoplasts

The ability of the protoplasts to regenerate was examined in CM, TB3 and Czapek-Dox broths. Briefly, 200 µl protoplasts (10^6/ml) were cultured in 5 ml broth and incubated at 25 °C for 18 h. Protoplasts were observed for regeneration with a light microscope at 20× magnification, and the percentage of regeneration was calculated by counting the number of regenerated protoplasts out of total protoplasts cultured. In order to isolate a single colony, the regenerated protoplast suspension was centrifuged at 2800 rpm, the supernatant discarded and pellet was resuspended in 200 µl CM broth, serially diluted and cultured on PDA for 5–7 days until the colonies appeared.

GFP plasmid and siRNAs

The GFP plasmid (pCH-sGFP, Additional file 1) was kindly provided by Professor Xie Bingyan of the Institute of Vegetables and Flowers, CAAS. Primers GFP-1 5′ CTTTCGACACTGAAATACGTCG3′ and GFP-2 5′ GCATCAGAGCAGATTGTACTGAGAG3′ were used to amplify the GFP cassette from the GFP plasmid.

The siRNAs targeting different regions of the *GFP* gene (siRNA-gfp1, siRNA-gfp2, siRNA-gfp3 and siRNA-gfp4) and the *Vta2* gene (siRNA-vtaNC, siRNA-vta1, siRNA-vta2, siRNA-vta3 and siRNA-vta4) were designed and synthesized by Oligobio, Beijing, China. The sequences of siRNAs are given in Table 1 and the positions of these siRNAs along the genes are shown in Additional file 1 and Additional file 2.

PEG-mediated transformation

For PEG-mediated transformation, an established protocol was followed with some modifications [32]. Briefly, 200 µl protoplasts (10^6/ml) was mixed with 12 µg GFP plasmid (12.2 kb, bearing the hygromycin resistance cassette (hph) as a selection marker) or linear GFP cassette (3.3 kb) in a 50 ml Falcon tube and incubated on ice for 30 min; 1.5 ml 60 % PEG solution in STC buffer was added to the tube dropwise, gently swirled and left at room temperature for 15 min followed by the addition

of 5 ml TB3 broth. The tubes were incubated at 25 °C for 18 h, and *GFP* expression was checked with a fluorescence microscope (Zeiss Axio Imager M1, Jena, Germany). Transformants, transformed with GFP plasmid, were selected on PDA media supplemented with hygromycin B (50mg/mL final concentration) after regeneration in TB3 broth. Further confirmation of the positive transformants was made by PCR using GFP-CF 5′AGCTGGACGGCGACGTAAAC3′ and GFP-CR 5′ GATGGGGGTGTTCTGCTGGT3′ primers.

Electroporation

Before using electroporation for transformation, protoplasts were shocked at different field strengths from 100-1000 V/cm to ensure that electroporation had no or a very low lethal effect on the regeneration of protoplasts.

The protoplasts were mixed with 12 µg of the GFP plasmid or linear GFP cassette as described above, using 1 M sorbitol as the buffer, and kept on ice for 10 min. The mixtures were transferred to a 0.2-cm gap cuvette (BioRad, Hercules, CA), and different voltages (300, 400 and 500 V) were applied for 5 ms using a GenPulser Xcell electroporation system (BioRad). Immediately after electroporation, the protoplasts were transferred to 5 ml TB3 broth and incubated at 25 °C for 18 h. After regeneration of protoplasts, the culture was centrifuged at 2800 rpm for 5 min. The supernatant was discarded, and the pellet was resuspended in 200 µl TB3 and observed with a fluorescence microscope.

siRNA inhibition assay for *GFP* and *Vta2* genes

For targeting the *GFP* gene, 200 µl Vd-GFP protoplasts (10^6/ml) were transformed with 10 µM siRNA-gfp1, siRNA-gfp2, siRNA-gfp3 or siRNA-gfp4 by PEG-mediated transformation and regenerated for 18 h as described. Inhibition of *GFP* expression was checked using fluorescence microscopy by counting the number of hyphae with fluorescence. The percentage of *GFP* inhibition was determined by dividing the number of hyphae with no fluorescence on the number of hyphae with fluorescence multiplied by 100.

Similarly, for silencing *Vta2* gene, protoplasts obtained from Vd-wt were treated with siRNA-vta1, siRNA-vta2, siRNA-vta3 and siRNA-vta4 independently in a final concentration of 10 µM or with 2.5 µM siRNAs mix using PEG-mediated transformation in RNase free environment. siRNA-vtaNC was used as a negative control. Briefly, protoplasts obtained from vd-wt (10^6/ml) were mixed with either of the siRNAs or with mixed siRNAs in 50 mL falcon tube and incubated on ice for 30 min; 1.5 ml 60 % PEG solution in STC buffer was added to the tube dropwise, gently swirled and left at room temperature for 15 min followed by the addition of 5 ml TB3 broth. After 18 h of incubation in TB3 broth, the

Table 1 siRNA sequences developed against *GFP* and *Vta2*

Name	Sense sequence	Antisense sequence
siRNA-gfp1	UCUUCAAGGACGACGGCAAUU	UUGCCGUCGUCCUUGAAGAUU
siRNA-gfp2	GCCACAACGUCUAUAUCAUU	AUGAUAUAGACGUUGUGGCUU
siRNA-gfp3	GCAUGGACGAGCUGUACAAUU	UUGUACAGCUCGUCCAUGCUU
siRNA-gfp4	UCAAGGAGGACGGCAACAUU	AUGUUGCCGUCCUCCUUGAUU
siRNA-vta1	CCAGGGCAUGUACUCUCAAUU	UUGAGAGUACAUGCCCUGGUU
siRNA-vta2	GCAUGUACUCUCAACACAAUU	UUGUGUUGAGAGUACAUGCUU
siRNA-vta3	CCACGCUCAACACCUCUAUU	AUAGAGGUGUUGAGCGUGGUU
siRNA-vta4	GGCGCAACAAGCAAGCAAUU	AUUGCUUGCUUGUUGCGCCUU
siRNA-vtaNC	UUCUCCGAACGUGUCACGUUU	ACGUGACACGUUCGGAGAAUU

cultures were centrifuged at 2500 rpm for 5 min, the supernatant was discarded, and the pellet was resuspended in 200 µl TB3 broth and pipeted onto the center of CM agar plates. The colony diameter was measured after 10 days at 25 °C.

qRT-PCR analysis of Vta2 expression level

In order to further confirm that the reduction in growth of *V. dahliae* was due to silencing of *Vta2* gene, we conducted qRT-PCR. After treating the protoplasts with siRNAs (siRNA-vtaNC, siRNA-vta1, siRNA-vta2, siRNA-vta3, siRNA-vta4 and siRNA-vtamix respectively), they were cultured in TB3 broth for 72 h and the mycelia were harvested for RNA extraction by RNA Extraction Kit (YPHBio, Tianjin, China). qRT-PCR was carried out in 7500 Real Time PCR System (ABI, Massachusetts, USA). Gene specific primers were used: vta2-F 5′GGCTTC CTCAAGGTCGGCTATG3′, vta2-R 5′GCTGCATGTCA TCCCACTTCTTC3′, Vdactin-F 5′GGCTTCCTCAAGG TCGGCTATG3′ and Vdactin-R 5′GCTGCATGTCATC CCACTTCTTC3′ *Vdactin* was used as a housekeeping gene [33]. The relative expression of the targeted gene was analyzed using the $2^{-\Delta\Delta Ct}$ method. The standard curve met experimental requirements ($R^2 > 0.99$, E > 95 %) [34].

Statistical analysis

All experiments were done in three independent replicates. Means ± standard deviation and significant differences were determined using Duncan's multiple range test and *t*-test with *p*-values < 0.05 in SPSS 17.0 software (SPSS Inc., Chicago, IL, USA).

Results

Isolation and regeneration efficiency of protoplasts

In order to select an efficient enzyme for protoplasts isolation from *V. dahliae*, we treated the mycelia with different enzymes (driselase, cellulase, snailase and lysozyme) and found that driselase resulted in maximum protoplasts yield (Additional file 3). While investigating the effect of driselase concentration on the yield of protoplast, 20 mg/mL was found the best (Additional file 3).

For selecting the optimal mycelial age and enzymolytic time to isolate protoplasts, spores were cultured for different times (16, 18, 20, 22 and 24 h) and then treated with driselase (0.5–3.5 h). The optimal mycelial age was found to be 20 h which yielded 5.5×10^7/ml ± 0.275 protoplasts when treated with driselase (Fig. 1a), while 2.5 h enzymolysis time was observed to produce the maximum number of protoplasts for all culture ages (Fig. 1b). When the three media were tested for regeneration efficiency, the efficiency was highest in TB3 (65 ± 3 %) (Fig. 1c).

Observation of fluorescence from GFP expression

Soon after PEG-mediated transformation and electroporation, the protoplasts were cultured for 18 h in TB3 broth to detect *GFP* expression as an indicator of transformation. Strong GFP fluorescence was observed in the transformed protoplasts, but none was seen in the Vd-wt (Fig. 2a). PEG-4000 gave the highest transformation of all the methods for both linear GFP (600 ± 20 transformants/µg DNA) and for GFP plasmid (250 ± 10 transformants/µg DNA) (Fig. 2b). Electroporation for both GFP plasmid (24 ± 1 transformants/µg DNA) and linear GFP cassette (29 ± 1 transformants/µg DNA) was significantly lower than the PEG-mediated (PEG-4000) transformation.

GFP transformants selection and stability of the transgene

GFP transformants derived from GFP plasmid transformation were selected after 5–7 days of culturing the regenerated protoplasts on PDA plates supplemented with hygromycin B. The selected GFP transformants fluoresced strongly when viewed with fluorescence microscopy. Further confirmation of the transformants was made by PCR (Additional file 4). Expression of GFP was observed from three generations of transformants, indicating stable *GFP* expression throughout these three generations.

Silencing of GFP gene in strain Vd-GFP with siRNAs

Protoplasts of strain Vd-GFP were treated with the different siRNAs targeting the *GFP* gene, and checked for inhibition of *GFP* expression after regeneration (Fig. 2c). We found out that siRNA-gfp4 gave the best silencing efficiency (up to 100 %) compared with 10 % or less with siRNA-gfp1, siRNA-gfp2 and siRNA-gfp3 (Fig. 2d). The silencing of the *GFP* gene lasted for at least 72 h.

Silencing of Vta2 gene

To validate whether *V. dahliae* genes can be silenced by siRNA using *Vta2* as a reference gene, the gene was successfully silenced with siRNAs. On CM plates, the colony diameter of siRNA-vta1 group (2.8 cm) was significantly smaller than that of the siRNA-vtaNC group (4.6 cm) (Fig. 3a and b). The colony diameters of siRNA-vta2, siRNA-vta3, siRNA-vta4 and siRNA-vtamix groups were 3.6 cm, 3.5 cm, 3.2 cm and 3.0 cm, respectively. As determined by colony diameter, siRNA-vta1 had the best silencing efficiency. To further confirm whether the reduction in colony diameter was due to silencing of *Vta2* gene, qRT-PCR was conducted to determine the relative expression level of this gene in all the groups. The data was in accordance with that obtained from colony assessment. Expression level of

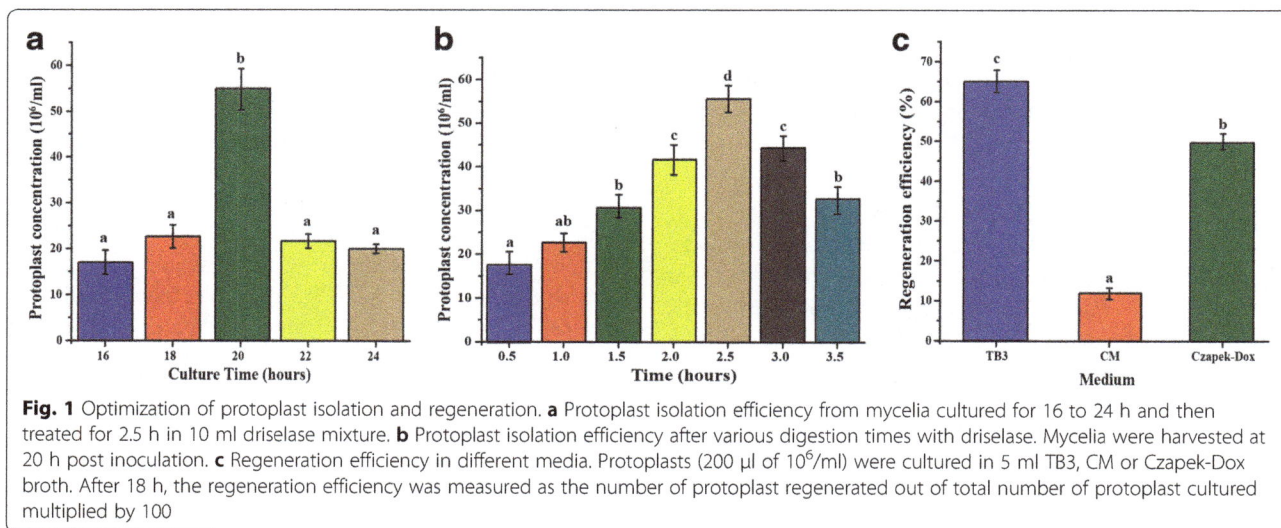

Fig. 1 Optimization of protoplast isolation and regeneration. **a** Protoplast isolation efficiency from mycelia cultured for 16 to 24 h and then treated for 2.5 h in 10 ml driselase mixture. **b** Protoplast isolation efficiency after various digestion times with driselase. Mycelia were harvested at 20 h post inoculation. **c** Regeneration efficiency in different media. Protoplasts (200 μl of 10^6/ml) were cultured in 5 ml TB3, CM or Czapek-Dox broth. After 18 h, the regeneration efficiency was measured as the number of protoplast regenerated out of total number of protoplast cultured multiplied by 100

Vta2 in siRNA-vta1 group was significantly lower than the other groups (Fig. 3c).

Discussion

In this study, we isolated and regenerated protoplasts from *V. dahliae*, then transformed the protoplasts with the GFP plasmid and linear GFP cassette using PEG-mediated transformation and electroporation, and silenced the *GFP* and *Vta2* genes using siRNAs.

Protoplasts have been isolated from many fungal species at various efficiencies depending on the species and conditions. Driselase has proved efficient for protoplast isolation from *Fusarium graminearum* (ca 10^9 g^{-1} wet mass) and *Ascosphaera apis* (98.36 × 10^5 mL^{-1} of protoplasts) [35, 36]. For *V. dahliae*, protoplasts were previously isolated using a combination of two enzymes [18, 19]. In our study, we isolated protoplasts from *V. dahliae* by using a single enzyme *driselase*, with up to 90 % efficiency. The main differences in our protocol and that developed in the previous study [19] are the number of spores they cultured for protoplasts isolation and the enzymolysis temperature. In our study $1.5 × 10^7$/ml spores were cultured for obtaining mycelia to be digested with the enzyme while they used 10^6/ml. We used 33 °C as the optimum enzymolysis temperature and 20 h old culture the best for protoplast isolations as compared to their 30 °C and 24 h old culture. There is a difference in the media used for the growth of mycelia as well between the two protocols that can also have a significant effect on the protoplast production [36–39].

The efficiency of protoplast isolation, in addition to the digestion enzyme and other factors, also depends on the age of the mycelia. Young and exponentially growing mycelia are the best choices for protoplast isolation [36], but the optimal age of the mycelia varies for different species, e.g., 60 h for *Blakeslea trispora* and 2 days for

Pleurotus pulmonarius and *Pleurotus florida* [40, 41]. For *V. dahliae*, previous studies have used 24-h-old mycelia for protoplast isolation [18, 19], but here we obtained more protoplasts from the 20-h culture than from the 24-h culture. The probable reason for this can be the sensitivity of the younger mycelia to the digesting enzymes, with the increase in growth the cell wall becomes thicker and the mycelia would be digested more difficultly leading to decreased protoplast yield [36, 38, 39]. In addition to the age of mycelia, the protoplast yield also depends on the duration of the enzyme digestion, which has ranged from 3 h to 16 h for maximum protoplast release in other fungal species [37, 40, 42]. The optimum time of enzymolysis in our study was 2.5 h. Less time is presumably not enough for all the mycelia to be digested by the enzyme, while prolonged enzymolyis can result in the breaking of protoplasts [37, 40–42].

Regeneration of protoplasts is a vital step and is the main limiting factor in a transformation experiment. Thus, a high frequency of regeneration is necessary for genetic manipulation of the particular fungus. Protoplasts from different fungi have been isolated with various regeneration frequencies, ranging from 3.3 to 77.5 % [37, 40–42], partly depending on the media used to culture the protoplast. For example, the protoplast regeneration efficiencies for *B. trispora* and *Nodulisporium sylviforme* were found to be 77.5 % and 72 % on PDA respectively while the regeneration frequencies decreased (for *B. trispora* 32.5 % on RM and for *N. sylviforme* 44 % on CM) with the use of other media [40, 42]. In our study, the frequency of regeneration was about 5-fold higher in TB3 broth than in CM broth.

To increase the transformation efficiency of *V. dahliae* protoplasts, we also tested a number of protocols to determine the best one. Transformation efficiency of

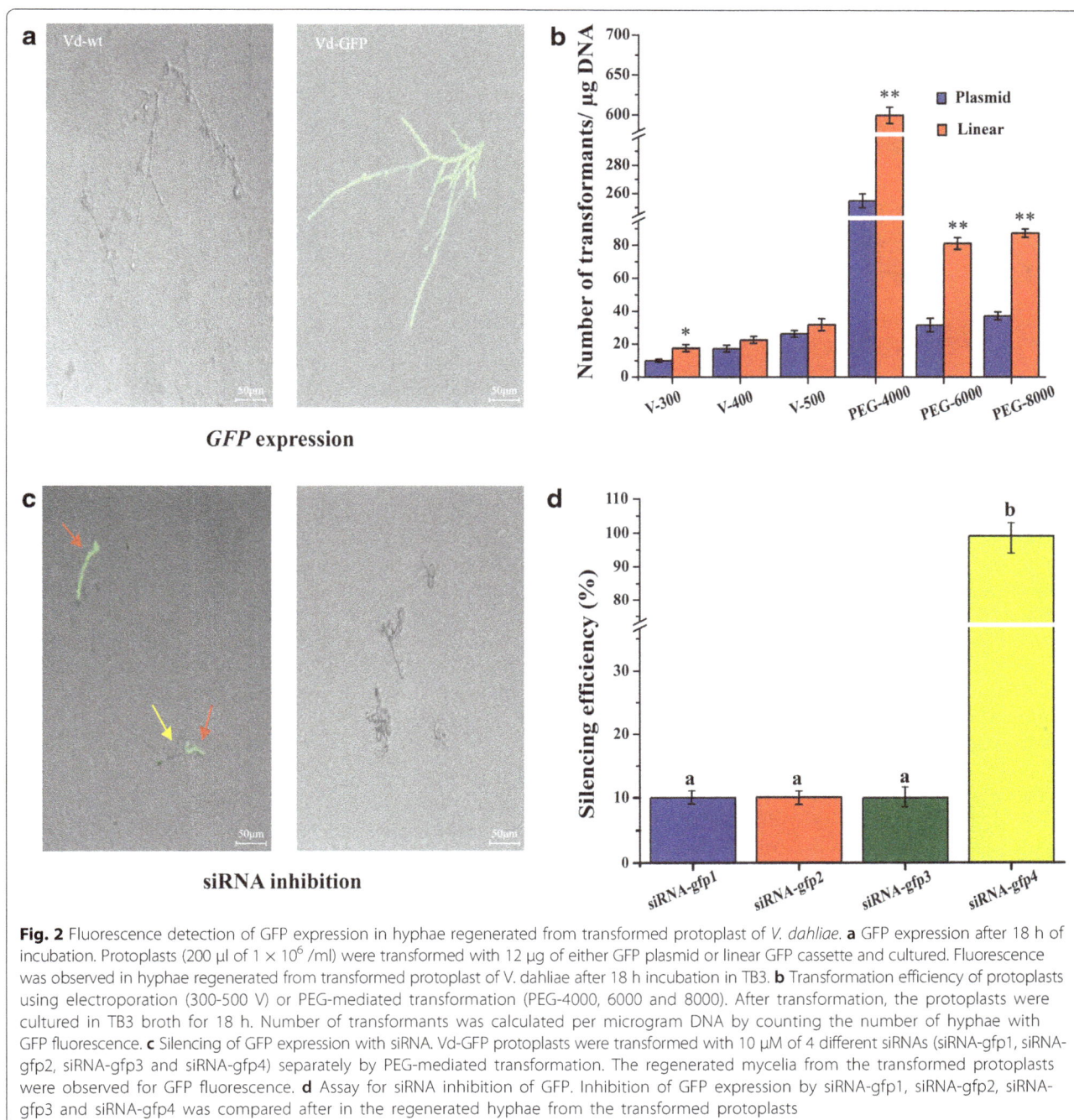

Fig. 2 Fluorescence detection of GFP expression in hyphae regenerated from transformed protoplast of *V. dahliae*. **a** GFP expression after 18 h of incubation. Protoplasts (200 µl of 1×10^6 /ml) were transformed with 12 µg of either GFP plasmid or linear GFP cassette and cultured. Fluorescence was observed in hyphae regenerated from transformed protoplast of V. dahliae after 18 h incubation in TB3. **b** Transformation efficiency of protoplasts using electroporation (300-500 V) or PEG-mediated transformation (PEG-4000, 6000 and 8000). After transformation, the protoplasts were cultured in TB3 broth for 18 h. Number of transformants was calculated per microgram DNA by counting the number of hyphae with GFP fluorescence. **c** Silencing of GFP expression with siRNA. Vd-GFP protoplasts were transformed with 10 µM of 4 different siRNAs (siRNA-gfp1, siRNA-gfp2, siRNA-gfp3 and siRNA-gfp4) separately by PEG-mediated transformation. The regenerated mycelia from the transformed protoplasts were observed for GFP fluorescence. **d** Assay for siRNA inhibition of GFP. Inhibition of GFP expression by siRNA-gfp1, siRNA-gfp2, siRNA-gfp3 and siRNA-gfp4 was compared after in the regenerated hyphae from the transformed protoplasts

various fungal species with PEG has been variable, e.g., 102 transformants/1 µg DNA for *V. fungicola*, 100-200 for the basidiomycete *Pleurotus ostreatus* [5, 43], and for *V. dahliae*, 10–50 transformants/1 µg DNA [18]. We obtained much higher yields in our experiments: 250 transformants/1 µg DNA using the GFP plasmid (12.2 kb) and 600 using the linear GFP cassette (3.3 kb) and PEG-4000. The higher frequency of transformation might be due to the high quality of protoplasts obtained. Plasmid size also plays a vital role in the transformation

efficiency. Previous studies have indicated that increasing plasmid size results in decreased transformation efficiency [44, 45]. The fewer transformants obtained with the GFP plasmid in comparison to the linear GFP cassette is thus probably due to the large size of the plasmid. On the other hand, transforming *V. dahliae* protoplast using electroporation did not yield promising results for either of the GFP plasmids. The main hurdle in electroporation is the regeneration of the protoplasts. As the voltage increases, the regeneration capacity of the

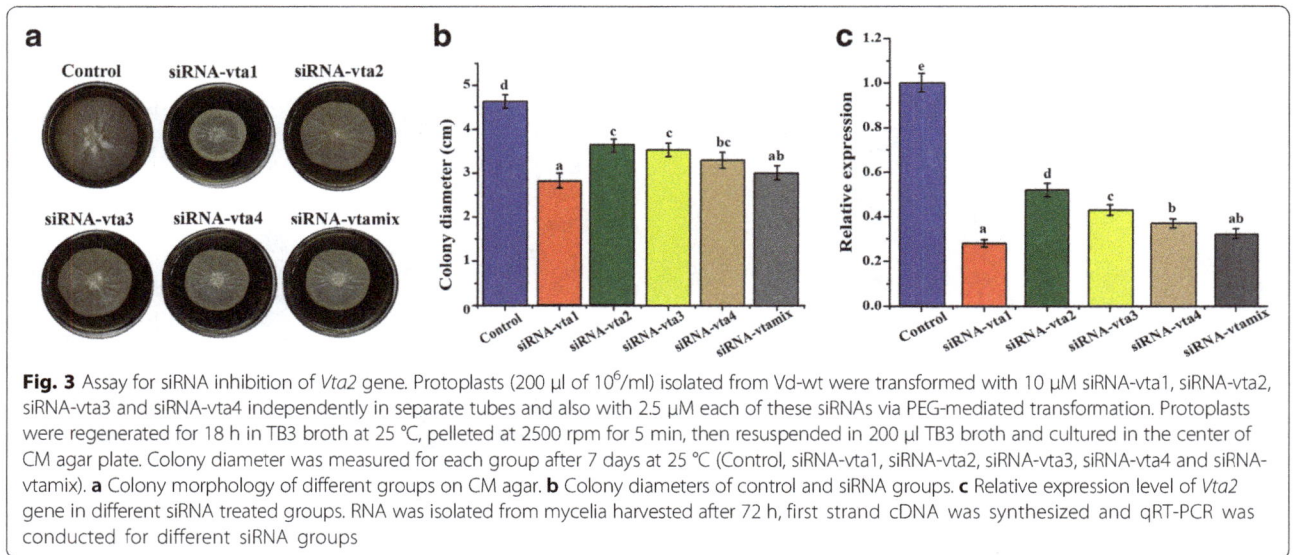

Fig. 3 Assay for siRNA inhibition of *Vta2* gene. Protoplasts (200 μl of 10^6/ml) isolated from Vd-wt were transformed with 10 μM siRNA-vta1, siRNA-vta2, siRNA-vta3 and siRNA-vta4 independently in separate tubes and also with 2.5 μM each of these siRNAs via PEG-mediated transformation. Protoplasts were regenerated for 18 h in TB3 broth at 25 °C, pelleted at 2500 rpm for 5 min, then resuspended in 200 μl TB3 broth and cultured in the center of CM agar plate. Colony diameter was measured for each group after 7 days at 25 °C (Control, siRNA-vta1, siRNA-vta2, siRNA-vta3, siRNA-vta4 and siRNA-vtamix). **a** Colony morphology of different groups on CM agar. **b** Colony diameters of control and siRNA groups. **c** Relative expression level of *Vta2* gene in different siRNA treated groups. RNA was isolated from mycelia harvested after 72 h, first strand cDNA was synthesized and qRT-PCR was conducted for different siRNA groups

protoplast decreases. At low voltage (300–500 V), the transformation efficiency ranged from 10 to 29 transformants/1 μg DNA for GFP plasmid and linear GFP cassette, respectively, much lower than with PEG-4000.

RNAi is a powerful reverse genetics tool for deciphering gene function in various organisms including fungi [46]. Characterizing gene function using gene deletion, disruption or insertion is a time-consuming process. Downregulation of a gene using RNAi is an alternative method in functional genomics as it is a rapid process as compared to the deletion or disruption of a gene. Moreover this approach is particularly helpful in studying essential genes or genes present in multiple copies within the genome that could compensate for each other's function. In fungi, synthetic siRNAs have been used to downregulate specific genes. For example, *hmgR* and *fpps* in *Fusarium* sp. HKF15 were silenced by delivering siRNAs designed against these genes into the protoplasts [31]. In another study, three siRNA sequences (Nor-Ia, Nor-Ib, Nor-Ic) targeting the mRNA sequence of the *aflD* gene were tested for controlling aflatoxin production in *Aspergillus flavus* and *Aspergillus parasiticus* [47]. Designing siRNAs that are more effective at downregulating is essential for gene silencing. Several siRNAs designed from different sites within the same gene can have striking differences in silencing efficiency [48, 49] as shown by the silencing of the *GFP* gene at various efficiencies using different siRNAs.

Vta2 gene was used as a reference gene for siRNA inhibition assay because its inhibition can easily be assessed from the colony growth [16]. After treating protoplasts obtained from Vd-wt transformed with different siRNAs designed for this gene, we observed a significant decrease in the colony diameter of the siRNA-treated groups compared with the control group. Differences in colony diameter were also observed among the siRNA groups, indicating that siRNA designed from different locations within the gene can have strikingly different silencing effects. The difference in the colony diameter among the siRNA groups and the control groups lasted for about 10 days, sufficient time for characterizing a gene.

Conclusion

Our improved method greatly increased the number of transformants per microgram of DNA over the others available. This method will be useful for elucidating gene functions by downregulating a particular gene of interest using siRNA and constructing gene deletion mutants of *V. dahliae* in a shorter time than required for ATMT.

Additional files

Additional file 1: Sketch of pCH-sGFP and position of siRNA along the *GFP* gene. (A) Diagram of GFP plasmid (pCH-sGFP). (B) Position of siRNAs along the *GFP* gene. siRNAs were designed and synthesized by Oligobio, Beijing, China.

Additional file 2: Position of siRNAs along the *Vta2* gene of *V. dahliae*. The position of different siRNAs designed to target this gene is shown in this figure. Sequence underlined with different colors shows different siRNAs.

Additional file 3: Selection of efficient enzyme and the effect of driselase concentration on the protoplasts isolation from *V. dahliae*. (A) Protoplasts isolation efficiency from the mycelia of Verticillium dahliae by treating with different enzymes, (B) The effect of driselase concentration on the release of protoplasts.

Additional file 4: Confirmation of GFP transformants by PCR. Single colony was selected and cultured in CM for 5-7 days. Mycelia were harvested and genomic DNA was isolated. PCR was carried out with gene specific primers.

Abbreviations

ATMT, *Agrobacterium tumefaciens*-mediated transformation; CM, complete medium; PDA, potato dextrose agar; PEG, polyethylene glycol; siRNA, short interfering RNA; *Vta*, *Verticillium* transcription activator of adhesion

Acknowledgments

The authors are thankful to Dr. Ijaz Ali and Mr. Adil Khan for revising the manuscript.

Funding

This work was supported by a grant from National Nonprofit Industry Research (201503109).

Authors' contributions

HC and HG designed the study and advised on protocols. LR and XS carried out the experimental procedures. XQ helped with experimental procedures and manuscript preparation. The manuscript was read and approved by all the authors.

Competing interests

The authors declare that they have no competing interests.

References

1. Fradin EF, Thomma BPHJ. Physiology and molecular aspects of *Verticillium* wilt diseases caused by *V. dahliae* and *V. albo-atrum*. Mol Plant Pathol. 2006;7:71–86.
2. Klosterman SJ, Atallah ZK, Vallad GE, Subbarao KV. Diversity, pathogenicity, and management of *Verticillium* species. Annu Rev Phytopathol. 2009;47:39–62.
3. Pegg GF, Brady BL. *Verticillium* Wilts. CABI Pulishing: New York, NY, USA; 2002.
4. Dobrowolska A, Staczek P. Development of transformation system for *Trichophyton rubrum* by electroporation of germinated conidia. Curr Genet. 2009;55:537–42.
5. Amey RC, Athey-Pollard A, Burns C, Mills PR, Bailey AM, Foster GD. PEG-mediated and *Agrobacterium*-mediated transformation in the mycopathogen *Verticillium fungicola*. Mycol Res. 2002;106:4–11.
6. Mullins ED, Chen X, Romaine P, Raina R, Geiser DM, Kang S. Agrobacterium-Mediated Transformation of *Fusarium oxysporum*: An Efficient Tool for Insertional Mutagenesis and Gene Transfer. Phytopathology. 2001;91:173–80.
7. Dobinson KF, Grant SJ, Kang S. Cloning and targeted disruption, via *Agrobacterium tumefaciens* -mediated transformation, of a trypsin protease gene from the vascular wilt fungus *Verticillium dahliae*. Curr Genet. 2004;45:104–10.
8. de Groot MJA, Bundock P, Hooykaas PJJ, Beijersbergen AGM. *Agrobacterium tumefaciens*-mediated transformation of filamentous fungi. Nat Biotechnol. 1998;16:839–42.
9. Chen X, Stone M, Schlagnhaufer C, Romaine CP. A Fruiting Body Tissue Method for Efficient *Agrobacterium*-Mediated Transformation of *Agaricus bisporus*. Appl Environ Microbiol. 2000;66:4510–3.
10. Tzima A, Paplomatas EJ, Rauyaree P, Kang S. Roles of the catalytic subunit of cAMP-dependent protein kinase A in virulence and development of the soilborne plant pathogen *Verticillium dahliae*. Fungal Genet Biol. 2010;47: 406–15.
11. Rauyaree P, Ospina-Giraldo MD, Kang S, Bhat RG, Subbarao KV, Grant SJ, et al. Mutations in *VMK1*, a mitogen-activated protein kinase gene, affect microsclerotia formation and pathogenicity in *Verticillium dahliae*. Curr Genet. 2005;48:109–16.
12. Klimes A, Dobinson KF. A hydrophobin gene, *VDH1*, is involved in microsclerotial development and spore viability in the plant pathogen *Verticillium dahliae*. Fungal Genet Biol. 2006;43:283–94.
13. He X-J, Li X-L, Li Y-Z. Disruption of Cerevisin via *Agrobacterium tumefaciens*-mediated transformation affects microsclerotia formation and virulence of *Verticillium dahliae*. Plant Pathol. 2015;64:1157–67.
14. Hoppenau CE, Tran V-T, Kusch H, Aßhauer KP, Landesfeind M, Meinicke P, et al. *Verticillium dahliae VdTHI4*, involved in thiazole biosynthesis, stress response and DNA repair functions, is required for vascular disease induction in tomato. Environ Exp Bot. 2014;108:14–22.
15. Qi X, Su X, Guo H, Qi J, Cheng H. A *ku70* null mutant improves gene targeting frequency in the fungal pathogen *Verticillium dahliae*. World J Microbiol Biotechnol. 2015;31:1889–97.
16. Tran VT, Braus-Stromeyer SA, Kusch H, Reusche M, Kaever A, Kuhn A, et al. *Verticillium* transcription activator of adhesion *Vta2* suppresses microsclerotia formation and is required for systemic infection of plant roots. New Phytol. 2014;202:565–81.
17. Dobinson KF. Genetic transformation of the vascular wilt fungus Verticillium dahliae. Can J Bot. 1995;73:710–5.
18. Wang Y, Xiao S, Xiong D, Tian C. Genetic transformation, infection process and qPCR quantification of *Verticillium dahliae* on smoke-tree *Cotinus coggygria*. Australas Plant Pathol. 2012;42:33–41.
19. Xiao S, Sun Y, Tian C, Wang Y. Optimization of factors affecting protoplast preparation and transformation of smoke-tree wilt fungus *Verticillium dahliae*. African J Microbiol Res. 2013;7:2712–8.
20. Liu Z, Friesen TL. Polyethylene Glycol (PEG)-Mediated Transformation in Filamentous Fungal Pathogens. In: Bolton MD, Thomma BPHJ, editors. Plant Fungal Pathog. Methods Protoc. Methods Mol. Biol. Totowa, NJ: Humana Press; 2012. p. 365–75.
21. Fjose A, Ellingsen S, Wargelius A, Seo H-C. RNA interference: mechanisms and applications. Biotechnol Annu Rev. 2001;7:31–57.
22. Maeda I, Kohara Y, Yamamoto M, Sugimoto A. Large-scale analysis of gene function in *Caenorhabditis elegans* by high-throughput RNAi. Curr Biol. 2001; 11:171–6.
23. Clemens JC, Worby CA, Simonson-Leff N, Muda M, Maehama T, Hemmings BA, et al. Use of double-stranded RNA interference in *Drosophila* cell lines to dissect signal transduction pathways. Proc Natl Acad Sci. 2000;97:6499–503.
24. Malhotra P, Dasaradhi PVN, Kumar A, Mohmmed A, Agrawal N, Bhatnagar RK, et al. Double-stranded RNA-mediated gene silencing of cysteine proteases (*falcipain-1 and -2*) of *Plasmodium falciparum*. Mol Microbiol. 2002;45:1245–54.
25. Nakayashiki H, Hanada S, Quoc NB, Kadotani N, Tosa Y, Mayama S. RNA silencing as a tool for exploring gene function in ascomycete fungi. Fungal Genet Biol. 2005;42:275–83.
26. Singh S, Braus-Stromeyer SA, Timpner C, Tran VT, Lohaus G, Reusche M, et al. Silencing of *Vlaro2* for chorismate synthase revealed that the phytopathogen *Verticillium longisporum* induces the cross-pathway control in the xylem. Appl Microbiol Biotechnol. 2010;85:1961–76.
27. Ullán RV, Godio RP, Teijeira F, Vaca I, García-Estrada C, Feltrer R, et al. RNA-silencing in *Penicillium chrysogenum* and *Acremonium chrysogenum*: validation studies using β-lactam genes expression. J Microbiol Methods. 2008; 75:209–18.
28. Shimizu T, Ito T, Kanematsu S. Functional analysis of a melanin biosynthetic gene using RNAi-mediated gene silencing in *Rosellinia necatrix*. Fungal Biol. 2014;118:413–21.
29. Khatri M, Rajam MV. Targeting polyamines of *Aspergillus nidulans* by siRNA specific to fungal ornithine decarboxylase gene. Med Mycol. 2007;45:211–20.
30. Mumbanza FM, Kiggundu A, Tusiime G, Tushemereirwe WK, Niblett C, Bailey A. In vitro antifungal activity of synthetic dsRNA molecules against two pathogens of banana, *Fusarium oxysporum* f. sp. *cubense* and *Mycosphaerella fijiensis*. Pest Manag Sci. 2013;69:1155–62.
31. Deshmukh R, Purohit HJ. siRNA mediated gene silencing in *Fusarium* sp. HKF15 for overproduction of bikaverin. Bioresour Technol. 2014;157:368–71.
32. Wang Y, Xiao S, Xiong D. Genetic transformation, infection process and qPCR quantification of *Verticillium dahliae* on smoke-tree *Cotinus coggygria*. Aust Plant Pathol. 2013;42:33–41.
33. Yang X, Ben S, Sun Y, Fan X, Tian C, Wang Y. Genome-Wide Identification, Phylogeny and Expression Profile of Vesicle Fusion Components in Verticillium dahliae. PLoS One. 2013;8, e68681.
34. Bustin SA, Benes V, Garson JA, Hellemans J, Huggett J, Kubista M, et al. The MIQE Guidelines: Minimum Information for Publication of Quantitative Real-Time PCR Experiments. Clin Chem. 2009;55:611–22.

35. Marilyn GW, Michaela N, Laurie M, Margaret LB, Geoffrey DR, Peter JP, et al. Protoplast production and transformation of morphological mutants of the Quorn® myco-protein fungus, Fusarium graminearum A3/5, using the hygromycin B resistance plasmid p AN7-1. Mycol Res. 1997;101:871-7.

36. Wubie AJ, Hu Y, Li W, Huang J, Guo Z, Xu S, et al. Factors Analysis in Protoplast Isolation and Regeneration from a Chalkbrood Fungus, *Ascosphaera apis*. Int J Agric Biol. 2014;16:89-96.

37. Dhar P, Kaur G. Optimization of different factors for efficient protoplast release from entomopathogenic fungus *Metarhizium anisopliae*. Ann Microbiol. 2009;59:183-6.

38. Shabana YM, Charudattan R. Preparation and Regeneration of Mycelial Protoplasts of *Alternaria eichhorniae*. J Phytopathol. 1997;145:335-8.

39. Li L, Yin Q, Liu X, Yang H. An efficient protoplast isolation and regeneration system in *Coprinus comatus*. African J Microbiol Res. 2010;4:459-65.

40. Li Y, Yuan Q, Du X. Protoplast from β-carotene-producing fungus *Blakeslea trispora*: Preparation, regeneration and validation. Korean J Chem Eng. 2008; 25:1416-21.

41. Eyini M, Rajkumar K, Balaji P. Isolation, Regeneration and PEG-Induced Fusion of Protoplasts of *Pleurotus pulmonarius* and *Pleurotus florida*. Mycobiology. 2006;34:73.

42. Zhao K, Zhou D, Ping W, Ge J. Study on the Preparation and Regeneration of Protoplast from Taxol-producing Fungus *Nodulisporium sylviforme*. Nat Sci. 2004;2:52-9.

43. Liu Y, Wang SX, Yin YG, Zhao S, Geng X, Xu F. Polyethylene glycol (PEG)-mediated transformation of the fused *egfp-hph* gene into *Pleurotus ostreatus*. African J Biotechnol. 2012;11:4345-53.

44. Vicky C, Lisa FD, Kerry AF, Sonja JL, Adrea AM. The effect of increasing plasmid size on transformation efficiency in *Escherichia coli*. J Exp Microbiol Immunol. 2002;2:207-23.

45. Ohse M, Takahashi K, Kadowaki Y, Kusaoke H. Effects of Plasmid DNA Sizes and Several Other Factors on Transformation of *Bacillus subtilis* ISW1214 with Plasmid DNA by Electroporation. Biosci Biotechnol Biochem. 1995;59: 1433-7.

46. Salame TM, Ziv C, Hadar Y, Yarden O. RNAi as a potential tool for biotechnological applications in fungi. Appl Microbiol Biotechnol. 2011; 89:501-12.

47. Abdel-Hadi A, Caley D, Carter D, Magan N. Control of Aflatoxin Production of *Aspergillus flavus* and Aspergillus parasiticus Using RNA Silencing Technology by Targeting aflD (*nor-1*) Gene. Toxins (Basel). 2011;3:647-59.

48. Holen T, Amarzguioui M, Wiiger MT, Babaie E, Prydz H. Positional effects of short interfering RNAs targeting the human coagulation trigger Tissue Factor. Nucleic Acids Res. 2002;30:1757-66.

49. Lam JK, Chow MY, Zhang Y, Leung SW. siRNA Versus miRNA as Therapeutics for Gene Silencing. Mol Ther Acids. 2015;4:e252.

4

Biochemical characterization and synergism of cellulolytic enzyme system from *Chaetomium globosum* on rice straw saccharification

Wanwitoo Wanmolee[1], Warasirin Sornlake[2], Nakul Rattanaphan[3], Surisa Suwannarangsee[2], Navadol Laosiripojana[1,4] and Verawat Champreda[2,4*]

Abstract

Background: Efficient hydrolysis of lignocellulosic materials to sugars for conversion to biofuels and chemicals is a key step in biorefinery. Designing an active saccharifying enzyme system with synergy among their components is considered a promising approach.

Results: In this study, a lignocellulose-degrading enzyme system of *Chaetomium globosum* BCC5776 (CG-Cel) was characterized for its activity and proteomic profiles, and synergism with accessory enzymes. The highest cellulase productivity of 0.40 FPU/mL was found for CG-Cel under the optimized submerged fermentation conditions on 1% (w/v) EPFB (empty palm fruit bunch), 2% microcrystalline cellulose (Avicel®) and 1% soybean meal (SBM) at 30 °C, pH 5.8 for 6 d. CG-Cel worked optimally at 50–60 °C in an acidic pH range. Proteomics analysis by LC/MS/MS revealed a complex enzyme system composed of core cellulases and accessory hydrolytic/non-hydrolytic enzymes attacking plant biopolymers. A synergistic enzyme system comprising the CG-Cel, a β-glucosidase (Novozyme® 188) and a hemicellulase Accellerase® XY was optimized on saccharification of alkaline-pretreated rice straw by a mixture design approach. Applying a full cubic model, the optimal ratio of ternary enzyme mixture containing CG-Cel: Novozyme® 188: Accellerase® XY of 44.4:20.6:35.0 showed synergistic enhancement on reducing sugar yield with a glucose releasing efficiency of 256.4 mg/FPU, equivalent to a 2.9 times compared with that from CG-Cel alone.

Conclusions: The work showed an approach for developing an active synergistic enzyme system based on the newly characterized *C. globosum* for lignocellulose saccharification and modification in bio-industries.

Keywords: Cellulase, *Chaetomium globosum*, Lignocellulose, Saccharification, Synergistic action

Background

The shift from petroleum-based industry to a greener bio-based platform is expedited by an increasing concern of global warming. Lignocellulosic plant biomass has attracted attention as a renewable resource for production of biofuels and commodity chemicals in biorefinery.

Lignocellulosic materials consist mainly of three different types of biopolymers: (i) cellulose, a linear homopolymer of D-glucose organized into highly crystalline microfibers which are intimately associated with an intricate network of (ii) hemicellulose, an amorphous branched polymer comprising various pentoses, hexoses, and sugar acids, and (iii) lignin, a heteropolymer of phenolic alcohols which shields the polysaccharide microstructure from external physical, chemical, and biological attacks. These biopolymers are organized into a complex and highly recalcitrant lignocellulosic structure [1].

In nature, lignocelluloses are degraded by the cooperation of various microorganisms, capable of producing

* Correspondence: verawat@biotec.or.th
[2]Enzyme Technology Laboratory, National Center for Genetic Engineering and Biotechnology (BIOTEC), 113 Thailand Science Park, Phahonyothin Road, Khlong Luang, Pathumthani 12120, Thailand
[4]BIOTEC-JGSEE Integrative Biorefinery Laboratory, Innovation Cluster 2 Building, 113 Thailand Science Park, Phahonyothin Road, Khlong Luang, Pathumthani 12120, Thailand
Full list of author information is available at the end of the article

an array of cellulolytic, hemicellulolytic, and ligninoloytic enzymes [2, 3]. Cellulases and hemicellulases are mainly composed of various hydrolytic enzymes in different glycosyl hydrolase (GH) families acting synergistically. Cellulases comprise three major groups of enzymes: (1) endoglucanases (EC 3.2.1.4), which attack regions of low crystallinity in cellulose fibers, creating free chain-ends; (2) exo-glucanases or cellobiohydrolases (EC 3.2.1.91) which further degrade the molecule by cleaving cellobiose from the free-chain ends; and (3) β-glucosidases (EC 3.2.1.21) which hydrolyze cellobiose to produce glucose [4]. In addition to the three major groups of cellulases, there are a number of endo- and exo-acting enzymes that attack the heterogeneous hemicelluloses, such as endo-β-1,4-xylanase, β-xylosidase, galactomannanase, glucomannanase, and acetylesterase. A challenge for development of a feasible biomass industry is identifying efficient lignocellulolytic microbes and developing an active enzyme system based on synergism of the major glycosyl hydrolases and the non-hydrolytic auxiliary components (e.g. expansins and lytic polysaccharide monooxygenases) [5, 6]. These molecular systems are required for efficient hydrolysis of lignocelluloses to sugars, the key intermediates for subsequent conversion to biorefinery products.

Aerobic fungi belonging to phylum Ascomycota are important plant biomass degraders capable of secreting an array of glycosyl hydrolases and non-hydrolytic auxiliary components. Several ascomycetes are commonly used for producing cellulases and hemicellulases in industry, e.g. *Trichoderma*, *Aspergillus*, and *Talaromyces* [7]. Enzymes from these fungi are different in their composite activities and enzyme components, which have different catalytic characteristics on lignocellulose decomposition. *Chaetomium* is a saprophytic fungus belonging to Ascomycota with high capability on degrading plant materials [8]. Most studies on *Chaetomium*, particularly on *C. globosum* have focused on laccases [9–11] while there have been few studies on their cellulolytic enzyme systems [12]. In this study, a biomass degrading enzyme system of the soft-rot fungus *C. globosum* BCC5776 was studied. The crude enzyme was characterized for its catalytic activities and its components identified by proteomics. The crude enzyme was used for formulation of an efficient biomass-degrading enzyme mixture by a mixture design approach and applied for hydrolysis of alkaline-pretreated rice straw. The work provides an alternative lignocellulose-degrading enzyme system potent for on-site enzyme production for the development of a feasible biorefinery industry.

Methods

Strain and media

Rice straw was obtained from a paddy field in Pathumthani province, Thailand according to the required national guideline. It was physically processed by a cutting mill (Retsch SM 200, Hann, Germany) and then sieved to particles smaller than 0.5 mm. The biomass was delignified with 10% (w/v) NaOH at 80 °C for 90 min at the solid/liquid ratio of 1:3, washed with water until neutral pH was obtained, and dried at 60 °C for overnight before use as the substrate for enzymatic hydrolysis. The alkaline-pretreated rice straw contained 74.3% cellulose, 13.4% hemicellulose, 1.36% lignin and 2.9% ash according to the standard NREL analysis method [13]. Empty palm fruit bunch (EPFB) was obtained from Kasetsart University, Thailand and physically processed by a chopping machine (Model JL800, Zhengzhou, China) and subsequently on a cutting mill and sieved through a 0.5 mm mesh. *C. globosum* BCC5776 was identified and obtained from the BIOTEC Culture Collection, Thailand (www.tbrcnetwork.org) and maintained on potato dextrose agar (PDA). Polysaccharides used as substrates in enzymatic activity analysis were obtained from Sigma-Aldrich. Accellerase®XY (hemicellulase from *Trichoderma reesei*) and Novozyme®188 (β-glucosidase from *Aspergillus niger*) were obtained from Dupont (Rochester, NY) and Sigma-Aldrich, respectively.

Optimization of cellulase production conditions

C. globosum BCC5776 was cultivated by submerged fermentation in 250-mL conical flasks. The inoculum was prepared from the culture grown on PDA by plunging four agar pieces covered with profuse mycelia using a cock borer no. 2 and inoculated into 50 mL of the production medium (4% (w/v) of microcrystalline cellulose Avicel® and 1% (w/v) of soybean meal in water). The culture was incubated at 30 °C for 6 d with shaking at 200 rpm. Culture media and conditions were varied as specified including concentration of Avicel® (2, 4, and 6% (w/v)), concentration of lactose (0, 0.05, and 0.1% (v/v)) and pH (5.8 and 7.0 using 50 mM potassium phosphate buffer). Effect on addition of 1% (w/v) empty palm fruit bunch (EPFB) as a co-carbon source for induction of cellulase was also studied. The cultures were collected periodically for determination of cellulase activity using the dinitrosalicylic acid (DNS) method [14]. The experiments were performed in triplicate and the average of cellulase activity was used as the response (dependent variable). The data were analyzed using SPSS 16.0 (StatSoft, Inc., Tulsa, OK).

Enzyme production and purification

A 10% (v/v) inoculum grown in the optimal production medium containing 2% Avicel®, 1% EPFB, 1% soybean meal (SBM), 0.1% lactose, and initial pH of 5.8 at 30 °C for 6 d was inoculated into a 5 L bioreactor (BIOSTAT B-DCU, Sartorius, Göttingen, Germany) with a 3 L working volume of the same medium. The culture was incubated at 30 °C for 6 d with constant mixing at 200 rpm and oxygen feed of 1 vvm. The fungal mycelia were separated by filtration on gauze and clarified by centrifugation at 10,000 × g for 10 min. The supernatant was then filtered through a 0.2 µm Supor®-200 membrane (Pall Corp, Ann Arbor, MI) followed by concentration (5×) using a Minimate™ tangential flow filtration (TFF) system equipped with a 10-kDa MWCO TFF membrane (Pall Corp, Ann Arbor, MI, USA). The enzyme was kept at 4 °C and used in subsequent experiments.

Enzyme activity assays

Polysaccharide-degrading enzyme activities were analyzed using the 3,5-dinitrosalicylic acid (DNS) method by measuring the amount of reducing sugars liberated [14] according to the standard procedure recommended by the Commission on Biotechnology, IUPAC [15] with modifications on the total reaction volume. Reactions of 3.5 mL contained 100 mM sodium acetate phosphate buffers, pH 5.5 with an appropriate dilution of the enzyme using a 1 × 6 cm Whatman no. 1 filter paper as the substrate and incubated at 50 °C for 60 min for determining the Filter paper activity (FPase) as filter paper unit (FPU). The carboxymethyl cellulase (CMCase), xylanase, mannanase, amylase, and pectinase activities were assayed using 1% (w/v) carboxymethyl cellulose, 1% (w/v) birchwood xylan, 0.5% (w/v) locust bean gum, 1% (w/v) soluble starch, and 0.5% (w/v) pectin from citrus peels as the substrates, respectively. The reactions were incubated at 50 °C for 30 min. The amount of reducing sugars was determined at the end of the reaction by measuring the absorbance at 540 nm using a UV-Vis spectrophotometer microplate reader (Multiskan Ascent, Thermo Scientific, Cambridge, MA) and interpolation from a standard curve prepared using dilutions of the corresponding sugar as standards. One enzyme activity unit (U) is defined as the amount of enzyme required to release 1 µmol of reducing sugars from a substrate in 1 min under the assay condition. The β-glucosidase and β-xylosidase activities were assayed using 0.1% (w/v) p-nitrophenyl-β-D-glucopyranoside (PNPG) and p-nitrophenyl-β-D-xylopyranoside (PNPX) as the substrates, respectively in 3 mL reactions containing an appropriate amount of enzyme in 100 mM sodium acetate buffer (pH 5.5). The reactions were incubated at 50 °C for 30 min and terminated by the addition of 2 mL of 1 M Na_2CO_3. The quantity of p-nitrophenolate was measured spectrophotometrically at 405 nm at the end of the reaction. One activity unit (U) is defined as amount of enzyme that produces 1 µmol p-nitrophenolate per minute under the experimental conditions. The total protein concentration of the crude enzyme extracts was determined using Bradford's method with the BioRad's Protein Assay reagent (BioRad, Hercules, CA) using bovine serum albumin (BSA) as the standard. The experiments were performed in triplicate.

Proteomic analysis

The enzyme preparation was applied to a 10% SDS-PAGE gel and separated using a MiniProtean II cell (Biorad, Hercules, CA, USA). The protein bands were visualized by staining with Coomassie blue R-250 and were manually excised into five fractions according to their apparent molecular weights (14.4–116.0 kDa). The polypeptides in gel were then digested with trypsin (Ettan Spot Handling Workstation User Manual 18-1153-55 Edition AC, GE Healthcare Biosciences, Uppsala, Sweden). The tryptic peptides were resuspended with 0.1% formic acid and analyzed on a Finigan LTQ linear ion trap mass spectrometer (Thermo Scientific, San Jose, CA, USA) according to a method described in Wongwilaiwalin et al., [16]. All MS/MS spectra were searched using the Mascot® search engine (Matrix Science, Boston, MA) against the NCBI-nr database following criteria: enzyme trypsin, static modification of Cys (+57.05130 Da), with differential modification of Met (+15.99940). The search results were filtered by cross-correlation versus charge state (+1 ≥ 1.5, +2 ≥ 2.0,+3 ≥ 2.5) and protein probability (minimum $1.00E^{-3}$). The candidate protein queries were mapped to the UniProt Knowledge base [17].

Synergistic action of *C. globosum* enzyme to commercial enzymes

Interactions among the *C. globosum* enzyme (CG-Cel), Novozyme®188 and Accellerase®XY were studied using experimental mixture design approach [18] with a fixed total enzyme volume [19]. The enzymatic hydrolysis reactions of 1 mL total volume contained 5% (w/v) alkaline-pretreated biomass in 100 mM sodium acetate buffer (pH 5.5) supplemented with 1 mM sodium azide and incubated at 50 °C for 48 h with rotary shaking at 200 rpm. An optimal enzyme mixture releasing the highest reducing sugar yield was defined by a {3,3}-augmented simplex lattice design using Minitab 16.0 software (Minitab Inc., State College, PA). The design contained 13 experimental points (see design summarized in Table 3), which were performed in quadruplicate with three components and a lattice degree of three. The three independent variables in the mixture design consisted of CG-Cel (X1), Novozyme®188 (X2), and Accellerase®XY (X3). The sum of all enzyme components in the reactions were 100% with the total enzyme volume fixed at 80 µL. The amount of released reducing sugar (Y1) was used as dependent variables for simulation of the respondent model equation. The amount of reducing sugars liberated at the end of the reactions was determined using the DNS method at the end of the reaction [14]. The sugar profile in the hydrolyzates was analyzed by high performance liquid chromatography (Waters e2695, Waters, Milford, MA) equipped with a differential refractometer using an

Aminex HPX-87H column (Bio-Rad, Hercules, CA). The column temperature was 65 °C. H_2SO_4 solution (5 mM) was used as the mobile phase at a flow rate of 0.5 mL/min. Concentration of sugar was determined from the calibration curve of standard solution. The experiments were performed in quadruplicate. Sugars from control reactions containing heat-inactivated enzymes were subtracted from the data.

Mixture design analysis

After regression analysis, the full cubic model was used to simulate the optimized ratio of the mixture components. The canonical correlation of the full cubic model is shown in Eq. (1):

$$Y = \sum_{i=1}^{3} \beta_i X_i + \sum\sum_{i<j}^{3} \beta_{ij} X_i X_j$$
$$+ \sum\sum_{i<j}^{3} \delta_{ij} X_i X_j (X_i - X_j) + \sum\sum\sum_{i<j<k}^{3} \beta_{ijk} X_i X_j X_k$$

$$(1)$$

where Y is a predicted response, β_i is a linear coefficient, β_{ij} is a quadratic coefficient, and β_{ijk} is a cubic coefficient. β_{ij} is a parameter of the model. $\beta_i X_i$ represents the linear blending portion, and the parameter β_{ij} represents either synergistic or antagonistic blending.

Results and discussion

Optimization of enzyme production conditions

Production of cellulolytic enzyme from *C. globosum* BCC5776 by submerged fermentation was studied. The optimization study included three variables: (1) concentration of Avicel®, (2) concentration of lactose, and (3) initial pH previously identified to show significant effects ($p < 0.05$) on cellulase level of the fungus in the preliminary screening of factors influencing enzyme production. According to Table 1, the basic fermentation condition containing Avicel® as the

sole carbon source (run no. 1) resulted in the CMCase and FPase activity of 6.37 and 0.21 U/mL, respectively. Further increases in Avicel® (run no. 6–9) did not enhance the target enzymatic activities. Addition of lactose (run no. 2–9) was found to significantly ($p < 0.05$) induce cellulase production level when compared with the same conditions in the absence of lactose. Higher cellulase productivity was found at acidic pH (5.8) compared with that observed under neutral conditions. The highest CMCase and FPase activities of 7.19 U/mL and 0.32 FPU/mL, respectively, were recorded after 144 h of fermentation in run no. 3 containing 1% lactose and an inducer with controlled pH at 5.8. This up-regulation of cellulase by lactose was previously reported in *C. papyrosolvens* [20] and *A. cellulolyticus* [21] where lactose was shown to act as an effective inducer of cellulase biosynthesis, which is controlled by a regulator protein responsive to the concentration of the target substrate [22]. The preference for acidic pH for production of cellulase and other lignocellulose degrading enzymes has been reported for most cellulase producing ascomycetes fungi e.g. *T. reesei* [23], *A. cellulolyticus* [24], *and C. globosum* [25]. This preference is related to pH for optimal growth and catalytic activity of the enzymes.

Addition of EPFB as a co-substrate at 1% (w/v) concentration under submerged fermentation was found to further improve the cellulase production by *C. globosum* BCC5776. This led to the enhancement of endo-glucanase (CMCase) and total cellulase (FPase) activities to 10.57 U/mL and 0.40 FPU/mL, equivalent to 47 and 25% increases, respectively, compared with those obtained in its absence (Fig. 1a and b). The use of agro-industrial wastes as the sole carbon source or in combination with pure cellulose as an effective and cost-efficient substrate for production of plant biomass degrading enzymes e.g. cellulase, hemicellulase, and pectinase production has been demonstrated in many fungal strains [26–28]. The cellulase activities produced by *C. globosum* in this study were in the same range (or higher in some cases, particularly for the CMCase activity) compared to several wild type fungi in genera *Aspergillus*, *Acremonium*, *Trichodema*, and *Schizophyllum* reported in many recent publications, which were in the range of 0.01–1.33 FPU/mL. The very high cellulase activities (up to five FPU/ml) were achieved by mutants or genetically modified fungal strains. Comparison of cellulase activities produced by different fungal strains recently reported is shown in the Additional file 1: Table S1.

The optimized enzyme production medium (2% Avicel®, 1% EPFB, 0.1% lactose, and 1% soybean meal at

Table 1 Effects of variables to cellulase activities in fermentation of BCC5776

No.	Concentration (%)		Initial pH	Activity	
	Avicel®	Lactose		CMCase (U/mL)	FPase (FPU/mL)
1	4	-	5.5	6.37 ± 0.01	0.21 ± 0.03
2	2	0.05	5.8	7.02 ± 0.01	0.22 ± 0.05
3	2	0.1	5.8	7.19 ± 0.00	0.32 ± 0.01
4	2	0.05	7.0	5.10 ± 0.03	0.20 ± 0.00
5	2	0.1	7.0	5.54 ± 0.02	0.17 ± 0.01
6	6	0.05	5.8	6.87 ± 0.04	0.18 ± 0.00
7	6	0.1	5.8	6.68 ± 0.01	0.20 ± 0.01
8	6	0.05	7.0	6.12 ± 0.05	0.17 ± 0.00
9	6	0.1	7.0	5.07 ± 0.04	0.17 ± 0.03

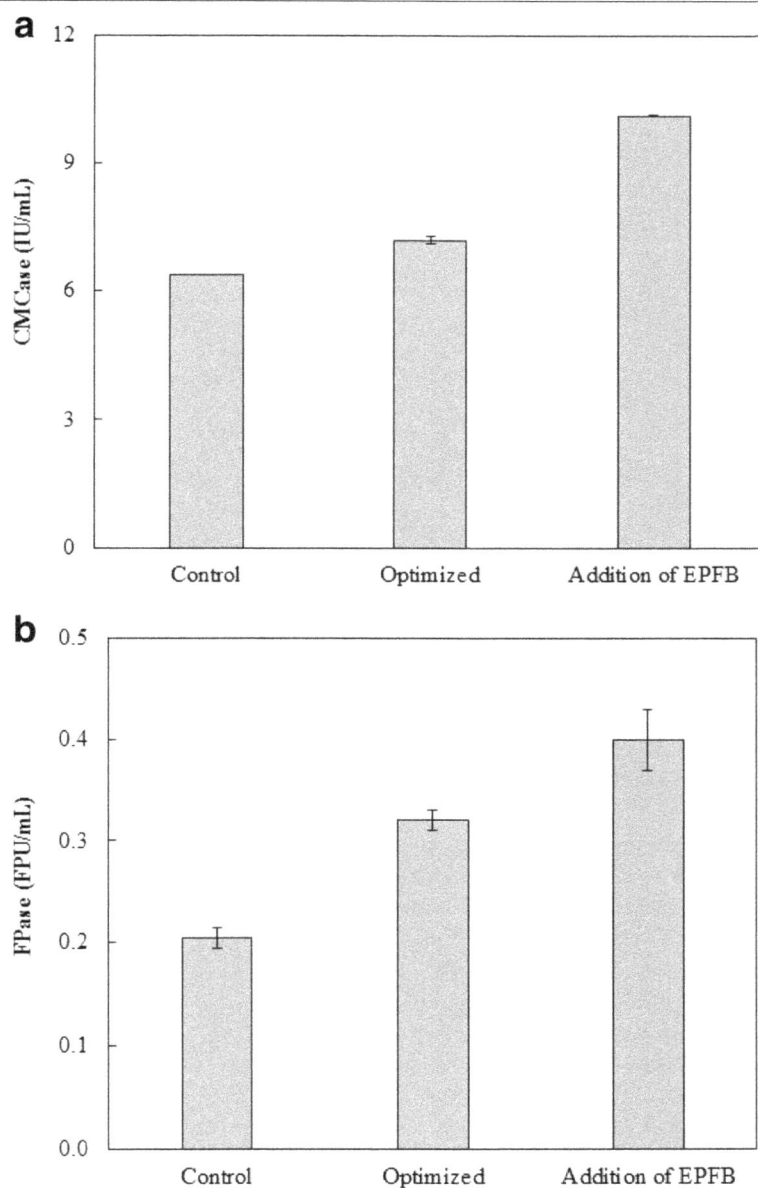

Fig. 1 Comparison of (**a**) CMCase and (**b**) FPase activities from *C. globosum* BCC5776 under various conditions. The cultures were incubated at 30 °C for 144 h with continuous shaking at 200 rpm. Control medium (4% (*w/v*) of microcrystalline cellulose Avicel® and 1% (*w/v*) of soybean meal in water); Optimized medium (2% (*w/v*) microcrystalline cellulose Avicel®, 0.1% (*v/v*) lactose, and 1% (*w/v*) soybean meal with 50 mM potassium phosphate, pH 5.8); Optimized medium + 1% (*w/v*) EPFB

pH 5.8) was used for production of the crude enzyme (CG-Cel) in a laboratory-scale fermenter. The enzyme showed similar composite activities to those obtained in the small scale experiments. The crude enzyme preparation (Table 2) contained a relatively high level of endo-glucanase as shown by CMCase activity (15.7 U/mL) with FPase activity (0.4 FPU/mL). Strong endo-β-1,4-xylanase (28.2 U/mL) and endo-β-1,4-mannanase (13.6 U/mL) activities were found together with other enzymes involved in plant polysaccharide degradation, such as pectinase (0.8 U/mL) and amylase (0.5 U/mL). However, regardless of the high activities of endo-acting glycosyl hydrolases, the crude enzyme exhibited substantially weak downstream activities i.e. β-glucosidase (1.5 U/mL) and β-xylosidase (0.04 U/mL) activities, which could limit degradation of cellulose and hemicelluloses to sugars. The 5× concentrated enzyme contained the cellulase

Table 2 The composite enzyme activity profiles of *C. globosum* BCC5776 enzyme extracts

Enzyme	Activity (U/mL)[a]
CMCase	15.70
FPase	0.40
β-glucosidase	1.50
Xylanase	28.20
β-xylosidase	0.04
Amylase	0.50
Mannanase	13.60
Pectinase	0.80

[a]Protein concentration = 1.30 mg/mL

activity of 2.0 FPU/mL while >90% activities of other enzymes were retained.

Synergistic action of CG-Cel and commercial enzymes

The mixture design approach was applied to study synergistic and cooperative interactions among CG-Cel and a hemicellulase, Accellerase®XY, and a β-glucosidase, Novozyme®188, chosen from our pre-screening study on identification of additive enzymes with complementing activities to the core cellulase (see the comparative enzyme profiles in Fig. 2). The measured response of this method was presumed to depend only on the enzyme proportion of the complements in the mixture. A {3,3}-simplex lattice model was applied which comprised 13 experimental points located inside the triangular graph, in which the sum of the three component loading for every experimental point was always 100% based on a volumetric basis. According to this design, the cellulase dosages as FPU loading/g substrate of each experimental point was varied according to different proportions of CG-Cel: Novozyme®188:Accellerase®XY. Each point was operated in quadruplicate to minimize the effect of experimental error.

The reducing sugar yield obtained from each experimental condition in the mixture design is summarized in Table 3. The standard deviation of all the data points was ≤ 5% of the mean. CG-Cel, equivalent to the FPase activity of 3.2 FPU/g (No. 1), demonstrated a higher reducing sugar yield (500.9 mg/g) than those obtained with the β-glucosidase Novozyme®188 (No. 7) and the hemicellulase Accellerase®XY (No. 10) alone. The binary combinations between CG-Cel and Accellerase®XY (No. 3 and 6) led to slight increases in reducing sugar yields (519.3–525.4 mg/g) compared with that of CG-Cel alone, despite the lower total FPase dosage in the reaction (1.17–2.18 FPU/g). Ternary mixtures of all three enzymes at specific compositions resulted in

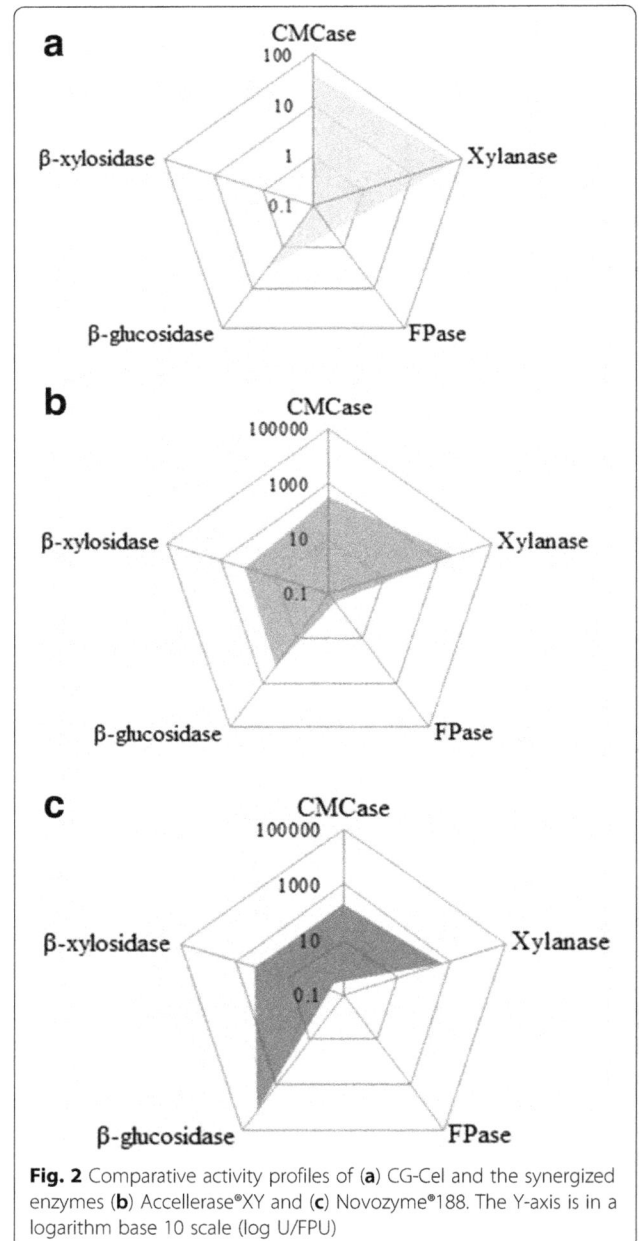

Fig. 2 Comparative activity profiles of (**a**) CG-Cel and the synergized enzymes (**b**) Accellerase®XY and (**c**) Novozyme®188. The Y-axis is in a logarithm base 10 scale (log U/FPU)

further increases in the total reducing sugar yields, with the maximal reducing sugar yield of 567.8 mg/g (No. 11) which contained a 4:1:1 ratio of BCC5776: Novozyme®188: Accellerase®XY, suggesting synergism of the three enzymes.

The response data for the reducing sugar were then analyzed using multiple regression analysis from the linear to full cubic model. The full cubic model was found to be the best fitted model for the reducing sugar (R^2 = 97.30%, P_{Model} < 0.01). The ANOVA analysis of the full cubic model is illustrated in

Table 3 Design of experiment for enzymatic hydrolysis of pretreated rice straw and the associated response data

Run no.	% composition			Reducing sugar (mg/g biomass)	
	BCC5776	Novozyme®188	Accellerase®XY	Average	SD
1	100.0	0.0	0.0	500.9	2.6
2	66.7	33.3	0.0	493.1	2.2
3	66.7	0.0	33.3	525.4	0.5
4	33.3	66.7	0.0	421.2	4.3
5	33.3	33.3	33.3	535.6	2.6
6	33.3	0.0	66.7	519.3	3.2
7	0.0	100.0	0.0	69.2	3.4
8	0.0	66.7	33.3	205.3	4.1
9	0.0	33.3	66.7	233.3	4.6
10	0.0	0.0	100.0	205.0	4.9
11	66.7	16.7	16.7	567.8	1.7
12	16.7	66.7	16.7	426.1	1.4
13	16.7	16.7	66.7	483.5	2.1

Reactions (1 mL) contained 5% (w/v) alkaline-pretreated rice straw in 100 mM sodium acetate buffer, pH 5.5 with a total enzyme volume of 80 μL and incubated at 50 °C for 48 h

Table 4. For a single factor, CG-Cel, Novozyme®188, Accellerase®XY coefficients showed a positive correlation with the reducing sugar yield. Marked synergisms between CG-Cel and Novozyme®188 or Accellerase®XY were observed (786.6 and 845.1), with a weak synergistic interaction between Novozyme®188 and Accellerase®XY. The highest coefficient was observed for the CG-Cel*Novozyme®188*Accellerase®XY, indicating strong interactions among these components. The fitted equation for the reducing sugar yield based on the significant terms is shown in Eq. (2):

$$
\begin{aligned}
\text{Reducing sugar (mg/g)} = {} & 4.82685 * \text{BCC5776} \\
& + 0.72036 * \text{Novozyme188} \\
& + 2.04999 * \text{AccelleraseXY} \\
& + 0.07866 * \text{BCC5776} * \text{Novozyme188} \\
& + 0.08451 * \text{BCC5776} * \text{AccelleraseXY} \\
& + 0.03772 * \text{Novozyme188} * \text{AccelleraseXY} \\
& + 0.00253 * \text{BCC5776} * \text{Novozyme188} \\
& \quad * \text{AccelleraseXY}(-) - 0.00059 * \text{BCC5776} \\
& \quad * \text{Novozyme188}(-) - 0.00053 * \text{BCC5776} \\
& \quad * \text{AccelleraseXY}(-)
\end{aligned}
$$

(2)

The responses of reducing sugar yield with respect to component combinations are represented by a ternary mixture contour plot (Fig. 3). The area that emphasized the greatest reducing sugar was in the middle of the CG-Cel and Accellerase®XY axes, and near the bottom of the Novozyme®188 vertex. This implies that a high level of reducing sugars could be achieved when CG-Cel was the major component with Novozyme®188 and Accellerase®XY in the minority. The optimal enzyme combination based on the maximal reducing sugar yield was determined to be 44.4% CG-Cel, 20.6% Novozyme®188, and 35.0% Accellerase®XY with a predicted reducing sugar yield of 591.2 mg/g pretreated rice straw with the total cellulase activity of 1.49 FPU/g. An experimental reducing experimental sugar yield of 572.7 mg/g was obtained, validating the model. The ternary enzyme mixture was found to increase the yields of all major composite sugars (i.e. glucose, xylose, and arabinose) from hydrolysis of the pretreated rice straw. The glucose yield (Additional file 2: Table S2) obtained using the optimal enzyme mixture was 381.2 mg/g, which was higher than that obtained using the individual enzymes (Fig. 4a). This led to the increase in glucose releasing efficiency from 88.7 mg glucose/FPU from CG-Cel alone by 2.9 fold to 256.4 mg glucose/FPU of the ternary optimal enzyme mixture. ($p < 0.05$ by T-test) [29]. Further increases in the

Table 4 The regression model analysis of the {3,3} full cubic model

Factor	Coefficient	SE	T	p-value
BCC5776	482.7	14.3	*	*
Novozyme188	72	14.3	*	*
AccelleraseXY	205	14.3	*	*
BCC5776*Novozyme188	786.6	63.94	12.3	0.000
BCC5776*AccelleraseXY	845.1	63.94	13.22	0.000
Novozyme188*AccelleraseXY	377.3	63.94	5.9	0.000
BCC5776*Novozyme188*Accellerase XY	2534.8	416.85	6.08	0.000
BCC5776*Novozyme188*(-)	−586.8	122.48	−4.79	0.000
BCC5776*AccelleraseXY*(-)	−528.2	122.48	−4.31	0.000
Novozyme188*AccelleraseXY*(-)	78.8	122.48	0.64	0.523
S = 28.6660	PRESS = 47389.9			
R² = 97.30%	R² (pred) = 96.29%		R² (adj) = 96.72%	

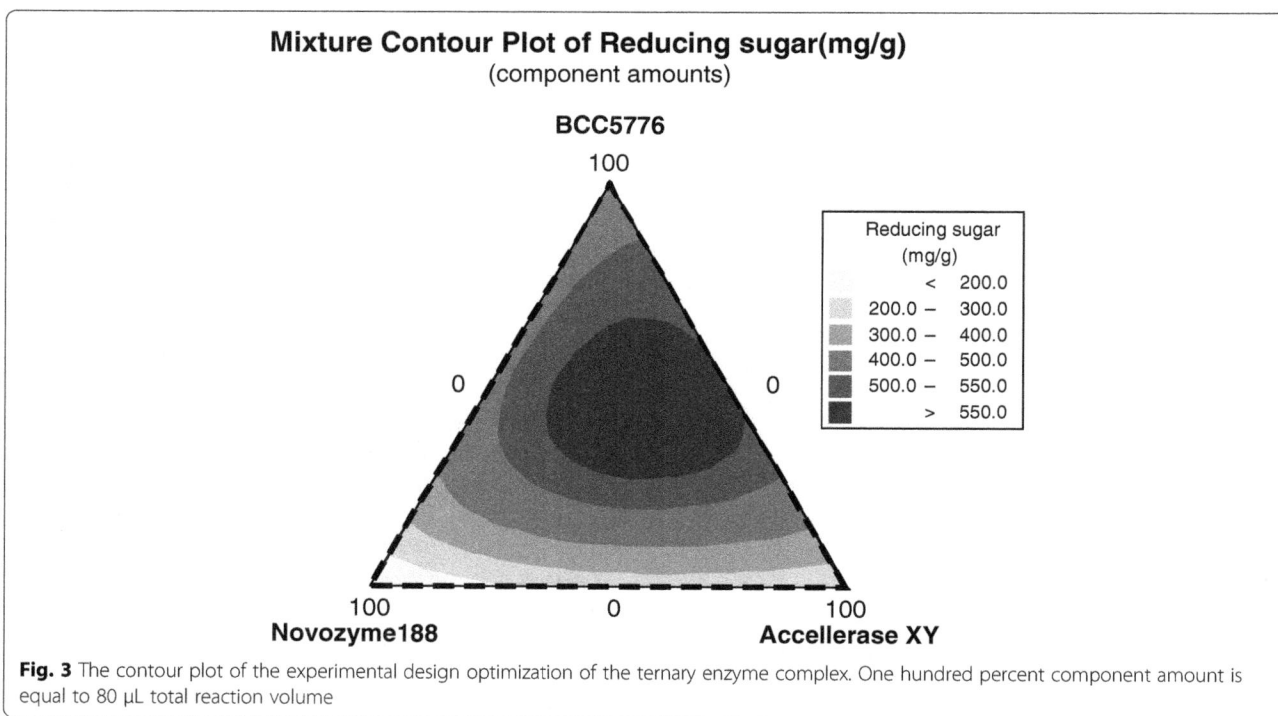

Fig. 3 The contour plot of the experimental design optimization of the ternary enzyme complex. One hundred percent component amount is equal to 80 μL total reaction volume

total enzyme loading from 1× (1.49 FPU/g) to 4× (5.96 FPU/g) led to a stepwise increase in reducing sugar yield to 764.7 mg/g and glucose released to 474.8 mg/g; however, respective decreases in glucose releasing efficiency were observed at higher enzyme loadings (Fig. 4b).

Proteomic analysis of BCC5776 crude enzyme

In order to analyze the composite glycosyl hydrolases and accessory non-hydrolytic enzymes with functions on attacking lignocellulosic materials in CG-Cel, the composite proteins in the secretome of BCC5776 (see Additional file 3: Figure S1) was analyzed using LC/MS/MS. Twenty seven different proteins were identified in CG-Cel, from which 81% were annotated as functional proteins while the rest were classified as hypothetical proteins. The majority of them are hydrolytic enzymes attacking cellulose and hemicelluloses, with a substantial fraction of non-cellulosic polysaccharide hydrolyzing enzymes (Table 5). Most of them are closely related to homologous enzymes in *Neofusicoccum parvum* UCRNP2 and other ascomycetes. The cellulose degrading enzymes were classified into various glycosyl hydrolase families and non-hydrolytic enzymes in different classes. The cellulolytic components are annotated as 5 endo-glucanases (GH5 and 7) and 8 cellobiohydrolases, including CBH-I (GH6 and 7), which attacks the reducing end of cellulose chain, and CBH-II, which attacks the non-reducing end, releasing cellobiose units, in addition to one carbohydrate binding module (CBM1).

A few hemicellulolytic enzymes were also identified in the BCC5776 secretome. These included GH10 and 11 endo-β-1,4-xylanase, a key endo-acting hemicellulase attacking xylan, the major component in hemicellulose in addition to exo-acting enzymes including a β-galactosidase (GH35) and α-mannosidase (GH47). Non-hydrolytic counterparts included a cellobiose dehydrogenase acting on oxidizing cellodextrins, an intermediate in cellulose hydrolysis to their corresponding lactones, and a polysaccharide deacetylase functioning on acetyl group branches in hemicelluloses. The functions of both enzymes on synergistic and cooperative actions with core cellulases have been demonstrated [30]. Other hydrolases in the secretome are mainly related to different classes of proteases, e.g. carboxypeptidase s1, leucyl aminopeptidase, and tripeptidyl-peptidase 1. No other non-cellulosic polysaccharide degrading enzymes e.g. amylases or pectinases were identified. The proteomic profiles of the secretome thus point to a lack of downstream cellulose degrading enzyme (i.e. a β-glucosidase) and various hemicellulolytic components. The lack of these activities could explain why CG-Cel is strongly synergistic with supplemented pure β-glucosidase and hemicellulases.

Synergism of core cellulases with hemicellulases and auxiliary components in various classes of carbohydrate processing enzymes according to the Carbohydrate Active Enzymes Database (CaZY) [31] has been demonstrated and could be applied for developing efficient enzyme systems for bioindustries. Their synergistic

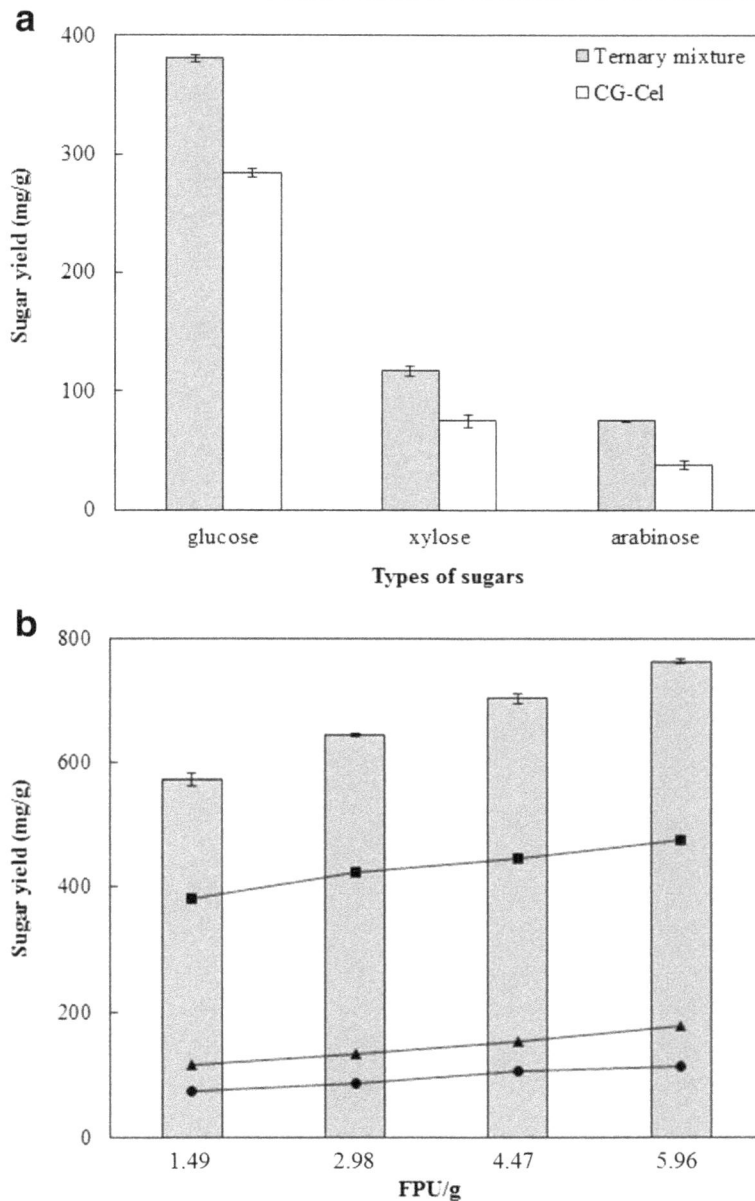

Fig. 4 Saccharification of pretreated rice straw using the optimal enzyme mixture compared with individual enzymes. The reactions (1 mL) contained 5% (*w*/*v*) alkaline-pretreated rice straw in 100 mM sodium acetate buffer, pH 5.5 with a fixed total enzyme volume of 80 μL (1×) and incubated at 50 °C for 48 h. (**a**) sugar profile of optimal enzyme mixture; (**b**) the total reducing sugars (bar) and individual sugar profile ((■) glucose; (▲) xylose; (♦) arabinose) versus FPU loading/g on pretreated rice straw hydrolysis. *1× enzyme loading equaled to 1.49 FPU/g

interactions can be explained by several mechanisms [32–34]. Activities of upstream enzymes can be enhanced by cooperative action with downstream enzymes acting on degrading smaller substrates, for example, alleviation of cellobiohydrolase (CBH) inhibition by a β-D-glucosidase which cleaves cellobiose, the released product with an inhibitory effect to CBHs to glucose molecules [35]. Synergism can also be resulted from the endo/exo effect between an endoglucanase which creates new, free cellodextrin chain ends for an exo-acting CBH [36] as well as synergic action between different exoglucanases attacking the reducing and non-reducing ends of cellulose chains [37]. Cooperative action of cellulases and hemicellulases can also increases accessibility to the target substrates of each other, providing synergisms between glycosyl hydrolases attacking different biopolymers in the plant cell wall [38]. Physical modification of the substrates e.g. loosening of the crystalline region of cellulose by auxiliary proteins (e.g. expansins) [32] and non-hydrolytic

Table 5 Identification of the polysaccharide degrading enzymes in the secretome of *Chaetomium globosum* BCC5776

GI number	GH Family	Protein name	Organism	Mascot Score	Prot Mass
343435330	GH7	Cellobiohydrolase I	*Uncultured fungus*	111	18482
343441380	GH7	Cellobiohydrolase I	*Uncultured fungus*	60	17704
517325505	GH5	Probable cellulase precursor	*Fusariam fujikuroi*	59	42508
343435350	GH7	Cellobiohydrolase I	*Uncultured fungus*	55	18629
485920895	GH6	Putative cellobiohydrolase II protein	*Neofusicocum parvum*	167	47768
310801037	GH7	Glycoside hydrolase family 7 endoglucanase	*Collectotrichum graminicola*	150	49663
46395332	GH7	Cellobiohydrolase	*Irpex lacteus*	111	56112
407917740	GH5	Glycosyl hydrolase family 5 endoglucanase	*Macrophomina phaseolina*	95	32163
477528227	GH7	Exoglucanase type c precursor	*Collectotrichum graminicola*	109	52197
367046256	GH7	Glycosyl hydrolase family 7 endoglucanase	*Theilavia terrestris*	84	56119
49333363	GH7	Cellobiohydrolase II	*Volvariella volvacea*	82	55320
343435720	GH7	Cellobiohydrolase I	*Uncultured fungus*	75	18463
575066085	-	Carbohydrate-binding module family 1 protein	*Heterobasidion irregulare*	68	35482
485929653	-	Putative cellobiose dehydrogenase protein	*Neofusicocum parvum*	99	897431
407926573	GH35	Glycoside hydrolase family 35 β-galactosidase	*Macrophomina phaseolina*	204	107902
407929733	GH47	Glycoside hydrolase family 47 α-mannosidase	*Macrophomina phaseolina*	122	57629
74664704	GH10	Endo-1,4-β-xylanase (Xylanase) (EC 3.2.1.8) (1,4-β-D-xylan xylanohydrolase)	*Aspergillus oryzae*	101	34903
407926897	GH10	Glycoside hydrolase family 10 endoxylanase	*Macrophomina phaseolina*	189	34499
485916263	GH10	Putative endo-β-xylanase protein	*Neofusicocum parvum*	185	35125
485916633	GH10	Putative extracellular endo-β-protein	*Neofusicocum parvum*	130	34638
485919833	GH11	Putative endo-β-xylanase protein	*Neofusicocum parvum*	87	23690
500259512	GH11	Putative endo-β-xylanase I protein	*Phaeoacremonium minimum*	84	23642
485916757	-	Putative carboxypeptidase s1 protein	*Neofusicocum parvum*	78	61265
485919267	-	Putative leucyl aminopeptidase protein	*Neofusicocum parvum*	147	40795
485924582	-	Putative tripeptidyl-peptidase 1 protein	*Neofusicocum parvum*	68	65131
407926489	-	Polysaccharide deacetylase	*Macrophomina phaseolina*	66	26890
325683994	GH7	Glycoside hydrolase family 7 endoglucanase	*Phialophora sp.*	161	50242

enzymes (e.g. lytic polysaccharide monoxygenases) [39] which in overall, results in increased accessibility of the hydrolytic enzymes to the substrates has also been demonstrated. However, they were not annotated in the secretome of *C. globosum* in this study.

Synergistic enzyme systems are developed by empirically determining combinations of different enzymes for optimal digestibility on specific lignocellulosic substrates in order to reduce enzyme usage without sacrificing the rate or yield from substrate hydrolysis. *T. reesei* cellulase, a well-known cellulase in biomass industry was shown to lack β-glucosidase activity, resulting in strong inhibition of the cellulase by accumulated cellobiose in the reaction. This inhibition can be overcome by addition of external β-glucosidase enzymes from different microbial origins [40]. The synergistic action of *T. reesei* cellulase with a crude enzyme mixture from *A. aculeatus* containing various cell wall polysaccharide degrading enzymes with strong downstream cellulolytic and hemicellulolytic activities and a non-catalytic bacterial expansin, acting on physical loosening of the crystalline region of cellulose fibers, on saccharification of alkaline pretreated rice straw was reported [19]. Glycosyl hydrolases from the metagenome of microflora present in sugarcane bagasse can enhance reducing sugar yield of *T. reesei* cellulase [41]. A recent work also showed strong synergistic action of a commercial cellulase, Acellerase®1500 with an endoxylanase, pectate lyase, and a hemicellulose side chain cleaving α-arabinofuranosidase [33]. These findings demonstrate the potential of formulating active synergistic enzyme systems from various microbial sources in order to maximize the hydrolysis efficiency on specific lignocellulosic substrates. The glucose releasing efficiency of the synergistic ternary enzyme system developed in this study was higher than that reported for Accellerase®1500 (81 mg glc/FPU) and the synergistic enzyme systems based on that enzyme, which was in the range of 122–173 mg/FPU [33]. The efficiency of the ternary enzyme mixture is in the same range to that reported for a synergistic system comprising *T. reesei* cellulase, a crude enzyme from *A. aculeatus* and a bacterial expansin [19] (229 mg glc/FPU) and between *T. reesei* cellulase and enzyme from *A. awamori* [42] (200 mg glc/FPU). The greater efficiency of the *C. globosum* enzyme system may depend though on the nature of the substrates and hydrolysis conditions. Elucidation of synergized enzyme activities in this study thus provides a basis for further modification of the CG-Cel enzyme either by complementation of lack activities by external in-house enzymes or genetic modification of the strain.

Conclusions

A biomass-degrading enzyme system from *C. globosum* BCC5776 (CG-Cel), a potent fungus for development of a cellulase producing strain for on-site enzyme production, was characterized. The primary cellulase activity can be augmented by other hydrolytic and non-hydrolytic accessory enzymes. The enzyme system shows a high degree of synergism with commercial β-glucosidase and hemicellulase. The optimal combination of CG-Cel, β-glucosidase and hemicellulase was found, leading to marked improvement in the sugar releasing efficiency. The work provides an approach for designing an effective enzyme system with specificity for saccharification of lignocellulosic materials, including agricultural biomass for the biorefinery industry.

Additional files

Additional file 1: Table S1. Comparison of cellulose production by *C. globosum* BCC5776 and other fungi in recent selected publications.

Additional file 2: Table S2. Sugar yields from hydrolysis of pretreated rice straw using various enzyme combinations and dosages.

Additional file 3: Figure S1. SDS-PAGE analysis of the BCC5776 crude enzyme extracts. Lane 1- Marker proteins; Lane 2 –the CG-Cel crude enzyme extracts stained with Coomassie brilliant blue.

Abbreviations

CG-Cel: Cellulase system of *Chaetomium globosum* BCC5776; CMCase: Cellulase activity on carboxymethyl cellulose; DNS: Dinitrosalicylic acid; EPFB: Empty palm fruit bunch; FPase: Cellulase activity on filter paper; GH: Glycosyl hydrolase; SBM: Soybean meal; TFF: Tangential flow filtration

Acknowledgements
The authors would like to thank Dr. Philip J. Shaw for manuscript editing.

Funding
The experimental works including the design of the study and collection, analysis and interpretation of data and manuscript preparation were financially supported by the National Science and Technology Development Agency (Grant number P-15-50502) and The Thailand Research Fund. WW was financially supported by The Petchra Pra Jom Klao Ph.D. Research Scholarship, King Mongkut's University of Technology Thonburi and The Joint Graduate School of Energy and Environment.

Authors' contribution
WW and WS carried out enzyme production, characterization, and mixture design study. NR participated in the upscale fermentation. SS advised on design of experiment. NL and VC contributed to the analysis of the results and manuscript writing. All authors read and approved the final manuscript.

Competing interests
The authors declare that they have no competing interests.

Author details
[1]The Joint Graduate School for Energy and Environment (JGSEE), King Mongkut's University of Technology Thonburi, Prachauthit Road, Bangmod, Bangkok 10140, Thailand. [2]Enzyme Technology Laboratory, National Center for Genetic Engineering and Biotechnology (BIOTEC), 113 Thailand Science Park, Phahonyothin Road, Khlong Luang, Pathumthani 12120, Thailand. [3]Bioprocess Laboratory, National Center for Genetic Engineering and Biotechnology (BIOTEC), 113 Thailand Science Park, Phahonyothin Road, Khlong Luang, Pathumthani 12120, Thailand. [4]BIOTEC-JGSEE Integrative Biorefinery Laboratory, Innovation Cluster 2 Building, 113 Thailand Science Park, Phahonyothin Road, Khlong Luang, Pathumthani 12120, Thailand.

References

1. Feldman D. Wood—chemistry, ultrastructure, reactions. J Polym Sci. 1985;23:601–2.
2. Ljungdahl LG. The cellulase/hemicellulase system of the anaerobic fungus Orpinomyces PC-2 and aspects of its applied use. Ann N Y Acad Sci. 2008; 1125:308–21.
3. Sánchez C. Lignocellulosic residues: Biodegradation and bioconversion by fungi. Biotechnol Adv. 2009;27:185–94.
4. Howard RL, Abotsi E, van Rensburg EL J, Howard SY. Lignocellulose biotechnology: issue of bioconversion and enzyme production. Afr J Biotechnol. 2003;2:602–19.
5. Woodward J. Enzymatic hydrolysis of cellulose synergism in cellulase systems. Bioresour Technol. 1991;36:67–75.
6. Arantes V, Saddler JN. Access to cellulose limits the efficiency of enzymatic hydrolysis: the role of amorphogenesis. Biotechnol Biofuels. 2010;3:4.
7. Dashtban M, Schraft H, Qin W. Fungal bioconversion of lignocellulosic residues; opportunities & perspectives. Int J Biol Sci. 2009;5:578–95.
8. Somrithipol S, Hywel-Jones NL, Jones EBG. Seed fungi. In: Jones EBG, Tanticharoen M, Hyde KD, editors. Thai fungal diversity. Thailand: BIOTEC; 2004. p. 129–40.
9. El Zayat S-A. Preliminary studies on laccase production by *Chaetomium globosum* an Endophytic fungus in *Glinus lotoides*. Am Eurasian J Agric Environ Sci. 2008;3:86–90.
10. Viswanath B, Chandra MS, Pallavi H, Reddy BR. Screening and assessment of laccase producing fungi from different environmental samples. Afr J Biotechnol. 2008;7:1129–33.
11. Abdel-Azeem AM, Gherbawy YA, Sabry AM. Enzyme profiles and genotyping of *Chaetomium globosum* isolates from various substrates. Plant Biosyst. 2014. doi:10.1080/11263504.2014.984791.
12. Kumar R, Singh S, Singh OV. Bioconversion of lignocellulosic biomass: biochemical and molecular perspectives. J Indus Microbiol Biotechnol. 2008;35:377–91.

13. Sluiter A, Hames B, Ruiz R, Scarlata C, Sluiter J, Templeton D, Crocker D. Determination of structural carbohydrates and lignin in biomass laboratory analytical procedure. Golden: National Renewable Energy Laboratory; 2008.

14. Miller GL. Use of dinitrosalicylic acid reagent for determination of reducing sugar. Anal Chem. 1959;31:426–9.

15. Wood TM, Bhat KM. Methods of measuring cellulase activities. Methods Enzymol. 1988;160:87–117.

16. Wongwilaiwalin S, Rattanachomsri U, Laothanachareon T, Eurwilaichitr L, Igarashi Y, Champreda V. Analysis of a thermophilic lignocellulose degrading microbial consortium and multi-species lignocellulolytic enzyme system. Enzym Microb Technol. 2010;47:283–90.

17. Magrane M, Consortium U. UniProt knowledgebase: a hub of integrated protein data. Oxford: Database; 2011.

18. Cornell J. Experiments with Mixtures: Designs, Models, and the Analysis of Mixture Data. 3rd ed. USA: Wiley; 2002.

19. Suwannarangsee S, Bunterngsook B, Arnthong J, Paemanee A, Thamchaipenet A, Eurwilaichitr L, Laosiripojana N, Champreda V. Optimisation of synergistic biomass-degrading enzyme systems for efficient rice straw hydrolysis using an experimental mixture design. Bioresour Technol. 2012;119:252–61.

20. Thirumale S, Swaroopa Rani D, Nand K. Control of cellulase formation by trehalose in Clostridium papyrosolvens CFR-703. Proc Biochem. 2001;37:241–5.

21. Fang X, Yano S, Inoue H, Sawayama S. Lactose enhances cellulase production by the filamentous fungus Acremonium cellulolyticus. J Biosci Bioeng. 2008;106:115–20.

22. Shiang M, Linden JC, Mohagheghi A, Grohmam K, Himmel ME. Characterization of engF, a gene for a non-cellulosomal Clostridium cellulovoras endoglucanase. Gene. 1991;182:163–7.

23. Li C, Yang Z, He Can Zhang R, Zhang D, Chen S, Ma L. Effect of pH on cellulase production and morphology of Trichoderma reesei and the application in cellulosic material hydrolysis. J Biotechnol. 2013;168:470–7.

24. Prasetyo EN, Kudanga T, Ostergaard L, Rencoret J, Gutierrez A, Rio JCD. Polymerization of lignosulfonates by the laccase-HBT (1-hydroxybenzotriazole) system improves dispersibility. Bioresour Technol. 2010;101:5054–62.

25. El-Said AHM, Saleem A. Ecological and Physiological Studies on Soil Fungi at Western Region, Libya. Mycobiology. 2008;36:1–9.

26. Golbeck R, Ramos MM, Pereira GAG, Maugeri-Filho F. Cellulase production from a new strain Acremonium strictum isolated from the Brazilian Biome using different substrates. Bioresour Technol. 2013;128:797–803.

27. Kilikian BV, Afonso LC, Souza TFC, Ferreira RG, Pinheiro IR. Filamentous fungi and media for cellulase production in solid state cultures. Braz J Microbiol. 2014;45:279–86.

28. Salihu A, Abbas O, Sallau AB, Alam MZ. Agricultural residues for cellulolytic enzyme production by Aspergillus niger: effects of pretreatment. 3. Biotech. 2015;5:1101–6.

29. R Development Core Team. R: A language and environment for statistical computing. Vienna: R Foundation for Statistical Computing; 2008.

30. Inoue H, Decker SR, Taylor LE, Yano S, Sawayama S. Identification and characterization of core cellulolytic enzymes from Talaromyces cellulolyticus (formerly Acremonium cellulolyticus) critical for hydrolysis of lignocellulosic biomass. Biotechnol Biofuels. 2014;7:151.

31. Lombard V, Golaconda Ramulu H, Drula E, Coutinho PM, Henrissat B. The carbohydrate-active enzymes database (CAZy) in 2013. Nucleic Acids Res. 2014;42:490–5.

32. Bunterngsook B, Eurwilaichitr L, Thamchaipenet A, Champreda V. Binding characteristics and synergistic effects of bacterial expansins on cellulosic and hemicellulosic substrates. Bioresour Technol. 2015;176:129–35.

33. Laothanachareon L, Bunterngsook B, Suwannarangsee S, Eurwilaichitr L, Champreda V. Synergistic action of recombinant accessory hemicellulolytic and pectinolytic enzymes to Trichoderma reesei cellulase on rice straw degradation. Bioresour Technol. 2015;198:682–90.

34. Leggio LL, Simmons TJ, Poulsen JCN, Frandsen KEH, Hemsworth GR, Stringer MA, Freiesleben PV, Tovborg M, Johansen KS, Maria LD, Harris PV, Soong CL, Dupree P, Tryfona T, Lenfant N, Henrissat B, Davies GJ, Walton PH. Structure and boosting activity of a starch-degrading lytic polysaccharide monooxygenase. Nature Commun. 2015;6:5961. doi:10.1038/ncomms6961.

35. Horn SJ, Vaaje-Kolstad G, Westereng B, Eijsink VG. Novel enzymes for the degradation of cellulose. Biotechnol Biofuels. 2012;5:45.

36. Yang M, Zhang KD, Zhang PY, Zhou X, Ma XQ, Li FL. Synergistic cellulose hydrolysis dominated by a multi-modular processive endoglucanase from Clostridium cellulosi. Front Microbiol. 2016;7:932.

37. Teeri TT. Crystalline cellulose degradation: new insight into the function of cellobiohydrolases. Trends Biotechnol. 1997;15:160–7.

38. Mohanram S, Amat D, Choudhary J, Arora A, Nain L. Novel perspectives for evolving enzyme cocktails for lignocellulose hydrolysis in biorefineries. Sustain Chem Proc. 2013;1:15.

39. Eibinger M, Ganner T, Bubner P, Rošker S, Kracher D, Haltrich D, Ludwig R, Plank H, Nidetzky B. Cellulose surface degradation by a lytic polysaccharide monooxygenase and its effect on cellulase hydrolytic efficiency. J Biol Chem. 2014;289:35929–38.

40. Dashtban M, Qin W. Overexpression of an exotic thermotolerant β-glucosidase in Trichoderma reesei and its significant increase in cellulolytic activity and saccharification of barley straw. Microb Cell Fact. 2012;11:1–15.

41. Kanokratana P, Eurwilaichitr L, Pootanakit K, Champreda V. Identification of glycosyl hydrolases from a metagenomic library of microflora in sugarcane bagasse collection site and their cooperative action on cellulose degradation. J Biosci Bioeng. 2015;119:384–91.

42. Gottschalk LMF, Oliveira RA, Bon EPDS. Cellulases, xylanases, β-glucosidase and ferulic acid esterase produced by Trichoderma and Aspergillus act synergistically in the hydrolysis of sugarcane bagasse. Biochem Eng J. 2010;51:72–8.

5

Optimization of laccase production from *Marasmiellus palmivorus* LA1 by Taguchi method of Design of experiments

Aiswarya Chenthamarakshan, Nayana Parambayil, Nafeesathul Miziriya, P. S. Soumya, M. S. Kiran Lakshmi, Anala Ramgopal, Anuja Dileep and Padma Nambisan[*]

Abstract

Background: Fungal laccase has profound applications in different fields of biotechnology due to its broad specificity and high redox potential. Any successful application of the enzyme requires large scale production. As laccase production is highly dependent on medium components and cultural conditions, optimization of the same is essential for efficient product production.

Results: Production of laccase by fungal strain *Marasmiellus palmivorus* LA1 under solid state fermentation was optimized by the Taguchi design of experiments (DOE) methodology. An orthogonal array (L8) was designed using Qualitek-4 software to study the interactions and relative influence of the seven selected factors by one factor at a time approach. The optimum condition formulated was temperature (28 °C), pH (5), galactose (0.8%*w/v*), cupric sulphate (3 mM), inoculum concentration (number of mycelial agar pieces) (6Nos.) and substrate length (0.05 m). Overall yield increase of 17.6 fold was obtained after optimization. Statistical optimization leads to the elimination of an insignificant medium component ammonium dihydrogen phosphate from the process and contributes to a 1.06 fold increase in enzyme production. A final production of 667.4 ± 13 IU/mL laccase activity paves way for the application of this strain for industrial applications.

Conclusion: Study optimized lignin degrading laccases from *Marasmiellus palmivorus* LA1. This laccases can thus be used for further applications in different scales of production after analyzing the properties of the enzyme. Study also confirmed the use of taguchi method for optimizations of product production.

Keywords: Laccase, Taguchi DOE, Solid state fermentation, *Marasmiellus palmivorus* LA1, Optimization

Background

Laccases (EC 1.10.3.2; benzenediol: oxygen oxidoreductases) are a major group of ligninolytic enzymes which are present in all the eukaryotic kingdoms described in the five kingdom classification by R.H Whittaker in 1969 [1–5]. Laccases non-specifically catalyse one-electron oxidation of four equivalent substrates concomitant with the four-electron reduction of molecular oxygen to water with the help of a copper containing catalytic apparatus [6, 7]. Physiologically, laccase fulfil diverse roles from plant lignin polymerisation [8] to fungal morphogenesis [9]. Being less substrate specific, energy-saving, and biodegradable, laccases were suitable in the development of highly effective, sustainable, and eco-friendly enterprises [10] in the areas of biofuel production [11], chemical transformation of xenobiotics [12], dye decolourisation [13], as biofuel cells [14], effluent treatment [15], pulp bleaching [16], as biosensors [17] and in general food quality improvement [18, 19]. Any application of laccase requires large scale production of the enzyme preferably in a cost effective manner.

Even though other enzyme production systems prefer submerged fermentation, enzyme production from fungi, especially filamentous fungi is better adapted to Solid state fermentation (SSF) as only SSF offers an adherence surface to filamentous fungi [20]. In SSF, growth and enzyme production occur in inert or natural solid

* Correspondence: plantbiotech1992@gmail.com
Department of Biotechnology, Cochin University of Science and Technology, Cochin-22, Kerala, India

material under near or complete absence of free flowing liquid. SSF have advantages like high volumetric productivity [21], effective utilization of agro industrial wastes as substrates that even mimic the natural living surface of fungi and economy [22] due to its static nature. SSF utilizes materials like orange peel [23], banana waste [24], barley bran [25] and pine apple leaves [26] for useful enzyme production, which otherwise pose solid waste disposal problems. This reutilization is appreciated in the context of sustainable development. However, robust control of parameters (both media composition and cultural conditions) in SSF is difficult particularly on an industrial scale, which explains the failure of adapting successful lab scale production systems to an industrial level in the past [27]. This can be overcome by the thorough optimization of the different factors that influence production. Classical single factor method of optimization is an inadequate choice as it is time consuming [28] and will not yield any outcome regarding the relative influence of any of the involving factors. Statistical methods which also accounts for variations in the production process would be appropriate for optimization. Taguchi method of design of experiment is an approach for optimization of parameters, where the production quality stands intact even in an altered environment [29].

The Taguchi method of Design of Experiments (DOE) was developed by Genechi Taguchi who was involved in modifying the Japanese telephone system [30]. The main aim of this method is to determine the optimal process characteristic that is weakly sensitive to noise factors [31]. The taguchi method operates systematically with fewer trials, thus reducing the time, cost and effort, but offer more quantitative information [32]. The method can work even if the parameters are discrete and qualitative. It functions by reducing the sensitivity of the system [33] through thorough parameter designing. For the purpose, taguchi employs a fractional factorial design in the form of an orthogonal array. This array includes representatives from all possible combinations of selected experimental parameters, which are apt to increase the efficiency and precision and simultaneously reducing any experimental errors [34]. Analysis of individual factor contribution along with their interactive effects eventually leads to the identification of finest factors which was further optimized through Analysis of variance (ANOVA). All these advantages contribute to its greater application in other fields of science especially biotechnology.

A newly isolated strain LA1 from rarely explored species palmivorus, is the laccase producing fungus that is selected in the present study. The strain was found to be utilizing pineapple leaves, an inexpensive, unused agro-residue, as substrate for laccase production. The

initial laccase activity expressed by *Marasmiellus palmivorus* LA1 was as good as or even higher than that of the initial activities of some of the other reported fungi [35–38]. The present study applies taguchi method for the optimization of extracellular laccase enzyme production in SSF from the fungi *Marasmiellus palmivorus* LA1. The experimental design comprises of seven different factors that proceeds at two levels with L8 (2^7) array layout for laccase production. This is the first attempt reported for the optimization of laccase production from any *Marasmiellus palmivorus*, which is generally viewed only in the context of palm pathogens.

Results and discussion
Determination of factors
Selection of the appropriate culture factors is the prime key for the success of any optimization process. Here the factors and levels selected were based on the preliminary studies of one factor at a time (OFAT) on laccase production by *Marasmiellus palmivorus* LA1. The selected factors do have an influential role in laccase production as it increases the laccase production from 38.53 to 627.7 IU/mL, which is 16.2 fold during OFAT. Previous studies on different fungal laccases also emphasise the requirement of temperature, pH [39], galactose [40], cupric sulphate [41], inoculum concentration [34], and substrate length [42] for increased laccase production. However, in OFAT only the individual factor contributions are taken into consideration, which may vary during factor interactions in an industrial scale scenario.

Designing of the matrix experiment
Taguchi method of DOE is an effective statistical plan for studying the optimization of laccase production involving several factors. It is reliable for parameter identification with the added advantage of sparing the cost. Implementation of taguchi through Qualitek-4 (QT4) windows version can be through any of the L-4 arrays with three factors at two levels to L-81 arrays with 40 factors at three levels. In the present study, the L-8 array was designed using Qualitek-4 applied in order to study seven different factors. In this orthogonal array, the control factors and the identified noise factors were varied in such a way to find out a combination where variations in noise no longer affect the overall production [43]. These were called the robust designs and the analysis is called the signal to noise ratio analysis. The signal to noise ratio is linked with quadratic loss function, which in turn assumes significant losses can happen within the specification limit [44]. Such losses within limits are expected and can easily be met. "Bigger is better" quality characteristic provides a single index for the measurable results from multiple criteria.

Experimentation of the designed matrix

All the 8 trials were carried out under SSF. On experimenting the matrix combination, trial 5, which comprises of temperature - 28 °C, pH - 3, ammonium dihydrogen phosphate -0.05% (w/v), galactose - 0.8% (w/v), cupric sulphate – 3 mM, inoculum concentration (number of mycelial agar pieces) - 4 Nos. and substrate length - 0.05 m yielded maximum production with 659 ± 12 IU/mL, while least production is for the trial 1. Trial 1 includes temperature - 26 °C, pH - 3, ammonium dihydrogen phosphate - 0.03% (w/v), galactose - 0.8% (w/v), cupric sulphate – 1 mM, inoculum concentration (number of mycelial agar pieces) - 4 Nos. and substrate length - 0.03 m (Table 1).

Data analysis

The average of obtained enzyme production, in which each factor is at given level, is described in Table 2. Difference between the average values, L2-L1 indicated the relative influence of the particular factor. Greater the difference in values, better the influence on production. The positive value indicates an increase in production as it moves from level 1 to level 2, while the negative value indicates production decrease during the course from L1 to L2. Thus among the selected factors, cupric sulphate increases the laccase production at level 2, followed by substrate length, inoculum concentration, pH, temperature, ammonium dihydrogen phosphate and galactose. Ammonium dihydrogen phosphate has very less or no effect on laccase production with very similar values at level 1 (54.741) and level 2 (54.746). Galactose on the other hand is showing a slight better production at level 1.

Individual factor interaction

Interaction analysis provides insight into the interaction of a factor with other factors considered during the experiment. The severity index (SI) represents the influence of two individual factors at different levels of interaction. Col. in Table 3 show the position to which interacting factors are allotted. Overall influences of the selected factors on laccase production were depicted graphically

Table 1 Experimental trial results of all the eight trials conducted

Trials	Laccase activity (IU/mL ± Standard deviation)
1	455.2 ± 1
2	639.9 ± 24
3	542 ± 17
4	557 ± 1
5	659 ± 12
6	458 ± 68
7	606.4 ± 8
8	505.8 ± 29

Table 2 Main effects of all the selected factors

Sl. No:	Factors	Level 1	Level 2	L2 - L1
1	Temperature (°C)	54.718	54.769	.05
2	pH	54.662	54.825	.163
3	$NH_4H_2PO_4$ (%)[a]	54.741	54.746	.005
4	Galactose (%)	54.967	54.52	-.447
5	Cupric sulphate (mM)	53.721	55.766	2.045
6	Inoculum concentration	54.626	54.861	.234
7	Substrate length (m)	54.185	55.302	1.116

[a]Percentage against buffer

(Fig. 1a-g). A perpendicular line represents full (100%) interaction while parallel line means no interaction between the given factors. On analysing the severity index, its noteworthy that ammonium dihydrogen phosphate, the least laccase production influencer interacts maximally with inoculum concentration to give higher severity index (89.72%, Col.5), while the high enzyme production influencing cupric sulphate shows modest interaction with inoculum concentration with low SI (0.19%, Col.3).

Individual factor contribution and ANOVA

Analysis of variance test was carried out to determine the significance of individual factors on total laccase production (Table 4). Test results showed that cupric sulphate has a significant impact (73.18%) on laccase production followed by substrate length (23.8%). The other factors cumulatively contribute about 4.98% only to laccase production. From the F-ratio of all selected parameters, it was noticed that ammonium dihydrogen orthophosphate has null effect on production thus its effect was pooled. Pooling also helps to avoid saturation of the designed system. All other factors and their interactions considered in the current design were statistically significant at 90% confidence interval indicating that their variability can be explained in terms of significant effects. Contribution of each factor on laccase enzyme production was represented in Fig. 2.

Optimum level determination and validation of the optimum

The taguchi method provided optimum culture conditions for each of the influencing factor. The optimum conditions estimated and their contribution are shown in Table 5. Cupric sulphate and substrate length were the major factors affecting laccase production from *Marasmiellus palmivorus* LA1 under solid state fermentation. Signal to noise ratio expected was 56.769 (Table 5), from which the expected production was calculated using the formula, square root (1/Mean Square Deviation (MSD)). MSD represents all the

Table 3 Predicted interactions of the given factors depicted via severity index

Sl. No:	Interacting factor pairs (Order based on SI)	Columns	SI (%)	Col.	Opt.
1	$NH_4H_2PO_4$ (%) x Inoculum concen.	3 × 6	89.72	5	[1,2]
2	Temperature (°C) x Inoculum concen.	1 × 6	82.66	7	[1,2]
3	Temperature (°C) x Galactose (%)	1 × 4	82.06	5	[2,1]
4	Temperature (°C) x $NH_4H_2PO_4$ (%)	1 × 3	76.01	2	[2,1]
5	$NH_4H_2PO_4$ (%) x Galactose (%)	3 × 4	71.42	7	[2,1]
6	pH x Inoculum concen.	2 × 6	65.61	4	[2,2]
7	pH x Substrate length (m)	2 × 7	64.67	5	[1,2]
8	pH x Cupric sulphate (mM)	2 × 5	35.32	7	[1,2]
9	Substrate length (m) x Galactose (%)	2 × 4	34.38	6	[2,1]
10	$NH_4H_2PO_4$ (%) x Substrate length (m)	3 × 7	28.57	4	[2,2]
11	pH x $NH_4H_2PO_4$ (%)	2 × 3	23.98	1	[2,1]
12	Galactose (%) x Inoculum concen.	4 × 6	23.92	2	[1,2]
13	Temperature (°C) x Cupric sulphate (mM)	1 × 5	17.93	4	[2,2]
14	Temperature (°C) x Substrate length (m)	1 × 7	17.33	6	[1,2]
15	$NH_4H_2PO_4$ (%) x Cupric sulphate (mM)	3 × 5	10.27	6	[1,2]
16	Cupric sulphate (mM) x Substrate length (m)	5 × 7	5.15	2	[2,2]
17	Inoculum concen. x Substrate length	6 × 7	3.8	1	[2,2]
18	Temperature (°C) x pH	1 × 2	2.08	3	[2,2]
19	Galactose (%) x Cupric sulphate (mM)	4 × 5	2.06	1	[1,2]
20	Galactose (%) x Substrate length (m)	4 × 7	.28	3	[1,2]
21	Cupric sulphate (mM) x Inoculum concen.	5 × 6	.19	3	[2,2]

Columns: Column locations to which the interacting factors are assigned, SI%: Interaction severity index, Col: Column that should be reserved if this particular interaction is to be studied, Opt.: indicates factor levels desirable for the optimum condition

variation around the given target and can be calculated from S/N, where S/N = - Log (MSD). The expected laccase production at optimum conditions was found to be 689.366 IU/mL.

Variation reduction plot
Variation reduction plot is a graphical representation of the current and improved production status within upper and lower control limits (UCL or LCL) (Fig. 3). Nominal value is 553.062 IU/mL while LCL and UCL being 317.737 and 789.397 IU/mL respectively. Reduced variation is represented by the steep peak in graph. From the graph it's deducted that the improved condition could cause a savings of 37.3%. This savings owes to the elimination of non necessitated media component.

Validation of the optimum
Tests were performed with the optimized factors with the recommended level. This resulted in the production of 667.4 ± 13 IU/mL of enzyme which is comparable to the predicted (689.366 IU/mL). Thus the taguchi method is validated for extracellular laccase production.

Fungal laccase production under solid state fermentation is influenced by various environmental (temperature) and cultural (pH, media components, substrate size, inoculum) conditions [45]. Involvement of many factors leads to optimization to improve the laccase enzyme production. Other than classical approaches, taguchi method of DOE offers a statistical design to create robustness in the process with low lost by considering only the main effects. The method also accounts multiple interaction possibilities between the parameters which is significant in industrial applications. Unlike the past decade, many works were currently relying on taguchi methods for culture parameter optimization [46], process optimization [47, 48], medium optimization [49, 50] and overall yield of enzyme production [51]. In this optimization process the most influencing factors affecting laccase production were found to be cupric sulphate, followed by substrate length and inoculum concentration. Increased production in the presence of copper can be attributed as the defensive response of fungi towards the induced metallic stress [52]. Similar increase in laccase production was observed in *Marasmius quercophilus* in addition of copper [53]. Copper can induce the production of laccase isozymes which

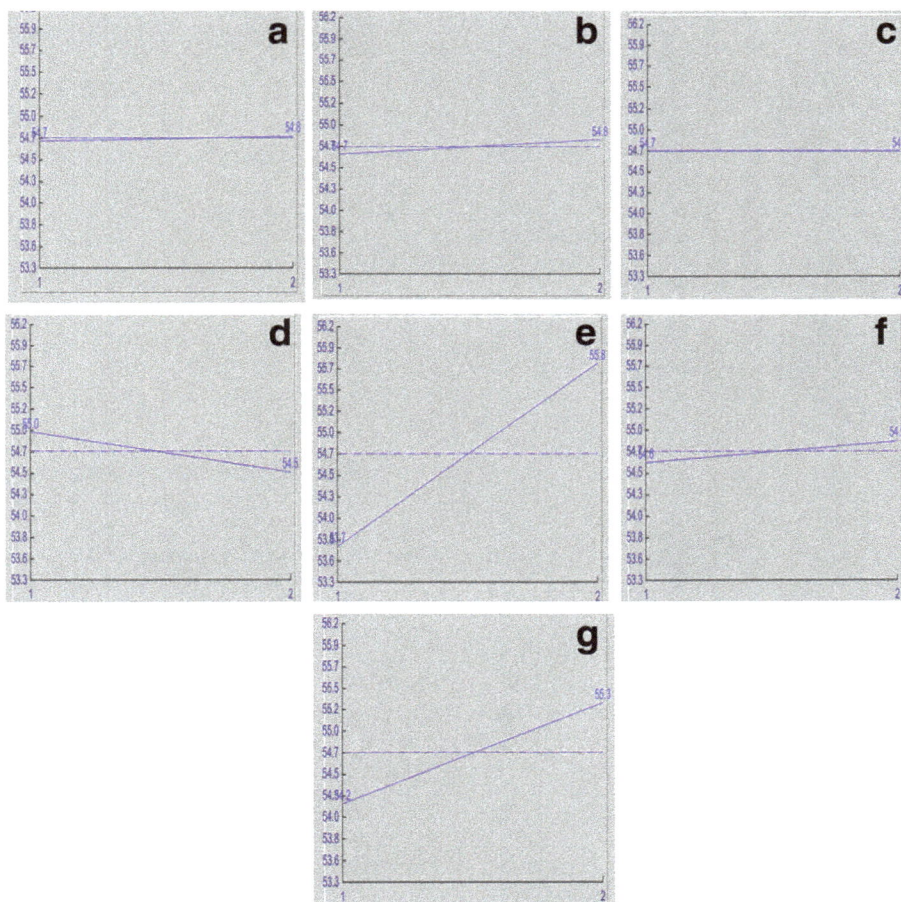

Fig. 1 Influence of the selected factors on laccase production by *Marasmiellus palmivorus* LA1. **a** Temperature (°C), **b** pH, **c** Ammonium dihydrogen phosphate (%), **d** Galactose (%w/v), **e** Cupric sulphate (mM), **f** Inoculum concentration (number of mycelial agar pieces), and **g** Substrate length (m). In all the graphs X-axis denotes the different levels (1 and 2) of the concerned factor and Y-axis average effect of the concerned factors

leads to increased production [54]. Influence of the pine apple leaf length in production is by offering more surface area and lignin for the basidiomycete, *Marasmiellus palmivorus* LA1 growth and laccase production [55]. Inoculum size does play an important role in establishing the culture in Erlenmeyer flasks, which explains its influential role. The insignificance of ammonium dihydrogen phosphate in the medium was stated through the statistical study, which leads to the elimination of the same from

Table 4 Analysis of variance (ANOVA)

Sl.No:	Factors	DOF	Sums of squares	Variance	F-Ratio	Pure Sum	Percent
1	Temperature (°C)	1	0.006	0.006	173.585	0.006	0.060
2	pH	1	0.053	0.053	1,325.232	0.053	0.464
3	$NH_4H_2PO_4$ (%)	1	0		POOLED	0	0
4	Galactose (%)	1	0.401	0.401	10,006.075	0.401	3.510
5	Cupric sulphate (mM)	1	8.366	8.366	208,581.269	8.366	73.181
6	Inoculum concentration	1	0.108	0.108	2,694.782	0.108	0.945
7	Substrate length (m)	1	2.496	2.496	62,237.074	2.496	21.835
	Other/Error	1	0	0			0.005
	Total	7	11.433				100.00%

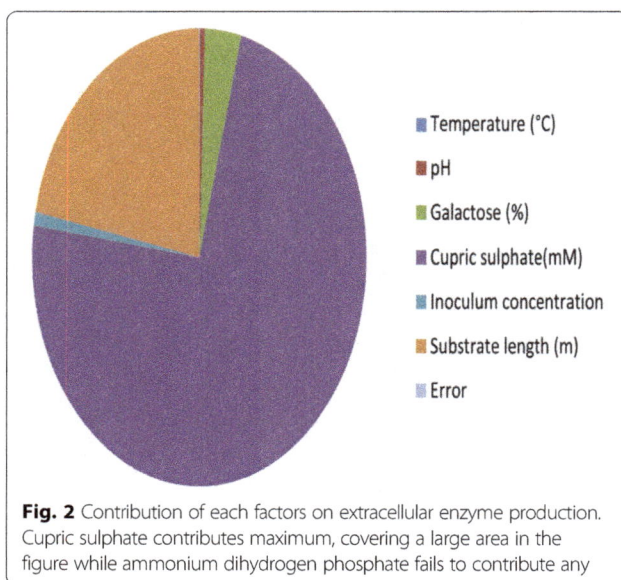

Fig. 2 Contribution of each factors on extracellular enzyme production. Cupric sulphate contributes maximum, covering a large area in the figure while ammonium dihydrogen phosphate fails to contribute any

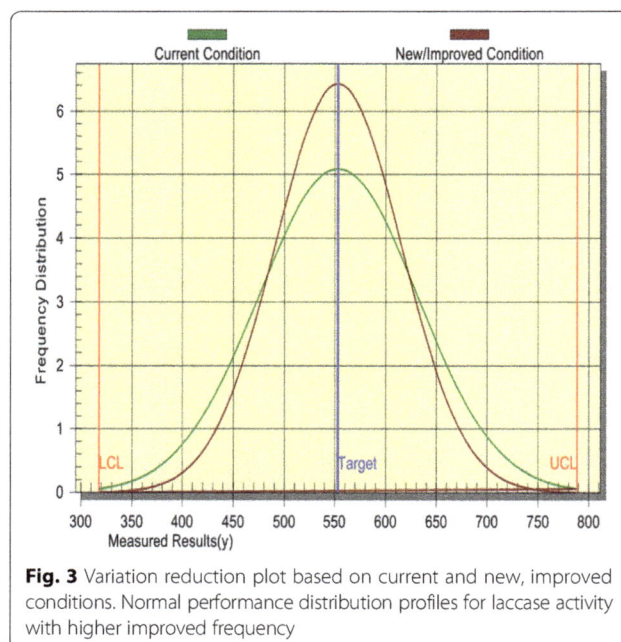

Fig. 3 Variation reduction plot based on current and new, improved conditions. Normal performance distribution profiles for laccase activity with higher improved frequency

medium, thus reducing the cost without compromising the production quality and quantity. The strain provides a satisfactory yield of laccase (Table 6) leading to the application of the enzyme directly or in immobilized form in industrial settings.

Conclusion

Optimization of laccase production under solid state fermentation by *Marasmiellus palmivorus* LA1 stain was done via taguchi method of DOE using Qualitek-4. The study aided in the understanding of individual factor contribution and interaction among factors. Elimination of unwanted factors significantly reduce the loss during the process, which otherwise needed to be met. Validation of optimized parameters provides an optimum set of conditions that are insensitive to noise factors which can be used in large scale bioprocess.

Table 5 Optimum culture condition predicted and their contributions on the selected levels

Sl. No:	Factors	Level description	Level	Contribution
1	Temperature (°C)	28	2	0.025
2	pH	5	2	0.081
3	Galactose (%)	0.8	1	0.223
4	Cupric sulphate (mM)	3	2	1.023
5	Inoculum concentration	6	2	0.117
6	Substrate length (m)	0.05	2	0.558
Total contribution from all factors				2.025
Current grand average of performance				54.743
Expected result at optimum condition				56.769

Methods

Microorganism

For the present study, the culture of fungi *Marasmiellus palmivorus* LA1 isolated from Palakkad district of Kerala, India was used for the production of extra cellular laccases. The strain was grown and then maintained on Potato dextrose agar (PDA) at 4 °C.

Solid state fermentation for enzyme production

Pineapple leaves of varying length were used as the substrate [26], onto which *Marasmiellus palmivorus* LA1 mycelial agar pieces (0.005 m × 0.005 m sized) were inoculated in 250 mL Erlenmeyer flask. The moisture content was adjusted to 10% with 0.1 M sodium citrate buffer of pH 5. The system was incubated for 5 days under static condition in appropriate temperatures.

Product extraction

Extracellular enzyme was extracted using 40 mL of 0.1 M sodium citrate buffer of pH 5. After incubation period, the mycelial-free supernatant was collected by gentle shaking followed by centrifugation at 9000 g for 10 min and used for further laccase activity assays.

Enzyme assay

The laccase assay was performed spectrophotometrically (Shimadzu 1601) at 420 nm using 2,2′-azino-bis (3-ethylbenzothiazoline-6-sulphonic acid) (ABTS) as substrate [56]. One unit (IU/mL) of laccase activity was defined as the amount of enzyme required for the conversion of one micromole of substrate per minute under assay conditions.

Table 6 Comparison of laccase yields of other fungi with the fungus of interest, grown under solid state fermentation

Sl. No:	Organism	Substrate	Enzyme activity (IU/mL)	Reference
1	Pleurotus ostreatus	Banana pseudostem	3	[58]
2	Pleurotus sajor-caju.	Banana pseudostem	3.6	[58]
3	Coprinellus disseminatus SW-1 NTCC 1165	Wheat bran	25.5	[59]
4	Aspergillus heteromorphus	Rice straw	6.6	[60]
5	Aspergillus heteromorphus	Sugarcane baggase	2.9	[60]
6	Schizophyllum commune IBL-06	Banana stalks	345	[61]
7	Ganoderma lucidum	Pineapple leaf	472.31 ± 41.2	[26]
8	Coculture of Pleurotus flabellatus and Pleurotus eous	Coffee pulp	8.8	[62]
9	Schyzophyllum commune	Corn stover	130.80	[63]
10	Pleurotus ostreatus IBL-04	Wheat straw	517 ± 1.05	[64]
11	Phanerochaete chrysosporium	Wheat straw	263.03	[65]
12	Trametes versicolor IBL-04	Corn cobs	869.65	[66]
13	Marasmiellus palmivorus LA1	Pine apple leaf	667.4 ± 13	Present study

Taguchi method for optimization

In the present study, the taguchi method of optimization moves in five stages: factors determination, matrix designing, experimentation of matrix, data analysis and optimum level validation. All these stages proceed in a stepwise manner to finally yield a valid output.

Determination of factors

Using the one factor at a time (OFAT) method for optimization seven different factors that were found to be crucial for laccase enzyme production in *Marasmiellus palmivorus* LA1 were listed out. Then these factors were used for further optimization using taguchi method. The significant influencing factors are temperature, pH, ammonium dihydrogen phosphate ($NH_4H_2PO_4$), galactose, cupric sulphate, inoculum concentration (number of mycelial agar pieces) and substrate length.

Designing of the matrix experiment

QUALITEK-4 software (Nutek Inc., MI, USA) was employed for the purpose [57]. Using an L8 (2^7) orthogonal array the seven major factors were studied in two levels (Table 7). "Bigger is better" was the quality characteristic preferred in the experimental studies. Signal to noise ratio analysis was used for result analysis.

Experimentation of the designed matrix

Based on the two levels mentioned (Table 7) 8 different trial sets of solid state fermentation were conducted with *Marasmiellus palmivorus* LA1 (Table 8). All the trials were performed in 250 mL Erlenmeyer flasks having pine apple leaves of length 0.03 or 0.05 m (depending on the assigned levels) wetted with pH 3 or 5 sodium citrate buffer (0.1 M). Ammonium dihydrogen phosphate (0.03% or 0.05% (*w/v*)) and galactose (0.8% or 1.2% (w/v)) were dissolved in the buffer and supplemented. Filter sterilized cupric sulphate was added after autoclaving of the flasks at 121 °C, 1.03 bar pressure for 20 min. Variation in the inoculum concentration was created by using different number of agar pieces (4 or 6). 26 °C or 28 °C temperature was maintained throughout the period of 5-day incubation. Enzyme extraction was performed

Table 7 Selected culture factors and their assigned levels

Sl.No:	Factors	Level 1	Level 2
1	Temperature (°C)	26	28
2	pH	3	5
3	$NH_4H_2PO_4$ (%)	0.03	0.05
4	Galactose (%)	0.8	1.2
5	Cupric sulphate (mM)	1	3
6	Inoculum concentration	4	6
7	Substrate length (m)	0.03	0.05

Table 8 L-8 orthogonal array design

Trials	Columns						
	1	2	3	4	5	6	7
Trial 1	1	1	1	1	1	1	1
Trial 2	1	1	1	2	2	2	2
Trial 3	1	2	2	1	1	2	2
Trial 4	1	2	2	2	2	1	1
Trial 5	2	1	2	1	2	1	2
Trial 6	2	1	2	2	1	2	1
Trial 7	2	2	1	1	2	2	1
Trial 8	2	2	1	2	1	1	2

as previously described. All the trials were performed in triplicates.

Data analysis

Analysis of the obtained results was done using Qualitek-4 software to infer the interactions between different factors and to give idea about the influence of each individual factor on enzyme production.

Optimum level determination and validation of the optimum

By analysing the interactions, the software predicts an optimum condition for maximum enzyme production. The software recommended optimum condition was validated by conducting solid state fermentation and assay testing in triplicates under the optimum condition.

Abbreviations

ABTS: 2, 2'-azino-bis (3-ethylbenzothiazoline-6-sulphonic acid); ANOVA: Analysis of variance; DOE: Design of experiments; LSL: Lower specification limit; MSD: Mean square deviation; OFAT: One factor at a time; PDA: Potato dextrose agar; SI: Severity index; SSF: Solid state fermentation; USL: Upper specification limit

Acknowledgment

The authors gratefully acknowledge Department of Biotechnology, Cochin University of Science and Technology, Cochin -22 for the facilities and infrastructure, and Kerala State Council for Science Technology and Environment (KSCSTE) for the financial assistance.

Funding

The fund to carry out the work was provided by Kerala state council for science, technology and environment (KSCSTE), Trivandrum, Kerala, India in the form of junior research fellowships. KSCSTE played no role in the design of the study, the collection, analysis, and interpretation of data or in writing the manuscript.

Authors' contributions

AC identified the fungus, participated in the design of the study, solid state fermentation experiments, statically interpretation and drafted revised the manuscript. NP, NM, SPS participated in helping solid state fermentation experiments. KLMS, AR participated in helping statistical interpretation. AD participated in helping the manuscript drafting and revision. PN conceived of the study, participated in its design, co-ordinated the work and carefully revised the manuscript. All authors read and approved the final manuscript.

Authors' information

Aiswarya Chenthamarakshan received her M.Sc Biotechnology from Department of Biotechnology, Cochin University of Science and Technology, Cochin 22 in 2013 and currently pursuing Ph.D under the guidance of Dr. Padma Nambisan in Plant Biotechnology Laboratory, Cochin University of Science and Technology. Her interests are in utilization of laccase producing fungi for industrial applications.
Nayana parambayil studied M.Sc Biotechnology in Department of Biotechnology, Cochin University of Science and Technology and presently doing Ph.D under the guidance of Dr. Padma Nambisan in Plant Biotechnology Laboratory. She mainly focuses on solid state fermentation of ligninases.
Nafeesathul Miziriya is currently studying M.Sc Biotechnology in Department of Biotechnology, Cochin University of Science and Technology and joins the group as a summer trainee. Her research interests are in the field of fungal enzymes.
Soumya P S pursued M.Sc. Biotechnology, from Bharathiar University, Tamilnadu and presently working in Plant Biotechnology Laboratory as a Ph.D scholar. Her area of study was laccases and their potential applications.
Kiran Lakshmi M S studied Gene Technology from J.J. College of Arts and Science, Tamil nadu, India where she got her master's degree, and joined Dr. Padma Nambisan in 2012 for her doctoral degree. Her area under study includes molecular biology of ligninolytes.
Anala Ramgopal is a Ph.D student working under Dr. Padma Nambisan in Plant Biotechnology Laboratory of Cochin University of Science and Technology. She completed her master's degree from Cochin University of Science and Technology in biotechnology. Biodegradation of polythene is her major area of interest.
Anuja Dileep after finishing her M.Sc. Biotechnology joined Plant Biotechnology Laboratory, Cochin University of Science and Technology in 2014. Her interests include effluent treatment using biological methods.
Dr. Padma Nambisan is currently working as a faculty (Professor and guide) in Department of Biotechnology, Cochin University of science and technology. She achieved Post Doctoral DBT Fellowship from Indian Institute of Science and Ph.D from Indian Agricultural Research Institute (I.A.R.I). Her major area of expertise includes Plant Biotechnology and Genetics and fungal biotechnology.

Competing interests

The authors declare that they have no competing interests.

References

1. Ranocha P, McDougall G, Hawkins S, Sterjiades R, Borderies G, Stewart D, Cabanes-Macheteau M, Boudet AM, Goffner D. Biochemical characterization, molecular cloning and expression of laccases–a divergent gene family–in poplar. Eur J Biochem. 1999;259(1-2):485–95.
2. Bourbonnais R, Paice MG, Reid ID, Lanthier P, Yaguchi M. Lignin oxidation by laccase isozymes from Trametes versicolor and role of the mediator 2, 2'-azinobis (3-ethylbenzthiazoline-6-sulfonate) in kraft lignin depolymerization. Appl Environ Microb. 1995;61(5):1876–80.
3. Sharma P, Goel R, Capalash N. Bacterial laccases. World J Microb Biot. 2007;23(6):823–32.
4. Arora DS, Sharma RK. Ligninolytic fungal laccases and their biotechnological applications. Appl Biochem Biotech. 2010;160(6):1760–88.
5. Tartar A, Wheeler MM, Zhou X, Coy MR, Boucias DG, Scharf ME. Parallel metatranscriptome analyses of host and symbiont gene expression in the gut of the termite Reticulitermes flavipes. Biotechnol Biofuels. 2009;2(1):1.
6. Solomon EI, Sundaram UM, Machonkin TE. Multicopper oxidases and oxygenases. Chem Rev. 1996;96(7):2563–606.
7. Messerschmidt A. Spatial structures of ascorbate oxidase, laccase and related proteins: implications for the catalytic mechanism. In: Messerschmidt A, editor. Multi-copper oxidases. Singapore: World Scientific; 1997. p. 23–80.
8. Sterjiades R, Dean JF, Eriksson KE. Laccase from sycamore maple (Acer pseudoplatanus) polymerizes monolignols. Plant Physiol. 1992;99(3):1162–8.
9. Leatham GF, Stahmann MA. Studies on the laccase of Lentinus edodes: specificity, localization and association with the development of fruiting bodies. Microbiology. 1981;125(1):147–57.
10. Osma JF, Toca-Herrera JL, Rodríguez-Couto S. Uses of laccases in the food industry. Enzyme Res. 2010;2010:1–8.
11. Kudanga T, Le Roes-Hill M. Laccase applications in biofuels production: current status and future prospects. Appl Microbiol Biot. 2014;98(15):6525–42.
12. Calvo AM, Copa-Patiño JL, Alonso O, González AE. Studies of the production and characterization of laccase activity in the basidiomycete Coriolopsis gallica, an efficient decolorizer of alkaline effluents. Arch Microbiol. 1998; 171(1):31–6.

13. Khlifi R, Belbahri L, Woodward S, Ellouz M, Dhouib A, Sayadi S, Mechichi T. Decolourization and detoxification of textile industry wastewater by the laccase-mediator system. J Hazard Mater. 2010;175(1):802–8.

14. Slomczynski D, Nakas J, Tanenbaum SW. Production and Characterization of Laccase from Botrytis cinerea 61-34. Appl Environ Microb. 1995;61(3):907–12.

15. Jaouani A, Guillén F, Penninckx MJ, Martínez AT, Martínez MJ. Role of Pycnoporus coccineus laccase in the degradation of aromatic compounds in olive oil mill wastewater. Enzyme Microb Tech. 2005;36(4):478–86.

16. Luisa M, Goncalves FC, Steiner W. Use of laccase for bleaching of pulps and treatment of effluents. In: Enzymes for pulp and paper processing. 1996. p. 197–206.

17. Montereali MR, Della Seta L, Vastarella W, Pilloton R. A disposable Laccase–Tyrosinase based biosensor for amperometric detection of phenolic compounds in must and wine. J Mol Catal B-Enzym. 2010;64(3):189–94.

18. Selinheimo E, Kruus K, Buchert J, Hopia A, Autio K. Effects of laccase, xylanase and their combination on the rheological properties of wheat doughs. J Cereal Sci. 2006;43(2):152–9.

19. Neifar M, Ellouze-Ghorbel RA, Kamoun A, Baklouti S, Mokni A, Jaouani A, Ellouze-Chaabouni SE. Effective clarification of pomegranate juice using laccase treatment optimized by response surface methodology followed by ultra filtration. J Food Process Eng. 2011;34(4):1199–219.

20. Pandey A. Solid-state fermentation. Biochem Eng J. 2003;13(2):81–4.

21. Duenas R, Tengerdy RP, Gutierrez-Correa M. Cellulase production by mixed fungi in solid-substrate fermentation of bagasse. World J Microb Biot. 1995;11(3):333–7.

22. Murugesan K, Nam IH, Kim YM, Chang YS. Decolorization of reactive dyes by a thermostable laccase produced by Ganoderma lucidum in solid state culture. Enzyme Microb Tech. 2007;40(7):1662–72.

23. Rosales E, Couto SR, Sanromán MA. Increased laccase production by Trametes hirsuta grown on ground orange peelings. Enzyme Microb Tech. 2007;40(5):1286–90.

24. Elisashvili V, Parlar H, Kachlishvili E, Chichua D, Bakradze M, Kokhreidze N, Kvesitadze G. Ligninolytic activity of basidiomycetes grown under submerged and solid-state fermentation on plant raw material (sawdust of grapevine cuttings). Adv Food Sci. 2001;23(3):117–23.

25. Couto SR, Gundín M, Lorenzo M, Sanromán MÁ. Screening of supports and inducers for laccase production by Trametes versicolor in semi-solid-state conditions. Process Biochem. 2002;38(2):249–55.

26. Hariharan S, Nambisan P. Optimization of lignin peroxidase, manganese peroxidase, and Lac production from Ganoderma lucidum under solid state fermentation of pineapple leaf. BioResources. 2012;8(1):250–71.

27. Cen P, Xia L. Production of cellulase by solid-state fermentation. In: Tsao GT, editor. Recent Progress in Bioconversion of Lignocellulosics. Berlin Heidelberg: Springer; 1999. p. 69–92.

28. Silva EM, Rogez H, Larondelle Y. Optimization of extraction of phenolics from Inga edulis leaves using response surface methodology. Sep Purif Technol. 2007;55(3):381–7.

29. Khuri AI, Mukhopadhyay S. Response surface methodology. WIREs Comp Stat. 2010;2(2):128–49.

30. Martínez-Lorente AR, Dewhurst F, Dale BG. Total quality management: origins and evolution of the term. TQM Mag. 1998;10(5):378–86.

31. Montgomery DC. Design and analysis of experiments. New Jersey: Wiley; 2008.

32. Stone RA, Veevers A. The Taguchi influence on designed experiments. J Chemometr. 1994;8(2):103–10.

33. Yang WP, Tarng YS. Design optimization of cutting parameters for turning operations based on the Taguchi method. J Mater Process Tech. 1998;84(1):122–9.

34. Prasad KK, Mohan SV, Rao RS, Pati BR, Sarma PN. Laccase production by Pleurotusostreatus 1804: Optimization of submerged culture conditions by Taguchi DOE methodology. Biochem Eng J. 2005;24(1):17–26.

35. Boehmer U, Suhardi SH, Bley T. Decoloring reactive textile dyes with white-rot fungi by temporary immersion cultivation. Eng Life Sci. 2006;6(4):417–20.

36. Stajić M, Persky L, Friesem D, Hadar Y, Wasser SP, Nevo E, Vukojević J. Effect of different carbon and nitrogen sources on laccase and peroxidases production by selected Pleurotus species. Enzyme Microb Tech. 2006;38(1):65–73.

37. Mohorčič M, Friedrich J, Pavko A. Decoloration of the diazo dye reactive black 5 by immobilized Bjerkandera adusta in a stirred tank bioreactor. Acta Chim Slov. 2004;51:619–28.

38. Gnanamani A, Jayaprakashvel M, Arulmani M, Sadulla S. Effect of inducers and culturing processes on laccase synthesis in Phanerochaete chrysosporium NCIM 1197 and the constitutive expression of laccase isozymes. Enzyme Microb Tech. 2006;38(7):1017–21.

39. Nyanhongo GS, Gomes J, Gübitz G, Zvauya R, Read JS, Steiner W. Production of laccase by a newly isolated strain of Trametes modesta. Bioresource Technol. 2002;84(3):259–63.

40. Elshafei AM, Hassan MM, Haroun BM, Elsayed MA, Othman AM. Optimization of laccase production from Penicillium martensii NRC 345. Adv life sci. 2012;2(1):31–7.

41. Revankar MS, Lele SS. Enhanced production of laccase using a new isolate of white rot fungus WR-1. Process Biochem. 2006;41(3):581–8.

42. Patel H, Gupte A, Gupte S. Effect of different culture conditions and inducers on production of laccase by a basidiomycete fungal isolate Pleurotus ostreatus HP-1 under solid state fermentation. BioResources. 2009;4(1):268–84.

43. Farnet AM, Tagger S, Le Petit J. Effects of copper and aromatic inducers on the laccases of the white-rot fungus Marasmius quercophilus. Cr Acad Sci IIIVIE. 1999;322(6):499–503.

44. Joglekar AM. Statistical methods for six sigma: in R&D and manufacturing. USA: Wiley; 2003.

45. Niku-Paavola ML, Karhunen E, Kantelinen A, Viikari L, Lundell T, Hatakka A. The effect of culture conditions on the production of lignin modifying enzymes by the white-rot fungus Phlebia radiata. J Biotechnol. 1990;13(2):211–21.

46. Sabarathinam S, Jayaraman V, Balasubramanian M, Swaminathan K. Optimization of culture parameters for hyper laccase production by Trichoderma asperellum by Taguchi design experiment using L-18 orthogonal array. Malaya j Biosci. 2014;1(4):214–25.

47. El Aty AA, Wehaidy HR, Mostafa FA. Optimization of inulinase production from low cost substrates using Plackett–Burman and Taguchi methods. Carbohyd Polym. 2014;102:261–8.

48. Mnif I, Sahnoun R, Ellouze-Chaabouni S, Ghribi D. Evaluation of B. subtilis SPB1 biosurfactants' potency for diesel-contaminated soil washing: optimization of oil desorption using Taguchi design. Environ Sci Pollut R. 2014;21(2):851–61.

49. Petlamul W, Prasertsan P. Medium optimization for production of Beauveriabassiana BNBCRC spores from biohydrogen effluent of palm oil mill using taguchi design. Int J Biosci Biochem Bioinforma. 2014;4(2):106.

50. Kamble R, Gupte A. Cyclodextrin glycosyltransferase production by alkaliphilic bacillus sp. Isolated from rice cultivated soil and media optimization using taguchi method. Int J Pharm Sci Res. 2014;5(7):2754.

51. Azin M, Moravej R, Zareh D. Production of xylanase by Trichoderma longibrachiatum on a mixture of wheat bran and wheat straw: Optimization of culture condition by Taguchi method. Enzyme Microb Tech. 2007; 40(4):801–5.

52. Fernandez-Larrea J, Stahl U. Isolation and characterization of a laccase gene from Podospora anserina. Molec Gen Genet. 1996;252(5):539–51.

53. Rao RS, Kumar CG, Prakasham RS, Hobbs PJ. The Taguchi methodology as a statistical tool for biotechnological applications: a critical appraisal. Biotechnol J. 2008;3(4):510–23.

54. Saparrat MC, Guillén F, Arambarri AM, Martínez AT, Martínez MJ. Induction, isolation, and characterization of two laccases from the white rot basidiomycete Coriolopsis rigida. Appl Environ Microb. 2002;68(4):1534–40.

55. Mishra S, Mohanty AK, Drzal LT, Misra M, Hinrichsen G. A review on pineapple leaf fibers, sisal fibers and their biocomposites. Macromol Mater Eng. 2004;289(11):955–74.

56. Bourbonnais R, Leech D, Paice MG. Electrochemical analysis of the interactions of laccase mediators with lignin model compounds. BBA-Gen Subjects. 1998;1379(3):381–90.

57. Montgomery DC. Introduction to statistical quality control. 2nd ed. Wiley; 1991.

58. Ghosh M, Mukherjee R, Nandi B. Production of extracellular enzymes by two Pleurotus species using banana pseudostem biomass. Acta Biotechnol. 1998;18(3):243–54.

59. Agnihotri S, Dutt D, Tyagi CH, Kumar A, Upadhyaya JS. Production and biochemical characterization of a novel cellulase-poor alkali-thermo-tolerant xylanase from Coprinellus disseminatus SW-1 NTCC 1165. World J Microb Biot. 2010;26(8):1349–59.

60. Singh A, Bajar S, Bishnoi NR, Singh N. Laccase production by Aspergillus heteromorphus using distillery spent wash and lignocellulosic biomass. J Hazard Ma Ter. 2010;176(1):1079–82.

61. Irshad M, Asgher M. Production and optimization of ligninolytic enzymes by white rot fungus Schizophyllum commune IBL-06 in solid state medium banana stalks. Afr J Biotechnol. 2011;10(79):18234–42.

62. Parani K, Eyini M. Production of ligninolytic enzymes during solid state fermentation of coffee pulp by selected fungi. Sci Res Rep. 2012;2(3):202–9.

63. Yasmeen Q, Asgher M, Sheikh MA, Nawaz H. Optimization of ligninolytic enzymes production through response surface methodology. BioResources. 2013;8(1):944–68.

64. Asgher M, Ahmad Z, Iqbal HM. Alkali and enzymatic delignification of sugarcane bagasse to expose cellulose polymers for saccharification and bio-ethanol production. Ind Crop Prod. 2013;44:488–95.

65. Koyani RD, Sanghvi GV, Sharma RK, Rajput KS. Contribution of lignin degrading enzymes in decolourisation and degradation of reactive textile dyes. Int Biodeter Biodegr. 2013;77:1–9.

66. Noreen S, Asgher M, Hussain F, Iqbal A. Performance Improvement of Ca-Alginate Bead Cross-Linked Laccase from Trametes versicolor IBL-04. North Carolina: BioResources. 2015;11(1):558–72.

6

Cell-bound lipases from *Burkholderia* sp. ZYB002: gene sequence analysis, expression, enzymatic characterization, and 3D structural model

Zhengyu Shu[1,2,3*†], Hong Lin[1,2,3†], Shaolei Shi[1,2,3], Xiangduo Mu[1,2,3], Yanru Liu[1,2,3] and Jianzhong Huang[1,2,3*]

Abstract

Background: The whole-cell lipase from *Burkholderia cepacia* has been used as a biocatalyst in organic synthesis. However, there is no report in the literature on the component or the gene sequence of the cell-bound lipase from this species. Qualitative analysis of the cell-bound lipase would help to illuminate the regulation mechanism of gene expression and further improve the yield of the cell-bound lipase by gene engineering.

Results: Three predictive cell-bound lipases, *lipA*, *lipC21* and *lipC24*, from *Burkholderia* sp. ZYB002 were cloned and expressed in *E. coli*. Both LipA and LipC24 displayed the lipase activity. LipC24 was a novel mesophilic enzyme and displayed preference for medium-chain-length acyl groups (C10-C14). The 3D structural model of LipC24 revealed the open Y-type active site. LipA displayed 96 % amino acid sequence identity with the known extracellular lipase. *lipA*-inactivation and *lipC24*-inactivation decreased the total cell-bound lipase activity of *Burkholderia* sp. ZYB002 by 42 % and 14 %, respectively.

Conclusions: The cell-bound lipase activity from *Burkholderia* sp. ZYB002 originated from a multi-enzyme mixture with LipA as the main component. LipC24 was a novel lipase and displayed different enzymatic characteristics and structural model with LipA. Besides LipA and LipC24, other type of the cell-bound lipases (or esterases) should exist.

Keywords: *Burkholderia* sp. ZYB002, Cell-bound lipase, Lipase LipC24, Lipase LipA

Background

Microbial lipase (triacylglycerol lipase, EC 3.1.1.3) catalyze hydrolysis of the long chain triglycerides, or the reverse reaction. Besides hydrolysis activity, lipases also displayed alcoholysis, aminolysis, interesterification, and esterification activity, etc. with rigorous regioselectivity, stereoselectivity, and chemoselectivity [1]. As one kind of non-aqueous enzymes, lipases kept high catalysis efficiency in organic solvent systems or micro-aqueous systems, and were widely used in many industrial fields [2].

Microbial strains can produce multiple types of lipases. *Candida rugosa* produced more than five types of lipase isoenzymes (Lip1-Lip5), which shared high sequence identity, but displayed significantly different enzymatic characteristics [3]. Different types of lipases produced by a specific microbial strain always were distributed to different cell compartments, respectively. Lipase LipA from *Pseudomonas aeruginosa* either was secreted into the culture medium, or interacted with the polysaccharide alginate and then anchored on the cell surface [4–6], while Esterase EstA from *P. aeruginosa* was located in the outer membrane [7, 8].

Cell-bound lipase could be directly used as a whole cell biocatalyst. Compared with the extracellular enzyme, whole cell biocatalysts displayed many advantages, including high stability in the long-term, inexpensive preparation, independence of the exogenous co-factor for redox reaction, etc. [9]. In previous research, *Burkholderia* sp. ZYB002 produced both extracellular lipase and cell-bound

* Correspondence: shuzhengyu@fjnu.edu.cn; hjz@fjnu.edu.cn
†Equal contributors
[1]National & Local United Engineering Research Center of Industrial Microbiology and Fermentation Technology, Ministry of Education, Fujian Normal University, Fuzhou 350117, China
Full list of author information is available at the end of the article

lipase [10, 11]. The cell-bound lipase from *Burkholderia cepacia* displayed excellent interesterification activity for biodiesel production and highly enantioselective hydrolysis activity for L-menthol synthesis [12, 13]. However, there wasn't any report on the type or the gene sequence of the cell-bound lipase from *B. cepacia*.

In this article, three predictive cell-bound lipase genes from *Burkholderia* sp. ZYB002, *lipA*, *lipC21* and *lipC24*, were cloned and expressed in *E. coli*, respectively. Furthermore, the component of the cell-bound lipase from *Burkholderia* sp. ZYB002 was analyzed.

Methods

Bacterial strains and plasmids

The bacterial strains and plasmids used in this study are listed in Table 1. Briefly, *E. coli* DH5α was used as the host strain for plasmid amplification, and *E. coli* BL21(DE3) and *E. coli* Origami2 (DE3) were used as the expression host strain for three lipase genes, *lipA*, *lipC21* and *lipC24*, respectively. *Burkholderia* sp. ZYB002 was the lipase-producing strain, which was isolated and identified

in our lab [14]. Antibiotics were added as required to the final concentrations of 60 μg/mL ampicillin, 35 μg/mL chloramphenicol, 50 μg/mL kanamycin, 100 μg/mL trimethoprim, 50 μg/mL gentamicin.

The cloning plasmid pMD18T-*lipAB*, pMD18T-*lipC21*, and pMD18T-*lipC24*, harbored the full length lipase gene of *lipA* and its chaperonin gene *lipB*, *lipC21* and *lipC24*, respectively. The expression plasmid pEDSF-*lipB-lipA*, pEDSF-*lipC21*, and pEDSF-*lipB-lipC24*, harbored the coding region for the mature LipA (lipase A) and its chaperonin LipB (the lipase-specific foldase), LipC21 (lipase C21), and LipC24 (lipase C24)/LipB, respectively.

Chemicals and biochemistry reagents

High-fidelity DNA polymerases, restriction enzymes, T_4-DNA ligases, PCR purification kits, the DNA Gel-Extraction Kits, DNA markers, and protein markers etc. were purchased from Takara Biotechnology Co. Ltd (Dalian, China). Primers synthesis and DNA sequencing was completed by Sangon Biotechnology Co. Ltd (Shanghai, Beijing). All antibiotics were purchased

Table 1 Strains and plasmids used in the current study

	Description	Source
Strains		
Burkholderis sp. ZYB002	Wild-type, lipase-producing strain with multiple antibiotic resistance	Shu et al., 2009 [14]
Burkholderis sp. ZYB002 -Δ*lipA*	*lipA*-inactivation mutant strain derived from *Burkholderis* sp. ZYB002; Tmp^r and *lipA*::gfp	This study
Burkholderis sp. ZYB002 -Δ*lipC24*	*lipC24*-inactivation mutant strain derived from *Burkholderis* sp. ZYB002; Tmp^r and *lipC24*::gfp	This study
E. coli DH5α	*fhuA2 lac*(del)*U169 phoA glnV44 Φ80' lacZ*(del)*M15 gyrA96 recA1 relA1 endA1 thi-1 hsdR17*	TAKARA
E. coli BL21(DE3)	Expression host strain for *lipA* and *lipC21*	Novagen
E. coli Origami2 (DE3)	Expression host strain for *lipC24*	Novagen
Plasmids		
pMD18T-*lipAB*	pMD18T containing the PCR-amplified *lipA* and *lipB*	This study
pMD18T-*lipC21*	pMD18T containing the PCR-amplified *lipC21*	This study
pMD18T-*lipC24*	pMD18T containing the PCR-amplified *lipC24*	This study
pEDSF-*lipB*	pACYCDuet-1 with insertion of *lipB* at MCS2	This study
pEDSF-*lipB-lipA*	pEDSF-*lipB* with insertion of *lipA* at MCS1	This study
pEDSF-*lipC21*	pET28a with insertion of *lipC21* at MCS	This study
pEDSF-*lipB-lipC24*	pEDSI-*lipB* with insertion of *lipC24* at MCS1	This study
pGro7	Expression plasmid containing *groES-groEL* gene, *araB* promoter	TAKARA
pBBR1TP	Broad host range cloning vector with the trimethoprim resistance gene	Yingrun Bio. Inc.
pJQ200SK	Suicide vector with gentamicin resistance gene	Yingrun Bio. Inc.
pRK2013	The helper plasmid with RK2 transfer genes and kanamycin resistance gene	Yingrun Bio. Inc.
pEGFP-N1	Expression vector with *gfp* gene and kanamycin resistance gene	Clontech.
pBCMB-S1	pJQ200SK containing the PCR-amplified *tmp* gene from pBBR1TP	This study
pBCMB-S2	pBCMB-S1 containing the PCR-amplified *lipA* gene fragment (*lipA'*)	This study
pBCMB-S3	pBCMB-S2 containing the PCR-amplified *gfp* gene from pEGFP-N1	This study
pBCMB-S4	pBCMB-S1 containing the PCR-amplified *lipC24* gene fragment (*lipC24'*)	This study
pBCMB-S5	pBCMB-S4 containing the PCR-amplified *gfp* gene from pEGFP-N1	This study

from Beijing dingguo changsheng biotechnology Co. Ltd (Beijing, China). Various 4-nitrophenyl fatty acid esters, triolein, oleic acid, 1, 3-diolein, 1, 2-diolein and 1-monoolein were purchased from Sigma-Aldrich. Silica gel GF254 was purchased from Haiyang Chemical Co. Ltd (Qingdao, China). Olive oil, *n*-hexane, chloroform and acetone were of analytical grade and purchased from Sinopharm Chemical reagent Co. Ltd (China).

Gene cloning and sequence alignment of *lipA*, *lipC21* and *lipC24*

The full lengths of three different lipase genes, *lipA/ lipB*, *lipC21* and *lipC24* were amplified by PCR using the genomic DNA from *Burkholderia* sp. ZYB002 as the templates. The primer pairs for PCR were listed in the Table 2. All PCR conditions and PCR procedures used in this research were given in the Additional file 1. PCR products were ligated into pMD18-T simple vector to construct the cloning plasmid pMD18T-*lipAB*, pMD18T-

lipC21, and pMD18T-*lipC24*, respectively. The three lipase genes were sequenced in full length.

The nucleotide sequences of *lipA/lipB*, *lipC21* and *lipC24* have been deposited in the GenBank database. To construct the phylogenetic tree, the deduced amino acid sequences of LipA, LipC21 and LipC24, were submitted to BLAST at the NCBI web site, respectively. The retrieved-sequences displaying over 30 % sequence identity to LipA, LipC21 or LipC24 were selected and then aligned using BioEdit editor (Version 7.0.1). The phylogenetic tree was constructed using the software MEGA4.

Construction of the expression plasmids for *lipA*, *lipC21* and *lipC24*

Plasmid pACYCDuet-1 was selected to functionally co-express *lipA* with *lipB* or *lipC24* with *lipB*, respectively. Plasmid pET28a was selected to functionally express *lipC21*. Primer pairs used for PCR amplification of *lipB*, *lipA*, *lipC21* and *lipC24* fragments were listed in the Table 2. The PCR products, plasmid pET28a, plasmid

Table 2 Oligonucleotide primers used in the current study[a]

Primers	Oligonucleotide sequence (5' to 3')	Annealing temperature (°C)	PCR products
lipACF	AAGGATCCTCGGCGTCGACAACGTGCTGAACAAG	52	Full length of *lipA* and the corresponding chaperonin *lipB*
lipACR	CGAAAGCTTCGCCAACACCATCGAGCAACATCTG		
lipC21CF	TCGATGGCTTGGGTGACGGACA	59	Full length of *lipC21*
lipC21CR	CGAAGTTGGCTGGCACTCTTTGGC		
lipC24CF	CTAGTGCAGCGTCTCGGGCGCGA	62	Full length of *lipC24*
lipC24CR	CACCATGTCCTCCAGACGTTTCATGATGG		
lipBEF	TATAGATCTCCCGCCGTCGCTCGCCGGCTCCAG	75	The coding region for the truncated LipB with deletion of N-terminal 70-amino acid residue.
lipBER	CTTCTCGAGCTGCATGCTGCCGGCCCCGCG		
lipAEF	TATGGATCCGGCCGATGGCTACGCGGCGACGC	73	The coding region for the mature LipA
lipAER	CTTAAGCTTTTACACGCCCGCCAGCTTCAG		
lipC21EF	CGCGGATCCGCTTCGCCCGGCCGCGTTCCC	60	The coding region for the mature LipC21
lipC21ER	CCCAAGCTTGCCGCGACACGGCCTGCTGCGC		
lipC24EF	CGCGGATCCGGCGCACCGGCCGTGTCCGA	60	The coding region for the mature LipC24
lipC24ER	CCCAAGCTTGGTGCAGCGTCTCGGGCGCGAG		
lipC24MF	GCTAT**GCA**GGCGGCGCGATCGCGAC	60	pEDSF-*lipB*-lipC24-Ser^{179}Ala
lipC24MR	GCCGCC**TGC**ATAGCCGATCATCGCGC		
tmpF	CTTAGATCTCACGAACCCAGTTGACATAAG	54	Full length of the trimethoprim resistance gene
tmpR	CTTAGATCTTTAGGCCACACGTTCAAG		
lipAIF	CTTGGATCCCGAGTATTGGTACGGCATCCAG	53	*lipA* gene fragment (named as *lipA'*)
lipAIR	CTTCTCGAGTTACACGCCCGCCAGCTTCAGC		
gfpF-lipA	CTTCTGCAGATGGTGAGCAAGGGCGAGGA	54	Full length of the *gfp* gene (construction pBCMB-S3)
gfpR-lipA	CTTCTGCAGTTACTTGTACAGCTCGTCCATG		
LipC24IF	TGCTCTAGAAATACGGGATGACCACGCTTGAT	66	*LipC24* gene fragment (named as *lipC24'*)
LipC24IR	CTTGGGCCCCGTTGAAACGGTCGTAGAGCCAC		
gfpF-lipC24	GGAATTCCATATGATGGTGAGCAAGGGCGAGGA	60	Full length of the *gfp* gene (construction pBCMB-S5)
gfpR-lipC24	GGAATTCCATTTACTTGTACAGCTCGTCCATG		

[a]Underlined nucleotides: restriction endonuclease site; Bold nucleotides: the mutated sites

pACYCDuet-1 and plasmid pEDSF-*lipB* (Table 1) were double digested by the restriction endonuclease, followed by the ligation reaction to yield the expression plasmid pEDSF-*lipB-lipA*, pEDSF-*lipC21* and pEDSF-*lipB-lipC24*, respectively (Table 1).

Expression of *lipA*, *lipC21* and *lipC24* in *E. coli*
E. coli BL21(DE3) was used as the expression host strain for pEDSF-*lipB-lipA* and pEDSF-*lipC21*. *E. coli* Origami2(DE3) was selected as the expression host strain for pEDSF-*lipB-lipC24*. Chaperone plasmid pGro7 was co-transformed with plasmid pEDSF-*lipC21* into *E. coli* BL21(DE3).

Same induction condition was adopted for *E. coli* BL21(DE3)-pEDSF-*lipB-lipA* and *E. coli* Origami2(DE3)-pEDSF-*lipB-lipC24*. When the cell density (OD_{600}) reached 0.6–0.9, IPTG was added to the culture medium to the final concentration of 1 mmol/L. Induction culture was lasted for 16 h at 25 °C and then the cells were collected by centrifugation.

Expression of *E. coli* BL21(DE3)-pEDSF-*lipC21*/pGro7 was induced by IPTG and L-arabinose, respectively, as described by Pérez et al. [15]. In brief, 0.5 mg/mL L-arabinose was initially added into the culture medium to induce expression of the chaperone gene of *groES-groEL*. IPTG (1 mmol/L final concentration) was not added into culture medium until the cell density (OD_{600}) reached 0.6. Induction incubation of IPTG was lasted for 16 h at 25 °C and the cells were then collected by centrifugation.

Purification of LipA, LipC21, and LipC24
E. coli cells were lysed using sonication and the supernatant was collected by centrifugation. The recombinant protein carrying a $(His)_6$-tag was purified from the supernatant using immobilized metal-affinity chromatography (HisTrap HP, 1 mL, GE Healthcare) followed by anion exchange chromatography (HiTrap DEAE F. F., 1 mL, GE Healthcare). Before loaded onto the HisTrap chromatography column, the supernatant was incubated with 2 mmol/L ATP for 10 min at 37 °C to dissociate the recombinant protein/chaperone complex.

The loading buffer for affinity chromatography column consisted of the following components, 20 mmol/L $Na_2HPO_4-NaH_2PO_4$ buffer (pH7.5),20 mmol/L imidazole,500 mmol/L NaCl. The recombinant protein was eluted using a linear concentration gradient from 20 mmol/L to 1 mol/L imidazole in the same buffer, and the active fractions (or the target fractions identified by anti-His Western-blot) were pooled and dialyzed against 50 mmol/L Tris–HCl buffer (pH7.5). The desalted protein solution was loaded on anion exchange chromatography column using 50 mmol/L Tris–HCl buffer (pH7.5) and eluted using a linear concentration gradient from 0 mol/L to 1 mol/L NaCl in the same buffer.

The homogeneity of the purified protein was determined by SDS-PAGE on a 12 % separating gel in the presence of 0.1 % SDS. The protein concentration was analyzed using the Bradford method, with bovine serum albumin as standard.

Biochemical characterization of LipC24
Temperature optimum and temperature stability The optimal temperature was determined by incubating the standard reaction mixture at different temperatures ranging from 30 °C to 60 °C, and the maximum lipase activity was considered 100 %. To analyze temperature stability, the LipC24 preparation was incubated at 40 °C and aliquots were continuously taken at 3-min interval to assay the residual activity. Inactivation process of LipC24 preparation was continued until 80 % of the activity was lost. Half-life of thermal inactivation was calculated using the method as described by Zhao and Arnold [16].

pH optimum and pH stability
The optimal pH for lipase activity was determined by incubating the lipase with substrate in a suitable buffer at various pH ranging from 4 to 9, and the maximum lipase activity was considered 100 %. The corresponding buffers were NaAc/HAc(pH4.0-5.0), Na2HPO4-NaH2PO4 (pH6.0-8.0), and Gly/NaOH (pH9.0), respectively, and the concentrations of all used buffers were 20 mmol/L. To determine the effect of pH on lipase stability at pH ranging from 6.0 to 8.5, aliquots of the concentrated LipC24 preparation were diluted five-fold in the corresponding buffer (pH 6.0, pH6.5, pH7.0, pH7.5, pH8.0 and pH8.5) and then incubated for 24 h at 4 °C. The residual lipase activity after incubation was determined and the lipase activity at the start was taken as 100 %.

Substrate specificity
The activities of LipC24 toward various 4-nitrophenyl fatty acid esters with varying chain length (C4, C8, C10, C12, C14 and C16) were investigated.

Kinetic parameters for hydrolysis of 4-nitrophenyl myristate (*p*NPM) The Michaelis-Menten constant (K_m) and maximal reaction rate (V_{max}) of LipC24 hydrolysis activity were determined at different *p*NPM concentration (0.1, 0.2, 0.4, 0.6, 0.8, 1.0, 1.2, 1.4, and 1.8 mmol/L, respectively) under identical conditions to the spectrophotometric assay. Data points were fitted by non-linear regression using Graphpad Prism6.

Positional specificity assay
Positional specificity was determined by analyzing lipolysis products of triolein by thin-layer chromatography (TLC) on silica gel GF254, following the procedure described by Rahman et al. [17]. In brief, the reaction

mixtures containing 0.1 mol/L of triolein, 1.3 mL Na_2HPO_4-NaH_2PO_4 (20 mmol/L, pH7.4), and 5 U(500 μL) LipC24 solution were shaken at 200 rpm at 40 °C for 3 h. The reaction products were extracted with n-hexane and then analyzed by TLC. The silica gel plate was developed in a mixture of chloroform and acetone (96:4).

The lipase activity of the purified LipC24 was measured using spectrophotometric assay under standard assay conditions, as described by Kordel et al. [18]. The spectrophotometric assay method was used in the whole experiment unless stated otherwise. All reactions were carried out at 40 °C and 20 mmol/L of Na_2HPO_4-NaH_2PO_4 buffer (pH7.5). One unit of lipase activity was defined as the amount of LipC24 that liberated 1 μmol 4-nitrophenol from 4-nitrophenyl fatty acid ester per min.

3D model of LipC24

Three dimensional (3D) structural model of LipC24 was generated and optimized using the software YASARA (version18.4.30; www.yasara.org) with default settings [19].

Site-directed mutagenesis of the lipC24 gene

Amino acid substitutions (Ser[179]Ala) in LipC24 were performed using the Quickchange site-directed mutagenesis method. The plasmid pEDSF-lipB-lipC24 was used as template and the complementary mutagenic oligonucleotide primers were listed in Table 2. PCR products were firstly hydrolyzed using DpnI restriction endonuclease to remove methylated parental template DNA, and then transformed into E. coli Origami2 (DE3). The desired nucleotide substitutions was confirmed by DNA sequencing.

Induction expression of lipC24-Ser[179]Ala and purification of LipC24-Ser[179]Ala was carried out using the same method as described above.

Construction of the lipA-inactivation mutation strain and the lipC24-inactivation mutation strain

To rapidly inactivate lipA gene or lipC24 gene of Burkholderia sp. ZYB002, the suicide plasmid pBCMB-S3 (for lipA-inactivation) and pBCMB-S5 (for lipC24- inactivation) were constructed, respectively. Construction of plasmid pBCMB-S3 and plasmid pBCMB-S5 was adopted the identical method and flow diagram (Additional file 2: Figure S1 and Figure S2). Details of the construction of the lipA-inactivation mutation strain was only given.

The suicide plasmid pBCMB-S3 was constructed as follows. Firstly, the full length of the trimethoprim resistance gene was cloned from plasmid pBBR1TP using tmpF/tmpR primers incorporated in the Bgl II restriction site. The PCR product was digested with Bgl II, and then ligated into the Bgl II-digested plasmid pJQ200SK. The resulting plasmid was designated

pBCMB-S1. Secondly, the fragment of the lipA gene was amplified using lipAMF/lipAMR primers, which were designed to add a BamH I site and a Xho I site at the 5'-terminal and the 3'-terminal of the lipA gene fragment, respectively. The PCR product (named as lipA') was digested with the respective enzymes and then ligated to the Xho I/BamH I-digested plasmid pBCMB-S1. The resulting plasmid was designated pBCMB-S2. Thirdly, the gfp gene was cloned from plasmid pEGFP-N1 using gfpF/gfpR primers incorporated the Pst I restriction site. The PCR product was digested with Pst I, and then ligated into the Pst I-digested plasmid pBCMB-S2. The resulting plasmid was designated pBCMB-S3.

The suicide plasmid pBCMB-S3 was delivered to Burkholderia sp. ZYB002 by triparental mating as described previously [20]. Candidate mutants were primarily selected on trimethoprim Luria–Bertani (LB) medium and identified by PCR, and further confirmed by Southern blot hybridization using the gfp gene fragment labeled with digoxigenin as a probe. The corresponding mutant strain was designated Burkholderis sp. ZYB002-ΔlipA.

Cell-bound lipase production and activity assay

Cell-bound lipase production was carried out as described by Shu et al. [11]. Cell-bound lipase activity was determined using alkali titration method as described by Saxena et al. [21]. All reactions were carried out at 40 °C and 20 mmol/L Na_2HPO_4-NaH_2PO_4 buffer (pH7.5) unless stated otherwise. One unit of cell-bound lipase activity was defined as 1 μmol of fatty acid produced from olive oil per min by the cell culture of 1 OD_{600} under the standard assay conditions.

Results

Sequence of LipA, LipC21 and LipC24

Nucleotide sequences of the lipA/B, lipC21 and lipC24 have been deposited in the GenBank database under the accession No. EU768869, No. KF192626, No. KF438175, respectively. Nucleotide sequence analysis revealed that the lipA ORF, lipC21 ORF and lipC24 ORF coded for a putative protein of 364 amino acids, 427 amino acids and 438 amino acids, respectively. Protein sequence analysis revealed that the LipA displayed 96 % identity with the known lipase from Pseudomonas sp. KWI-56 [22], while all homologous protein sequences of LipC21 and LipC24 were the putative and uncharacterized lipases from the whole genomic DNA sequences. Moreover, protein sequence alignment through BLAST did not reveal any sequence identity among LipA, LipC21 and LipC24 (Fig. 1).

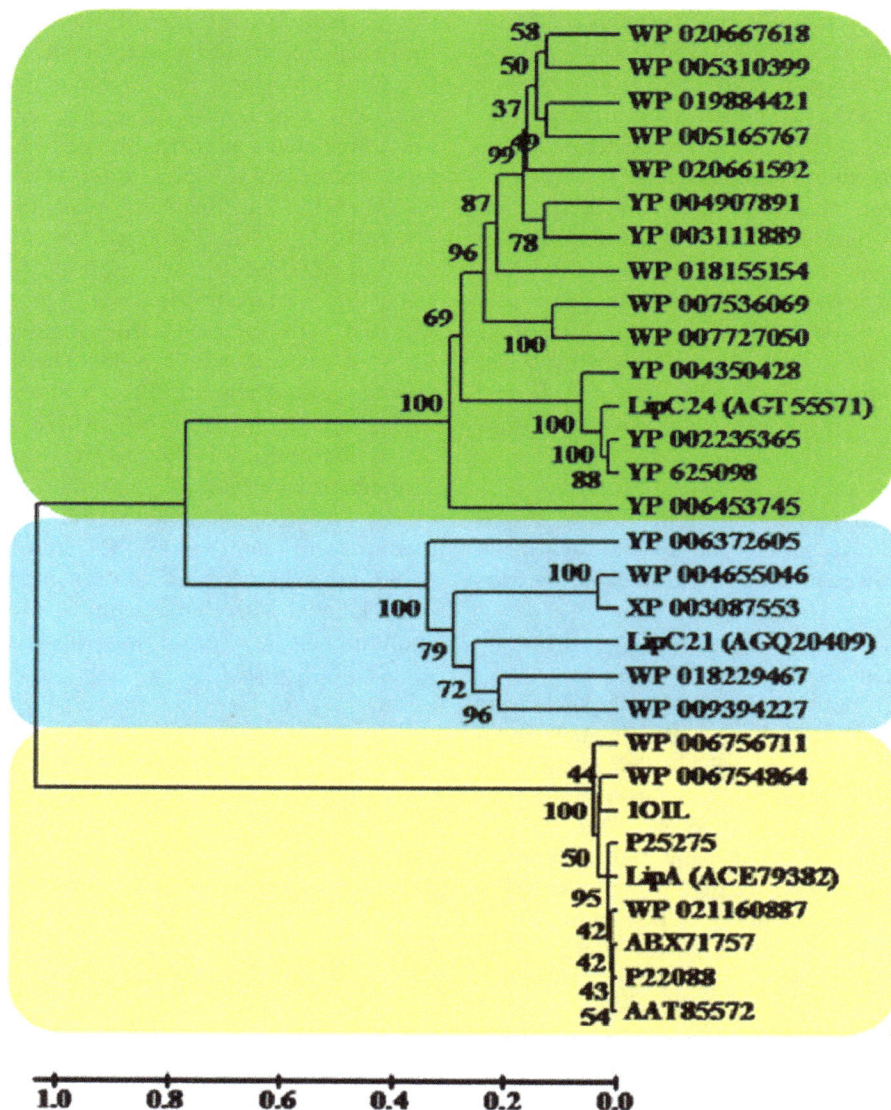

Fig. 1 Phylogenetic tree of LipA cluster, LipC21 cluster and LipC24 cluster. The amino acid sequences included the putative, uncharacterized lipases showing over 30 % identity to LipA, LipC21 and LipC24

Expression and enzymatic characterization of LipA, LipC21 and LipC24

Soluble LipA, LipC21 and LipC24 could be obtained only when *lipA*, *lipC21* and *lipC24* were co-expressed with their corresponding chaperone genes (Fig. 2). The soluble expression of *lipA* and *lipC24* required the assistance of the lipase-specific folding gene, *lipB*. However, co-expression of the chaperone *GroEL-GroES* gene is the prerequisite for the soluble expression of *lipC21*. Moreover, the soluble expression level of LipC24 could be significantly increased when *E. coli* Origami2 (DE3) was used as the expression host strain (Data not shown).

Except for LipC21, both LipA and LipC24 displayed lipase activity. As most reports on the enzymatic

characterization of the LipA from *B. cepacia*, the relative molecular weight of the LipA from *Burkholderia* sp. ZYB002 was 34 kDa. LipA was an alkaline mesothermal-active lipase [23]. The optimum temperature and pH of LipA for hydrolysis activity were 40 °C and 8.0, respectively [23]. The enzymatic characterization of LipC24 was totally different from that of LipA. The LipC24 was purified 17.7-fold from the supernatant of the *E. coli* cell lysate and yielded 21.49 % of the initial activity. The specific activity of LipC24 was 15.63 U/mg using 4-nitrophenyl palmitate as substrate (Table 3), which was far lower than that of LipA (253.82 U/mg for 4-nitrophenyl palmitate) [23]. SDS/PAGE analysis of LipC24 displayed a single band, which corresponded to

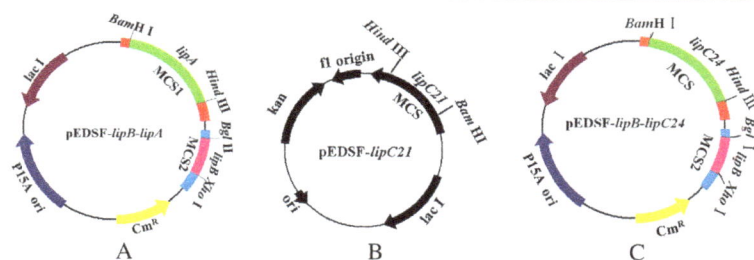

Fig. 2 Map of the expression plasmid for *lipA*, *lipC21* and *lipC24*. **a** pEDSF-*lipB*- *lipA* was derived from pACYCDuet-1, which was inserted *lipA* gene at the MCS1 site and the chaperone *lipB* gene at the MCS2 site; **b** pEDSF-*lipC21* was derived from pET28a, which was inserted *lipC21* gene at the MCS site. To obtain the soluble expression of *lipC21*, plasmid pEDSF-*lipC21* and plasmid pGro7 must be co-transformed into *E. coli* BL21(DE3); **c** pEDSF-*lipB-lipC24* was derived from pACYCDuet-1, which was inserted *lipC24* gene at the MCS1 site and the chaperone *lipB* gene at the MCS2 site

a molecular mass of 45 kDa (Fig. 3). The optimum temperature and pH of LipC24 for hydrolysis activity were found to be 40 °C and 7.5, respectively. The LipC24 could be kept stable in the pH range 7.0-8.0 for 24 h at 4 °C, while the half-time of the LipC24 was only 16 min at 40 °C (Fig. 4). LipC24 was less stable than LipA. LipA displayed excellent thermostability up to 65 °C and could keep stability over a broad pH range from 3.0 to 10 [23]. The LipC24 indicated a clear preference for esters with the medium acyl chain length (C10-C14) when assayed using 4-nitrophenyl derivatives (Table 4). The LipC24 exhibited a simple Michaelis-Menten kinetics for *p*NPM hydrolysis. The values of K_m and V_{max} of LipC24 were 0.37 ± 0.07 mmol/L and 138.8 ± 7.90 $\mu mol \cdot min^{-1} \cdot mg^{-1}$, respectively (Fig. 5). Michaelis constan K_m of LipC24 is less than that of lipase from *Bacillus* sp.. Accordingly, the maximum reaction velocity V_{max} of LipC24 was higher than that of *Bacillus* sp. [24]. LipC24 cleaved not only the 3-positioned ester bonds, but also the 2-positioned ester bond of triolein (Fig. 6). Thus, LipC24 could nonspecifically hydrolyze the ester bonds of triolein. The same experiment results were verified with other *Pseudomonas* sp. lipases [17, 25].

Protein sequence and structural model analysis of LipC24

There were several conserved sequence blocks between the deduced amino acid sequence of LipC24 and the other putative homologous lipases. In block 3, there was a conserved pentapeptide Gly-Tyr-Ser-Gly-Gly, in which the catalytic serine residue was embedded in most lipases. Besides the conserved serine residue, there were three conserved aspartate residues in block 1, block 2

and block 5, and a conserved histidine residue in block 5 (Additional file 2: Figure S3). The 3D homology model of LipC24 presented the characteristics of a canonical α/β-hydrolase fold, in which parallel or mixed β sheet in the molecular center was surrounded (or connected) by helices (Fig. 7a). A hydrogen bond network was formed among Ser^{179}, Asp^{336}, and His^{367}, which constituted the catalytic triad (Fig. 7b). Mutant of LipC24-Ser^{179}Ala lost 100 % lipase activity, which confirmed the function of Ser^{179} in the active site. The oxyanion hole consisted of Ala^{82} and Gly^{180}, which stabilized the transient state of LipC24-ethyl acetate complex (Fig. 7c). The substrate-binding pocket of LipC24 displayed the distinct open Y-type structure (Fig. 7d).

Component of the cell-bound lipase

The cell-bound lipase activity of *Burkholderis* sp. ZYB002-Δ*lipA* and *Burkholderis* sp. ZYB002-Δ*lipC24* significantly decreased to 58 % and 86 % of its original activity, respectively (Fig. 8a). The cell-bound lipase activity originated from a multi-enzyme mixture in which LipA was the main component. Besides LipA and LipC24, other type of lipases could exist on the cell surface of *Burkholderis* sp. ZYB002.

Discussion

Different types of lipase produced by a specific microbial strain always displayed a totally different 3D structure and enzymatic characterization [26–28]. In previous research, cell-bound lipase from *B. cepacia* displayed excellent catalytic activity for organic synthesis [12, 13]. However, there was not any report on the gene sequence

Table 3 Purification of LipC24 from *E. coli* Origami 2(DE3)-pEDSF-*lipB-lipC24*

Steps	Total activity (U)	Total protein (mg)	Specific activity (U/mg)	Yield (%)	Purification (fold)
Cell-free extract	245.39	89.78	2.73	100	1
HisTrap HP	57.96	1.8	32.20	23.62	11.79
HiTrap DEAE FF	52.74	1.09	48.31	21.49	17.70

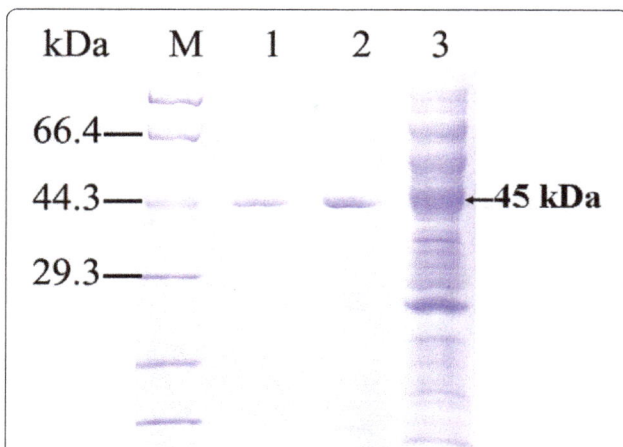

Fig. 3 SDS-PAGE analysis of LipC24 in different purification steps. M: protein marker; 1: the purified LipC24 by HiTrap DEAE FF anion-exchange chromatography column; 2: the purified LipC24 by HisTrap HP affinity chromatography column; 3: cell-free extract of *E. coli* Origami 2(DE3)-pEDSF-*lipB-lipC24*

BCAM0949] and the chaperone gene *lipB* [gene locus: BCAM0949]. Moreover, two predictive lipases, LipC21 [gene locus: BCAL1969] and LipC24 [gene locus: BCAM2764], were distributed on the cytoplasmic membrane. The genes, *lipC21* and *lipC24* were situated on the gene cluster for ammonia metabolism and pilus synthesis, respectively (Fig. 8b). In *P. aeruginosa*, part of the extracellular lipase could be anchored on the cell surface and act as cell-bound lipase [4–6].

Three speculative cell-bound lipase genes (*lipA*, *lipC21* and *lipC24*) from *Burkholderia* sp. ZYB002 were cloned and expressed in *E. coli*. Among LipA, LipC21 and LipC24, only LipA displayed a high sequence identity with the known extracellular lipase from *Pseudomonas* sp. KWI-56 [22], which suggested that LipA was the authentic triacylglycerol lipase. The protein sequences of LipC21 and LipC24 had not any sequence identity with known lipases or esterases, which could lead to speculations that LipC21 and LipC24 were the novel lipases (Fig. 1).

The expression soluble lipases required different chaperone protein genes. It was necessary for *lipA* and *lipC24* to be co-expressed with the *lipB* gene. Due to the strong hydrophobicity, a 70-amino acid residue fragment at the N-terminal of LipB had to be truncated when *lipB* was heterogeneously co-expressed in *E. coli*

nor any structural investigation of the cell-bound lipase genes from *B. cepacia*. From the whole genomic DNA sequence of *B. cepacia* J2315 (www.burkholderia.com), more than 10 gene sequences were predicted as lipase genes, including the extracellular lipase *lipA* [gene locus:

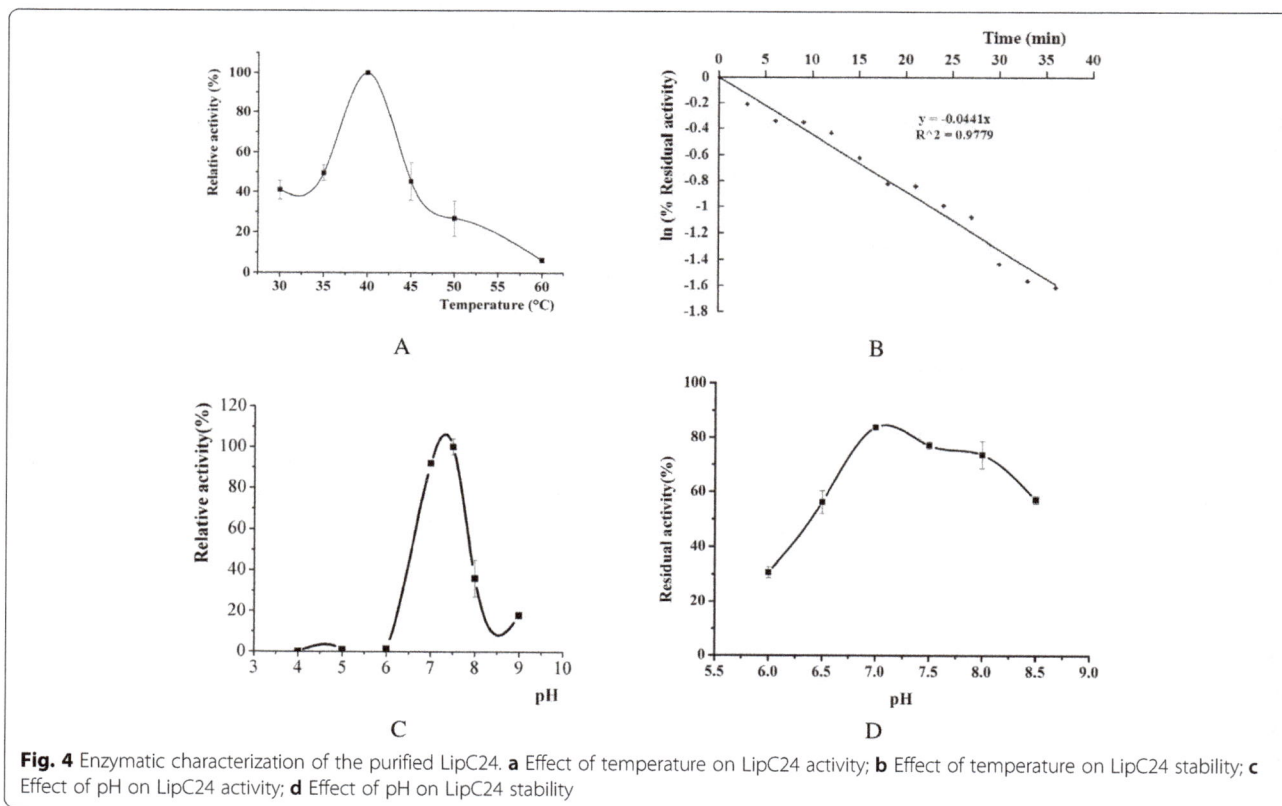

Fig. 4 Enzymatic characterization of the purified LipC24. **a** Effect of temperature on LipC24 activity; **b** Effect of temperature on LipC24 stability; **c** Effect of pH on LipC24 activity; **d** Effect of pH on LipC24 stability

Table 4 The specific activity of LipC24 towards 4-nitrophenyl esters

Substrate	Specific activity (U/mg)
4-nitrophenyl palmitate (C16)	15.63 ± 1.08
4-nitrophenyl myristate (C14)	55.49 ± 1.87
4-nitrophenyl laurate (C12)	31.14 ± 2.59
4-nitrophenyl decanoate (C10)	48.31 ± 2.06
4-nitrophenyl octanoate (C8)	18.54 ± 1.67
4-nitrophenyl butyrate (C4)	0.51 ± 0.12

[29]. Among *lipB*, *groES-groEL* gene, *dnaK-dnaJ-grpE* gene and *tig* gene, it was only the chaperone *groES-groEL* gene that improved the soluble expression level of *lipC21*. GroEL-GroES was also reported to be necessary for the soluble expression of the family VIII lipase *lipBL* from *Marinobacter lipolyticus* [15].

Enzymatic characterization and 3D structure of LipC24 was totally different from that of LipA. LipC24 displayed high activity in the neutral buffer (pH 7.0-7.5) and mesothermal reaction conditions. Furthermore, LipC24 would sharply abolish the lipolytic activity when LipC24 was kept at high temperature, alkaline solution, or acid solution, respectively. On the contrary, LipA was thermostable, alkaline-tolerant, and organic solvent-resistant [30, 31]. The open Y-type active site of LipC24 was totally different from the funnel-shaped active site of LipA [32]. In the molecular model of LipC24, a predictive intramolecular disulfide bond was formed between Cys^{352} and Cys^{395} (PredictProtein 2013 server, https://www.predictprotein.org/), which corresponded to the requirement of the host strain, *E. coli* Origami2 (DE3) for the soluble expression of *lipC24*.

Although both titrimetric assay method and colorimetric assay method were widely used for lipase activity

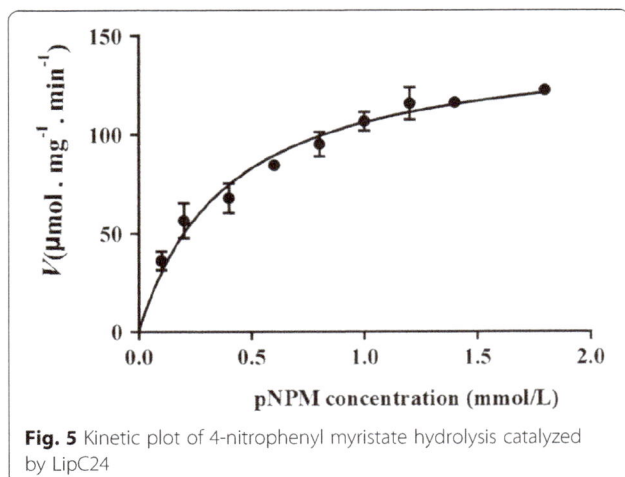

Fig. 5 Kinetic plot of 4-nitrophenyl myristate hydrolysis catalyzed by LipC24

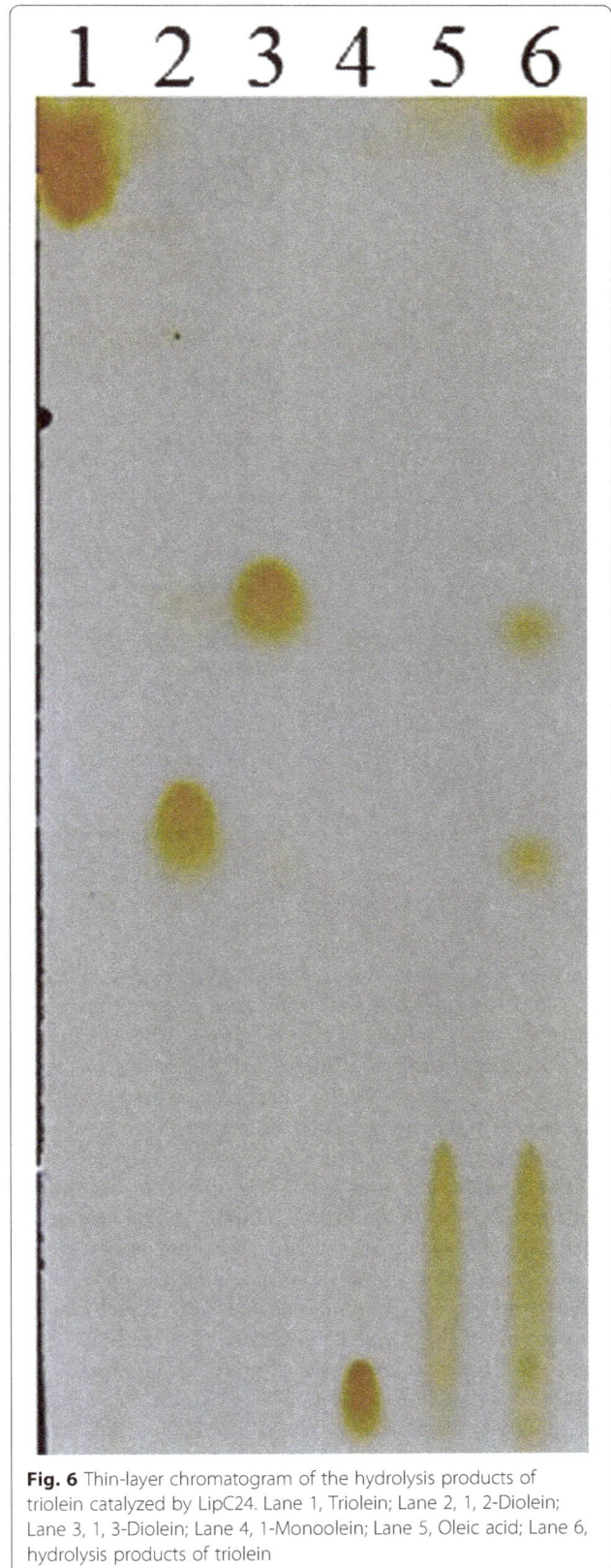

Fig. 6 Thin-layer chromatogram of the hydrolysis products of triolein catalyzed by LipC24. Lane 1, Triolein; Lane 2, 1, 2-Diolein; Lane 3, 1, 3-Diolein; Lane 4, 1-Monoolein; Lane 5, Oleic acid; Lane 6, hydrolysis products of triolein

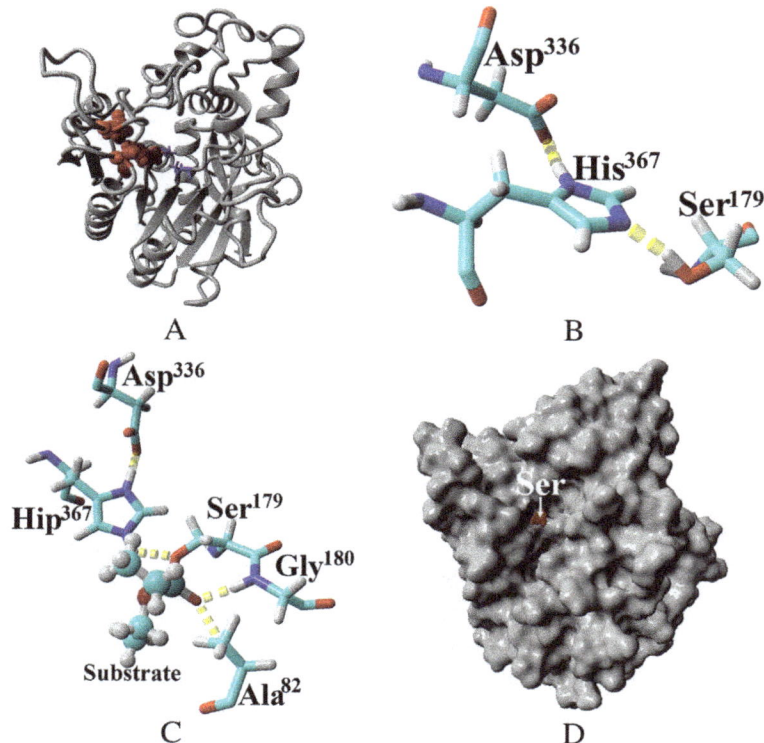

Fig. 7 The 3D model of LipC24. **a** The overall three-dimensional structure of LipC24. β-strands were represented as arrows and surrounded by the helices; **b** Ser^{179}, Asp^{336}, and His^{367} formed the catalytic triad within the range of H-bond interactions; **c** The transient state model of LipC24-ethyl acetate complex, which was stabilized by Ala^{82} and Gly^{180}. Hip^{367} originated from His^{367}, which accepted a proton from the hydroxyl group of Ser^{179}; **d** The open Y-type substrate-binding pocket of LipC24

determination, there were obvious differences between the two assay methods [33, 34]. Olive oil or other triacylglycerol was used as the substrate in the titrimetric assay method, while 4-nitrophenyl esters were always used as the substrate in the colorimetric assay method. However, 4-nitrophenyl esters could be permeated into the cytoplasm [35–37], and was hydrolyzed by the intracellular lipolytic enzymes (including lipase and esterase). 4-nitrophenyl esters could not be used as the substrates for activity determination of the cell-bound lipase. In the present work, membrane-impermeable olive oil and the alkali titration assay method was used for the activity determination of the whole cell lipase.

The cell-bound lipase activity of *Burkholderis* sp. ZYB002-Δ*lipA* decreased by 42 % of the total cell-bound lipase activity *Burkholderis* sp. ZYB002 (Fig. 8a), which demonstrated that LipA was the main component of cell-bound lipases. It had been reported that microbial strains could simultaneously produce extracellular lipases and various kinds of biosurfactants (rhamnolipid, lipopolysaccharide, polysaccharide alginate, etc.) when various oils or lipids were used as the inducer or carbon

source [4, 38]. Part of biosurfactants were firmly associated with the outer membrane of the host strain and could interact with lipases by electrostatic interaction, which resulted in cell surface anchoring of the extracellular lipases [5, 6]. LipC24 contributed 14 % of the total cell-bound lipase activity *Burkholderis* sp. ZYB002 (Fig. 8a). Besides LipA and LipC24, other type of the cell-bound lipases (or esterases) could exist. Further analysis of the whole genome DNA sequence of *Burkholderia cepacia* J2315 predicted several esterase gene sequences, including two novel family VIII esterase genes, *estVIII-C11* [gene locus: BCAL2802] and *estVIII-C21* [gene locus: BCAM0511] (Fig. 8b).

Conclusions

The cell-bound lipase activity of *Burkholderia* sp. ZYB002 was shown to be a multi-enzyme mixture, which at least consisted of LipA and LipC24. LipA was the main component of the cell-bound lipase. LipC24 was a novel lipase, which displayed a totally different enzymatic characterization and 3D structure to that of LipA.

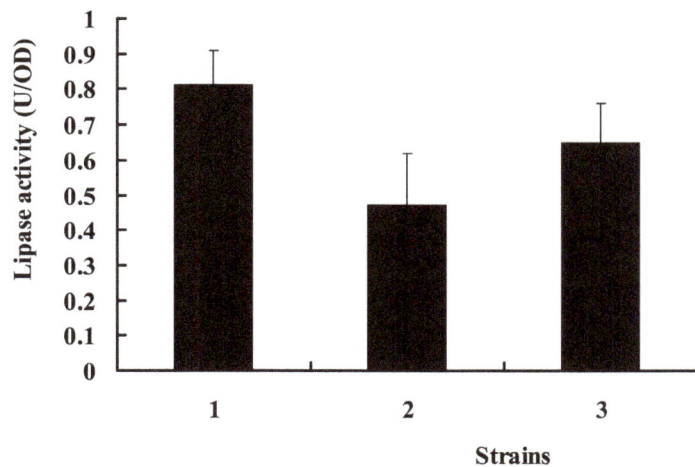

A

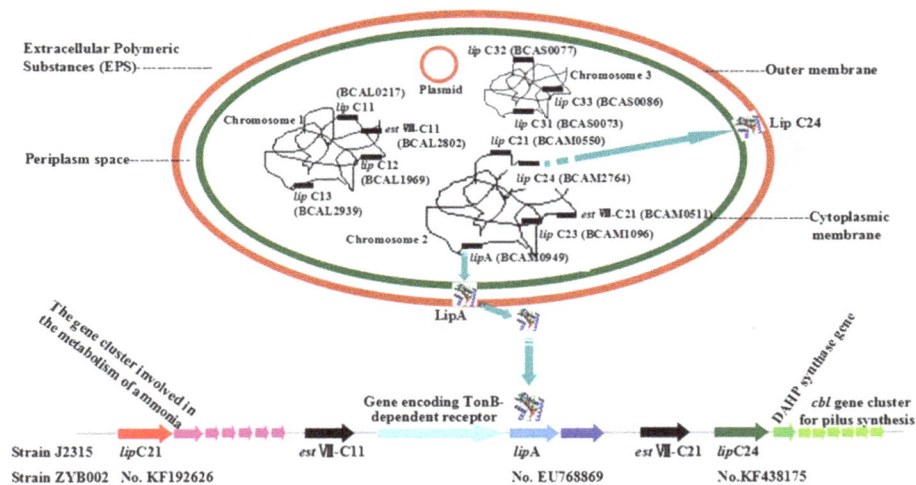

B

Fig. 8 Component of the cell-bound lipase from *Burkholderia* sp. ZYB002. **a** Comparsion analysis of the cell-bound lipase activity from *Burkholderis* sp. ZYB002, *Burkholderis* sp. ZYB002-ΔlipA, and *Burkholderis* sp. ZYB002-ΔlipC24. 1 *Burkholderis* sp. ZYB002 strain; 2 *Burkholderis* sp. ZYB002-ΔlipA strain; 3 *Burkholderis* sp. ZYB002-ΔlipC24 strain. **b** The predictive lipase gene family and two family VIII esterase genes (*estVIII-C11* and *estVIII-C21*) from *B.cepacia* J2315

Besides LipA and LipC24, other type of the cell-bound lipases (or esterases) may exist.

Additional files

Additional file 1: All PCR conditions and PCR procedures used in this research.

Additional file 2: Figure S1. The construction flow diagram for the suicide plasmid pBCMB-S3, which was used to construct the *lipA*-inactivation mutation strain. Figure S2. The construction flow diagram for the suicide plasmid pBCMB-S5, which was used to construct the *lipC24*-inactivation mutation strain. Figure S3. Blocks of sequences conserved between LipC24 and other putative homologous lipases.

Abbreviations
gfp: green fluorescent protein-encoding gene; IPTG: isopropyl-beta-d-thiogalactopyranoside; pNPM: 4-nitrophenyl myristate; SDS-PAGE: sodium lauryl sulfate-polyacrylamide gel electrophoresis; TLC: thin-layer chromatography.

Competing interests
The authors declare that they have no competing interests.

Authors' contributions
ZS designed and supervised the work. HL participated in the gene cloning, expression of the three predictive lipase genes, and purification of the recombinant lipase. SS constructed the lipase gene-inactivation mutation strain. XM contributed to the simulation of the 3D structural model of LipC24 and enzymatic characterization analysis. YL provided advice for the discussion. This manuscript was drafted and revised by ZS. JH gave final approval of the version to be published. All authors have read and approved the final manuscript.

Authors' information

Hong Lin is Ph.D. student of College of Life Sciences, Fujian Normal University.
Shaolei Shi and Xiangduo Mu are Master students of College of Life
Sciences, Fujian Normal University.
Zhengyu Shu, Yanru Liu, and Yanru Liu are staffs of College of Life Sciences,
Fujian Normal University.

Acknowledgments

We gratefully thank Karl Hult, emeritus professor of school of Biotechnology,
KTH Royal Institute of Technology, for the review of this manuscript and for
the English language editing.

Funding

This work was supported by the National Natural Science Funds of P. R.
China (No. 31370802 and No. 30870545), by the Key Project from Science
and Technology Bureau of Fujian province (No. 2013H0021) and by the
Natural Science Funds for Distinguished Young Scholar of Fujian Province
(No. 2009 J06013).

Author details

[1]National & Local United Engineering Research Center of Industrial
Microbiology and Fermentation Technology, Ministry of Education, Fujian
Normal University, Fuzhou 350117, China. [2]Engineering Research Center of
Industrial Microbiology, Ministry of Education, Fujian Normal University,
Fuzhou 350117, China. [3]College of Life Sciences, Fujian Normal University
(Qishan campus), Fuzhou 350117, China.

References

1. Reetz MT. Lipases as practical biocatalysts. Curr Opin Chem Biol. 2002;6:
   145–50.
2. Hasan F, Shah AA, Hameed A. Industrial applications of microbial lipases.
   Enzyme Micro Technol. 2006;39:235–51.
3. de Domínguez María P, Sánchez-Montero JM, Sinisterra JV, Alcántara AR.
   Understanding Candida rugosa lipases: an overview. Biotechnol Adv. 2006;
   24:180–96.
4. Stuer W, Jaeger KE, Winkler UK. Purification of extracellular lipase from
   Pseudomonas aeruginosa. J Bacteriol. 1986;168:1070–4.
5. Wingender J. Interactions of alginate with exoenzymes. In: Gacesa P, Russell
   NJ, editors. Pseudomonas infection and alginates-Biochemistry, genetics and
   pathology. London: Chapman and Hall; 1990. p. 160–80.
6. Tielen P, Kuhn H, Rosenau F, Jaeger KE, Flemming HC, Wingender J.
   Interaction between extracellular lipase LipA and the polysaccharide
   alginate of Pseudomonas aeruginosa. BMC Microbiol. 2013;13:159.
7. Wilhelm S, Tommassen J, Jaeger KE. A novel lipolytic enzyme located in the
   outer membrane of Pseudomonas aeruginosa. J Bacteriol. 1999;181:6977–86.
8. van den Berg B. Crystal structure of a full-length autotransporter. J Mol Biol.
   2010;396:627–33.
9. Ishige T, Honda K, Shimizu S. Whole organism biocatalysis. Curr Opin Chem
   Biol. 2005;9:174–80.
10. Shu ZY, Wu JG, Chen D, Cheng LX, Zheng Y, Chen JP, et al. Optimization of
    Burkholderia sp. ZYB002 lipase production for pitch control in
    thermomechanical pulping (TMP) processes. Holzforschung. 2012;66:341–8.
11. Shu ZY, Wu JG, Cheng LX, Chen D, Jiang YM, Li X, et al. Production and
    characteristics of the whole-cell lipase from organic solvent tolerant
    Burkholderia sp. ZYB002. Appl Biochem Biotechnol. 2012;166:536–48.
12. Liu Y, Li C, Wang SH, Chen WY. Solid-supported microorganism of
    Burkholderia cenocepacia cultured via solid state fermentation for biodiesel
    production: optimization and kinetics. Appl Energy. 2014;113:713–21.
13. Yu LJ, Xu Y, Wang XQ, Yu XW. Highly enantioselective hydrolysis of dl-
    menthyl acetate to l-menthol by whole-cell lipase from Burkholderia cepacia
    ATCC 25416. J Mol Catal B-Enzym. 2007;47:149–54.
14. Shu ZY, Lin RF, Jiang H, Zhang YF, Wang MZ, Huang JZ. A rapid and
    efficient method for directed screening of lipase-producing Burkholderia
    cepacia complex strains with organic solvent tolerance from rhizosphere.
    J Biosci Bioeng. 2009;107:658–61.
15. Pérez D, Kovacic F, Wilhelm S, Jaeger KE, García MT, Ventosa A, et al.
    Identification of amino acids involved in the hydrolytic activity of lipase
    LipBL from Marinobacter lipolyticus. Microbiology. 2012;158:2192–203.
16. Zhao H, Arnold FH. Directed evolution converts subtilisin E into a functional
    equivalent of thermitase. Protein Eng. 1999;12:47–53.
17. Rahman RN, Baharum SN, Basri M, Salleh AB. High-yield purification of an
    organic solvent-tolerant lipase from Pseudomonas sp. strain S5. Anal
    Biochem. 2005;341:267–74.
18. Kordel M, Hofmann B, Schomburg D, Schmid RD. Extracellular lipase of
    Pseudomonas sp. strain ATCC 21808: purification, characterization,
    crystallization, and preliminary X-ray diffraction data. J Bacteriol. 1991;173:
    4836–41.
19. Krieger E, Joo K, Lee J, Lee J, Raman S, Thompson J, et al. Improving
    physical realism, stereochemistry, and side-chain accuracy in homology
    modeling: four approaches that performed well in CASP8. Proteins. 2009;77:
    114–22.
20. Köthe M, Antl M, Huber B, Stoecker K, Ebrecht D, Steinmetz I, et al. Killing of
    Caenorhabditis elegans by Burkholderia cepacia is controlled by the cep
    quorum-sensing system. Cell Microbiol. 2003;5:343–51.
21. Saxena RK, Davidson WS, Sheoran A, Giri B. Purification and characterization
    of an alkaline thermostable lipase from Aspergillus carneus. Process Biochem.
    2003;39:239–47.
22. Iizumi T, Nakamura K, Shimada Y, Sugihara A, Tominaga Y, Fukase T.
    Cloning, nucleotide sequencing, and expression in Escherichia coli of a
    lipase and its activator genes from Pseudomonas sp. KWI-56. Agric Biol
    Chem. 1991;55:2349–57.
23. Liu YR, Qiu FJ, Shu ZY, Wang ZZ, Qiu LQ, Li X, et al. Purification and
    enzymatic characterization of the lipase from Burkholderia sp. ZYB002.
    J Fujian Normal Univ (Natur Sci Ed). 2014;30:100–5.
24. Dosanjh NS, Kaur J. Biochemical analysis of a native and proteolytic
    fragment of a high-molecular-weight thermostable lipase from a mesophilic
    Bacillus sp. Protein Expr Purif. 2002;24:71–5.
25. Gaur R, Gupta A, Khare SK. Purification and characterization of lipase from
    solvent tolerant Pseudomonas aeruginosa PseA. Process Biochem. 2008;43:
    1040–6.
26. Uppenberg J, Hansen MT, Patkar S, Jones TA. The sequence, crystal structure
    determination and refinement of two crystal forms of lipase B from Candida
    antarctica. Structure. 1994;2:293–308.
27. Ericsson DJ, Kasrayan A, Johansson P, Bergfors T, Sandström AG, Bäckvall JE,
    Mowbray SL. X-ray structure of Candida antarctica lipase A shows a novel
    lid structure and a likely mode of interfacial activation. J Mol Biol. 2008;376:
    109–19.
28. Fickers P, Marty A, Nicaud JM. The lipases from Yarrowia lipolytica: genetics,
    production, regulation, biochemical characterization and biotechnological
    applications. Biotechnol Adv. 2011;29:632–44.
29. Rosenau F, Tommassen J, Jaeger KE. Lipase-specific foldases. Chembiochem.
    2004;5:152–61.
30. Wang XQ, Yu XW, Xu Y. Homologous expression, purification and
    characterization of a novel high-alkaline and thermal stable lipase from
    Burkholderia cepacia ATCC 25416. Enzyme Micro Technol. 2009;45:94–102.
31. Dandavate V, Jinjala J, Keharia H, Madamwar D. Production, partial
    purification and characterization of organic solvent tolerant lipase from
    Burkholderia multivorans V2 and its application for ester synthesis. Bioresour
    Technol. 2009;100:3374–81.
32. Kim KK, Song HK, Shin DH, Hwang KY, Suh SW. The crystal structure of a
    triacylglycerol lipase from Pseudomonas cepacia reveals a highly open
    conformation in the absence of a bound inhibitor. Structure. 1997;5:173–85.
33. Hasan F, Shah AA, Hameed A. Methods for detection and characterization
    of lipases: a comprehensive review. Biotechnol Adv. 2009;27:782–98.
34. Beisson F, Tiss A, Rivière C, Verger R. Methods for lipase detection and assay:
    a critical review. Eur J Lipid Sci Technol. 2000;102:133–53.
35. Zha D, Xu L, Zhang H, Yan Y. Molecular identification of lipase LipA from
    Pseudomonas protegens Pf-5 and characterization of two whole-cell
    biocatalysts Pf-5 and Top10lipA. J Microbiol Biotechnol. 2014;24:619–28.
36. Martinez MB, Flickinger M, Higgins L, Krick T, Nelsestuen GL. Reduced outer
    membrane permeability of Escherichia coli O157:H7: suggested role of
    modified outer membrane porins and theoretical function in resistance to
    antimicrobial agents. Biochemistry. 2001;40:11965–74.
37. Cotterell D, Whittam R. The uptake and hydrolysis of p-nitrophenyl
    phosphate by red cells in relation to ATP hydrolysis by the sodium pump.
    J Physiol. 1972;223:773–802.
38. Colla LM, Rizzardi J, Pinto MH, Reinehr CO, Bertolin TE, Costa JA.
    Simultaneous production of lipases and biosurfactants by submerged and
    solid-state bioprocesses. Bioresour Technol. 2010;101:8308–14.

7

Identification of two transcription factors activating the expression of OsXIP in rice defence response

Yihua Zhan[1], Xiangyu Sun[1], Guozeng Rong[2], Chunxiao Hou[3], Yingying Huang[1], Dean Jiang[1] and Xiaoyan Weng[1*]

Abstract

Background: Xylanase inhibitors have been confirmed to be involved in plant defence. OsXIP is a XIP-type rice xylanase inhibitor, yet its transcriptional regulation remains unknown.

Results: Herbivore infestation, wounding and methyl jasmonate (MeJA) treatment enhanced mRNA levels and protein levels of OsXIP. By analyzing different 5′ deletion mutants of *OsXIP* promoter exposed to rice brown planthopper *Nilaparvata lugens* stress, a 562 bp region (−1451 − −889) was finally identified as the key sequence for the herbivores stress response. Using yeast one-hybrid screening, coupled with chromatin immunoprecipitation analysis, a basic helix-loop-helix protein (OsbHLH59) and an APETALA2/ETHYLENE RESPONSE FACTOR (AP2/ERF) transcription factor OsERF71 directly binding to the 562 bp key sequence to activate the expression of *OsXIP* were identified, which is further supported by transient expression assay. Moreover, transcriptional analysis revealed that mechanical wounding and treatment with MeJA resulted in an obvious increase in transcript levels of *OsbHLH59* and *OsERF71* in root and shoot tissues.

Conclusions: Our data shows that two proteins as direct transcriptional activators of *OsXIP* responding to stress were identified. These results reveal a coordinated regulatory mechanism of OsXIP, which may probably be involved in defence responses via a JA-mediated signaling pathway.

Keywords: 5′ deletion, OsbHLH59, OsERF71, Plant defence, Rice xylanase inhibitor, Transcriptional regulation

Background

Xylanase inhibitors (XIs) are a kind of plant-produced proteinaceous inhibitor that inhibits the activity of xylanase [1]. Recently, RIXI, riceXIP, OsXIP and OsHI-XIP xylanase inhibitors have been identified in rice plants [2–5], which all belong to XIP-type XIs. XIs have been thought to be involved in plant defence mainly for the reason that XIs only inhibit xylanases of microbial origin but not of plant origin. And many data provide evidence that XIs do indeed participate in plant defense [1].

Xylanase inhibitor genes act as defence-responsive genes in stress-induced signal transduction pathways. *Taxi-Ia* expression was induced 2.5 times and the transcripts of *Taxi-Ib/III* and *Taxi-IIb/IV* rose up to 20-fold by *F. graminearum* infection of wheat lemma, palea and ovary [6]. Infestation of wheat leaves by the powdery mildew fungus *B. graminis* induced the expression of *Taxi-Ib/III* and *Taxi-IIb/IV* [6]. The transcripts of *OsXIP* and *riceXIP* were drastically induced by wounding and methyl jasmonate (MeJA) treatment in the root [2]. Our previous study also revealed that pathogens can induce the expression of the rice xylanase inhibitor gene *RIXI* [7]. In planta direct evidence for this role has not been reported until Moscetti et al. [8] found that constitutive expression of the xylanase inhibitor TAXI-III delayed *Fusarium* head blight symptoms. Furthermore, overexpression of the RIXI xylanase inhibitor improved disease resistance of rice to the fungal pathogen, *Magnaporthe oryzae* [9]. In addition, overexpression of *OsHI-XIP* enhanced resistance in rice to herbivores, which is also the first time that a xylanase inhibitor has been demonstrated to play a role in resistance among rice herbivores [5]. However, the molecular basis underlying the regulation of XIs in plant defense is poorly understood.

* Correspondence: xyweng@zju.edu.cn
[1]College of Life Science, Zhejiang University, Hangzhou 310058, China
Full list of author information is available at the end of the article

A number of biotic and abiotic stress-responsive elements were observed by comparative analysis of cis-elements of xylanase inhibitors gene promoter by bioinformatics softwares PLACE and PlantCARE. The promoter region of a gene can provide valuable information about the factors inducing expression. For instance, cis-acting elements implicated in pathogen- and wound-inducible gene expression, i.e., GCC-box and W-box sequences could be recognized in the promoter region of *TAXI-III* [6]. Also investigation of the durum wheat *Xip-II* upstream region revealed the presence of a number of cis-acting elements controlling the expression of defense-related genes such as several W-boxes and a Myb-binding element, supporting its role in plant defense against pathogens [10]. The importance of these promoters regions has not yet been confirmed by promoter deletion analyses.

OsXIP is a XIP-type rice xylanase inhibitor, which was induced by various stresses such as MeJA treatment and wounding. And the expression patterns of OsXIP and riceXIP resemble each other and the induction of their expression by wounding may occur via a JA-mediated signaling pathway [2]. However, whether OsXIP plays an important role in resistance to invaders via a JA-mediated signaling pathway remains unclear.

Despite all these observations, there have been no reports on in planta functional characterization of the promoter region of xylanase inhibitor gene and its transcriptional regulation pattern so far. In this study, the promoter of *OsXIP* was cloned and analyzed, and a 562 bp region (−1451 to −889) was identified as the key sequence for the herbivores stress response by promoter deletion analyses. Using this 562 bp sequence as the bait, OsbHLH59 [11] and OsERF71 [12] proteins as direct transcriptional regulators of *OsXIP* responding to stress were identified. Collectively, our results, for the first time, reveal a transcriptional regulatory mechanism of OsXIP involved in defence responses.

Methods

Plant materials, growth conditions and stress treatments

The rice genotypes used in this study were Nipponbare wild-type (WT) and transgenic lines (see below). Rice seeds were sown in water and grown in normal culture solution in a greenhouse with natural day length extended to light/dark cycle of 14/10 h using high-pressure sodium lamp, with heating or ventilation used to maintain temperature at 28 °C and 18 °C during day and night respectively.

For wounding stress, 14-day-old seedlings were cut into 5–10 mm width and floated on distilled water. For phytohormone treatment, 14-day-old rice seedlings were submerged in 200 μM MeJA solution for 0, 2, 6, 12 and 24 h, and then shoots and roots were harvested separately. For BPH treatment, plants were individually infested with

20 adult BPH confined in a glass cylinder [diameter 4 cm, height 8 cm, with 48 small holes (diameter 0.8 mm)], the top of which was covered with a piece of sponge. One empty cylinder was used for control plants (non-infested).

Construction of the *OsXIP* promoter vectors and rice transformation

The full-length *OsXIP* gene promoter named OP1 (−2070 bp to +52 bp) was amplified from the genomic DNA of WT with OP1-U (forward primer) and OP1-L (reverse primer). Then a series of nested 5' deletions of OP1 fragments OP2 (−1451 bp to +52 bp), OP3 (−889 bp to +52 bp), OP4 (−569 bp to +52 bp), OP5 (−380 bp to +52 bp), OP6 (−172 bp to +52 bp), OP7 (−90 bp to +52 bp) were amplified by PCR from pMD19T-OP1 using the common reverse primer OP1-L and either the forward primers OP2-U, OP3-U, OP4-U, OP5-U, OP6-U, or OP7-U, respectively. The primers are shown in Additional file 1: Table S1. The full-length promoter and 5'-deletion derivatives were cloned into the pBI101.3-GUS upstream of *GUS* (β-glucuronidase). Empty vector pBI101.3-GUS was used as a negative control (VC). All constructs were mobilized into *Agrobacterium tumefaciens* EHA105 and transformed into calli derived from mature seeds of rice according to a previously described protocol [13, 14]. Approximately 90 calli were co-cultured for each vector, with the number of putative independent transformed plants being regenerated being 25, 19, 20, 16, 22, 13, 17 and 10 respectively for OP1-OP7 and a vector control. Primary transformants (T0) were raised, transferred to soil and allowed to grow in a greenhouse. Seeds were harvested and used for analysis in the next generation.

RNA extraction and quantitative RT-PCR

Total RNA was isolated from roots and shoots of rice seedlings using the RNAprep pure plant kit (Tiangen) according to the manufacturer's protocol. RNA (1 μg) was used to synthesize the first strand complementary DNA (cDNA) with an oligo (dT) primer according to the instruction of the PrimeScript first-strand cDNA synthesis kit (Takara). The qRT-PCR assay was performed on LightCycler480 instrument (Roche) using a SYBR® Premix Ex TaqTM kit (Takara). A rice actin gene *Osactin* (GenBank: Os03g50885) was used as an internal standard to normalize cDNA concentrations. The primers for qRT-PCR are listed in Additional file 2: Table S2. The relative quantification of gene expression was analyzed by the comparative method ($2^{-\Delta\Delta Ct}$) [15] with some modifications. Using the $2^{-\Delta\Delta Ct}$ method, data were presented as the fold-change in mRNA expression normalized to the endogenous reference gene (*Osactin*) and relative to the control.

Western blot analysis

OsXIP-specific polyclonal antibody was produced against a 15-residue synthetic peptide sequence of OsXIP (CGGRRNGVYRPFGDA) by GenScript USA Inc (China). Anti-OsXIP rabbit polyclonal antibody or anti-β-actin mouse monoclonal antibody (Beijing ComWin Biotech Co.,Ltd, CW0096) was used as the primary antibody. Immunoblot analysis was performed as described by Akulinkina et al. [16]. Samples were prepared from leaves of 2-weeks old rice plants treated with BPH for 24 h, wounding for 12 h and MeJA for 12 h, respectively. Protein fractions were separated by SDS-PAGE, and then transferred onto nitrocellulose membrane. Finally, immune complexes on a membrane were detected with BCIP/NBT. The reaction was stopped after 3 minutes of incubation by rinsing the membrane with water.

Quantitative GUS analysis

Quantitative GUS activity was measured according to the method described by Jefferson et al. [17] with some modifications. Briefly, the shoots or roots of 14-days seedlings that carried different fragments of OP1 were homogenized in GUS extraction buffer (50 mM PBS, pH 7.0, 10 mM EDTA, pH 8.0, 20% methanol, 0.1% Triton X-100, 0.1% sodium lauryl sarcosine, and 10 mM β-mercaptethanol). Crude protein extract (50 μl) was added to 450 μl of extraction buffer containing 2 mM 4-methylumbelliferyl-β-D-glucuronide (MUG) at 37 °C for 30 min or 60 min, and thereafter 200 μl of the reaction mixture was added to 800 μl of stop buffer (0.2 M Na_2CO_3). The 4-methylumbelliferone fluorescence was measured using a spectrofluorophotometer (RF-5301PC, Shimadzu) at 460 nm with excitation at 355 nm. Protein concentration was quantified by methods described by Bradford [18]. GUS activity was calculated as pmol of 4-methylumbelliferon (4-MU) min per minute and per milligram of total soluble proteins and presented as GUS activity relative to the VC.

Yeast one-hybrid (Y1H) screening

The Y1H screening used the Matchmaker Gold One-Hybrid Library Screening System (Clontech, Cat. Nos. 630491). The bait sequence (562 bp fragment) was cloned into the pAbAi vector that harbors the AUR1-C gene, conferring resistance to Aureobasidin A (AbA, a cyclic depsipeptide antibiotic used as a yeast selection marker). The resulting pAbAi-Bait construct was then linearized and integrated into the genome of the Y1HGold yeast strain by homologous recombination to generate a bait-specific reporter strain. The minimal inhibitory concentration of Aureobasidin A for the bait-specific reporter strain was determined. And the strain was used to screen a cDNA library generated from the

leaves of WT treated by BPH for 24 h. The transformants were initially screened on selective medium (SD/−Leu/AbA^{100}) and the positive colonies were identified by PCR and DNA sequencing.

For the re-transformation assay, the full-length CDSs of candidate genes were amplified from cDNA using the primers 59-AD-U/L and 71-AD-U/L as listed in Additional file 1: Table S1. The PCR products were then cloned into the pGADT7 vector and the resulting constructs were transferred into the bait reporter yeast strain mentioned above, respectively. The cells were grown on SD/−Leu and (SD/−Leu/AbA^{100}) plates at 30 °C for 3 days, and resuspended in liquid media to OD_{600} of 0.1 (10^{-1}) and diluted in a 10× dilution series (10^{-2} to 10^{-3}). Of each dilution, 7 μl was spotted on media selecting for both plasmids (SD/−Leu) and selecting for interaction (SD/−Leu/AbA^{100}), supplemented with 100 ng ml^{-1} to suppress background growth. The empty vector pGADT7 was used as a negative control.

Chromatin immunoprecipitation (ChIP)-PCR analysis

The 35Sp::OsbHLH59:GFP and 35Sp::OsERF71:GFP expression vectors were constructed by subcloning the full-length CDSs without terminators of OsbHLH59 and OsERF71 into the pCAMBIA1300-sGFP vector under the control of the 35S promoter [19], respectively. The primers 59-GFP-U/L and 71-GFP-U/L are listed in Additional file 1: Table S1. The resulting constructs were then introduced into Agrobacterium tumefaciens strain EHA105, and transformed into WT rice.

2–3 g of 3−weeks-old 35S:OsbHLH59-GFP or 35S:OsERF71-GFP transgenic seedlings were used for ChIP-PCR experiments as described in Haring et al. [20]. In brief, transgenic rice was fixed with 60 mL of 1.0% formaldehyde by vacuuming for 10 min. The chromatin DNA was sheared to 200–500 bp fragments by sonicating. Sheared DNA was incubated with GFP antibody (Biogot) (ChIP). Chromatin before immunoprecipitation was used as an input control. The primers for PCR of the target DNA are listed in Additional file 3: Table S3.

Transient expression in *Nicotiana benthamiana* leaves

The construction of vectors and transient expression in *Nicotiana benthamiana* leaves were performed as described by Ding et al. [21]. Briefly, for construction of effector vectors, the full length ORFs of OsbHLH59 and OsERF71 were amplified and cloned into the pCAMBIA1300-sGFP vector under the control of the 35S promoter. The 5'−deleted OP2 and OP3 promoters were constructed into the reporter vector pGreenII0800-LUC [22]. The recombinant plasmids were transferred into the Agrobacterium EHA105 lines. Then the EHA105 lines were co-infiltrated into the N. benthamiana leaves as described previously [23]. The Firefly and Renilla luciferase activities were quantified using a Dual Luciferase assay kit (Promega, http://www.promega.com/).

Subcellular localization

The 35Sp::OsbHLH59:GFP and 35Sp::OsERF71:GFP expression vectors were infiltrated into the *N. benthamiana* leaves by *Agrobacterium*-mediated transformation [24]. The two constructs were also transfected into rice protoplasts according to the protocol of Yoo et al. [25] with some modifications. The rice protoplasts were isolated from stems of 12-day-old WT seedlings by enzyme hydrolysis. Then 10 μg plasmid DNA was polyethylene glycol/calcium-transfected into these protoplasts. Empty vector was used as a control. The cells were observed with a confocal microscope (Zeiss LSM 710).

Results

Herbivore infestation, wounding and methyl jasmonate (MeJA) treatment enhanced mRNA levels and protein levels of OsXIP

qRT-PCR analysis revealed herbivore infestation, mechanical wounding, and MeJA treatment, especially wounding, resulted in an obvious increase in transcript levels of *OsXIP* (Fig. 1a). *OsXIP* expression was induced 3.5 times by BPH infestation and five-fold by wounding and MeJA treatment. To determine whether these stress

also affected the protein levels of OsXIP, we analyzed the protein levels of OsXIP by quantitative GUS activity of transgenic rice that carried the full-length *OsXIP* gene promoter named OP1 (−2070 bp to +52 bp) and western blotting. The expression levels of GUS all significantly rose up (Fig. 1b). Immunoblot analysis showed that the level of OsXIP also increased upon these stress (Fig. 1c). These results demonstrate that *OsXIP* is a stress-responsive gene.

Identification of herbivore-responsive promoter region

Given the observation that the expression of OsXIP was induced by different stress (Fig. 1), we used the PLACE and PlantCARE to analyse the promoter sequence of *OsXIP*. And as expected, a series of biotic and abiotic stress-responsive cis-regulatory elements exist in the promoter region (Additional file 4: Figure S1), such as W-box (TGACY) element, ABRE (CACGTG) element, MYB-binding (CGGTCA) site. To further determine the key sequences of the *OsXIP* promoter responding to herbivory, the transgenic plants that carried a set of 5' deletion promoter reporter constructs, OP2, OP3, OP4, OP5, OP6 or OP7 were obtained (Fig. 2) and their quantitative GUS activity was measured.

Without BPH treatment (Fig. 2), about 110- and 54-fold higher expression of GUS was detected with OP1

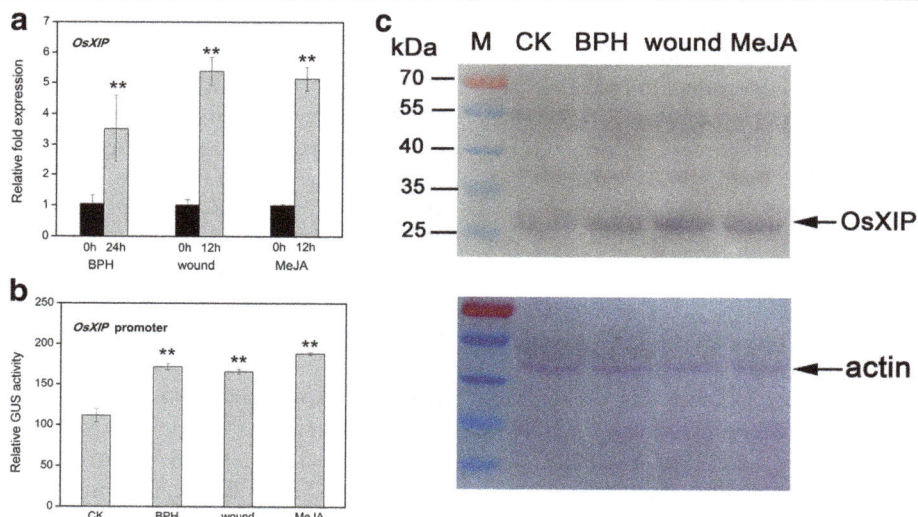

Fig. 1 Inducible expression of OsXIP after different treatments. **a** Transcript levels of *OsXIP* in rice leaves after different treatments were analyzed by qRT-PCR. Two-weeks-old WT seedlings were treated with BPH for 24 h, wounding for 12 h and MeJA for 12 h, respectively. The data represent means ± SD of three independent replicates. Asterisks indicate statistically significant differences compared with CK (0 h) (**$P < 0.01$; Student's t test). **b** GUS activity of OP1 transgenic rice after different treatments was measured by quantitative fluorescence method. Two-weeks-old transgenic rice seedlings were treated with BPH for 24 h, wounding for 12 h and MeJA for 12 h, respectively. The data represent means ± SD of three independent replicates. Asterisks indicate statistically significant differences compared with CK (**$P < 0.01$; Student's t test). **c** Western blot analysis of OsXIP. Protein fractions were isolated from leaves of WT treated with BPH for 24 h, wounding for 12 h and MeJA for 12 h, respectively and subjected to immunoblot with anti-OsXIP antibody (top panel). M, markers of proteins, the sizes of the markers are indicated at the left of the picture. Fractions corresponding to 10 μg were loaded into each lane and equal loading was confirmed by anti-actin antibody (bottom panel)

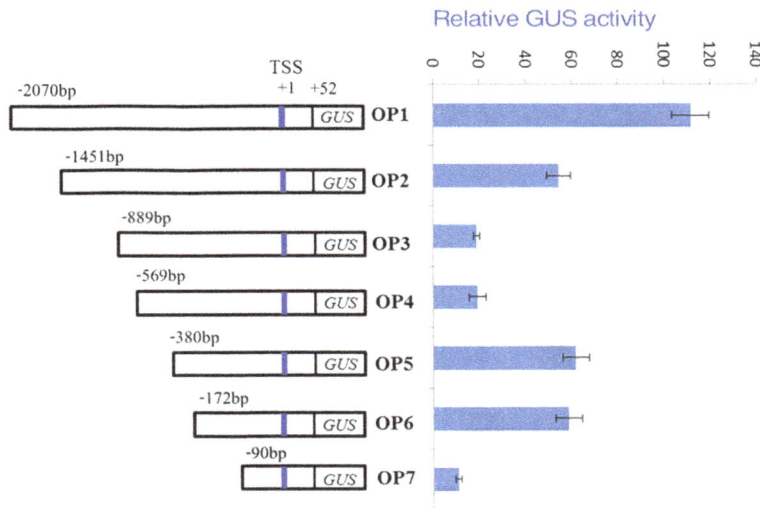

Fig. 2 Schematic representation of 5′ deletion promoter constructs and their GUS expression in shoot tissues. TSS, transcriptional start site. For every construct, three independent T2 transgenic lines were measured, and similar results were obtained. The data represent means ± SD of three independent replicates

and OP2 in the shoots, respectively, compared with vector control (VC). It was interesting to observe a sudden drop in expression with OP3, which was 18-fold higher compared with VC and similar to OP4. Surprisingly, the relative GUS expression for OP5 and OP6 are both three-fold greater than OP3 and OP4, despite having a greater deletion in the promoter region. Both OP5 and OP6 retain restore the expression level observed in OP2, whereas OP7 with the greatest promoter deletion has very minimal expression.

After infestation with BPH for 24 h (Fig. 3), GUS expression increased significantly in both the shoots and roots of transgenic lines OP1 and OP2. Elevated expression of GUS was also observed in the shoots of OP5, OP6 and OP7 and root of OP6. However, no significant change in expression was observed in the shoots of OP3 and OP4 and roots of OP4, OP5 and OP7. A sudden decrease in GUS expression was observed in the root of OP3. As mentioned above, the GUS expression of OP1 and OP2 was induced after BPH infestation, but OP3 and OP4 were not induced. Comparing the promoter regions upstream from the *GUS* gene in the constructs OP2 and OP3, OP2 had 562 bp region [−1451 bp to −889 bp relative to the transcriptional start site (TSS)] that was deleted from OP3. And the 562 bp fragment contains some stress-responsive cis-acting elements, such as W-box, ARR1AT, MBS. So we speculated that the 562 bp fragment was the key sequence involved in herbivore stress response. While the 290 bp fragment (−380 bp to −90 bp) was involved in herbivore stress response, this 562 bp fragment was then used for further research in this paper.

Proteins bound to the herbivores-responsive promoter region

To determine the proteins bound to the herbivores-responsive promoter region, we used the 562 bp fragment mentioned above as a bait to screen transformants from a cDNA library generated from the leaves of rice plants infested with BPH for 24 h by yeast one-hybrid (Y1H) screening system. Through initial screening, DNA sequencing and BLAST analysis, some proteins including functional proteins and regulatory proteins showing interaction with *OsXIP* promoter were obtained. The functional proteins included fructose 1, 6-bisphosphatase, HSP, photosystem, Tify domain containing protein. While only two genes LOC_Os02g02480 and LOC_Os06g09390, according to the rice genome annotation of The Institute for Genomic Research (TIGR; http://rice.plantbiology.msu.edu/), were the candidates of regulatory proteins. Based on BLAST analysis and literatures, LOC_Os02g02480 is a basic helix-loop-helix protein (OsbHLH59) [11]; LOC_Os06g09390 is an APETALA2/ETHYLENE RESPONSE FACTOR (AP2/ERF) transcription factor (OsERF71) [12].

The interaction between herbivores-responsive promoter region and the corresponding complete encoding products of the two genes were re-tested by Y1H assay. The analysis showed that the two proteins interacted specially with the promoter region (Fig. 4a). Meanwhile, we performed ChIP-PCR analysis by transforming 35S:OsbHLH59-sGFP and 35S:OsERF71-sGFP into rice to determine whether OsbHLH59 and OsERF71 regulate gene expression by binding to the 562 bp region in vivo. Our results showed that the promoter fragments of OsXIP were detected in the ChIP assays (Fig. 4b),

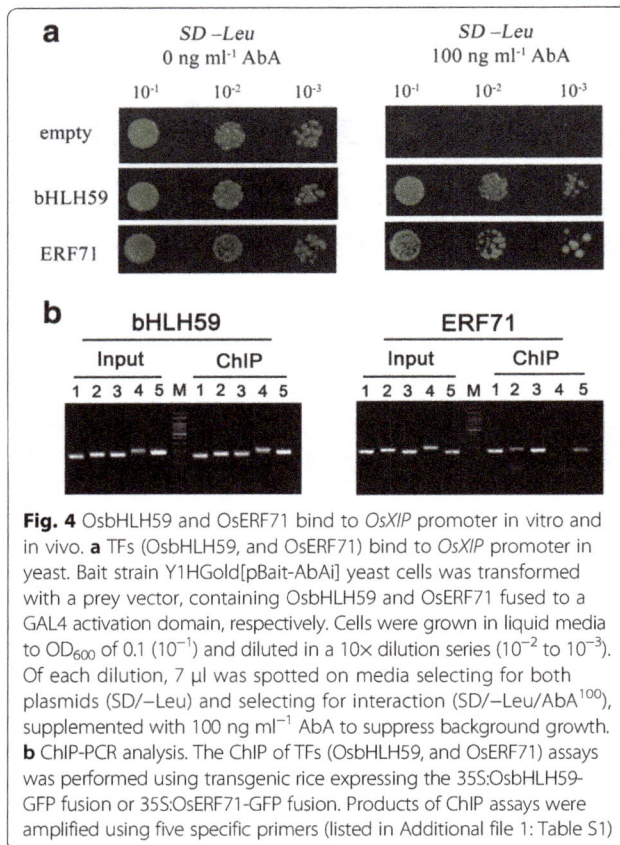

Fig. 3 GUS expression of 5′ deletion promoter constructs infested by BPH in T2 transgenic plants. Relative (to VC) GUS activity in shoots (**a**) and roots (**b**) of different transgenic lines (OP1–OP7) infested by BPH for 24 h. For every construct, three independent T2 transgenic lines were measured, and similar results were obtained. The data represent means ± SD of three independent replicates. Asterisks indicate statistically significant differences compared with control (Non-infested) (**P < 0.01; Student's t test)

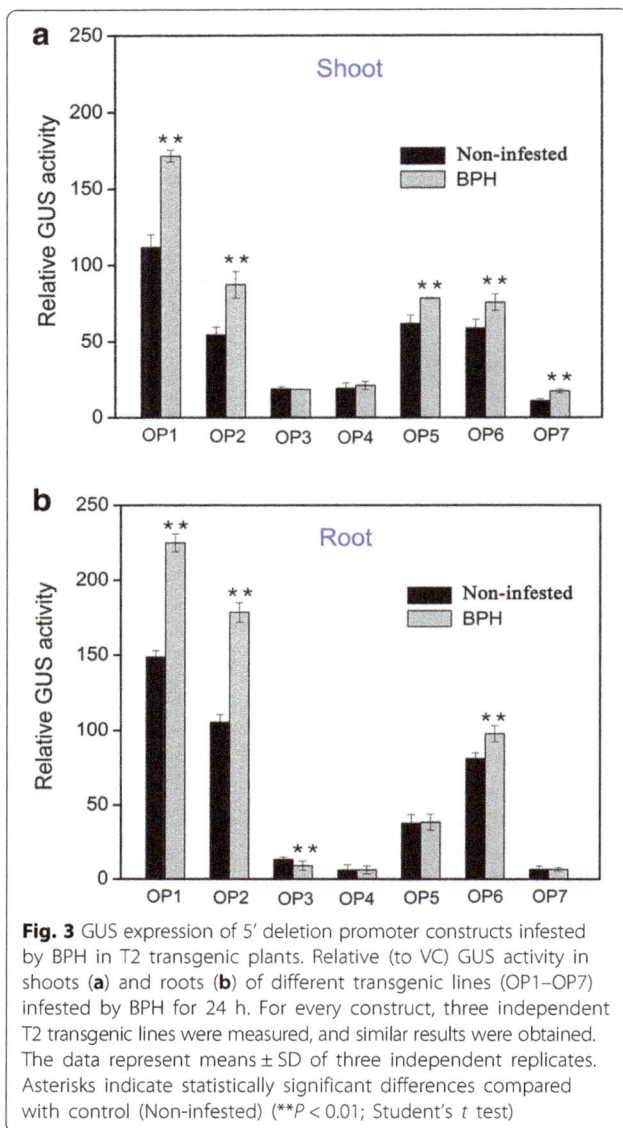

Fig. 4 OsbHLH59 and OsERF71 bind to *OsXIP* promoter in vitro and in vivo. **a** TFs (OsbHLH59, and OsERF71) bind to *OsXIP* promoter in yeast. Bait strain Y1HGold[pBait-AbAi] yeast cells was transformed with a prey vector, containing OsbHLH59 and OsERF71 fused to a GAL4 activation domain, respectively. Cells were grown in liquid media to OD_{600} of 0.1 (10^{-1}) and diluted in a 10× dilution series (10^{-2} to 10^{-3}). Of each dilution, 7 μl was spotted on media selecting for both plasmids (SD/−Leu) and selecting for interaction (SD/−Leu/AbA^{100}), supplemented with 100 ng ml^{-1} AbA to suppress background growth. **b** ChIP-PCR analysis. The ChIP of TFs (OsbHLH59, and OsERF71) assays was performed using transgenic rice expressing the 35S:OsbHLH59-GFP fusion or 35S:OsERF71-GFP fusion. Products of ChIP assays were amplified using five specific primers (listed in Additional file 1: Table S1)

further confirming that OsbHLH59 and OsERF71 can directly bind to the promoter motifs in vivo.

Transcriptional activation of the *OsXIP* promoter by OsbHLH59 and OsERF71

We also performed the tobacco transient expression assay to further clarify the DNA-binding activities of the two proteins. The dual luciferases vector was used as a reporter system following Hellens et al. [22]. The full length ORFs of OsbHLH59 and OsERF71 were amplified and cloned into the effector vector. The OP2 and OP3 promoters were constructed into the reporter vector as P1 and P2 (Fig. 5a), which infiltrated *N. benthamiana* leaves alone or co-expressed with the corresponding effector vectors, respectively. It was obvious that co-expression of OsbHLH59 remarkably increased LUC

expression driven by the OP2 promoter, as did expression of OsERF71 (Fig. 5b), suggesting OsbHLH59 and OsERF71 proteins promote the transcription of *OsXIP*. These results indicated that the two proteins may function as positive transcriptional regulators of *OsXIP* expression.

Subcellular localization of OsbHLH59 and OsERF71

To further evaluate the role of the OsbHLH59 and OsERF71 proteins, their subcellular localization were determined. We constructed 35Sp::OsbHLH59:GFP and 35Sp::OsERF71:GFP fusion genes, and transiently expressed the constructs in *N. benthamiana* leaves (Additional file 5: Figure S2) and rice protoplasts (Fig. 6), respectively. Fluorescence analysis revealed that the proteins localized only in the nucleus (Fig. 6). Cells infiltrated with GFP construct (control) yielded fluorescence both in the cytosol and the nucleus. These results further indicate that the two proteins may be transcriptional factors that function in the nucleus to regulate *OsXIP* expression.

Influence of abiotic stress on the expression of transcription factors

The expression patterns of the two genes that encode transcription factors bound to the rice herbivore-responsive cis-elements were examined by qRT-PCR in

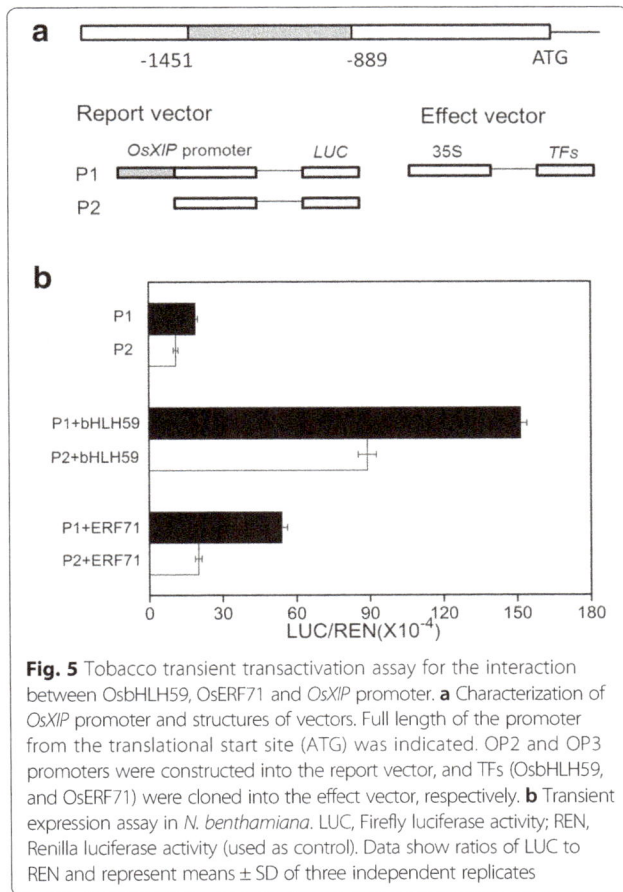

Fig. 5 Tobacco transient transactivation assay for the interaction between OsbHLH59, OsERF71 and *OsXIP* promoter. **a** Characterization of *OsXIP* promoter and structures of vectors. Full length of the promoter from the translational start site (ATG) was indicated. OP2 and OP3 promoters were constructed into the report vector, and TFs (OsbHLH59, and OsERF71) were cloned into the effect vector, respectively. **b** Transient expression assay in *N. benthamiana*. LUC, Firefly luciferase activity; REN, Renilla luciferase activity (used as control). Data show ratios of LUC to REN and represent means ± SD of three independent replicates

wild-type rice plants treated with MeJA or wounding at varying time intervals (2–24 h). The expression of these genes appeared to be responsive upon abiotic stress and was similar (Fig. 7). In shoot tissues, transcript expression of *OsbHLH59* and *OsERF71* increased concomitantly with time under MeJA treatment to 6 h, while decreased at 2 h and thereafter increased under wounding stress. The expression levels of *OsbHLH59* and *OsERF71* in root tissues were maximally induced approximately six- and seven-fold, and seven- and eight-fold after MeJA and wounding treatment, respectively. These results further suggest that the genes encoding these transcription factors may be involved in defence responses against herbivores by a JA- mediated pathway.

Discussion

The phytohormone jasmonic acid (JA) play a vital role in plant defense when plants are exposed to invaders [26]. In general, herbivorous insects and necrotrophic pathogens are more sensitive to JA-induced defenses [27, 28]. Wounding that caused by mechanical injury or insect feeding, leads to the accumulation of JA. Subsequently, the JA pathway is activated, which induces JA-responsive gene expression [29]. Our results revealed that herbivore infestation, mechanical wounding and MeJA treatment enhanced the expression of OsXIP at the transcriptional and protein levels (Fig. 1), which is consistent with OsHI-XIP [5]. To date, there is one

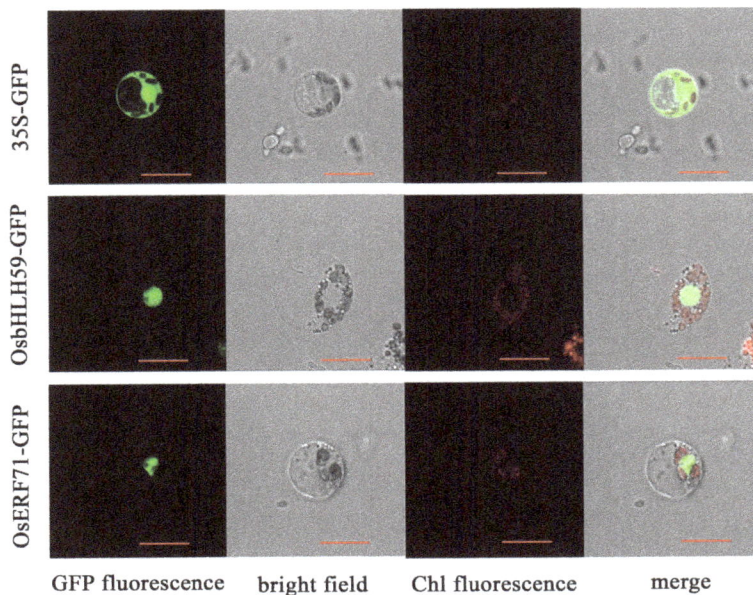

Fig. 6 Subcellular localization of OsbHLH59 and OsERF71 in rice protoplasts. The rice protoplasts were transformed with 35Sp::OsbHLH59:GFP, 35Sp::OsERF71:GFP or pCAMBIA1300-GFP. The transformed cells were observed under a confocal microscope. The photographs were taken under detecting GFP fluorescence, bright field, chloroplast auto-fluorescence, and merged microscope images, respectively. Empty vector (pCAMBIA1300-GFP) was used as a control. *Bars*, 10 μm

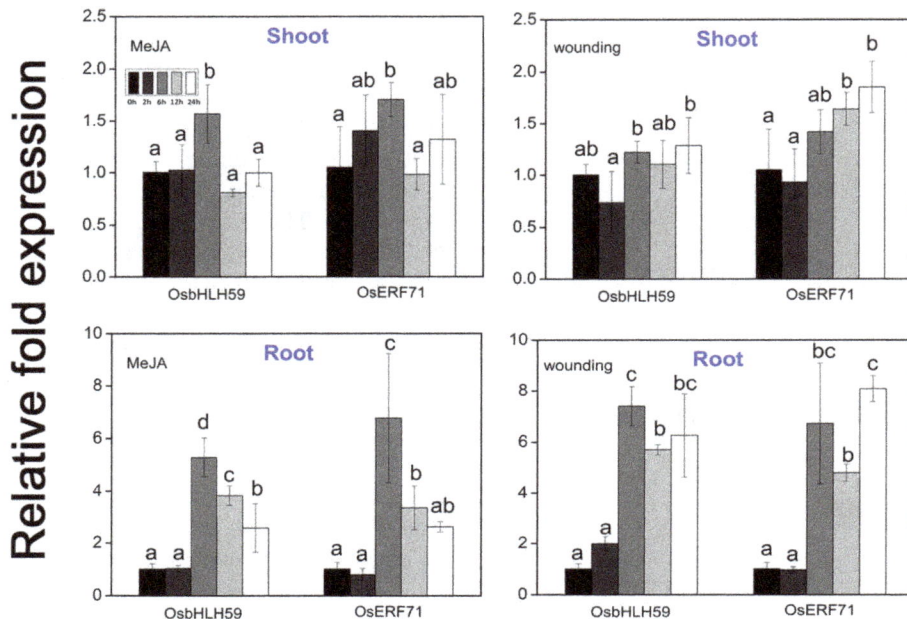

Fig. 7 Expression patterns of the genes encoding transcription factors under abiotic stress analysed by qRT-PCR. Two-weeks-old WT seedlings were treated with 200 μM MeJA or wounding. Samples were collected immediately after treatment (0 h) and at 2 to 24 h after treatment. The data represent means ± SD of three independent replicates. Letters indicate significant differences between means using Duncan's multiple range mean comparisons (5% α)

report that the xylanase inhibitor XIP-I could inhibit a xylanase from the digestive tract of the coffee berry borer [30]. *OsXIP* is a wound stress-responsive gene in rice [31]. When the herbivore feeds on the plants, wounding generates and then induces the accumulation of jasmonic acid (JA), which in turn activates wounding-associated defense pathways [32]. In our study, the expression level of OsXIP was up-regulated by BPH and the *N. lugens* induction process was very similar to that caused by wounding. When the rice was infested by BPH, wounding generated first and then activated the wound-responsive gene *OsXIP*, thus enhancing resistance against herbivores.

Bioinformatic analysis of the promoter of *OsXIP* revealed that it was a stress-induced promoter. *OsXIP* promoter contained stress-responsive cis-acting elements, such as ARR1AT, W-box, G-box and TGACG-motif. The W-box (TGACY), has been confirmed to be present upstream of salicylic acid or wound signal responsive genes; The G-box (CACGTG) and G-box-like (CANNTG, also called the E-box) are known to the binding sites of basic helix-loop-helix (bHLH) transcription factors [33]; TGACG motif has been reported to be essential for responsiveness to MeJA; ARR1AT and WRKY71OS exist in the promoter of growth regulators responsive genes. Thus, *OsXIP* gene responded to different stress (Fig. 1).

OsXIP promoter deletion analyses showed that GUS expression of OP3 reduced sharply compared with OP2,

while OP4 slightly enhanced and OP5 reached a maximum (Fig. 2). This suggested there were at least an enhancer (*e*) and a repressor (*r*) motif between OP2 and OP3 (−1451 bp to −889 bp) and OP4 and OP5 (−569 bp to −380 bp), respectively. So a regulation model of the *OsXIP* gene was proposed: an enhancer site (*e*) may exist in −1451 bp to −889 bp, and a repressor site (*r*) may be located between −569 bp and −380 bp in the promoter. GUS expression in OP2 was high due to the binding of enhancer (E) to its site (*e*). The enhancer might preferentially bound to the promoter and facilitate the gene expression on the grounds that binding of enhancer E at *e* site prevents the repressor R from binding to *r* site by either binding with R or obstructing *r* site. GUS expression of OP3 and OP4 decreased because of deletion of enhancer site (*e*). GUS expression of OP5 and OP6 increased after deletion of the *r* site. Serial deletion promoter constructs between −1451 and −889 bp and between −569 and −380 bp can be constructed to determine the locations and sequences of *e* and *r* by transient expression assay and electrophoretic mobility shift assay (EMSA).

After BPH stress, a 562 bp region (−1451 to −889) in the *OsXIP* promoter were finally identified as the key sequences involved in the herbivores stress response (Fig. 3). By Y1H screening, positive DNA-protein binding in vitro and in vivo (Fig. 4) and nuclear localization (Fig. 6), OsbHLH59 and OsERF71 proteins were confirmed as the direct regulators of *OsXIP* expression. The two proteins activated the expression of *OsXIP* gene by transient trans-activation assay (Fig. 5). Moreover, the two genes encoding corresponding OsbHLH59 and OsERF71 proteins showed differential expression upon MeJA and wounding stress (Fig. 7), further suggesting that they may be involved in positively regulating *OsXIP* expression via JA-mediated defence responses.

In the present study, two proteins (OsbHLH59, and OsERF71) as transcription activators of *OsXIP* were found. OsbHLH59 is a member of bHLH transcription factors, which are known to bind to G-box or E-box [11]. We analyzed the bait sequence and found that several E-box elements were present, including CAGTTG, CACTTG and CAATTG. And whether OsbHLH59 binds to these E-box elements of the bait can be further determined by EMSA. Some bHLH transcription factors in rice are involved in stress responses. For instance, RERJ1 (OsbHLH006) responded to wound and drought [34, 35]; OsBP-5 (OsbHLH102) is related to transcriptional regulation of the rice *Wx* gene [36]; OsbHLH094 forms a complex with RSS3 and JASMONATE ZIM-DOMAIN (JAZ) proteins to modulate the expression of JA-responsive genes [37]; DPF (OsbHLH25) positively regulates the accumulation of diterpenoid phytoalexins [38].

In *Arabidopsis*, the bHLH transcription factor MYC2 is a key positive transcriptional regulator of JA signaling pathway, which is inhibited by JAZ transcriptional repressors [39, 40]. Similarly, OsbHLH062 interacted with OsJAZ9 to regulate JA-responsive genes expression in rice [41]. Furthermore, the JA signaling pathway in *Arabidopsis* composes of the two major branches: the ERF branch and MYC branch [29]. And the ERF branch is controlled by AP2/ERF transcription factors, such as ORA59 [42]. Thus, the OsERF71 protein, an AP2/ERF transcription factor, may belong to the ERF branch to positively regulate the expression of *OsXIP*. In addition, transcriptional analysis revealed mechanical wounding and MeJA induced transcriptional expression of *OsbHLH59* and *OsERF71* in rice (Fig. 7). These results reinforce the possibility that the induction expression of *OsXIP* by wounding may be regulated by the JA-mediated signaling pathway [2]. So OsbHLH59 and OsERF71 may belong to the MYC branch and the ERF branch, respectively, to positively regulate the expression of *OsXIP*. This speculation is further supported by the findings that *OsXIP* was independent of growth and development in rice plants [30] and its expression was not regulated by phytohormones associated with growth [2].

Conclusions

In summary, we reveal the transcriptional regulatory mechanism of OsXIP and its involvement in defense response in rice. In response to herbivore infestation, mechanical wounding or MeJA stress, OsbHLH59 and OsERF71 transcription factors promote this process by activating the expression of *OsXIP* via directly binding to its promoter. Our discovery contributes to clarify the regulatory mechanism of OsXIP and gives us a better understanding of the function of OsXIP in plant defence.

Additional files

Additional file 1: Table S1. List of primers used for gene constructs.

Additional file 2: Table S2. Primers for quantitative real-time (qRT) PCR.

Additional file 3: Table S3. Primers for chromatin immunoprecipitation (ChIP)-PCR.

Additional file 4: Figure S1. The sequence of the putative promoter region (−2,070/+52) of *OsXIP*. The transcription start site is indicated as +1, and the putative start codon is underlined; All potential cis-acting elements are boxed; All the primers (forward OP1-U- OP7-U and reverse OP-L) are indicated by red arrows.

Additional file 5: Figure S2. Subcellular localization of OsbHLH59 and OsERF71 in *N. benthamiana*. The full length ORFs without terminators of OsbHLH59 and OsERF71 were cloned into the pCAMBIA1300-sGFP vector under the control of the 35S promoter. Then *N. benthamiana* cells were transformed with 35Sp::OsbHLH59:GFP, 35Sp::OsERF71:GFP or pCAMBIA1300-GFP. After incubating for 48 h, the transformed cells were observed under a confocal microscope. The photographs were taken under detecting GFP fluorescence, bright field, and in combination (merge), respectively. Empty vector (pCAMBIA1300-GFP) was used as a control. *Bars*, 10 µm.

Abbreviations
BPH: Brown planthopper; ChIP: Chromatin immunoprecipitation; GFP: Green fluorescent protein; GUS: β-glucuronidase; JA: Jasmonic acid; JAZ: JASMONATE ZIM-DOMAIN; MeJA: Methyl jasmonate; XIs: Xylanase inhibitors

Acknowledgments
We thank Dr. Jian Xue (College of Agriculture and Biotechnology, Zhejiang University, Hangzhou, China) for BPH infestation.

Funding
This work was supported by the National Natural Science Foundation of China (Grant Nos. 30971702, 31271632 and 31672462), and by research grants from the Science and Technology Department of Zhejiang Province, China (2016C32086).

Authors' contributions
XW, YZ designed the project. YZ, XS, GR, CH and YH performed the experiments and data analysis. YZ wrote the manuscript. DJ guided the experiments. All authors read and approved the final manuscript.

Competing interests
The authors declared that they have no competing interests.

Author details
[1]College of Life Science, Zhejiang University, Hangzhou 310058, China. [2]Cixi Agricultural Technology Promotion Center, Cixi 315300, China. [3]The Institute of Rural Development and Information Institute, Zhejiang Academy of Agricultural Sciences, Hangzhou 310021, China.

References

1. Dornez E, Croes E, Gebruers K, De Coninck B, Cammue BPA, Delcour JA, Courtin CM. Accumulated evidence substantiates a role for three classes of wheat xylanase inhibitors in plant defense. Crit Rev Plant Sci. 2010;29:244–64.

2. Tokunaga T, Esaka M. Induction of a novel XIP-type xylanase inhibitor by external ascorbic acid treatment and differential expression of XIP-family genes in rice. Plant Cell Physiol. 2007;48:700–14.

3. Goesaert H, Gebruers K, Courtin CM, Delcour JA. Purification and characterization of a XIP-type endoxylanase inhibitor from Rice (Oryza sativa). J Enzyme Inhib Med Chem. 2005;20:95–101.

4. Durand A, Hughes R, Roussel A, Flatman R, Henrissat B, Juge N. Emergence of a subfamily of xylanase inhibitors within glycoside hydrolase family 18. FEBS J. 2005;272:3227.

5. Xin ZJ, Wang Q, Yu ZN, Hu LC, Li JC, Xiang CY, Wang BH, Lou YG. Overexpression of a xylanase inhibitor gene, OsHI-XIP, enhances resistance in rice to herbivores. Plant Mol Biol Rep. 2014;32:465–75.

6. Igawa T, Ochiai-Fukuda T, Takahashi-Ando N, Ohsato S, Shibata T, Yamaguchi I, Kimura M. New TAXI-type xylanase inhibitor genes are inducible by pathogens and wounding in hexaploid wheat. Plant Cell Physiol. 2004;45:1347–60.

7. Hou CX, Zhan YH, Jiang DA, Weng XY. Functional characterization of a new pathogen induced xylanase inhibitor (RIXI) from rice. Eur J Plant Pathol. 2014;138:405–14.

8. Moscetti I, Tundo S, Janni M, Sella L, Gazzetti K, Tauzin A, Giardina T, Masci S, Favaron F, D'Ovidio R. Constitutive expression of the xylanase inhibitor TAXI-III delays Fusarium head blight symptoms in durum wheat transgenic plants. Mol Plant Microbe In. 2013;26:1464–72.

9. Hou CX, Lv T, Zhan YH, Peng YY, Huang YY, Jiang D, Weng XY. Overexpression of the RIXI xylanase inhibitor improves disease resistance to the fungal pathogen, Magnaporthe oryzae, in rice. Plant Cell Tiss Org. 2015; 120:167–77.

10. Elliott G, Durand A, Hughes RK, Kroon PA, D'Ovidio R, Juge N. Isolation and characterisation of a xylanase inhibitor Xip-II gene from durum wheat. J Cereal Sci. 2009;50:324–31.

11. Li XX, Duan XP, Jiang HX, Sun YJ, Tang YP, Yuan Z, Guo JK, Liang WQ, Chen L, Yin JY, Ma H, Wang J, Zhang DB. Genome-wide analysis of basic/helix-loop-helix transcription factor family in rice and Arabidopsis. Plant Physiol. 2006;141:1167–84.

12. Nakano T, Suzuki K, Fujimura T, Shinshi H. Genome-wide analysis of the ERF gene family in Arabidopsis and rice. Plant Physiol. 2006;140:411–32.

13. Hiei Y, Ohta S, Komari T, Kumashiro T. Efficient transformation of rice (Oryza sativa L.) mediated by Agrobacterium and sequence analysis of the boundaries of the T-DNA. Plant J. 1994;6:271–82.

14. Mei CS, Qi M, Sheng GY, Yang YN. Inducible overexpression of a rice allene oxide synthase gene increases the endogenous jasmonic acid level, PR gene expression, and host resistance to fungal infection. Mol Plant Microbe In. 2006;19:1127–37.

15. Livak KJ, Schmittgen TD. Analysis of relative gene expression data using real-time quantitative PCR and the 2(T)(−Delta Delta C) method. Methods. 2001;25:402–8.

16. Akulinkina DV, Bolychevtseva YV, Elanskaya IV, Karapetyan NV, Yurina NP. Association of high light-inducible HliA/HliB stress proteins with photosystem 1 trimers and monomers of the cyanobacterium Synechocystis PCC 6803. Biochemistry (Mosc). 2015;80:1254–61.

17. Jefferson RA, Kavanagh TA, Bevan MW. GUS fusions: beta-glucuronidase as a sensitive and versatile gene fusion marker in higher plants. EMBO J. 1987;6:3901–7.

18. Bradford MM. A rapid and sensitive method for the quantitation of microgram quantities of protein utilizing the principle of protein-dye binding. Anal Biochem. 1976;72:248–54.

19. Zhou G, Qi J, Ren N, Cheng J, Erb M, Mao B, Lou Y. Silencing OsHI-LOX makes rice more susceptible to chewing herbivores, but enhances resistance to a phloem feeder. Plant J. 2009;60:638–48.

20. Haring M, Offermann S, Danker T, Horst I, Peterhansel C, Stam M. Chromatin immunoprecipitation: optimization, quantitative analysis and data normalization. Plant Methods. 2007;3:11.

21. Ding ZJ, Yan JY, Xu XY, Li GX, Zheng SJ. WRKY46 functions as a transcriptional repressor of ALMT1, regulating aluminum-induced malate secretion in Arabidopsis. Plant J. 2013;76:825–35.

22. Hellens RP, Allan AC, Friel EN, Bolitho K, Grafton K, Templeton MD, Karunairetnam S, Gleave AP, Laing WA. Transient expression vectors for functional genomics, quantification of promoter activity and RNA silencing in plants. Plant Methods. 2005;1:13.

23. Yang YN, Li RG, Qi M. In vivo analysis of plant promoters and transcription factors by agroinfiltration of tobacco leaves. Plant J. 2000;22:543–51.

24. Wroblewski T, Tomczak A, Michelmore R. Optimization of Agrobacterium-mediated transient assays of gene expression in lettuce, tomato and Arabidopsis. Plant Biotechnol J. 2005;3:259–73.

25. Yoo SD, Cho YH, Sheen J. Arabidopsis mesophyll protoplasts: a versatile cell system for transient gene expression analysis. Nat Protoc. 2007;2:1565–72.

26. Pieterse CM, Van der Does D, Zamioudis C, Leon-Reyes A, Van Wees SC. Hormonal modulation of plant immunity. Annu Rev Cell Dev Biol. 2012; 28:489–521.

27. Glazebrook J. Contrasting mechanisms of defense against biotrophic and necrotrophic pathogens. Annu Rev Phytopathol. 2005;43:205–27.

28. Howe GA, Jander G. Plant immunity to insect herbivores. Annu Rev Plant Biol. 2008;59:41–66.

29. Van der Does D, Leon-Reyes A, Koornneef A, Van Verk MC, Rodenburg N, Pauwels L, Goossens A, Korbes AP, Memelink J, Ritsema T, Van Wees SCM, Pieterse CMJ. Salicylic acid suppresses jasmonic acid signaling downstream of SCF^COI1-JAZ by targeting GCC promoter motifs via transcription factor ORA59. Plant Cell. 2013;25:744–61.

30. Padilla-Hurtado B, Florez-Ramos C, Aguilera-Galvez C, Medina-Olaya J, Ramirez-Sanjuan A, Rubio-Gomez J, Acuna-Zornosa R. Cloning and expression of an endo-1,4-beta-xylanase from the coffee berry borer, Hypothenemus hampei. BMC Res Notes. 2012;5:23.

31. Tokunaga T, Miyata Y, Fujikawa Y, Esaka M. RNAi-Mediated knockdown of the XIP-type endoxylanase inhibitor gene, OsXIP, has no effect on grain development and germination in rice. Plant Cell Physiol. 2008;49:1122–7.

32. Erb M, Meldau S, Howe GA. Role of phytohormones in insect-specific plant reactions. Trends Plant Sci. 2012;17:250–9.

33. Atchley WR, Fitch WM. A natural classification of the basic helix-loop-helix class of transcription factors. Proc Natl Acad Sci U S A. 1997;94:5172–6.

34. Kiribuchi K, Jikumaru Y, Kaku H, Minami E, Hasegawa M, Kodama O, Seto H, Okada K, Nojiri H, Yamane H. Involvement of the basic helix-loop-helix transcription factor RERJ1 in wounding and drought stress responses in rice plants. Biosci Biotech Bioch. 2005;69:1042–4.

35. Kiribuchi K, Sugimori M, Takeda M, Otani T, Okada K, Onodera H, Ugaki M, Tanaka Y, Tomiyama-Akimoto C, Yamaguchi T, Minami E, Shibuya N, Omori T, Nishiyama M, Nojiri H, Yamane H. RERJ1, a jasmonic acid-responsive gene from rice, encodes a basic helix-loop-helix protein. Biochem Biophys Res Commun. 2004;325:857–63.

36. Zhu Y, Cai XL, Wang ZY, Hong MM. An interaction between a MYC protein and an EREBP protein is involved in transcriptional regulation of the rice Wx gene. J Biol Chem. 2003;278:47803–11.

37. Toda Y, Tanaka M, Ogawa D, Kurata K, Kurotani K, Habu Y, Ando T, Sugimoto K, Mitsuda N, Katoh E, Abe K, Miyao A, Hirochika H, Hattori T, Takeda S. RICE SALT SENSITIVE3 forms a ternary complex with JAZ and class-C bHLH factors and regulates jasmonate-induced gene expression and root cell elongation. Plant Cell. 2013;25:1709–25.

38. Yamamura C, Mizutani E, Okada K, Nakagawa H, Fukushima S, Tanaka A, Maeda S, Kamakura T, Yamane H, Takatsuji H, Mori M. Diterpenoid phytoalexin factor, a bHLH transcription factor, plays a central role in the biosynthesis of diterpenoid phytoalexins in rice. Plant J. 2015;84: 1100–13.

39. Chini A, Fonseca S, Fernandez G, Adie B, Chico JM, Lorenzo O, Garcia-Casado G, Lopez-Vidriero I, Lozano FM, Ponce MR, Micol JL, Solano R. The JAZ family of repressors is the missing link in jasmonate signalling. Nature. 2007;448:666–U664.

40. Melotto M, Mecey C, Niu Y, Chung HS, Katsir L, Yao J, Zeng W, Thines B, Staswick P, Browse J, Howe GA, He SY. A critical role of two positively charged amino acids in the Jas motif of Arabidopsis JAZ proteins in mediating coronatine- and jasmonoyl isoleucine-dependent interactions with the COI1 F-box protein. Plant J. 2008;55:979–88.

41. Wu H, Ye HY, Yao RF, Zhang T, Xiong LZ. OsJAZ9 acts as a transcriptional regulator in jasmonate signaling and modulates salt stress tolerance in rice. Plant Sci. 2015;232:1–12.

42. Pre M, Atallah M, Champion A, De Vos M, Pieterse CM, Memelink J. The AP2/ERF domain transcription factor ORA59 integrates jasmonic acid and ethylene signals in plant defense. Plant Physiol. 2008;147:1347–57.

8

The antibacterial activity and mechanism of ginkgolic acid C15:1

Zhebin Hua[1,3], Caie Wu[1,2*], Gongjian Fan[1,2], Zhenxing Tang[2] and Fuliang Cao[1,3]

Abstract

Background: The present study investigated the antibacterial activity and underlying mechanisms of ginkgolic acid (GA) C15:1 monomer using green fluorescent protein (GFP)-labeled bacteria strains.

Results: GA presented significant antibacterial activity against Gram-positive bacteria but generally did not affect the growth of Gram-negative bacteria. The studies of the antibacterial mechanism indicated that large amounts of GA (C15:1) could penetrate GFP-labeled *Bacillus amyloliquefaciens* in a short period of time, and as a result, led to the quenching of GFP in bacteria. In vitro results demonstrated that GA (C15:1) could inhibit the activity of multiple proteins including DNA polymerase. In vivo results showed that GA (C15:1) could significantly inhibit the biosynthesis of DNA, RNA and *B. amyloliquefaciens* proteins.

Conclusion: We speculated that GA (C15:1) achieved its antibacterial effect through inhibiting the protein activity of *B. amyloliquefaciens*. GA (C15:1) could not penetrate Gram-negative bacteria in large amounts, and the lipid soluble components in the bacterial cell wall could intercept GA (C15:1), which was one of the primary reasons that GA (C15:1) did not have a significant antibacterial effect on Gram-negative bacteria.

Keywords: GA, Green fluorescent protein, Antibacterial activity

Background

Plants can synthesize over 200,000 compounds through various metabolic pathways [1]. Secondary metabolites in plants are derived from primary metabolites, and their categories and chemical structures are complex and diverse, including nitrogen-containing organic compounds, terpenoids, phenols and polyacetylenes, of which alkaloids, terpenoids and phenols are the most common. Secondary metabolites are widely involved in plant growth, development and defense as well as other physiological and biological processes [2]. Plant secondary metabolites provide many useful natural organic compounds for human use. Because traditional chemical pesticides contaminate soil and water, the development of environmentally friendly bio-pesticides has become a popular research focus. However, the development of synthetic pesticides has many problems, such as a low successful rate, long cycle and

huge cost etc. Therefore, discovering lead compounds (plant-derived antibacterial reagents) from natural plant products with improved biological activity has become an effective method to develop new biological pesticides. Self-defense mechanisms have been evolved in plants, and many secondary plant metabolites are natural antibacterial agents [3–5]. Wilkins et al. [6] reported that 1389 plants could be used as sources of plant antibacterial agents including ingredients that could kill or inhibit bacteria, such as antibiotics, flavonoids, organic acids, polyphenols and specific proteins. Wilson et al. [7] studied the inhibition of *Botrytis cinerea* by 345 crude plant extracts and 49 essential oils, found that 13 crude extracts and 4 essential oils provided antibacterial activities.

Resorcinolic lipids are widely distributed plant secondary metabolites produced in large numbers. Recent studies have shown that they have extraordinarily high antibacterial activity. Resorcinolic lipids produced by *Pseudomonas carboxydoflava* can inhibit the growth of many bacteria species, such as *Micrococcus lysodeictius* and *Bacillus subtilis* [8, 9]. Resorcinolic lipids isolated from cashew apple have strong antibacterial effects on Gram-positive bacteria, including methicillin-resistant *S. aureus* strains [10, 11].

* Correspondence: wucaie@njfu.edu.cn
[1]Co-Innovation Centre for Sustainable Forestry in Southern China, Nanjing Forestry University, Nanjing 210037, China
[2]College of Light Industry Science and Engineering, Nanjing Forestry University, Nanjing 210037, China
Full list of author information is available at the end of the article

Sixteen phenolic compounds have been isolated from the cashew *Anacardium occidentale* (Anacardiaceae) nut shell oil, including various C15 phenolic compounds. Their antimicrobial activity has been tested against four typical microorganisms, *Bacillus subtilis*, a Gram-positive bacterium; *Escherichia coli*, a Gram-negative bacterium; *Saccharomyces cereuisiae*, a yeast; and *Penicillium chrysogenum*, a mold. Most of them exhibited potent antibacterial activity against only Gram-positive bacteria [12].

Ginkgo is a Chinese-specific rare relict species that is well known as a "living fossil of gymnosperms" [13]. The fruit and leaves of ginkgo have relatively high economic and medicinal values. However, its sarcotestas is usually discarded, causing secondary pollution of the environment [14]. GA, which is in high level in sarcotestas, is a natural plant-derived active substance contained in ginkgo, and it belongs to long-chain phenolic compounds that are derivatives of sumac acid [15]. Current studies have shown that the biological activities of GA include anti-tumor activity, neuroprotective activity, anxiolytic and antibacterial activity [16–20]. These biological activities may make a possibility that increases the utilization of ginkgo sarcotestas and reduces environmental pollution. The potential uses of Ginkgo have been attracted many concern. Studies of GA antibacterial activity have found that although GA could inhibit the activity of bacteria and plant pathogens, it just showed selective antibacterial activity, with strong inhibition towards to Gram-positive bacteria and almost no inhibition to Gram-negative bacteria [21–24].

The present study employed GFP-labeled strains and analyzed the antibacterial activity and mechanisms of GA C15:1 monomer, high amounts of which was in ginkgo sarcotestas and had relatively high antibacterial activity. Investigations of the selective antibacterial activity of GA could provide a scientific and theoretical basis for the development of new plant-derived pesticides using ginkgo sarcotestas as the raw material.

Results

Antibacterial activity of GA (C15: 1)

The antibacterial activity of GA (C15:1) is shown in Table 1. GA (C15:1) had strong antibacterial activity against Gram-positive bacteria, the MIC values of all of the tested Gram-positive bacteria were not greater than 10 µg mL^{-1}..In this study, all of the tested Gram-negative bacteria could grow well after the addition of large doses of GA (C15:1) (final concentration 500 µg mL^{-1}), and no differences were observed compared with the controls supplemented with salicylic acid, indicating that GA (C15:1) did not have significant antibacterial action against Gram-negative bacteria.

The effect of GA (C15:1) on GFP in bacteria

Using a GFP-labeled strain as the target, we studied effect of GA (C15:1) on GFP fluorescence in bacteria, and the results are show in Fig. 1-a). GA (C15:1) could significantly affect GFP fluorescence in the Gram-positive bacteria *B. amyloliquefaciens* SQR9-gfp within 1 min. Compared with the results for the CK (bacteria only containing DMSO), GA (C15:1) at the concentration of 5 µg mL^{-1} could reduce GFP fluorescence intensity in SQR9 bacteria by more than 50% within 1 min, and GA (C15:1) at higher concentrations could almost completely quench GFP fluorescence in SQR9 bacteria within 1 min.

Although GA (C15:1) could significantly affect GFP fluorescence in *B. amyloliquefaciens* SQR9-gfp bacteria within 1 min, it did not have a significant effect on GFP fluorescence in Gram-negative bacteria *E. coli* DH5α-gfp and *P. putida* KT2440-gfp. Within 1 min, a significant decrease of fluorescence intensity was not detected in the studied Gram-negative bacteria, and fluorescence intensity values in the CK were close to the fluorescence intensity value in bacteria supplemented with GA.

We extended the contact time of Gram-negative bacteria *E. coli* DH5α-gfp and *P. putida* KT2440-gfp with GA (C15:1) to 4 h. The results (Fig. 1-b) showed that

Table 1 Antibacterial activities of GA (C15:1) and salicylic acid

Strains		Control		GA (C15:1)		Salicylic acid	
		MIC (µg mL^{-1})	MBC (µg mL^{-1})	MBC (µg mL^{-1})	MBC (µg mL^{-1})	MIC (µg mL^{-1})	MBC (µg mL^{-1})
G$^-$	*E. coli* DH5α	>500	-	>500	-	>500	-
	E. coli O157:H7	>500	-	>500	-	>500	-
	P. putida KT2440	>500	-	>500	-	>500	-
	P. aeruginosa PAO1	>500	-	>500	-	>500	-
	R. solanacearum	>500	-	>500	-	>500	-
G$^+$	*B. amyloliquefaciens* SQR9	>500	-	5	60	500	>500
	R. jostii RHA1	>500	-	10	20	500	>500
	S. thermophilus ND03	>500	-	10	20	500	>500
	S. aureus	>500	-	10	20	500	>500

- Not measured, *GA* ginkgolic acid, *MIC* the minimum inhibitory concentration, *MBC* the minimum bactericidal concentration

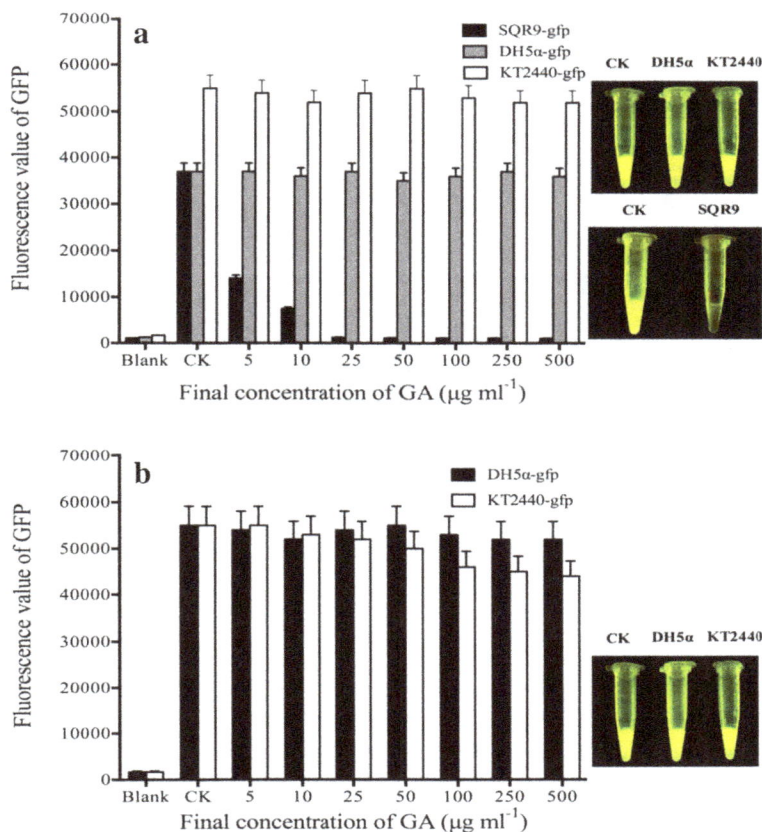

Fig. 1 Effect of GA (C15:1) on GFP fluorescence in bacteria. Three independent experiments were conducted ($n = 3$); the error bars indicate one standard error. Three individual tubes were collected from LB plates, and each tube was performed in triplicate. Each bar in the gram represents means of three individual tubes (mean + − non-log transfoemed SE.). **a**: Bacteria were incubated at 30 °C for 1 min before GFP fluorescence was measured.10 μL of DMSO without the drug was used as a control. The blank was the *E. coli* bacteria solution without GFP. One-way ANOVA was used for analyzing the data ($F_{7,16} = 656.9$ $P < 0.001$(SQR9-gfp); $F_{7,16} = 0.208$ $P > 0.05$(DH5α-gfp); $F_{7,16} = 0.357$ $P > 0.05$ (KT2440-gfp)); (**b**) Bacteria were incubated at 30 °C for 4 h before GFP fluorescence was measured. 10 μL of DMSO without the drug was used as a control. The blank was the *E. coli* bacteria solution without GFP. One-way ANOVA was used for analyzing the data ($F_{7,16} = 0.178$ $P > 0.05$(DH5α-gfp); $F_{7,16} = 1.412$ $P > 0.05$ (KT2440-gfp))

even with longer incubation times, GA could only reduce GFP fluorescence in Gram-negative bacteria by a small amount. GA (C15:1) at the concentration of 500 μg mL^{-1} had the most significant effect on fluorescence in *E. coli* DH5α-gfp, causing approximately 30% fluorescence reduction. The fluorescence reduction values at other concentrations were all less than 25%.

The scanning electron microscopy examination showed that after the addition of GA (C15:1), the cells of the three bacteria still remained intact without apparent cell lysis (Fig. 2). Because GFP protein was only present in the bacteria, we speculated that GFP fluorescence decay in Gram-positive bacteria *B. amyloliquefaciens* SQR9-gfp was caused by a large amount of GA that entered the bacteria within a short time, whereas the reason that GFP fluorescence in both Gram-negative bacteria did not show decay was that GA (C15:1) did not enter these bacteria in a large amount. The lack of a significant reduction in GFP fluorescence in the two Gram-negative bacteria was caused by a limited amount of GA entering the cells.

Effect of GA (C15:1) on GFP in bacteria crude extracts

To verify the hypothesis that "GFP fluorescence decay was related to GA entering the bacteria cells", the bacteria cells of Gram-negative bacteria *E. coli* DH5α-gfp and *P. putida* KT2440-gfp were lysed and centrifuged, and the crude lysate supernatants which contained GFP, were collected. GA (C15:1) was directly added to the supernatant, and the GFP fluorescence intensity was examined. The same procedure was performed on Gram-positive bacteria *B. amyloliquefaciens* SQR9-gfp. The results showed that GA (C15:1) could significantly affect GFP fluorescence in the crude lysates of *E. coli* DH5α-gfp and *P. putida* KT2440-gfp within 1 min (Fig. 3). Compared with the results for the CK, GA (C15:1) at a final concentration > 25 μg mL^{-1} could completely quench GFP fluorescence in the crude lysates of *E. coli* DH5α-gfp and *P. putida* KT2440-gfp within 1 min.

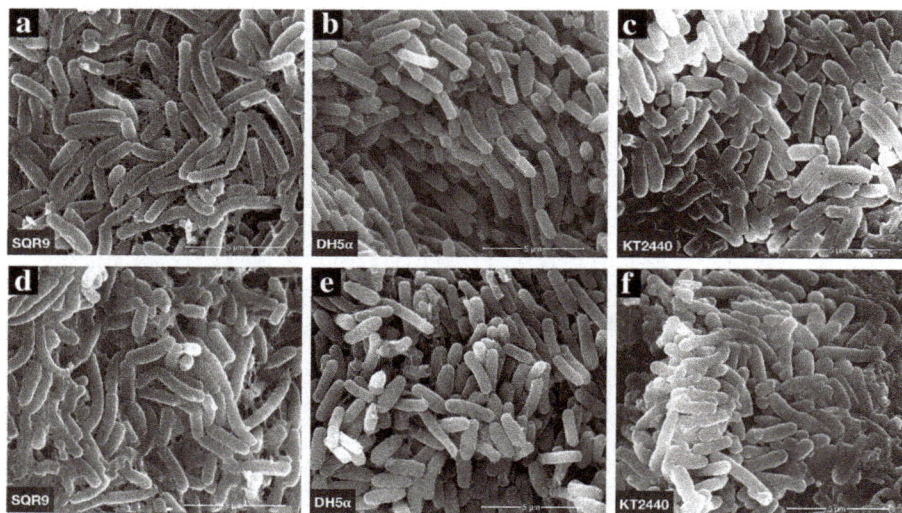

Fig. 2 SEM observation of bacteria cell morphology. respectively, (**a**, **b**, **c**) represent Bacteria cell morphology before the addition of GA; (**d**, **e**, **f**) represent Bacteria cell morphology after the addition of GA at a final concentration of 100 μg mL^{-1} for 1 min

Similar fluorescence decay results were also showed in the crude cell lysates of Gram-positive bacteria *B. amyloliquefaciens* SQR9-gfp (Fig. 3). When the concentration of GA (C15:1) was >> 10 μg mL^{-1}, the fluorescence in the crude lysates of *B. amyloliquefaciens* SQR9-gfp was completely quenched within 1 min. This results were consistent with the results of the GFP fluorescence

decay experiment in *B. amyloliquefaciens* SQR9-gfp cells, and suggested that GA could enter the cells of Gram-positive bacteria *B. amyloliquefaciens* SQR9-gfp GFP within a short period of time.

Effect of GA on the activity of a variety of proteins

According to the PCR results (Fig. 4-a), GA (C15:1) significantly inhibited the PCR reaction. The addition of 1 μg mL^{-1} and 5 μg mL^{-1} GA (C15:1) could interfere with PCR reactions, resulting in decreased specific bands, although these concentrations would not completely inhibit PCR reactions. However, when the concentration of GA was ≥ 10 μg mL^{-1}, the PCR reaction was completely inhibited, and electrophoresis could not detect specific target bands. This result suggested that lower concentration of GA (C15:1) could inhibit Taq DNA polymerase activity and interfere with DNA replication.

Restriction digestion electrophoresis results showed that GA (C15:1) could significantly inhibit the enzymatic activity of Kpn I, Hind III and EcoR I (Fig. 4-b, c, d). GA (C15:1) at a final concentration of 1 μg mL^{-1} could partially inhibit the enzymatic activity of the three restriction enzymes. When the concentration of GA (C15:1) was > 5 μg mL^{-1}, the enzymatic activity of the three restriction enzymes were inhibited almost completely, and the super coiled pUC19 plasmid was barely digested.

GA (C15:1) could significantly inhibit SOD enzyme activity and β-galactosidase activity (Fig. 5-a, b). When the final concentration of GA (C15:1) was 1 μg mL^{-1}, both SOD enzyme activity and β-galactosidase activity were decreased by 50% compared with that of the control. When the final concentration of GA was > 5 μg mL^{-1},

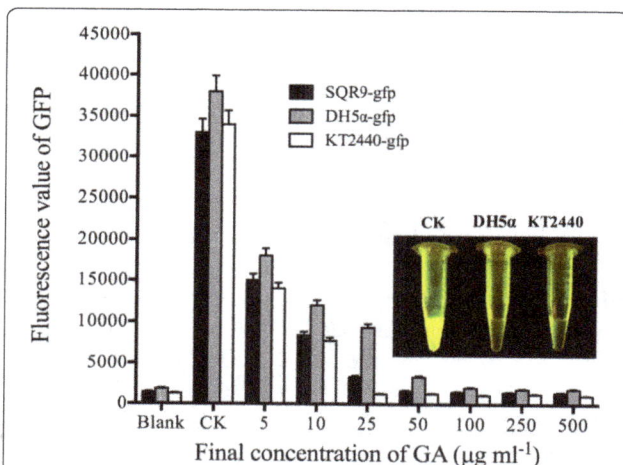

Fig. 3 Effect of GA (C15:1) on GFP fluorescence in bacteria crude lysates (1 min). Three independent experiments were conducted (*n* = 3). The error bars indicate one standard error. All of the tests were incubated at 30 °C for 1 min before GFP fluorescence was measured. 10 μL of DMSO without the drug was used as a control. The blank was the *E. coli* bacteria solution without GFP. One-way ANOVA was used for analyzing the data (F$_{7,16}$ = 281.4 *P* < 0.001 (SQR9-gfp); F$_{7,16}$ = 246.3 *P* < 0.001 (DH5α-gfp); F$_{7,16}$ = 304.0 *P* < 0.001 (KT2440-gfp)). Three individual tubes were collected from LB plates, and each tube was performed in triplicate. Each bar in the gram represents means of three individual tubes (mean + − non-log transfoemed SE.)

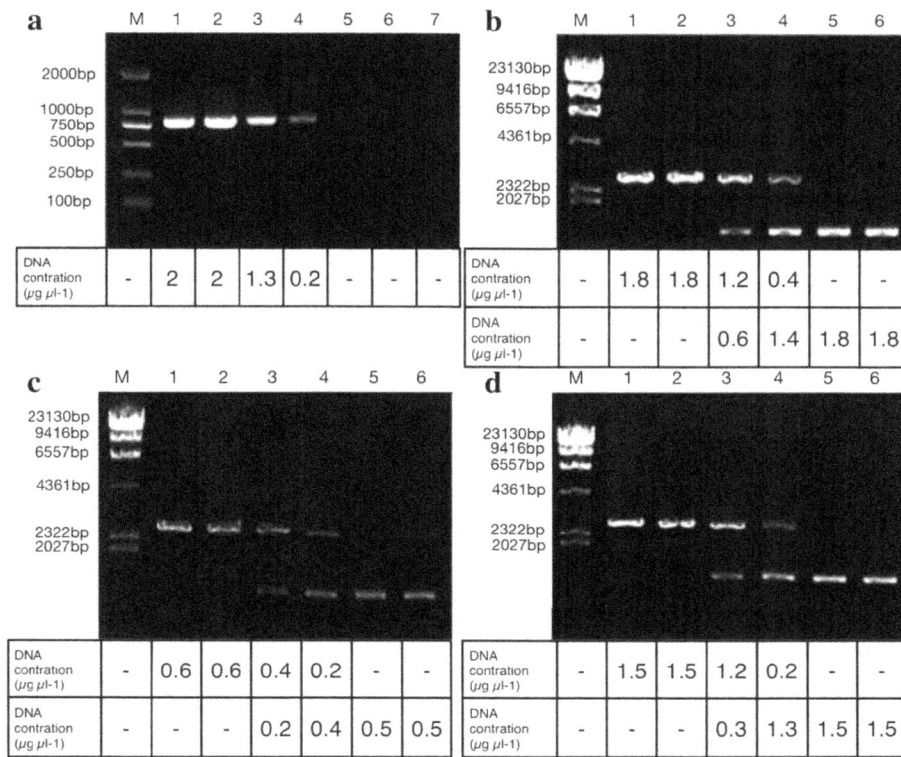

Fig. 4 Effect of GA (C15:1) on a variety of proteins. respectively :(**a**) represents the effect of GA (C15:1) on the activity of *Taq* DNA Polymerase. The DNA contration at the bottom of fig represents the PCR reaction was inhibited by GA (C15:1). **b** represents the effect of GA (C15:1) on the activity of *Kpn* I. The DNA contration at the bottom of fig represents the enzymatic activity *Kpn* I was inhibited by GA (C15:1) (**c**) represents the effect of GA (C15:1) on the activity of *Hind* III. The DNA contration at the bottom of fig represents the enzymatic activity *Hind* III was inhibited by GA (C15:1) (**d**) represents the effect of GA (C15:1) on the activity of *EcoR* I. The DNA contration at the bottom of fig represents the enzymatic activity *EcoR* I was inhibited by GA (C15:1) lane 1, CK; lane 2, addition of DMSO into PCR or digestion system; lane 3, addition of 1 µg mL^{-1} GA (C15:1) into PCR or digestion system; lane 4, addition of 5 µg mL^{-1} GA (C15:1) into PCR or digestion system; lane 5, addition of 10 µg mL^{-1} GA (C15:1) into PCR or digestion system; lane 6, addition of 25 µg mL^{-1} GA (C15:1) into PCR or digestion system; lane 7, addition of 50 µg mL^{-1} GA (C15:1) into PCR or digestion system

SOD enzyme activity and β-galactosidase enzyme activity were almost undetectable.

These proteins had different sources and were selected randomly. Thus, the results of this study suggested that the inhibition of GA (C15:1) on protein activities was non-selective.

Inhibition of isotope incorporation experiments

The aforementioned experiments showed that GA could inhibit DNA polymerase function in vitro. However, it was unclear whether GA had a similar function in bacteria, including whether GA (C15:1) could inhibit DNA polymerase activity in vivo, which would inhibit DNA replication. In addition, it is unclear whether GA could inhibit RNA polymerase and ribosome activities, which would inhibit transcription and translation. To further clarify the mechanism of GA (C15:1), we used the method of the inhibition of isotope incorporation to verify the effects of GA (C15:1) in vivo. (Methy-^{3}H) thymine ([^{3}H] TdR), ^{3}H-uridine ([^{3}H] UR) and ^{3}H-tyrosine ([^{3}H]

Tyr) were used as precursors to determine effect of GA (C15:1) on the biosynthesis of DNA, RNA and *B. amyloliquefaciens* SQR9 proteins. The results are showed in Fig. 6. Compared with that of the control, GA (C15:1) could inhibit DNA replication, RNA synthesis and protein synthesis to different extents under all three concentrations (25 µg mL^{-1}, 10 µg mL^{-1} and 5 µg mL^{-1}). When the concentration of GA (C15:1) reached 25 µg mL^{-1}, the inhibition of [^{3}H] TdR incorporation was approximately 99%, of [^{3}H] UR incorporation was approximately 90%, and of protein precursor [^{3}H] tyrosine was approximately 85%. These data indicated that GA (C15:1) could inhibit DNA replication in vivo as well as RNA transcription and protein synthesis.

The interception of GA by Gram-negative bacteria cell walls

Using *E. coli* as the target, lysozyme was used to destroy the peptidoglycan structure in the cell wall and thus *E. coli* DH5α-gfp protoplasts were obtained. Effect of GA (C15:1)

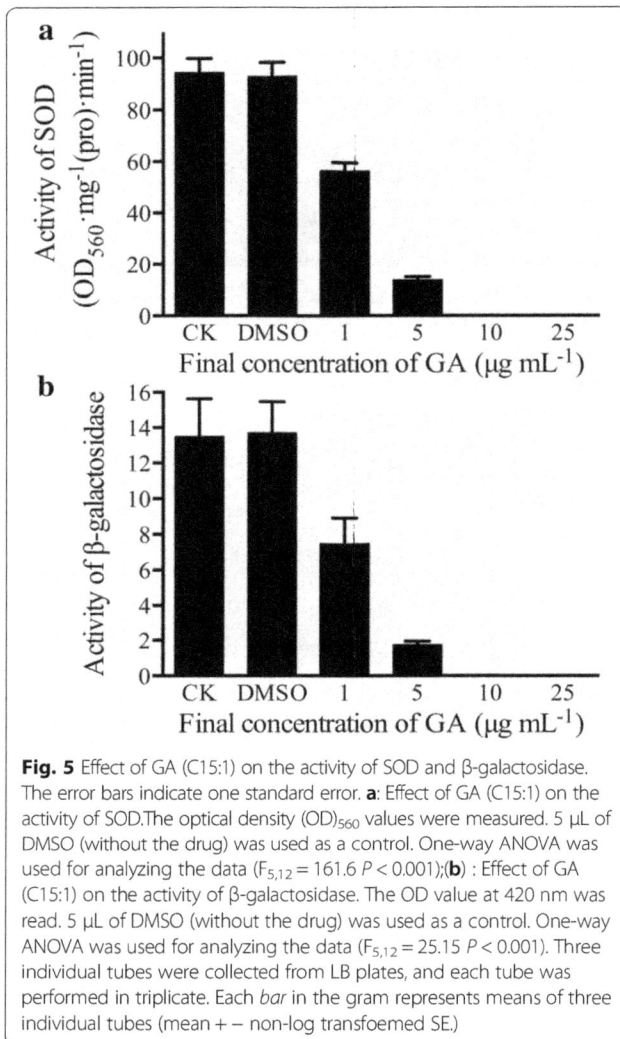

Fig. 5 Effect of GA (C15:1) on the activity of SOD and β-galactosidase. The error bars indicate one standard error. **a**: Effect of GA (C15:1) on the activity of SOD. The optical density (OD)$_{560}$ values were measured. 5 μL of DMSO (without the drug) was used as a control. One-way ANOVA was used for analyzing the data ($F_{5,12} = 161.6$ $P < 0.001$);(**b**) : Effect of GA (C15:1) on the activity of β-galactosidase. The OD value at 420 nm was read. 5 μL of DMSO (without the drug) was used as a control. One-way ANOVA was used for analyzing the data ($F_{5,12} = 25.15$ $P < 0.001$). Three individual tubes were collected from LB plates, and each tube was performed in triplicate. Each *bar* in the gram represents means of three individual tubes (mean + − non-log transfoemed SE.)

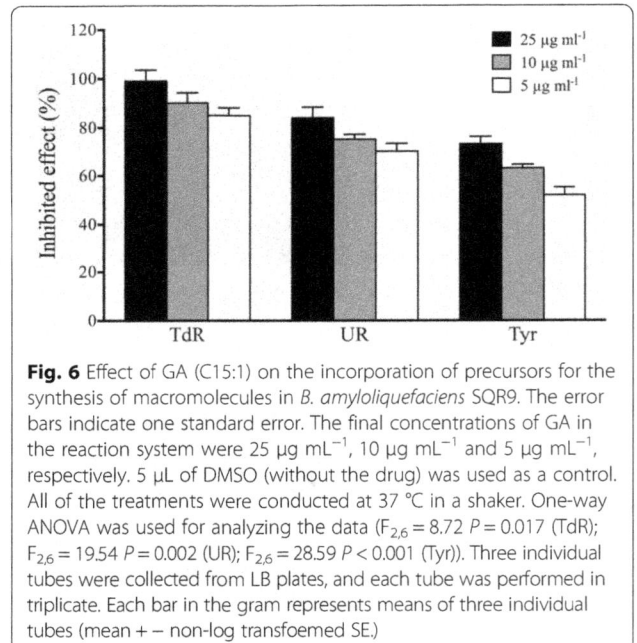

Fig. 6 Effect of GA (C15:1) on the incorporation of precursors for the synthesis of macromolecules in *B. amyloliquefaciens* SQR9. The error bars indicate one standard error. The final concentrations of GA in the reaction system were 25 μg mL^{-1}, 10 μg mL^{-1} and 5 μg mL^{-1}, respectively. 5 μL of DMSO (without the drug) was used as a control. All of the treatments were conducted at 37 °C in a shaker. One-way ANOVA was used for analyzing the data ($F_{2,6} = 8.72$ $P = 0.017$ (TdR); $F_{2,6} = 19.54$ $P = 0.002$ (UR); $F_{2,6} = 28.59$ $P < 0.001$ (Tyr)). Three individual tubes were collected from LB plates, and each tube was performed in triplicate. Each bar in the gram represents means of three individual tubes (mean + − non-log transfoemed SE.)

(C15:1) on GFP fluorescence in the *E. coli* cells that did not have lipid-soluble components in their cell walls, was measured. The results (Fig. 8) showed that when the final concentration of GA reaches was 5 μg mL^{-1}, a large degree decrease of GFP fluorescence in *E. coli* occurred. When the final concentration of GA was above

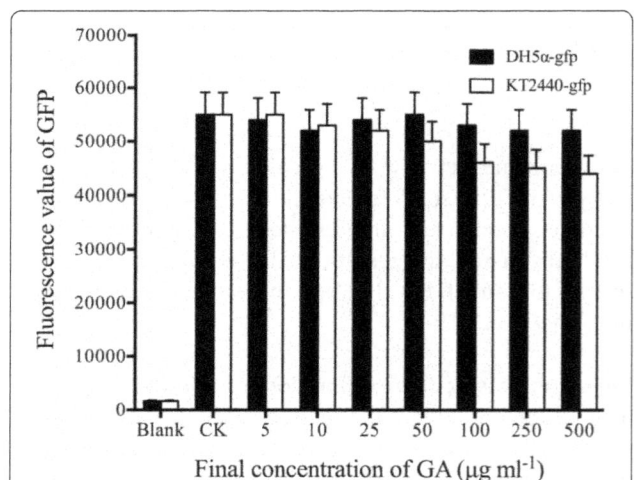

Fig. 7 Effect of GA (C15: 1) on GFP fluorescence in the protoplasts of *E. coli* DH5α-gfp (1 min). Three independent experiments were conducted ($n = 3$); The error bars indicate one standard error. 5 μL of DMSO (without the drug) was used as a control. The blank was *E. coli* protoplast solution without GFP. All the tests were incubated at 30 °C for 1 min. One-way ANOVA was used for analyzing the data ($F_{7,16} = 0.106$ $P > 0.05$ (DH5α-gfp); $F_{7,16} = 1.412$ $P > 0.05$ (KT2440-gfp)). Three individual tubes were collected from LB plates, and each tube was performed in triplicate. Each bar in the gram represents means of three individual tubes (mean + − non-log transfoemed SE.)

on GFP fluorescence in the protoplast within 1 minute was measured. The results (Fig. 7) showed that when the final concentration of C15:1 was lower than 25 μg mL^{-1}, it did not have significant effect on GFP fluorescence in protoplasts. When the final concentration of GA was ≥ 25 μg mL^{-1}, the decreased GFP fluorescence intensity in the protoplasts was found. It was positively correlated with increasing concentration of GA. When the final concentration of GA reached 500 μg mL^{-1}, the GFP fluorescence intensity in the protoplasts decreased to approximately 80% of the control. This result suggested that after the peptidoglycan structure in the Gram-negative bacteria cell wall was destroyed, a small amount of high concentration GA could enter the Gram-negative bacteria cell and produce a low level of GFP fluorescence decay.

By soaking *E. coli* cells in ethanol solution for a short period of time, the lipid-soluble components (mainly included lipopolysaccharide and phospholipids) in the cell wall were removed/partially removed. Effect of GA

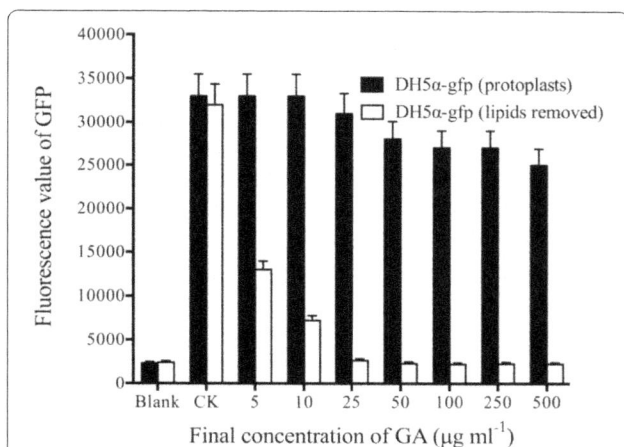

Fig. 8 Effect of GA (C15: 1) on GFP fluorescence in *E. coli* DH5α-gfp (1 min), in which the lipids have been removed from the strains' cell wall. Three independent experiments were conducted (*n* = 3). The error bars indicate one standard error. 5 µL of DMSO (without the drug) was used as a control. The blank was E. coli bacteria solution without GFP (after ethanol solubilization). All the tests were incubated at 30 °C for 1 min. One-way ANOVA was used for analyzing the data ($F_{7,16}$ = 2.116 P > 0.05 (*Black*); $F_{7,16}$ = 121.6 P < 0.001 (*White*)). Three individual tubes were collected from LB plates, and each tube was performed in triplicate. Each bar in the gram represents means of three individual tubes (mean + − non-log transfoemed SE.)

10 µg mL^{-1}, it could completely quench the GFP fluorescence in *E. coli*. However, if the lipid-soluble components were not removed from the *E. coli* bacteria cell wall, even the final concentration of GA at 500 µg mL^{-1} could not quench the GFP fluorescence (Fig. 1-a). These results showed that the lipid-soluble components in Gram-negative bacteria cell walls could intercept GA.

Discussion

In this study, *E. coli* DH5α, *E. coli* O157: H7, *P. putida* KT2440, *P. aeruginosa* PAO1, *R. solanacearum*, *Rhodococcus* RHA1, *S. thermophilus* ND03, *S. aureus* and other common strains were used to study the antibacterial activity of GA (C15:1), and GA was found to have significant antibacterial activity against Gram-positive bacteria but little effect on the growth of Gram-negative bacteria. A relatively strong selective antibacterial mechanism of GA was observed. The MIC value of *B. amyloliquefaciens* SQR9 was the smallest among all of the tested Gram-positive bacteria. However, its MBC value (60 µg mL^{-1}) was the largest among all of the tested Gram-positive bacteria. These results might be caused by small amount of endospores that were generated when *B. amyloliquefaciens* SQR9 was cultured. Endospores had relatively strong resistance, and could withstand higher concentrations of GA without being killed. Therefore, *B. amyloliquefaciens* SQR9 had significantly higher MBC values than other Gram-positive bacteria. The antibacterial activity of GA has been reported these years. Himejima and Kubo [12]

found that 2-hydroxy-6-(8-pentadecenyl) salicylic (another name of ginkgolic acid C15:1) showed lower MICs (about 10 µg mL^{-1}) against Gram-positive bacteria and higher MICs (>100 µg mL^{-1}) against Gram-negative bacteria. Choi et al. [23] also showed that GA (C15:1) had significant antibacterial activity against 18 g-positive vancomycin-resistant. The results of the present study are consistent with above studies.

Additional studies on antibacterial mechanisms using GFP fluorescence-labeled Gram-positive bacteria *B. amyloliquefaciens* SQR9 and GFP-labeled Gram-negative bacteria *E. coli* DH5α and *P. putida* KT2440 showed that GA (C15:1) could significantly affect GFP fluorescence in the cells of Gram-positive *B. amyloliquefaciens* SQR9-gfp, whereas it had no significant effect on GFP fluorescence in the cells of Gram-negative bacteria *E. coli* DH5α-gfp and *P. putida* KT2440-gfp. The green fluorescent protein (GFP) has been widely used as a highly useful tool in the fluorescence studies of living cells, which is found in cell cytoplasm of jellyfish and is an extremely stable protein with 238 amino acids [25, 26]. The fluorescence produced by GFP was caused by its protein conformation. In general, as long as the protein conformation of GFP did not change, the fluorescence would not decay or disappear. Previous reports showed that GA and sumac acids, which had a similar structure, could affect the activity of numerous enzymes, including protein phosphatase, lipoxygenase and histone acetyltransferase [27–29]. In addition, GA affected in vivo regulation mechanism of small ubiquitin-related modifier (SUMO) and altered protein conformation, thereby affecting protein expression [30].

According to the above test, we suggested that the mechanism by which GA (C15:1) decayed GFP fluorescence was through conformation changes in the GFP protein. In addition, the mechanism by which GA promoted antibacterial activity against Gram-positive bacteria was through conformational changes of the proteins in the bacteria that inactivated the proteins and inhibited the growth of Gram-positive bacteria.

The results of crude cell lysate experiments showed that GFP fluorescence decay might be related to the interaction between GFP and GA. The GFP fluorescence in both Gram-negative and Gram-positive bacteria crude lysates was quenched by GA in a short period of time, which indicated that GFP fluorescence would be quenched as long as it had contact with GA and was not related to the micro-organism tagged with GFP. Because the structure between Gram-negative and Gram-positive bacteria was similar and results showed that peptidoglycan in Gram-positive bacteria could not prevent GA (C15:1) from entering the cell, we suggested that the peptidoglycan structure of Gram-negative bacteria also could not block GA (C15:1) from entering the cell. In the protoplast experiment, a small amount of GA molecules could enter the cells after the peptidoglycan structure in *E. coli* cell wall was destroyed by lysozyme, which might be the result of the action of lysozyme. After the peptidoglycan structure was destroyed by lysozyme, pores might be present on the surface of the peptidoglycan layer that allowed GA molecules to pass through. However, only a small amount of GA molecules could enter the cells because the number of pores generated on the surface of the peptidoglycan layer was low, and the surface of Gram-negative bacteria was still covered by a large amount of lipids (including lipopolysaccharides and phospholipids), which could intercept a large amount of GA molecules. In order to further confirm lipid-soluble components in the cell wall of Gram-negative bacteria intercept the majority of GA molecules, the studies use high resolution electron microscopy to observe membrane change or other methods to study transport of GA through membrane will be carried out.

Some studies demonstrated that GA markedly inhibited the biofilm formation of *S. mutans* and *Escherichia coli* O157:H7, and disrupted biofilm integrity [24, 31]. Therefore, we speculate that the GA may affect the secondary metabolism of Gram-positive and Gram-negative bacteria. Due to the secondary metabolism of bacteria, such as the formation of biofilm, fluorescence formation and synthesis of antibiotics are regulated by quorum-sensing, further studies on this section will be investigated.

Conclusions

GA (C15:1) has a relatively strong selective antibacterial mechanism, which significant antibacterial activity against Gram-positive bacteria but little effect on the growth of Gram-negative bacteria. Additional studies on antibacterial mechanisms showed that GA (C15:1) could inhibit the activities of the selected proteins to a certain degree, and non-selectively induce protein conformational changes. GA (C15:1) also inhibit DNA replication in vivo as well as RNA transcription and protein synthesis. Thus, we suggested that the mechanism by which GA (C15:1) promoted antibacterial activity against Gram-positive bacteria was through conformational changes of the proteins in the bacteria that inactivated the proteins and inhibited the growth of Gram-positive bacteria. The research results indicated that lipid-soluble components (including lipopolysaccharide and phospholipids) in the cell wall of Gram-negative bacteria intercepted the majority of GA molecules, whereas the peptidoglycan layer in the cell wall showed a reduced capacity to intercept GA molecules.

Methods
Media and reagents
Lysogeny broth (LB) medium (g L^{-1}) was composed as follows: peptone 10.0 g L^{-1}, yeast extract 5.0 g L^{-1}, NaCl 10.0 g L^{-1}, pH 7.2, (solid, addition of 1.5% agar), and deionized water 1000 mL, which was sterilized at 121 °C for 20 min.

Proteinase K, lysozyme, ampicillin (Amp), kanamycin (Km), gentamicin (Gm), isopropyl-β-D-thiogalactopyranoside (IPTG), o-nitrophenyl β-D-galactopyranoside (ONPG), and o-nitrophenol (ONP) were purchased from Shanghai Sangon Biotech (Sangon Biotech, Shanghai, China). GA C15:1 standard was purchased from Shanghai Tauto Biotechnology (Shanghai, China). Other chemical reagents were analytical grade.

Strains
Escherichia coli DH5α (*E. coli* DH5α) (ATCC53338), *E. coli* O157: H7 (*E. coli* O157: H7) (ATCC43895), *Pseudomonas putida* KT2440 (ATTC47054), *Pseudomonas aeruginosa* PAO1 (ATCC15692), *Ralstoniasolanacearum* (ATCC11 696), *Rhodococcusjostii* RHA1 [32], *Streptococcus thermophilus* ND03 [33], and *S. aureus* (ATCC25923) were from our laboratory.

Bacillus amyloliquefaciens SQR9 (CGMCC 5808; China General Microbiology Culture Collection Center) [34], *E. coli*-gfp (*E. coli* DH5α-gfp), *P. putida* KT2440-gfp, and *B. amyloliquefaciens* SQR9-gfp were provided by the Environmental Microbiology Lab at the College of Resources and Environment, Nanjing Agricultural University.

Determination of antibacterial activity of GA
A conventional broth-dilution method was adopted [35]. GA and salicylic acid were dissolved into dimethylsulfoxide (DMSO) to prepare stock solutions with different concentrations, respectively. The final concentrations of the drug (GA or salicylic acid) in the medium (500 µg mL^{-1}, 250 µg mL^{-1}, 100 µg mL^{-1}, 80 µg mL^{-1}, 60 µg mL^{-1}, 40 µg mL^{-1}, 20 µg mL^{-1}, 10 µg mL^{-1}, 5 µg mL^{-1}, 2 µg mL^{-1}, 1 µg mL^{-1}, 0.5 µg mL^{-1} and 0.1 µg mL^{-1}) were obtained. 10 µL stock solutions were added to 3 mL liquid LB medium that was inoculated with bacteria. All of the tests were incubated at 200 rpm. *E. coli* was cultured at 37 °C for 2 d, whereas other bacteria were cultured at 30 °C for 2 d. The lowest concentration without turbidity was defined as the minimum inhibitory concentration (MIC) of that substance. The medium without turbidity (50 µL) was inoculated with 3 mL of fresh LB liquid medium, and cultured in a shaker. *E. coli* was cultured at 37 °C for 2 d, and all other bacteria were cultured at 30 °C for 2 d. The lowest concentration without turbidity was defined as the minimum bactericidal concentration (MBC) of that substance. Same concentration of DMSO without GA and salicylic acid was added in control groups.

Measurement of fluorescence decay of GFP in bacteria

Single colonies of *E. coli* DH5α-gfp, *P. Putida* KT2440-gfp and *B. amyloliquefaciens* SQR9-gfp were extracted from solid LB plates, inoculated in liquid LB medium, and then 50 μg mL^{-1} Amp, Gm and Km was added to the medium to maintain the normal replication of plasmids in each strain. The medium was then centrifuged, and the bacteria were collected. The supernatant was discarded, and phosphate buffer was added to the collected bacteria to obtain the bacteria concentration of 10^8 CFU mL^{-1}. GA was then dissolved in DMSO to prepare stocks with different concentrations.

(1) 10 μL of GA stock solution at different concentrations was added to 1 mL of bacteria solution. The final concentrations of GA in the solution were 500 μg mL^{-1}, 250 μg mL^{-1}, 100 μg mL^{-1}, 50 μg mL^{-1}, 25 μg mL^{-1}, 10 μg mL^{-1} and 5 μg mL^{-1}, respectively. 10 μL of DMSO without the drug was used as a control. All of the tests were incubated at 30 °C for 1 min before GFP fluorescence was measured. The morphology of the bacteria was monitored under a scanning electron microscopy. For strains of *E. coli* DH5α-gfp and *P. Putida* KT2440-gfp, the incubation was extended to 4 h and additional samples were collected to measure GFP fluorescence. The morphology of the bacteria was measured using a scanning electron microscopy. Bacteria GFP fluorescence was initially observed by the naked eye using an LB16 Maestrogen UltraSlim no-damage blue LED transilluminator. Measurements of bacterial GFP fluorescence were performed using a Spectra ax M5 multifunctional microplate reader. The samples' fluorescence intensity was measured when the excitation wavelength was 488 nm and the emission wavelength was 509 nm. The blank was the *E. coli* bacteria solution without GFP.

(2) 1 mL of bacteria solution was lysed by sonication. The broken bacteria were centrifuged at 12,000 × g, 4 °C for 20 min, and then the supernatant was collected. GA stock solution was added to the supernatant. The final concentrations of GA in the bacteria solution were 500 μg mL^{-1}, 250 μg mL^{-1}, 100 μg mL^{-1}, 50 μg mL^{-1}, 25 μg mL^{-1}, 10 μg mL^{-1} and 5 μg mL^{-1}, respectively. 10 μL of DMSO (without the drug) was used as a control. All of the tests were incubated at 30 °C for 1 min before GFP fluorescence was measured. The control was the crude enzyme solution of the corresponding bacteria without GFP.

Effect of GA on the activities of a variety of proteins

Effect of GA on the activity of Taq DNA polymerase

Stock solutions at different concentrations were prepared by dissolving GA in DMSO. GA stock solutions (0.5 μL) at different concentrations were then added into each PCR reaction system. The final concentrations of GA in the reaction system were 25 μg mL^{-1}, 10 μg mL^{-1}, 5 μg mL^{-1} and 1 μg mL^{-1}, respectively. 0.5 μL of DMSO (without the drug) was as a control. The PCR reaction conditions were adopted the reference of Chester and Marshak [36]. After the reaction was finished, 3 μL of PCR product was analyzed on an appropriate agarose gel. GelRed nucleic acid dye was used to stain the gel.

Effect of GA on the activity of restriction enzymes

Three common restriction enzymes (Kpn I, Hind III and EcoR I) were selected as the targets. GA was dissolved in DMSO to prepare stock solutions with different concentrations. A single colony of *E. coli* that carried the pUC19 plasmid, was extracted and inoculated in 3 mL of LB medium containing Amp, which was then cultured at 37 °C overnight with vigorous shaking. 41.5 μL of pUC19 (200 ng μg^{-1}), 5 μL of 10 × restriction enzyme buffer, 3 μL of corresponding restriction enzyme, and 0.5 μL of GA stock solution with different concentrations were mixed. The final concentrations of GA in the reaction system were 25 μg mL^{-1}, 10 μg mL^{-1}, 5 μg mL^{-1} and 1 μg mL^{-1}, respectively. 5 μL of DMSO (without the drug) was used as a control. All of the components in the reaction systems were digested at 37 °C for 4 h, and then the temperature was increased to 75 °C for 15 min to stop the enzyme digestion. Electrophoresis was then carried out to examine the enzyme digestion.

Effect of GA on the activity of superoxide dismutase

According to the method by Beauchamp et al. [37], GA was dissolved in DMSO to prepare stock solutions with different concentrations. 80 μmol L^{-1} riboflavin, 77 μmol L^{-1} nitro blue tetrazolium (NBT), 13 mmol L^{-1} methionine, 0.1 mmol L^{-1} EDTA, 20 μL superoxide dismutase (SOD) enzyme solution (1 μg mL^{-1}), and 5 μL of the different concentrations of GA stock solution (final concentrations of GA in the reaction system of 25 μg mL^{-1}, 10 μg mL^{-1}, 5 μg mL^{-1} and 1 μg mL^{-1}) were mixed. 5 μL of DMSO (without the drug) was used as a control. After the samples were exposed to light at 4500 Lux light intensity for 15 min, the reaction was stopped by shielding the light. The optical density (OD)$_{560}$ values were measured. One active unit (U) occurred when NBT was inhibited by 50%, and enzyme activity = $(\triangle A \times N \times 60)/(W \times T \times V \times 50\%)$, where $\triangle A$ represented the difference in OD values between the control and sample, N represented the total volume of enzyme solution, W represented the protein mass, T represented the light reaction time and V represented the volume of enzyme solution added. Enzyme activity was represented as OD$_{560}$ • mg^{-1} (pro) • min^{-1}.

Effect of GA on the activity of β-galactosidase

2 μg mL^{-1} the enzyme β-galactosidase was prepared in 10 mM pH 7.0 phosphate buffer. GA was dissolved in DMSO to prepare the stock solutions with different concentrations. Enzyme solution (1 mL) was incubated at 37 °C for 5 min, then 1 mL of phosphate buffer (pH 7.0) containing 20 mM ONPG preheated to 37 °C was added, finally 5 μL of the different concentrations of GA stock solution were added. The final concentrations of GA in the reaction system were 25 μg mL^{-1}, 10 μg mL^{-1}, 5 μg mL^{-1} and 1 μg mL^{-1}, respectively. 5 μL of DMSO (without the drug) was used as a control. All of the samples were incubated in a 37 °C for 10 min, and then 3 mL of 0.5 mol L^{-1} Na$_2$CO$_3$ was added to stop the reaction. The OD value at 420 nm was read. One enzyme unit was defined as the amount of enzyme required to release 1 μmol of ONP per minute at 37 °C.

Inhibition of isotopic precursor incorporation

According to the method by Aspedon and Groisman [21], [^{3}H] TdR, [^{3}H] UR and [^{3}H] Tyr were incorporated into precursors to investigate effect of GA (C15:1) on the biosynthesis of DNA, RNA and B. amyloliquefaciens SQR9 protein. The logarithmic growth phase SQR9 culture was diluted with sterile water. In a 96-well plate, 0.9 mL of bacterial suspension (OD$_{600}$ = 0.1) was added to each well. The isotope-labeled precursor (final concentration was 0.5 μCi mL^{-1}) and 5 μL of the different concentration of GA (C15: 1) were also added into the well. The final concentrations of GA in the reaction system were 25 μg mL^{-1}, 10 μg mL^{-1} and 5 μg mL^{-1}, respectively. 5 μL of DMSO (without the drug) was used as a control. All of the treatments were conducted at 37 °C in a shaker. When examining effect of GA (C15:1) on the synthesis of DNA and RNA, culture time for bacteria was limited to one generation. After growing the culture for 30 min, the culture was centrifuged at 12,000 × g, 4 °C for 5 min to harvest the bacteria pellet. Because protein synthesis was relatively slow, the bacteria culture time was extended to 2 h. OD$_{600}$ was also adjusted so that the cultures had the same bacteria concentration. The bacteria pellet was washed three times with phosphate buffer and then placed in an oven overnight to dry it. Scintillation solution was directly added to the Eppendorf tubes containing bacteria, and the tubes were then transferred to a scintillation counter (LS3801, Beckman) to determine the counts per minute (CPM) values. The average CPM values of the experimental group and control group were compared. The incorporation inhibition rate was calculated. Incorporation inhibition rate = control group CPM - experimental group CPM/control group CPM × 100%.

Effect of GA (C15:1) on GFP fluorescence in the protoplasts of E. coli

E. coli protoplasts were prepared according to the method described by Weiss [38]. GA was dissolved in DMSO, and its stock solutions with different concentrations were prepared. E. coli protoplasts (1 mL) were diluted with sucrose-magnesium-maleate (SMM) buffer (protoplast number > 10^7 CFU mL^{-1}), and 5 μL of the different concentration GA stock solutions were added. The final concentrations of GA in the reaction system were 500 μg mL^{-1}, 250 μg mL^{-1}, 100 μg mL^{-1}, 50 μg mL^{-1}, 25 μg mL^{-1}, 10 μg mL^{-1} and 5 μg mL^{-1}, respectively. 5 μL of DMSO (without the drug) was used as a control. All the tests were incubated at 30 °C for 1 min, and then the samples were analyzed for GFP fluorescence. GFP fluorescence was initially observed by the naked eye using an LB-16 Maestrogen UltraSlim no-damage blue LED transilluminator. GFP fluorescence was measured with a Spectra ax M5 multifunctional microplate reader. The samples' fluorescence intensity was measured when the excitation wavelength was 488 nm and the emission wavelength was 509 nm. The blank was E. coli protoplast solution without GFP.

The interception of GA by the lipid-soluble component in E. coli cell walls

E. coli DH5α-gfp was cultured in 100 mL of liquid LB medium at 37 °C, 200 rpm overnight. After centrifugation at 12,000 × g for 5 min, the supernatant was discarded to collect the bacteria. The bacteria pellet was then re-suspended in the same volume of phosphate buffer. The half was retained for the experiments and the left half was centrifuged at 12,000 × g for 5 min. The supernatant was then discarded. A small volume of phosphate buffer was used to re-suspend the bacteria pellet, and then 10 mL of 70% ethanol solution was added, mixed well, and kept at room temperature for 30 s before centrifuging at 12,000 × g for 1 min. The supernatant ethanol solution was then discarded. Phosphate buffer (50 mL) was used to re-suspend the bacteria pellet, which was centrifuged at 12,000 × g for 5 min. The supernatant was then discarded. Finally, 50 mL of phosphate buffer was used to re-suspend the bacteria pellet for the experiments.

GA was dissolved in DMSO, and its stock solutions with different concentrations were prepared. GA stock solutions (5 μL) at different concentrations were added to 1 mL of bacteria solution before/after ethanol solubilization of the lipids. The final concentrations of GA in the reaction system were 500 μg mL^{-1}, 250 μg mL^{-1}, 100 μg mL^{-1}, 50 μg mL^{-1}, 25 μg mL^{-1}, 10 μg mL^{-1} and 5 μg mL^{-1}, respectively. 5 μL of DMSO (without the drug) was used as a control. All the tests were incubated at 30 °C for 1 min, and then the samples were examined for GFP fluorescence. GFP fluorescence was initially observed by the naked eye using an LB-16 Maestrogen UltraSlim no-damage blue LED transilluminator. GFP fluorescence was measured using a Spectra ax M5 multifunctional microplate reader. The samples' fluorescence intensity was measured when the excitation wavelength was 488 nm and the emission wavelength was 509 nm. The blank was *E. coli* bacteria solution without GFP (after ethanol solubilization).

Statistical analyses

For all the experiments throughout the study, we collected three individual tubes from LB plates, and each tube was performed in triplicate. In analysis, comparisons were carried out using the fluorescent means of each triplicate, then, finally each bar in the gram represents means of three individual tubes. The data of each strain was analyzed using one-way ANOVA (i.e., *B. amyloliquefaciens* SQR9-gfp *E. coli* DH5α-gfp and *P. putida* KT2440-gfp), with final concentration of GA as the fixed factors ($P <$ 0.05). The experimental data were log transformed to meet the homogeneity of variance or a normal distribution of residuals. All statistical analyses were conducted using SPSS 13.0 (SPSS, Chicago, IL, USA).

Abbreviations

[³H] TdR: (Methy-³H) thymine; [³H] Tyr: ³H-tyrosine; [³H] UR: ³H-uridine; Amp: Ampicillin; CFU: Colony-Forming Units; CK: Control check; CPM: Counts per minute; DMSO: Dimethylsulfoxide; DNA: Deoxyribonucleic acid; EDTA: Ethylenediaminetetraacetic acid; GA: Ginkgo acid; GFP: Green fluorescent protein; Gm: Gentamicin; IPTG: Isopropyl-β-D-thiogalactopyranoside; Km: Kanamycin; LB: Lysogeny broth; LED: Light emitting diode; MBC: Minimum bactericidal concentration; MIC: Minimum inhibitory concentration; NBT: Nitro blue tetrazolium; OD: Optical density; ONP: O-nitrophenol; ONPG: O-nitrophenyl-β-D-galactopyranoside; PCR: Polymerase chain reaction; RNA: Ribonucleic acid; SMM: Sucrose-magnesium-maleate; SOD: Superoxide dismutase

Acknowledgements

It is a great pleasure to thank research fund for key projects in the national science & technology pillar program (2012BAD21B04) and the science and technology project of jiangsu province (BE2015315). The authors would like to acknowledge the support of the project funded by the priority academic program development (PAPD) of jiangsu higher education institutions and co-innovation entre for sustainable forestry in southern china. The authors also would like to thank Dr. Xie lulu of University of Rochester, Dr. Xiedong of Nanjing Forestry University for guidance in the data analysis.

Funding

This work was supported by research fund for key projects in the national science & technology pillar program (2012BAD21B04) and the science and technology project of jiangsu province (BE2015315).

Authors' contributions

HZB did the experimental, analyses of experimental data, figures drawing, critical reading of the manuscript and writing of the manuscript; WCE and CFL were responsible for study conception and design, analysis and interpretation of data work; FGJ contributed to study conception, to analysis of experimental data and to the writing of the manuscript; TZX analyzed the data, supervised the statistical analyses and contributed to the writing of the manuscript. All authors discussed and commented the results and gave their final approval for submission.

Competing interests

The authors declare that they have no competing interests.

Author details

[1]Co-Innovation Centre for Sustainable Forestry in Southern China, Nanjing Forestry University, Nanjing 210037, China. [2]College of Light Industry Science and Engineering, Nanjing Forestry University, Nanjing 210037, China. [3]College of Forestry, Nanjing Forestry University, Nanjing 210037, China.

References

1. Dixon RA, Strack D. Phytochemistry meets genome analysis, and beyond. Phytochemistry. 2003;62(6):815–6.
2. Toni M, Kutchan. Ecological arsenal and developmental dispatcher. The paradigm of secondary metabolism. Plant Physiol. 2001;125(1):58–60.
3. Cowan MM. Plant products as antimicrobial agents. Clin Microbiol Rev. 1999;12(4):564–82.
4. Li JWH, Vederas JC. Drug discovery and natural products: end of an era or an endless frontier? Science. 2009;325(5937):161–5.
5. Zhao J, Davis LC, Verpoorte R. Elicitor signal transduction leading to production of plant secondary metabolites. Biotechnol Adv. 2005;23(4):283–333.
6. Wilkins KM, Board RG, Gould GW: Mechanisms of action of food preservation procedures. London: Elservierapplied Science EA; 1989.
7. Wilson CL, Solar JM, El-Ghaouth A, Wisniewski ME. Rapid evaluation of plant extracts and essential oils for antifungal activity against *Botrytis cinerea*. Plant Dis. 1997;81(2):204–10.
8. Erin AN, Davitashvili NG, Prilipko LL, Boldyrev AA, Lushchak VI, Batrakov SG, Pridachina NN, Serbinova AE, Kagan VE. Influence of alkylresorcin on biological membranes during activation of lipid peroxidation. Biokhimiia. 1987;52:1180–5.
9. Kaprelyants AS, Suleimenov MK, Sorokina AD, Deborin GA, El-Registan GI, Stoyanovich FM, Lille YE, Ostrovsky DN. Structural functional changes in bacterial and model membranes induced by phenolic lipids. Biol Membr (Moscow). 1987;4:254–61.
10. Kubo I, Komatsu S, Ochi M. Molluscicides from the cashew *Anacardium occidentale* and their large-scale isolation. J Agr Food Chem. 1986;34(6):970–3.
11. Muroi H, Kubo I. Antibacterial activity of anacardic acid and totarol, alone and in combination with methicillin, against methicillin-resistant *Staphylococcus aureus*. J Appl Microbiol. 1996;80(4):387–94.
12. Himejima M, Kubo I. Antibacterial Agents from the Cashew *Anacardium occidentale* (Anacardiaceae) Nut Shell Oil. J Agric Food Chem. 1991;39:418–21.
13. Boonkaew T, Camper ND. Biological activities of Ginkgo extracts. Phytomedicine. 2005;12(4):318–23.
14. Chen JJ, Zhang T, Jiang B, Mu WM, Miao M. Characterization and antioxidant activity of *Ginkgo biloba* exocarp polysaccharides. Carbohydr Polym. 2012;87(1):40–5.
15. van Beek TA, Montoro P. Chemical analysis and quality control of *Ginkgo biloba* leaves, extracts, and phytopharmaceuticals. J Chromatogr A. 2009;1216(11):2002–32.
16. Ahlemeyer B, Krieglstein J. Neuroprotective effects of *Ginkgo biloba* extract. Cell Mol Life Sci. 2003;60(9):1779–92.
17. Satyan KS, Jaiswal AK, Ghosal S, Bhattacharya SK. Anxiolytic activity of ginkgolic acid conjugates from Indian *Ginkgo biloba*. Psychopharmacology (Berl). 1998;136(2):148–52.
18. Yang XM, Ye YR, Wang P, Chen J, Guo T. Study on anti-bacterium activities of extract of *Ginkgo biloba* leaves (EGbs) and Ginkgolic Acids (GAs). Food Sci. 2004;25(4):68–71.
19. Itokawa H, Totsuka N, Nakahara K, Takeya K, Lepoittevin JP, Asakawa Y. Antitumor principles from Ginkgo biloba L, vol. 35. Tokyo: JAACC; 1987.

20. Ni XW, Wu MC. Study on isolation identification and the antibacterial activity of ginkgolic acids. Food Sci. 2004;25(9):59–63.

21. Aspedon A, Groisman EA. The antibacterial action of protamine: evidence for disruption of cytoplasmic membrane energization in *Salmonella typhimurium*. Microbiology. 1996;142(pt12):3389–97.

22. Begum P, Hashidoko Y, Islam MT, Ogawa Y, Tahara S. Zoosporicidal activities of anacardic acids against *Aphanomyces cochlioides*. Z Naturforsch C: Biosci. 2002;57(9–10):874–82.

23. Choi JG, et al. Antibacterial activity of hydroxyalkenyl salicylic acids from sarcotesta of *Ginkgo biloba* against vancomycin-resistant Enterococcus. Fitoterapia. 2009;80:18–20.

24. He J, Wang S, Wu T, Cao Y, Xu X, Zhou X. Effects of ginkgoneolic acid on the growth, acidogenicity, adherence, and biofilm of Streptococcus mutans in vitro. Folia Microbiol (Praha). 2013;58:147–53.

25. Jefferson RA, Kavanagh TA, Bevan MW. GUS fusions: beta-glucuronidase as a sensitive and versatile gene fusion marker in higher plants. EMBO J. 1987;20:3901–7.

26. Wang S, Hazelrigg T. Implications for bcd mRNA localization from spatial distribution of exu protein in *Drosophila oogenesis*. Nature. 1994;369:400–3.

27. Ahlemeyer B, Selke D, Schaper C, Klumpp S, Krieglstein J. Ginkgolic acids induce neuronal death and activate protein phosphatase type-2C. Eur J Pharmacol. 2001;430(1):1–7.

28. Balasubramanyam K, Swaminathan V, Ranganathan A, Kundu TK. Small molecule modulators of histone acetyltransferase p300. J Biol Chem. 2003; 278(19):134–40.

29. Grazzini R, Hesk D, Heininger E, Hildenbrandt G, Reddy CC, Cox-Foster D, Medford J, Craig R, Mumma RO. Inhibition of lipoxygenase and prostaglandin endoperoxide synthase by anacardic acids. Biochem Bioph Res Co. 1991;176(2):775–80.

30. Fukuda I, Ito A, Hirai G, Nishimura S, Kawasaki H, Saitoh H, Kimura K, Sodeoka M, Yoshida M. Ginkgolic acid inhibits protein SUMOylation by blocking formation of the E1-SUMO intermediate. Chem Biol. 2009;16(2):133–40.

31. Lee JH, Kim YG, Ryu SY, Cho MH, Lee J. Ginkgolic acids and Ginkgo biloba extract inhibit *Escherichia coli* O157:H7 and Staphylococcus aureus biofilm formation. Int J Food Microbiol. 2014;174:47–55.

32. Seto M, Kimbara K, Shimura M, Hatta T, Fukuda M, Yano K. A novel transformation of polychlorinated biphenyls by Rhodococcus sp. strain RHA1. Appl Environ Microbiol. 1995;61:3353–8.

33. Sun Z, et al. Identification and characterization of the dominant lactic acid bacteria from kurut: the naturally fermented yak milk in Qinghai, China. J Gen Appl Microbiol. 2010;56:1–10.

34. Rosado A, Duarte GF, Seldin L. Optimization of electroporation procedure to transform B. polymyxa SCE2 and other nitrogen-fixing Bacillus. J Microbiol Methods. 1994; 19(1–11). doi: 10.1016/0167-7012(94)90020-5.

35. Kubo I, Muroi H, Kubo A. Structural functions of antimicrobial long-chain alcohols and phenols. Bioorgan Med Chem. 1995;3(7):873–80.

36. Chester N, Marshak DR. Dimethyl sulfoxide-mediated primer T_m reduction: a method for analyzing the role of renaturation temperature in the polymerase chain reaction. Anal Biochem. 1993;209(2):284–90.

37. Beauchamp C, Fridovich I. Superoxide dismutase: improved assays and an assay applicable to acrylamide gels. Anal Biochem. 1971;44(1):276–87.

38. Weiss RL. Protoplast formation in *Escherichia coli*. J Bacteriol. 1976;128(2):668–70.

9

The role of community engagement in the adoption of new agricultural biotechnologies by farmers: the case of the *Africa harvest* tissue-culture banana in Kenya

Sunita V. S. Bandewar[1*], Florence Wambugu[2], Emma Richardson[3,4] and James V. Lavery[3,5*]

Abstract

Background: The tissue culture banana (TCB) is a biotechnological agricultural innovation that has been adopted widely in commercial banana production. In 2003, Africa Harvest Biotech Foundation International (AH) initiated a TCB program that was explicitly developed for smallholder farmers in Kenya to help them adopt the TCB as a scalable agricultural business opportunity. At the heart of the challenge of encouraging more widespread adoption of the TCB is the question: what is the best way to introduce the TCB technology, and all its attendant practices and opportunities, to smallholder farmers. In essence, a challenge of community or stakeholder engagement (CE).

Results: In this paper, we report the results of a case study of the CE strategies employed by AH to introduce TCB agricultural practices to small-hold farmers in Kenya, and their impact on the uptake of the TCB, and on the nature of the relationship between AH and the relevant community of farmers and other stakeholders. We identified six specific features of CE in the AH TCB project that were critical to its effectiveness: (1) adopting an empirical, "evidence-based" approach; (2) building on existing social networks; (3) facilitating farmer-to-farmer engagement; (4) focusing engagement on farmer groups; (5) strengthening relationships of trust through collaborative experiential learning; and (6) helping farmers to "learn the marketing game". We discuss the implications of AH's "values-based" approach to engagement, and how these guiding values functioned as "design constraints" for the key features of their CE strategy. And we highlight the importance of attention to the human dimensions of complex partnerships as a key determinant of successful CE.

Conclusion: Our findings suggest new ways of conceptualizing the relationship between CE and the design and delivery of new technologies for global health and global development.

Keywords: Community engagement, Commercialization, Biotechnology, Agricultural biotechnology, Tissue culture, Bananas, Africa, Global health, Global development, Stakeholder engagement

* Correspondence: sunita.bandewar@utoronto.ca; james.v.lavery@emory.edu
[1]McLaughlin-Rotman Centre for Global Health, Toronto, Canada
[3]Centre for Ethical, Social & Cultural Risk, Li Ka Shing Knowledge Institute of St. Michael's Hospital, Toronto, Canada
Full list of author information is available at the end of the article

Background

"TCB Orchard is my second husband. It fetches me money as and when I want! ...I prefer TCBs because they don't have disease, they grow faster, they give me money quicker than the traditional bananas. So I decided to plant more and more.".[Farmer (18)].

The tissue culture banana (TCB) is a biotechnological agricultural innovation that has been adopted widely in commercial banana production. Tissue culture (TC) is based on the ability of many plant species to regenerate a whole plant from a single shoot tip. A single shoot tip is dissected and the pieces are placed in a sterile growth medium. Various hormones are added at different stages to promote shoot initiation, multiple shoot formation, and rooting induction. Within 6 months, up to 2,000 individual plantlets can be produced from a single shoot. TC technology enables rapid and large-scale proliferation of bananas. TC plants grow faster than plants grown from nursery "suckers", are free from pests and diseases, are uniform, produce fruits earlier, and the second generation crop also matures earlier. TC bananas are ready for harvest340 days after planting, compared to 420 days for plants grown from nursery "suckers" [1].

The TCB was introduced to Kenyan farmers initially in 1997 through a partnership between the Kenyan Agricultural Research Institute (KARI) (now known as Kenya Agriculture and Livestock Research Organization) and the International Service for the Acquisition of Agri-Biotech Applications (ISAAA). The first two phases of TCB introduction were carried out between 1997–2003, during which the feasibility and appropriateness of the technology were tested and systems of production and distribution of the TCB were piloted [2]. Commercial adoption presented a challenge in Kenya because bananas have traditionally been grown by the smallholder farmers for home and local consumption [3, 4]. In 2003, Africa Harvest Biotech Foundation International (AH), a Nairobi-based non-governmental organization (NGO) that applies agricultural biotechnologies and management practices to improve the viability of small-holder agriculture [4], initiated a TCB program that was explicitly developed for smallholder farmers and adopted a 'whole value chain' approach that aimed to help them adopt the TCB as a scalable agricultural business opportunity [2].

Adopting TCB has been shown to increase farm and household income and reduce relative food insecurity in Kenya [2, 5]. Despite the advantages of TCB, and the importance of banana-growing in general in East Africa, TCB adoption has remained relatively low and uneven in Kenya, amounting to only 7% of total banana acreage by 2012 [3, 6]. Most TCB production takes place in 12 counties in the central and eastern provinces of Kenya, where dissemination programs, such as AH, have been concentrated [3]. One reason for the modest adoption rates of the TCB is the special agricultural management practices and higher input intensities that TCB cultivation requires. AH's 'whole value chain' approach resulted from a recognition of these challenges, along with the demonstrated advantages of improved yields over a shorter growing period, a lower incidence of diseases, and greater uniformity of banana bunches, which translates into better marketability [3, 6]. At the heart of the challenge of encouraging more widespread adoption of the TCB is the question: what is the best way to introduce the TCB technology, and all its attendant practices and opportunities, to smallholder farmers. In essence, a challenge of community or stakeholder engagement (CE).

In this paper, we report the results of a case study of the CE strategies employed by AH to introduce TCB agricultural practices to small-hold farmers in Kenya, and their impact on the uptake of the TCB, and on the nature of the relationship between AH and the relevant community of farmers and other stakeholders.

CE has been recognized as an important factor in the successful introduction and adoption of new technologies [7]. But despite a growing appreciation of its importance, what makes CE effective remains poorly understood. A recent UK Parliamentary Report on lessons from the Ebola outbreak of 2015 in West Africa [8] emphasizes the need for substantial improvement to our ability to evaluate the impact of CE in global health and global development to improve effective employment of these practices.

We studied the TCB case as part of a research program on CE [9] undertaken by the Ethical, Social and Cultural (ESC) Program for the Bill and Melinda Gates Foundation's Grand Challenges in Global Health (GCGH) initiative [10], which aimed to gain a better understanding of what makes CE effective in the context of global health and global development research. This case study was one of several that examined a CE strategy that was already considered to be successful in an attempt to describe and explain the factors that contributed to its success.

Methods

We conducted a retrospective qualitative case study informed by grounded theory, an approach we have used successfully in other CE case studies [9, 11, 12]. The case study was conducted in collaboration with Africa Harvest (AH) in its TCB intervention areas in Murang'a County, Kenya. We made an exploratory visit to the TCB project in October 2007 to determine the scope of the case study, followed by an 8-weeks data collection visit in May-June, 2008.

We used a retrospective qualitative case study approach [13] to understand the history of the TCB Project over the previous 10 years and the role of CE practices. We conducted 16 in-depth semi-structured interviews with 17 individuals: TCB smallholder farmers ($n = 7$), AH staff ($n = 2$), field trainers/facilitators ($n = 2$); a plant biotechnology scientist from a Kenyan public agriculture research institute ($n = 1$); an agricultural NGO staff member ($n = 1$); a staff member from a private agricultural nursery ($n = 1$); a senior business consultant ($n = 1$); a consultant in community partnerships and group dynamics ($n = 1$); and a TCB extension officer from a private, farmer-owned company ($n = 1$). These interviews were complemented by non-participant observation by SB of three field training school activities at the AH TCB project intervention areas in the Thika Murang'a and Nairobi districts. Subsequently, SB conducted focus group discussions with three farmer groups at three different sites within these districts—Kairi, Muirigo and Ngorongo— on various aspects of the TCB project, with a central focus on CE activities. (Table 1) We also studied AH documents relating to TCB interventions: periodic progress reports, study reports of its formative research, training materials, and press releases and media coverage.

The goal of sampling in qualitative studies is not to construct a sample that mirrors major demographic features of the target population, but rather to identify key informants with unique experiences and personal knowledge of the phenomenon in question who can provide useful descriptions, insights and explanations of events relevant to the research questions. Prospective interviewees were identified initially by AH staff members and complemented by sequential referral sampling. We approached the prospective participants at their farms or respective work places with prior appointments which were facilitated by the AH staff.

The interview guide and the consent forms were translated into Kiswahili, although many farmers in Kenya understand and speak English. The interviews were simultaneously translated by facilitators, allowing participants to switch between Kiswahili and English. Facilitators were the field trainers associated with AH, belonged to the farmers' community, and were involved in the TCB project. Farmers' interviews were conducted at their own farms. Interviews with other interviewees were conducted in English at their own work places. Interviews lasted between 30 and 60 min. In one instance, two farmers chose to be interviewed together. Interviews were recorded, translated and transcribed verbatim.

The analytic approach combined techniques of grounded theory and qualitative description [14–16]. Two main rationales informed our choice of method: First, grounded theory emphasizes the experiences of participants, the meaning of these experiences to participants and their understanding of events [17], as opposed to seeking confirmation of investigators' hypotheses. Second, the grounded theory method aims to generate a theory of the phenomenon in question. The goal in this case was to produce an explanatory account that combines rich description of CE in the AH process with explanations of how various actions and structures lead to specific outcomes of interest [18].

In keeping with qualitative methodology, data collection and analysis were done in parallel [14, 19]. Preliminary data analysis was conducted by SB during the data collection phase in Nairobi to integrate initial insights into subsequent interviews. Later, upon the completion of the field work, SB conducted initial line-by-line coding of interview transcripts using the qualitative data software ATLAS.ti version 5.2., to identify key concepts, categories, and patterns. A constant comparative approach was used to compare findings within and across interviews and between categories [14, 18]. This was followed by focused coding and creation of analytical memos, as described by Charmaz [14]. This process produced a set of thematic and conceptual categories. We developed network views in ATLAS.Ti to explore relationships among the emerging concepts. Techniques for ensuring analytic rigor and trustworthiness included discussing coding between analysts (SB and JL), seeking alternative explanations for the data, and interrogating the coherence of interpretations through deliberations among the analysts [16]. We drafted a preliminary manuscript

Table 1 Characteristics of focus group participants

Group name (by site)	Number of group members who participated in the focus group discussion	Gender		Age	Land ownership	No of TCB plants owned
		Women	Men			
Kairi	8	5	3	33–75	1–4 acres	10–20
Muirigo	17	5	12	39–80 years	½–5 acres with most of them with 1 acre	5–30
Ngorongo	21	3	18	30–74 years	½–10 acres with most owning 1–2 acres	3–70
Total	46	13	33			

as a "best fit" for the data and refined it through several subsequent iterations.

The case study research protocol was approved by the respective Research Ethics Boards at the University of Toronto, and the Ministry of Agriculture, Nairobi, Kenya.

Results

The context: Introducing the TCB in the aftermath of 'breaches of trust'

During 1970s and 1980s tea and coffee cash crops offered small-hold farmers (smallholders) in Kenya a reliable, if modest, living. Farming cooperatives offered a buffer against economic hardship, particularly for poorer farmers, and played a powerful role in shaping farming communities in the country.

> "...back in the 70s and 80s...we took pride in ourselves on having been taken to school by coffee. Because, what a farmer would do if he didn't have school fees, coffee would give him a loan to pay school fees for his children. If he didn't have food, the coffee societies would buy maize...get food on their table... all to be deducted when farmers received their proceeds.". [Agricultural NGO staff member (Interview # 8)].

In the early 1990s, a crash in coffee and tea prices caught smallholders off guard [20]. The economic crisis had immediate implications for the normal functioning of the tea and coffee cooperatives, as falling revenues increasingly pitted stakeholders' interests against one another.

> "There were issues of corruption...the coffee board was not operational...Farmers suffered the most since they relied so much on the coffee [plantation].".[Agricultural NGO staff member (8)]

Inevitably, various schemes were proposed by dubious NGOs, promising farmers better markets and prices for their produce. Farmers recounted several stories of how such schemes for french beans, pineapples, avocado, maize, vanilla, agricultural water supply and collection of chameleons for export for medicinal purposes resulted in disappointment and betrayals of the farmers' trust.

> "It was bad because I remember we had planted about 2 acres of French beans...we were to harvest at least 60 cartons per day. They would come and probably [take only] 10 or 5 cartons. So we used to feed the rest to the animals...So the idea was just to sell their crops and then they go. So we have that experience, a bitter one, yes.". [Farmers, joint interview (13)].

Struggling for subsistence, and with a reinforced fear of new initiatives and a deep distrust of 'outsiders', the farmers increasingly turned to bananas, known locally as a 'back yard crop', a 'subsistence crop' and a 'woman's crop'. Some varieties of banana, such as the imported Cavendish, proved to be viable in Kenya and offered some economic opportunity for small landholders. But infestation by pests and diseases substantially limited yields and jeopardized the viability of the banana as a cash crop [2, 21].

Community engagement towards rebuilding 'relationships of trust'

We identified six specific features of CE in the AH TCB project which were critical to its effectiveness.

1. Determining the local relevance of an "imported" innovation: An evidence based approach

   As part of the foundation for its CE strategy, AH chose to carry out empirical research—surveys, rapid appraisals, and qualitative studies—with farmers to gain an understanding of their perspectives and share the findings with farmers and other key constituents, such as the MoA. This approach was thought to be necessary because of the unfamiliarity of smallhold farmers with TCB. When TCB was introduced to Kenya by AH in 1997 its suitability to Kenyan climatic and soil conditions had not been tested and some preliminary field trials were conducted at Jomo Kenyatta University of Agriculture and Technology to address these gaps in knowledge. Perhaps more importantly, its acceptability to Kenyan farmers and consumers was not well established. In fact, at the outset of the program, farmers viewed the TCB as "alien" and of limited relevance as a cash-crop option.

> "Initially when we first took the plant to the field, the other farmers, and the neighbours were laughing at them and telling them what they were planting! They had never seen anything like that because a TCB plantlet looks like a small flower with just a few leaves.". [Plant biotechnology scientist (7)]

   Understanding these initial attitudes through their empirical research helped AH to shape the content of the learning opportunities it was designing for farmers. And the discipline of seeking the perspectives of farmers also signalled AH's desire for a respectful relationship with farmers.

> "...if your decisions are top-bottom a lot of time they either don't work or nobody really trusts them. But most of our ideas are actually bottom-top whereby we start with the baseline. We go to the community and find out what they want...So by the time we are

writing our proposal and even writing our operational strategy, we already have an idea of what people want. I think the MoA (Ministry of Agriculture) likes that, they really appreciate it.". [AH staff (11)].

AH's empirical research also helped the AH team to identify community based resources, which later proved to be valuable social capital during CE: "We also wanted to find out or establish the available resources on the ground that we can work with...to identify the ground networks like community-based organizations (CBOs), field based organizations that we can work with.". [AH staff (A)]. AH's emphasis on generating relevant evidence also proved convincing for other key players such as the Ministry of Agriculture (MoA), which had important interests in understanding the commercial potential of the TCB.

THE AH invested in a pilot project to assess the promise of the TCB innovation in real world context which proved critical to the TCB initiative.

So the project went on very well and actually we proved a concept that for sure the TCB technology is workable. Farmers can adopt TCB which are actually profitable...And we found that for sure, for every one dollar you invest in the TCB, it fetches about 3 or 4 dollars back. So we were able to prove the concept...[AH staff (12)].

2. Introducing the TCB innovation: Building on existing social networks

AH used several channels to introduce the TCB innovation, capitalizing first on those most trusted by farmers: "Because most of the time when you are approaching a community it's usually easy when you use the existing networks than going to start afresh...Because people trust their own...[AH staff (12)]. For example, given the high levels of religious affiliation within the farming communities, AH engaged religious leaders and churches as one of their first points of contact with farmers.

"First of all when AH came and talked to members at church some of them decided that they are going to plant the bananas. So they were picked and their names were known. Then AH requested them to go on spreading the gospel about planting the bananas.". [Farmer (4)].

AH leveraged the support of the Ministry of Agriculture (MoA) by sharing endorsement letters from the MoA with stakeholders during their community entry activities: "And we have the letters from the agricultural offices because that is

our first point. You go through the agricultural offices to make it official." [AH staff (B)]. This strategy helped to authenticate and legitimize AH's messages and helped AH establish relationships with key stakeholders, such as local agricultural non-government organisations (NGOs), and various crop or livestock-based financing or management cooperatives. During our field work, we observed that farmers generally interacted comfortably and constructively with MoA staff, likely due to the positive legacy of MoA extension staff engaging with farmers throughout the country. These extension staff were also important stakeholders for AH as a result of their deep knowledge of the farming communities.

Not all relationships between farmers and stakeholders were equally trusting. Particularly for the identification of implementation sites, AH devoted considerable time and energy to assessing levels of trust with prospective partners, and prioritized those communities where productive partnerships could be built upon a relatively trusting base.

3. Facilitating farmer-to-farmer engagement

The TCB plantlet is vastly different from traditional bananas in its appearance. Unfamiliar appearance has been an impediment to the successful uptake of other agricultural innovations by farmers and consumers, such as with the introduction of Golden Rice in South East Asia [22] and this was true also for the TCB. But farmers' scepticism ran deeper than the appearance of the plantlet. They were concerned, in particular, about the need for trained personnel to carefully grow TCB plantlets in nurseries, and several other unfamiliar agricultural practices that made the TCB's commercial potential more difficult to understand.

"...here is a technology, those are scientists and they want to reach the community and that is beyond science. So we were dealing with all the social issues... if you want to adopt a technology, it is not just about the banana, there are perceptions and attitudes... So the first thing we had to tackle, of course, was the perceptions [of farmers and communities about TCB].". [Consultant in community partnerships and group dynamics (6)]

AH's claims about the viability of TCB required validation for farmers to even consider adopting the crop. AH adopted the practice of exposure visits or travelling workshops, to provide farmers with opportunities to experience TCB farming practices first-hand. 'Exposure visits' typically involve farmers visiting demonstration or pilot agricultural sites developed by private companies, NGOs, or government agricultural research institutes. Farmers take part in demonstrations and have an opportunity to ask questions of the organizers and peer farmers who have been early adopters of the new practices. These strategies were used by the Kenyan Ministry of Agriculture (MoA) and government institutes such as Kenya Agricultural Research Institute (KARI), to introduce new practices to Kenyan farmers.

"So it was easier because a few farmers had walked through the whole cycle. So when it was the time for mass adoption, we gravitated around pilot sites. So that became easier because there was something they could see...just creating a desire and in the process their attitude would change. They would say 'oh, I will try'...Exposure trips worked wonders...So, in fact, that was the most effective. So they were saying this is someone we know, it is so and so's father but look, his is also doing well...".[Consultant in community partnerships and group dynamics (6)]

"So we were convinced that if those farmers are able, just the way I am talking now, I am talking from experience. They talked from experience. And having been farmers all the way through when somebody talks we could actually understand.". [Farmers, joint interview (13)]
Exposure visits made the potential benefits of the TCB innovation evident to interested farmers and offered them engagement with other farmers who had already adopted the TCB, sometimes from within their own social networks, offering an opportunity to assess the merits of AH's claims. One of the farmers' genuine concerns was about viable markets for their bananas, a concern rooted in past bitter disappointments. An AH staff member recounted the farmers' scepticism:

"...you have been sent from Nairobi to come and tell us to plant bananas and then after we plant, you go back to Nairobi. Where are we going to take the produce?" But we [AH team] kept on assuring them, we are with you here 3 days in a week and we will be listening to your concerns and we are going to try and see how you can work together.". [AH staff (11)]

One of the persistent challenges for AH was to convince farmers that bananas that had been growing effortlessly in their backyards for generations would now require more intensive and technical agronomics if they adopted TCBs. The travelling workshops and legitimate business partnerships were instrumental in guiding farmers through this transition, in stark contrast to the disreputable operators from the past.

4. Focusing engagement on farmer groups
From the mid 1980s, contract farming evolved in Kenya to include more cooperative "schemes" whereby farmers would group together and establish their own terms of engagement with contracting companies to maintain better control of their stakes in the process [23]. These farmer groups also became eligible for various government subsidies and loan facilities, and registered groups were visited by the agriculture extension officers from the Ministry of Agriculture, who shared new information about a wide range of agricultural practices, at no cost to the individual farmers. AH was well aware of the advantages of farmer groups and adopted them as an explicit focus of their engagement strategy, encouraging independent farmers to consider the merits of joining with other farmers to pool risk and to better prepare for opportunities that they would be introduced to in the 'whole value chain' model [4].

"Once these farmers are aware about the TC technology, about commercial banana farming, then we move on and talk about forming groups or use the existing ones which might be doing another crop, such as, avocado or maize and looking for another enterprise.". [Field trainer/facilitator (16)]

Professionals in the field of group dynamics and psychology from the public and private sectors, along with experts in TCB agronomy, trained the AH trainers who then engaged with farmer groups.

"Mainly the lecturers were from outside...these trainings were very exhaustive. They handled the TCB innovation from a broad perspective...The group dynamics that we were trained in was broad...We were taught about the various methods of approaching a group. It was all broad, not specific... It was very helpful...[to learn] about group dynamics and the way groups behave.". [Field trainer/facilitator (2)]

A formal approach to group formation encouraged group members to develop their own "group constitutions" emphasizing transparency, equity in opportunities, and participatory decision making. These constitutions established the grounding principles for group governance. These included, among others: setting eligibility criteria for nominations for specific positions of responsibility within the group, such as chairperson, secretary, treasurer and banana grader, and procedures for matters related to finance and sales of the TCB produce.

"It is not one person who makes decisions...We discuss the issues at hand and voice our opinions. We discuss the issues thoroughly and see what we stand to gain. ... In this group we have some rules. For example, as a member of the group you cannot sell your bananas to people out there without the group deciding...You cannot as an individual farmer decide to sell yours to someone on your own. If you do so, you are asked to leave the group.". [Ngorongo FGD] (Table 1)

These group training activities emphasized the rationale for collective action. "You must first have a reason for forming the group. Also everyone joining the group must be interested in that reason for which you are forming the group. You must also have a vision for the future of the group for it to succeed." [Ngorongo FGD] (Table 1). They also focused on issues of leadership within the group. Group leaders were elected by group members, which fostered a sense of legitimacy within the group, especially since they knew that they also had the power to remove leaders who were not living up to the group's expectations. The growing collective sense of the group's power also helped in the transition to a real sense of ownership of the TCB technology, in both a symbolic and economic sense.

"Yes, because unless they take it as their activity, their project...we cannot say it is community development, unless they own it as their project, their bananas...So they say with or without a market, we are going to do banana farming and we are going to do it as a commercial venture. So those things I think are very important. The ownership of the project also means a lot to them.". [Field trainer/facilitator (16)]

5. Strengthening relationships of trust through collaborative experiential learning
    Adoption of appropriate agronomy practices was crucial for the realisation of TCB's commercial potential. Without sufficient TCB yields, AH would have run the risk of being perceived as simply another outside organization that failed farmers.

"Once farmers planted, and when we began talking about good management and adopting a technology, that was difficult...If it was only the technology and everything else was the same it would have been easier. But now one is bringing a new technology and one is also changing the agronomy. I think that change was a lot!". [Consultant in community work and group dynamics (6)]

TCB farmer groups were trained in TCB-specific agronomic practices at demonstration plots, which they developed with the kinds of training described above. These plots served as 'field schools' for group members.

"After forming the group, the idea [of demonstration plots] was brought by AH so that we can have a central place where we can all be coming to be taught as to how to plant...That is how we started the field school or demonstration land...Drawing upon these learning at the field school, we started doing it on our own farms where we did exactly what they had taught us...".[Farmer (4)]

"The AH has done an unusual thing here in the area. They have trained us on good farming practices. And they don't ask us to pay them, they pay the teacher [Trainer from the community (16)] to come and teach us. All they need is our presence to attend the training sessions. We like this aspect of them training us." [Ngorongo FGD]
"After a lot of training they told us to try and plant the TCB...So we saw the difference between the traditional and the TC bananas.". [Farmer (P15)]
    The exposure visits and demonstration plots required the AH team to be in the fields with farmers in their own environment and communities. Farmers found the sustained presence of AH in their communities unusual and reassuring.

"Because I thought for myself that they can't work all that hard to come and visit the farmers while they are cheating...".[Farmer (9)]
"What made us believe is that we thought that these people cannot come all the way here to just cheat us. Because we were all grown-ups who could think. Like me personally I don't believe someone can come from wherever to just tell you to plant something bad. So that is why I believed them.". [Farmer (15)]
6. Learning the marketing game
    "The problems we were having before were relating to marketing. Our marketing was not that good... No customers, the bananas are ripening in the *shamba* [orchard] and trucks are not coming to collect them. There was that difficulty but nowadays not much.". [Farmer (3)]
    The traditional bananas, particularly those of smallholders, rarely made it to the local retail market. It was imperative that farmers develop the required skills and understanding to have better

control over the commercial opportunities offered by TCB. Failure to enact improvements in supply chain and marketing practices would have caused enormous damage to farmers' trust: "Because you see they would start doubting, if you are telling us to grow bananas and if they had seen bananas rotting somewhere or bananas were being bought at 50 shillings." [AH staff (11)]

Because of its limited shelf-life, significant management skills are required to maintain banana quality until delivery to market. Farmers had little marketing knowledge, access to non-local markets or buffers to protect against price variations, making them vulnerable to exploitative middle men. Furthermore, public sector supports are scarce, as the banana market is largely unorganized with no government policies: "... the government actually had a very streamlined policy for marketing tea and coffee which are considered cash crops but banana has always been considered a subsistence crop which did not have this advantage." [AH staff (12)]

The primary goals for the commercialisation of the TCB were: (1) to develop a pool of customers with sustained interest in bulk produce; (2) consistent and timely supply of quality and quantity TCB produce to win bulk buyers' confidence; and, consequently, (3) the establishment, over time, of a more stable demand–supply chain. For smallholders with limited financial resources, maintaining both quantity and quality of production was a formidable challenge, with many critical factors beyond their control, such as their complete reliance on rainfall for adequate irrigation.

"Because with supermarkets, once you go into a contract with them, if they say we are going to supplying you a ton per week, it is a ton per week for 52 weeks a year! I have told you our farmers are using rain fed agriculture and under this [condition] farmers cannot achieve this [quality and quantity]...And if you are going into a contract [with bulk buyers] you must ensure that you are going to meet your part of the commitment." [Agricultural NGO staff member (8)]

To respond to these challenges, AH integrated marketing training for farmers in field schools, collaborating with a private company, TechnoServe, and with Tissue Culture Banana Enterprise Ltd. (TCBEL), a farmer-led marketing company whose establishment was facilitated by AH.

"And from there I have been working as an extension officer training farmers...the post-harvest handling and organizing the market for them...So that is why I

was saying that they need some retraining on the marketing...They never knew the optimum number they needed to plant and how much they can get from it... I also train them on the concept of collection centers, that is, TCB farmers have to take their produce to collection centers instead of someone to go around to individual farmers for collection...We also want to train them on financial management...".[Field trainer/facilitator (2)]

"TCBEL really played a very big role. So they [farmers] can now see where to sell their bananas... the market opening is a very big and very significant thing for us.". [AH staff (12)]

"Probably we can replicate TCBEL as a success story in other areas.". [Agricultural NGO staff member (8)]

AH also expanded the exposure visits to include marketing partners

"Well, this time we had no problem because they [TechnoServe and TCBEL] came together with the AH in whom we had confidence. These two entities told us that they would take over the marketing side. They also took the trouble to take us around to other farmers where they were actually active, they were helping farmers in marketing...So we knew how they were sold. So it was something which was well arranged, I have never seen.". [Farmers, joint interview (13)]

TCBEL was established as a member based organisation with affordable membership fees for farmers in the neighbourhood, but has evolved to a shareholder based company run by the farmers [24]. Given the corruption-fraught history of tea and coffee cooperatives, TCBEL has set itself apart in many concrete ways, including a commitment to fair and transparent practices, such as purchasing by weight and not by bunch count, systematic paper work, periodic payment to farmers, and the complete elimination of middlemen. It has also trained staff in a wide range of skills to ensure the consistent quality of the TCB produce.

AH's systematic approach to TCB marketing and commercialisation has shown results. But, satisfaction with the TCB's commercial development is uneven amongst farmers. Also, scepticism about availability of sustained markets for TCB continues.

"But with horticulture or dairy the first market creates demand. There will be challenges. The farmers will not know how to meet the requirements of the buyers...Those problems are always there. But if there is continuity they will be handled and....So it works as long as you build trust and the moment you build trust the demand–supply chain is maintained between traders and farmers.". [Senior Business Consultant (5)].

Limitations

Our study has two specific limitations. First, because of its retrospective nature, participants' accounts were vulnerable to normal limits of recall. Second, we had to rely on AH staff to help us identify participants, since there were no natural sampling frames to identify TCB farmers and external stakeholders. As a result, we might have failed to capture a fully representative range of stakeholder perspectives, including more critical and unfavourable views of AH or their CE strategies than are represented in our final sample of individual interviews and focus groups. Throughout our analyses we remained conscious of this limitation and sought to take appropriate care with the inferences we drew from our data.

Discussion

A values-based approach

AH's approach to community engagement is grounded in an explicit commitment to certain core values and guiding principles. These are publicly accessible (http://africaharvest.org/) and their substance was reflected in many of the comments and actions of AH staff and other stakeholders we interviewed. These core values included a commitment to "excellence", "institutional and scientific integrity and accountability", "service to farm families, especially small landholders", among others, and their guiding principles included a "commitment to partnerships that strengthen African agriculture", a "programmatic approach based on developing the whole value chain", "reaching out and empowering [their] stakeholders", "ensuring gender equality and benefit sharing from the development interventions", and "focus on impact and tangible results to the beneficiaries", among others. This explicit values-based approach served as a public declaration by AH that it was mindful of the farmers' previous experiences with unfair partnerships and was committed to not perpetuating these practices. Of more immediate concern for our case study, the guiding values served as *design constraints* for its CE strategy; an internally-generated framework for assessing the implications of its activities for its stakeholders. Although our findings do not permit strong conclusions about its efficacy, we believe the explicit application of values as design constraints for CE strategies is a very promising avenue for improving CE planning and management. In this case, the approach appears to have facilitated strong relationships that helped to restore farmers' willingness to place trust in new partners, with one another, and in a novel technology and its associated agricultural practices.

Key features of the CE strategy

Four key features of the AH's CE approach appear to be critical to its effectiveness. First, the decision to introduce the TCB was informed by empirical evidence that was collected through explicit processes of formative research that were part of the CE strategy. Second, CE was geared towards systematically transferring skills to farmers and other relevant players, a foundational aspect of 'technology transfer'. Exposure visits, group learning, inputs by trainers drawn from within the communities, and field schools were critical contexts for engagement among stakeholders and provided opportunities for authentic relationships to develop, which formed the basis for all the critical points of cooperation and trust that are required throughout the scale-up of the TCB operation. Third, by seeking partnerships with both public and private sector SHs, AH effectively expanded the skills, expertise and experience at its disposal to engage farmers, and help them engage with one another. In the process, it also created opportunities to enhance the capabilities of other relevant SHs, such as academic institutions, public and private research laboratories, and agricultural nurseries, while remaining front-and-centre in establishing and maintaining their trustworthiness for farmers, in particular.

Fourth, AH staff made concerted efforts to keep the promises they made to farmers. In the context of eroded trust, farmers began their journey with AH with scepticism. Kept promises—a commitment to "match words with action"—gradually won AH the farmers' confidence. As well, the model for the TCB scale-up put decision-making in the hands of the farmers and their collectives, which represented a fundamental departure from the "old ways" and fuelled a sense of ownership of the innovation. Importantly, the AH engagement strategy required farmers to become knowledgeable and skilled through the full value chain of the TCB, breaking old patterns established by previous partners of cultivating farmers' dependencies on them for key aspects of the production and delivery processes.

Attention to the human dimensions of complex partnerships

Training and capacity building are typically viewed as separate categories of activity from "core" Research & Development (R&D) activities for new technologies. Although this mind-set has begun to shift [25], there are few clear examples of how training and capacity building activities have been designed and effectively integrated with the introduction of a new technology. AH's TCB CE strategy demonstrates how training and capacity building efforts contribute directly to the development of a human infrastructure [26] to further the aims of the initiative across the TCB value chain. Given the farmers' collective experience with exploitative partnerships in the past, AH staff understood clearly that their trustworthiness, i.e., the farmers' willingness to place trust in them and in the technology [27] could be the difference between the success and failure of the TCB. More generally, this attention to trustworthiness requires intensive engagement and listening to stakeholders, even prior to the execution of the project, to understand where they are starting from—what experiences have they had, and how have these experiences shape their attitudes about the specific project at hand. And it requires sensitivity and thoughtful consideration of how all the actions, behaviours, and commitments the SHs will experience in the process of being "engaged" will enhance, or detract from, their confidence in placing trust in their new partners. The success of the AH TCB introduction demonstrates the critical importance of follow-through and maintaining a sustained presence in the lives of the collaborating SHs and committing to authentic relationships.

The AH results demonstrate how CE and the complex web of relationships it produces between implementers and SHs [26] can facilitate a wide range of complex interactions including extensive training, the establishment of new business partnerships and regulatory relationships. The scope of these activities/opportunities and the relationships that facilitate them, stand in contrast to the more mechanistic approaches to CE (e.g., community advisory boards (CABs)), which typically have a more remote and limited influence on the day to day details of implementation. The findings suggest that effective CE must "break the plane" of the typical advisory, or 'message-delivery' models of CE to create significant relationships and shared experience. It is through these that sound judgements about trustworthiness can be made, not simply through the proceedings of advisory committees and other structural mechanisms. AH offers a useful example of what is required to make this happen.

Challenges

In cultural terms, the re-imagining of banana cultivation from a 'back-yard' and 'woman's crop' to a scalable commercial enterprise introduces new incentives for men to 'take over' roles that have traditionally been performed by women. Although the creation of new economic opportunities for men is not, in itself, a threat to the well-being of women, avoidance of this type of displacement and disruption of women's economic opportunities has been recognized by AH as a central plank in its mission and an on-going challenge. As well, there are inherent challenges for the scale-up of TCB as an economic venture, such as the heavy demand for water in an environment that is increasingly dry. Anticipating, and fairly accounting for, this type of externality is a chronic challenge for CE and fair research partnerships

Conclusions

We undertook this case study, in part, to help explain the perceived success of AH's CE strategy. Our findings suggest that AH's explicit articulation of its guiding values provided critical design constraints for its CE strategy. The resulting CE efforts helped to restore and build relationships of trust between farmers, AH and a wide range of other stakeholders. AH's attention to the interests of farmers across the 'whole value chain' gave rise to training and capacity building efforts that were effectively integrated into the development of the commercial enterprise. Our findings suggest new ways of conceptualizing the relationship between CE and the design and delivery of new technologies for global health and global development.

Abbreviations

AH: Africa Harvest Biotech International Foundation; CAB: Community advisory board; CE: Community engagement; ISAAA: International Service for the Acquisition of Agri-Biotech Applications; KARI: Kenyan Agricultural Research Institute; R&D: Research & Development; TCB: Tissue culture banana; TCBEL: Tissue Culture Banana Enterprise Ltd.

Acknowledgments

We express our deep gratitude to the participants, who made time for us and provided valuable insights. We are grateful for the support from a number of staff members of the Africa Harvest Biotech International Foundation Inc., in particular, Jane Ndiritu, Victoria Ndungu, Kiragu Wangari, Josephine Songa, and Michael Njuguna for their support.

Funding

This research was supported by the Bill & Melinda Gates Foundation through the Grand Challenges in Global Health Initiative (grant 39673).

Authors' contributions

SVSB and JVL originated the study as part of the larger international community engagement research initiative of the Ethical, Social, Cultural Program for the Grand Challenges in Global Health initiative. SVSB was involved in all aspects of its implementation - data collection, preliminary data analysis during the data collection phase, detailed analysis, developing the draft manuscript as a "best fit" for the data, and revising and editing the manuscript. ER conducted supplementary data analysis, and contributed to the revision and final editing of the manuscript. FW helped to coordinate the data collection visits and provided critical insights throughout the development of the manuscript. JVL supervised all aspects of the study's implementation, participated in the detailed analysis, drafting and editing of the final manuscript. All authors helped to conceptualize ideas, interpret findings, and review drafts of the article. All authors read and approved the final manuscript.

Competing interests

The authors declare that they have no completing interests.

Author details

[1]McLaughlin-Rotman Centre for Global Health, Toronto, Canada. [2]Africa Harvest Biotech Foundation International Inc., Nairobi, Kenya. [3]Centre for Ethical, Social & Cultural Risk, Li Ka Shing Knowledge Institute of St. Michael's Hospital, Toronto, Canada. [4]Clinical Epidemiology & Biostatistics Department, Faculty of Health Sciences, McMaster University, Hamilton, Canada. [5]Dalla Lana School of Public Health and Joint Centre for Bioethics, University of Toronto, Toronto, Canada.

References

1. Africa Harvest Biotech Foundation International. A Decade of Dedication: How tissue culture banana as improved rural livelihoods in Kenya. 2008.

2. Acharaya SS, Mackey MGA. Socio-economic impact assessment of the tissue culture banana industry in Kenya. Africa Harvest Biotech Foundation International (AHBFI). Nairobi: Africa Harvest; 2009. https://issuu.com/africaharvest/docs/socio-economic. Accessed 6 July 2016.

3. Kabunga NS, Dubois T, Qaim M. Heterogeneous information exposure and technology adoption: the case of tissue culture bananas in Kenya. Agric Econ. 2012;43(5):473–86.

4. Vuylsteke DR, Ortiz R. Field performance of conventional vs. in vitro propagules of plantain (Musa spp., AAB group). HortSci. 1996;31(5):862–5.

5. Kabunga NS, Dubois T, Qaim M. Impact of tissue culture banana technology on farm household income and food security in Kenya. Food Policy. 2014; 45:25–34.

6. Dubois T, Dusabe Y, Lule M, Van Asten P, Coyne D, Hobayo JC, Nkurunziza S, Ouma E, Kabunga N, Qaim M, Kahangi E. Tissue culture banana (Musa spp.) for smallholder farmers: lessons learnt from east Africa. Acta Hort. 2013;986:51–9.

7. Singer PA, Berndtson K, Tracy SC, Cohen ERM, Masum H, Daar AS, Lavery JV. A tough transition. Nature. 2007;449:160–3.

8. Department of Health, United Kingdom. Government response to the House of Commons Science and Technology Committee 2nd Report of Session 2015–2016: Science in Emergencies: UK lessons from Ebola. Cm 9236. London: Department of Health, 2016. https://www.gov.uk/government/uploads/system/uploads/attachment_data/file/516829/DH_Cm_9236_Ebola_Print.PDF. Accessed 6 July 2016.

9. Tindana PO, Singh JA, Tracy CS, Upshur REG, Daar AS, Singer PA, et al. Grand challenges in global health: community engagement in research in developing countries. PLoS Med. 2007;4(9):e273.

10. Singer PA, Taylor AD, Daar AS, Upshur REG, Singh JA, Lavery JV. Grand challenges in global health: the ethical, social and cultural program. PLoS Med. 2007;4(9):1440–4.

11. Tindana PO, Rozmovits L, Boulanger RF, Bandewar SV, Aborigo RA, et al. Aligning community engagement with traditional authority structures in global health research: a case study from northern Ghana. Am J Public Health. 2011;101:1857–67. doi:10.2105/AJPH.2011.300203. PMID: 21852635.

12. Bandewar SVS, Kimani J, Lavery JV. The origins of a research community in the Majengo ob- servational cohort study, Nairobi, Kenya. BMC Public Health. 2010;10:630–40. doi:10.1186/1471-2458-10-630. PMID: 20964821.

13. Yin RK. Case study research: design and methods. 3rd ed. Thousand Oaks: Sage Publications; 2003.

14. Charmaz K. Constructing grounded theory. 2nd ed. Thousand Oaks: Sage Publications; 2006.

15. Wolcott HF. Transforming qualitative data: Description, analysis, and interpretation. Thousand Oaks: Sage Publications; 1994.

16. Patton MQ. Qualitative research and evaluation methods. 4th ed. Thousand Oaks: Sage Publications; 2014.

17. Wuest J. Grounded theory: The method. In: Munhall P, editor. Nursing research: a qualitative perspective. 4th ed. Toronto: Jones & Bartlett Learning; 2007. p. 239–72.

18. Corbin J, Strauss A. Basics of qualitative research: Techniques and procedures for developing grounded theory. 3rd ed. Thousand Oaks: Sage Publications; 2007.

19. Creswell JW. Qualitative inquiry and research design: Choosing among five traditions. 3rd ed. Thousand Oaks: Sage Publications; 2013.

20. Bevan D, Collier P, Gunning JW. Trade shocks in developing countries: consequences and policy responses. Eur Econ Rev. 1993;37(2):557–65.

21. Kubiriba J, Tushemereirwe WK. Approaches for the control of banana Xanthomonas wilt in East and Central Africa. African J Plant Sci. 2014; 8(8):398–404.

22. Nestel P, Bouis HE, Meenakshi JV, Pfeiffer W. Symposium: food fortification in developing countries. J Nutr. 2006;136:1064–7.

23. Ochieng C. The Importance of Contract Farming and Its Prospects for Contributing to Poverty Reduction in Africa. Key Note Paper, NEPAD Workshop "Contract Farming: Expanding Agri-Business Links with Smallholder Farmers in Africa", November 21–25, 2005, Entebbe, Uganda.

24. Njuguna MM, Wambugu FM. Towards optimizing the impact of tissue culture banana in Kenya. In: Wambugu FM, Kamanga D, editors. Biotechnology in Africa. Geneva: Springer International Publishing; 2014. p. 115–31. doi:10.1007/978-3-319-04001-1_7.

25. Zachariah R, Guillerm N, Berger S, Kumar AMV, Satyanarayana S, Bissell K, Edginton M, Hinderaker SG, Tayler-Smith K, Van den Bergh R, Khogali M, Manzi M, Reid AJ, Ramsay A, Reeder JC, Harries AD. Research to policy and practice change: is capacity building in operational research delivering the goods? Tropical Med Int Health. 2014;19(9):1068–75.

26. King KF, Kolopack P, Merritt MW, Lavery JV. Community engagement and the human infrastructure of global health research. BMC Medical Ethics. 2014;15:84. http://www.biomedcentral.com/1472-6939/15/84.

27. O'Neil O. A question of trust. Lecture 1: Spreading Suspicion. BBC Radio Reith Lectures 2002. London: BBC Radio. http://www.bbc.co.uk/radio4/reith2002/. Accessed 6 July 2016.

10

Improving the active expression of transglutaminase in *Streptomyces lividans* by promoter engineering and codon optimization

Song Liu[1]*, Miao Wang[4], Guocheng Du[1,3]* and Jian Chen[1,2]

Abstract

Background: Transglutaminases (TGase), which are synthesized as a zymogen (pro-TGase) in *Streptomyces* sp., are important enzymes in the food industry. Because this pro-peptide is essential for the correct folding of *Streptomyces* TGase, TGase is usually expressed in an inactive pro-TGase form, which is then converted to active TGase by the addition of activating proteases in vitro. In this study, *Streptomyces hygroscopicus* TGase was actively produced by *Streptomyces lividans* through promoter engineering and codon optimization.

Results: A gene fragment (*tg1*, 2.6 kb) that encoded the pro-TGase and its endogenous promoter region, signal peptide and terminator was amplified from *S. hygroscopicus* WSH03-13 and cloned into plasmid pIJ86, which resulted in pIJ86/*tg1*. After fermentation for 2 days, *S. lividans* TK24 that harbored pIJ86/*tg1* produced 1.8 U/mL of TGase, and a clear TGase band (38 kDa) was detected in the culture supernatant. These results indicated that the pro-TGase was successfully expressed and correctly processed into active TGase in *S. lividans* TK24 by using the TGase promoter. Based on deletion analysis, the complete sequence of the TGase promoter is restricted to the region from −693 to −48. We also identified a negative element (−198 to −148) in the TGase promoter, and the deletion of this element increased the TGase production by 81.3 %, in contrast to the method by which *S. lividans* expresses pIJ86/*tg1*. Combining the deletion of the negative element of the promoter and optimization of the gene codons, the yield and productivity of TGase reached 5.73 U/mL and 0.14 U/mL/h in the recombinant *S. lividans*, respectively.

Conclusions: We constructed an active TGase-producing strain that had a high yield and productivity, and the optimized TGase promoter could be a good candidate promoter for the expression of other proteins in *Streptomyces*.

Keywords: Transglutaminase, Endogenous promoter, Codon optimization, *Streptomyces hygroscopicus*, *Streptomyces lividans*

Background

Transglutaminase (TGase, EC 2.3.2.13) is an enzyme that exhibits several catalytic activities: the crosslinking of proteins by forming N^ε-(γ-glutamyl) lysine bonds, the incorporation of polyamines into protein, and the deamidation of protein-bound glutamines [1]. Because of

these catalytic abilities, TGase has been widely used in industrial processing, especially in food processing, for improving the functional properties of various proteins, including meat, soy, myosin, globulin, casein, peanut, and whey proteins [2]. TGase is widely distributed in various organisms, including plants [3], mammals [4], and microorganisms [5]. Among the TGases, the TGase from *Streptomyces* is Ca^{2+}-independent and is advantageous for industrial applications because it has a higher reaction rate, broad substrate specificity for an acyl

* Correspondence: liusong@jiangnan.edu.cn; gcdu@jiangnan.edu.cn
[1]Key Laboratory of Industrial Biotechnology, Ministry of Education, School of Biotechnology, Jiangnan University, Wuxi, China
Full list of author information is available at the end of the article

donor, and a smaller molecular size [6, 7]. The development of an efficient and easy-to-use expression system for the production of *Streptomyces* TGase is therefore highly desirable.

Streptomyces TGase is secreted as pro-TGase and becomes active after the cleavage of the pro-peptide by endogenous activating proteases [5]. Because the pro-peptide is essential for the correct folding of TGase, direct expression of mature TGase yields insoluble inclusion bodies [8] or inactive enzyme [9]. Thus, TGase is usually expressed in a pro-TGase form [10, 11]. Due to the absence of activating protease in the host strain, co-expression of heterologous proteases is required to convert pro-TGase into active TGase [12]. Because of the ability to convert the pro-protein into the active enzyme with its own proteases, *Streptomyces* hosts became ideal hosts for producing active TGase. TGases from *Streptoverticillium mobaraense*, *Streptoverticillium ladakanum*, and *Streptomyces platensis* have been heterologously expressed in *Streptomyces lividans* as an active enzyme [13–15]. However, the secretion level of TGase in *S. lividans* 3131 is less than 0.01 U/mL [13]. When *S. lividans* JT46 was used as the host strain, the yield of TGase reaches only 1.23–2.22 U/mL after 3–6 days of fermentation [14, 15]. Overall, both the yield and productivity of TGase as expressed in *Streptomyces* hosts are still low.

The *ermE* and *tipA* promoters have proven to be highly successful for the over-expression of *Streptomyces* genes [16]. However, the *ermE* promoter improved TGase production by 0.8 U/mL [17], and there are no reports for TGase expression with the other strong promoters. It has been found that the endogenous promoter of TGase is recognized in *S. lividans*, and the yield of the recombinant *S. platensis* TGase reached 2.22 U/mL [15], which suggests that the endogenous promoter of different TGases or its modified versions could be more efficient for TGase expression by *S. lividans* in contrast to heterologous strong promoters. In addition, the *Streptomyces* genome has a high (>70 %) GC content, and rare codons such as TTA could significantly reduce the protein expression in *S. lividans* [18]. However, the *Streptomyces* TGase gene contains rare codons such as TTA, although it was found in *Streptomyces* [14, 15, 19]. Thus, codon optimization could also benefit TGase expression in *S. lividans*.

Previously, we cloned the DNA fragment (GenBank No: HM231108) that contained the TGase gene with a flanking region sequence from the *S. hygroscopicus* genome, and a putative promoter region was found upstream to TGase [10]. In this study, the *S. hygroscopicus* TGase gene was expressed in *S. lividans* TK24 by using its putative endogenous promoter. Then, the putative promoter was partially deleted, and the effects of the

deletions on the expression of TGase in *S. lividans* TK24 were analyzed. In addition, the codons of TGase were optimized to further enhance the level of TGase expression. Finally, a relatively high level of TGase expression in *S. lividans* was achieved.

Results

Expression of TGase in *S. lividans* using its endogenous promoter

To express the TGase in *S. lividans* using its endogenous promoter, a gene fragment (*tg1*, 2.6 kb) was amplified from the *S. hygroscopicus* genome (Fig. 1a) and cloned into pIJ86, which resulted in the plasmid pIJ86/*tg1* (Fig. 1b). The *tg1* encoded the TGase ORF (1257 bp), the upstream sequence (893 bp) and the downstream sequence (458 bp) (Fig. 1a). As analyzed previously, the *S. hygroscopicus* TGase ORF was composed of a secretory signal peptide gene, a pro-peptide gene, and the mature TGase gene; the upstream and downstream sequence of the ORF contain a putative promoter and a putative terminator, respectively [10]. The expression vector was transformed into *S. lividans* TK24, yielding *S. lividans* TK24/pIJ86/*tg1*.

When cultivated for 48 h, *S. lividans* TK24/pIJ86/*tg1* obtained 1.8 U/mL of extracellular TGase, which was approximately 1.5-fold of that achieved in the wild strain *S. hygroscopicus* WSH03-13 under the same cultivation conditions (Fig. 1c). TGase activity was not detected in the culture supernatants of the control strains *S. lividans* TK24/pIJ86 (*S. lividans* TK24 carrying pIJ86) and *S. lividans* TK24 (Fig. 1c). After treatment with TGase-activating protease dispase [10], the culture supernatants of the control strains still did not exhibit TGase activity (data not shown). Then, the culture supernatants of *S. lividans* TK24/pIJ86/*tg1*, *S. hygroscopicus*, and the control strains were subjected to SDS-PAGE analysis. As shown in Fig. 1d (lane 1), the *S. lividans* TK24/pIJ86/*tg1* showed a remarkable band that had a size of 38 kDa, which corresponds to the molecular weight of *S. hygroscopicus* TGase [20]. In the case of control strains, a small number of TGase/pro-TGase-like bands was detected in the culture supernatants (Fig. 1d, lanes 2 and 3). For failing to detect TGase activity in the control samples (Fig. 1c), these TGase/pro-TGase-like bands could correspond to the endogenous extracellular proteins of *S. lividans* TK24. Two proteins with approximate molecular weights of pro-TGase and TGase were detected in the culture supernatants of *S. hygroscopicus* (Fig. 1d, lane 4), which indicates that pro-TGase is not fully processed [10]. Because the *ermE* promoter was removed in pIJ86/*tg1*, our results indicated that the upstream sequence (893 bp) contains the endogenous promoter, which could drive the expression of TGase in *S. lividans* TK24. Moreover, the pro-TGase is correctly

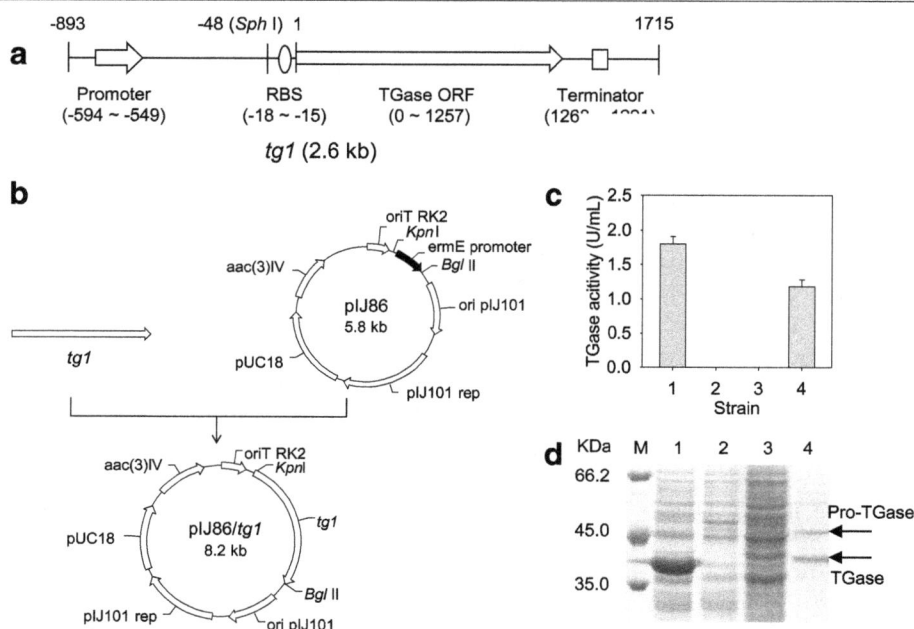

Fig. 1 Production of TGase by *S. lividans* TK24/pIJ86/*tgl*. **a** The gene structure of *tg1*. **b** Construction of TGase expression plasmid pIJ86/*tg1*. **c** TGase activity assay of the culture supernatants of *Streptomyces* strains. **d** SDS-PAGE analysis of the TGases in the culture supernatants of *Streptomyces* strains. Labeling for (**a**): The numbers in the illustration indicate the base positions. Labeling for (**c**): 1: *S. lividans* TK24/pIJ86/*tgl* (the recombinant strain that expresses the *S. hygroscopicus* TGase ORF using the TGase endogenous promoter), 2: *S. lividans* TK24/pIJ86 (the control strain that carries pIJ86), 3: *S. lividans* TK24 (the control strain without the expression plasmid), 4: *S. hygroscopicus* WSH03-13 (the wild type strain that produces TGase). Labels for (**d**): M: protein marker, 1: *S. lividans* TK24/pIJ86/*tgl*, 2: *S. lividans* TK24/pIJ86, 3: *S. lividans* TK24, 4: *S. hygroscopicus* WSH03-13. The recombinant *S. lividans* TK24 were inoculated into 30 mL of medium (which contained 50 μg/mL apramycine) and cultured at 30 °C and 200 rpm for 48 h

processed by the host proteases, which suggests that *S. lividans* TK24 is an ideal host for the active expression of TGase.

Deletion analysis of the TGase promoter

As shown in Fig. 1a, the putative core promoter was located in the upstream sequence (between −594 bp and −549 bp) of the TGase ORF. To identify the TGase promoter, we analyzed the effect of the upstream sequence deletions on the expression of TGase in *S. lividans* TK24.

First, deletions at the 5′-end of the upstream sequence were conducted. Deleting the upstream down to −793 (pTGU2) or −693 (pTGU3) had no significant effect on the expression level of the TGase gene (Fig. 2a). However, deleting the upstream down to −593 (pTGU4) resulted in a significant decrease in the TGase production, approximately 57.6 % of the activity of non-deletion (Fig. 2a). Deleting down to −493 (pTGU5) resulted in the complete loss of the TGase activity (Fig. 2a). These results suggest that the region from −693 to −493 contains important components of the TGase promoter.

Second, deletions at the 3′-end of the upstream sequence were conducted. Because the putative ribosome-binding site was located in the upstream sequence

between −18 and −15, the 3′-end deletion was initiated at −48. Deleting upstream from −48 up to −98 (pTGD2) did not have a significant effect on the expression level of the TGase gene (Fig. 2b). Deletion of up to −148 (pTGD2) resulted in a decrease in the TGase activity, approximately 82.1 % of the activity of the non-deletion (pIJ86/*tg1*) (Fig. 2b). Interestingly, deleting up to −198 (pTGD4) increased the TGase activity by 27.3 % (Fig. 2b). However, the deletion mutant pTGD5 resulted in a decrease in the TGase activity (91.2 %), and further deletion (pTGD6-pTGD10) caused a significant decrease in the TGase activity (less than 40 % of the activity of pIJ86/*tg1*) (Fig. 2b). Last, TGase activity could not be detected with the deletion mutant pTGD10 (−498) (Fig. 2b). These results suggest that the region from −498 to −198 and the region from −148 to −98 could be the positive elements for the TGase promoter, while the region from −198 to −148 was the negative element for the TGase promoter.

Based on the deletion analysis (Fig. 2), the complete promoter of TGase could be restricted to the sequence from −693 to −48 in *tg1*. Because the region from −198 to −148 negatively affected the expression, this region was deleted from the complete promoter (−693 to −48), which yielded the TGase expression plasmid pTGO. As

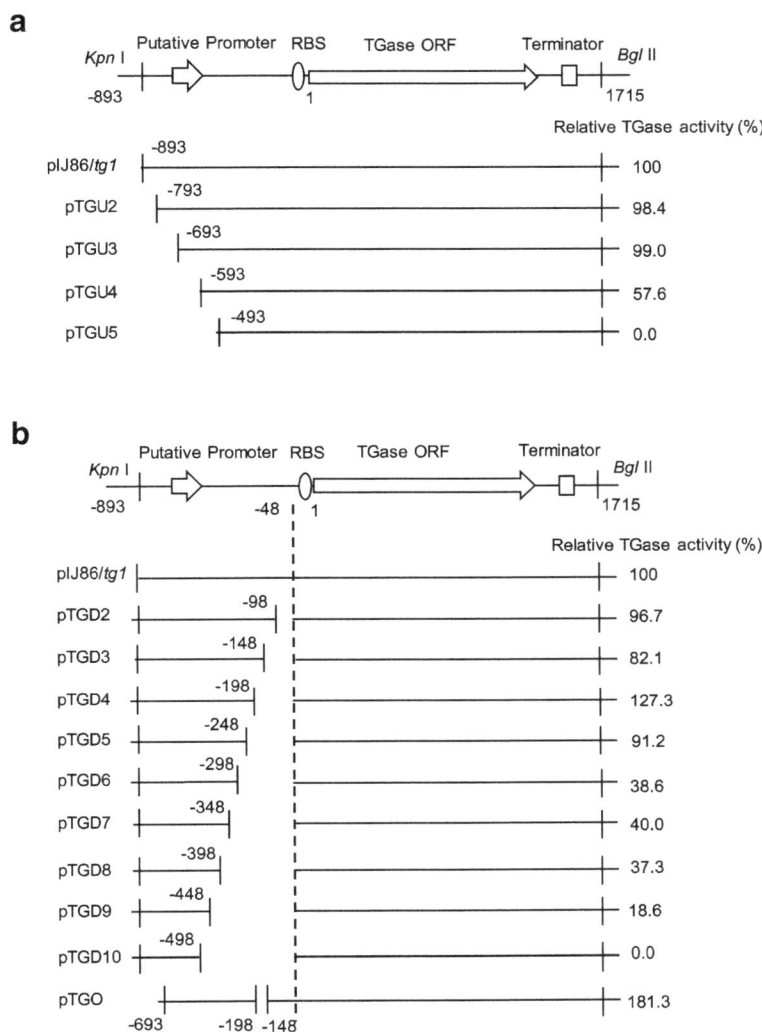

Fig. 2 The effect of the endogenous promoter modification on the expression of TGase in *S. lividans*. **a** Partially deleting the 5'-end of the TGase promoter region. **b** Partially deleting the 3'-end of the TGase promoter region. The recombinant *S. lividans* TK24 were inoculated into 30 mL of medium (which contained 50 μg/mL apramycine) and cultured at 30 °C and 200 rpm for 48 h

indicated by Fig. 2b, cells that carried pTGO achieved 3.3 U/mL of TGase, which was 81.3 % higher than that obtained by pIJ86/*tg1* (Fig. 2b).

Codon optimization of the TGase gene in *S. lividans*

To improve the TGase expression in *S. lividans*, the gene sequence of TGase ORF was optimized according to the gene codon bias of *Streptomyces* and was chemically synthesized (Fig. 3). The codon-optimized TGase ORF along with the intact upstream (−48 to −1) and downstream (1258 to 1715) was then cloned into the *Sph* I-*Bgl* II sites of pTGO, which yielded pTGOm. As shown in Fig. 4, when *S. lividans* expressed pTGOm, the highest yield of TGase (5.73 U/mL) was obtained, which was 73.6 % higher than that produced by *S. lividans* when it harbored pTGO. Moreover, the former

recombinant strain achieved the highest yield of TGase at 42 h, while for the latter strain, the highest yield was obtained at 48 h. Consequently, the productivity of *S. lividans* when it expressed pTGOm was 0.14 U/mL/h, which was twofold higher than that of *S. lividans* while it harbored pTGO.

Discussion

Although TGase from *Sv. ladakanum* [14] and *S. platensis* [15] has been expressed in *S. lividans* JT46 by TGase promoters, the yields of TGase reach only 1.23–2.22 U/mL after 3–6 days of fermentation, and the productivities are less than 0.03 U/mL/h [14, 15]. Recently, another recombinant *S. lividans* has obtained only 0.07 U/mL/h of TGase productivity by using the *Streptomyces cinnamoneus* phospholipase D promoter and signal

```
I    ATG TAC AAG CGT CGG AGT TTA CTC GCC TTC GCC ACT GTG AGT GCG GCG ATA TTC ACC GCT GGA GTC ATG CCG  72
II   --- --- --- CGG --- AGC CTC CTG --- --- GCG ACG GTC TCG --- --- ATC --- ACG GCG GGC --- --- CCC
III   M   Y   K   R   R   S   L   L   A   F   A   T   V   S   A   A   I   F   T   A   G   V   M   P      24

I    TCG GTC AGC CAT GCC GCC AGC GGC GGC GAC GGG GAA AGG GAG GGG TCC TAC GCC GAA ACG CAC GGT CTG ACG  144
II   --- GTG TCC CAC GCG --- TCC --- --- --- GGC GAG CGG --- GGC AGC --- --- GAG ACC --- GGC --- ---
III   S   V   S   H   A   A   S   G   G   D   G   E   R   E   G   S   Y   A   E   T   H   G   L   T      48

I    GCG GAG GAC GTC AAG AAC ATC AAC GCA CTG AAC AAA AGG GCC CTG ACT GCG GGT CAA CCT GGC AAT TCT CTG  216
II   --- --- --- --- --- --- --- --- GCC --- --- AAG CGC --- CTC --- --- GGC CAG CCC --- AAC TCC ---
III   A   E   D   V   K   N   I   N   A   L   N   K   R   A   L   T   A   G   Q   P   G   N   S   L      72

I    GCG GAA TTG CCG CCG AGC GTC AGT GCG CTC TTC CGG GCC CCC GAC GCT GCC GAC GAG AGG GTG ACC CCT CCT  288
II   --- GAG CTC --- CCC TCG --- TCC GCC CTG --- --- GCC --- --- GCC --- --- --- --- ACG CCG CCC
III   A   E   L   P   P   S   V   S   A   L   F   R   A   P   D   A   A   D   E   R   V   I   P   P      96

I    GCC GAG CCG CTC AAC CGG ATG CCT GAC GCG TAC CGG GCC TAC GGA GGT AGG GCC ACT ACG GTC GTC AAC AAC  360
II   --- --- CCC --- --- --- --- CCG --- GCC --- --- --- --- GGC GGC CGG --- ACC --- --- GTG --- ---
III   A   E   P   L   N   R   M   P   D   A   Y   R   A   Y   G   G   R   A   T   T   V   V   N   N     120

I    TAC ATA CGC AAG TGG CAG CAG GTC TAC AGT CAC CGC GAC GGC ATC CAA CAG CAA ATG ACC GAA GAG CAG CGA  432
II   --- ATC --- --- --- --- --- --- --- AGC --- --- --- --- --- CAG --- CAG --- --- GAG --- --- CGC
III   Y   I   R   K   W   Q   Q   V   Y   S   H   R   D   G   I   Q   Q   M   T   E   E   Q   R       144

I    GAA AAG CTG TCC TAC GGC TGC GTC GGC GTC ACC TGG GTC AAT TCG GGC CCC TAC CCG ACG AAC AAA TTG GCG  504
II   CAG --- --- TCG --- --- --- --- --- GTG --- --- GTG AAC TCC --- CCG --- CCC --- --- AAG CTG GCC
III   E   K   L   S   Y   G   C   V   G   V   T   W   V   N   S   H   P   Y   P   T   N   K   L   A     168

I    TTC GCG TTC TTC GAC GAG GAC AAG TAC AAG AGT GAC CTG GAA AAC AGC AGG CCA CGC CCC AAT GAG ACG CAA  576
II   --- --- --- --- --- --- --- --- --- --- TCC --- --- GAG --- --- CGC CCG CGG GCC AAC --- ACC CAG
III   F   A   F   F   D   E   D   K   Y   K   S   D   L   E   N   S   R   P   R   P   N   E   T   Q     192

I    GCC GAG TTT GAG GGG CGC ATC GTC AAG GAC AGT TTC GAC GAG GGG AAG GGC TTC AAG GCG GCG CGT GAT GTG  648
II   --- --- TTC --- --- --- --- --- --- --- TCC --- --- --- GGC --- --- --- --- CGC GCC CGG GAC ---
III   A   E   F   E   G   R   I   V   K   D   S   F   D   E   G   K   G   F   K   R   A   R   D   V     216

I    GCG TCC ATC ATG AAC AAG GCC CTG GAT AGT GCG CAC GAC GAG GGG ACT TAC ATC GAC AAC CTC AAG AAA GAG  720
II   --- AGC --- --- --- --- --- CTC GAC TCG --- --- --- --- GGC ACC --- --- --- --- CTG --- AAG ---
III   A   S   I   M   N   K   A   L   D   S   A   H   D   E   G   T   Y   I   F   N   L   K   K   E     240

I    CTC GCG AAC AAA AAT GAC GCT CTG CGC TAC GAG GAC AGT CGC TCG AAC TTT TAC TCG GCG CTG AGG AAT ACG  792
II   --- GCC --- AAG AAC --- GCG --- --- --- --- --- AGC --- --- --- TTC --- TCC GCC CTC --- AAC ---
III   L   A   N   K   D   A   L   R   Y   E   D   S   R   S   N   F   Y   S   A   L   R   N   T       264

I    CCG TCC TTC AAG GAA AGG GAT GGA GGC AAC TAC GAC CCA TCC AAG ATG AAG GCG GTG GTC TAC TCG AAA CAC  864
II   CCC AGC --- --- GAG CGC GAC GGT --- --- --- --- CCG AGC --- --- --- GCC GTC GTG --- --- AAG ---
III   P   S   F   K   E   R   D   G   G   N   Y   D   P   S   K   M   K   A   V   V   Y   S   K   H     288

I    TTC TGG AGC GGG CAG GAC CAG CGG GGC TCC TCT GAC AAG AGG AAG TAC GGC GAC CCG GAT GCC TTC CGC CCC  936
II   --- --- TCC TCG --- --- --- --- --- --- AGC --- --- CGC --- --- --- --- CCC GAC --- --- --- CCG
III   F   W   S   G   Q   D   Q   R   G   S   S   D   K   R   K   Y   G   D   P   D   A   F   R   P     312

I    GAC CAG GGC ACA GGC CTG GTC GAC ATG TCG AAG GAC AGG AAT ATT CCG CGC AGT CCC GCC CGA CCT GGC GAA  1008
II   --- --- --- ACC --- --- GTG --- --- --- --- --- CGC AAC ATC --- CGG TCG --- --- CGC CCG --- GAG
III   D   Q   G   T   G   L   V   D   M   S   K   D   R   N   I   P   R   S   P   A   R   P   G   R     336

I    AGT TGG GTG AAT TTC GAC TAC GGC TGG TTT GGG GCT CAG ACG GAA GCG GAT GCC GAC AAA ACC ATA TGG ACC  1080
II   TCG --- --- AAC --- --- --- --- --- TTC GGG GCG --- ACC GAG GCC --- GCG --- AAG --- ATC --- ACG
III   S   W   V   N   F   D   Y   G   W   F   G   A   Q   T   E   A   D   A   D   K   T   I   W   T     360

I    CAC GCC AAC CAC TAT CAC GCG CCC AAC GGC GGC GTG GGC CCC ATG AAC GTA TAC GAG AGC AAG TTC CGG AAC  1152
II   --- --- --- --- TAC --- --- --- --- --- --- GTC --- CCG --- --- GTG --- --- --- --- --- --- ---
III   H   A   N   H   Y   H   A   P   N   G   G   V   G   P   M   N   V   Y   E   S   K   F   R   N     384

I    TGG TCT GCC GGG TAC GCG GAT TTC GAC CGC GGA ACC TAC GTC ATC ACG TTC ATA CCC AAG AGC TGG AAC ACC  1224
II   --- TCG --- GGC --- --- GAC --- --- --- GGC --- --- --- --- ACC --- ATC --- --- TCG --- --- ---
III   W   S   A   G   Y   A   D   F   D   R   G   T   Y   V   I   T   F   I   P   K   S   W   N   T     408

I    GCC CCC GCC GAG GTG AAG CAG GGC TGG TCG TAA  1257
II   --- --- --- --- GTC --- --- --- --- --- TGA
III   A   P   A   E   V   K   Q   G   W   S   *
```

Fig. 3 Codon optimization of *S. hygroscopicus* WSH03-13 TGase. The Greek numerals (left side) indicated the sequence type of TGase: I: original sequence of *S. hygroscopicus* TGase ORF; II: *Streptomyces* preferred gene sequence of the TGase ORF; III: amino acid sequence of the TGase ORF. The grey shadows and "—" indicated the mutated positions and invariant positions. The numbers (right side) indicated the nucleotide sequences

peptide [21]. In this study, *S. lividans* TK24 that harbored pTGM obviously obtained a higher yield (5.73 U/mL) and productivity (0.14 U/mL/h) for the TGase (Fig. 4). It has been reported that different proteases showed variant activation efficiencies against *Streptomyces* pro-TGase in vitro [22]. Because all of these TGases are expressed in pro-TGase form, *S. lividans* TK24 could have those proteases that are more favorable for the pro-TGase activation in contrast to *S. lividans* JT46 [23].

To improve the production of TGase, the endogenous promoter of *S. hygroscopicus* TGase was engineered. Previously, we isolated a TGase-producing strain *S. hygroscopicus* WSH03-13 and cloned the TGase ORF with a flanking sequence [10, 20]. According to the sequence

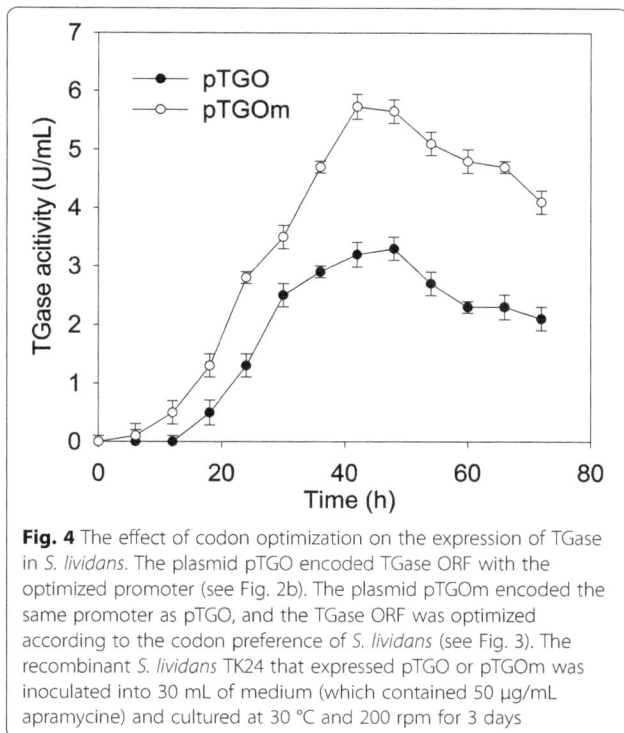

Fig. 4 The effect of codon optimization on the expression of TGase in *S. lividans*. The plasmid pTGO encoded TGase ORF with the optimized promoter (see Fig. 2b). The plasmid pTGOm encoded the same promoter as pTGO, and the TGase ORF was optimized according to the codon preference of *S. lividans* (see Fig. 3). The recombinant *S. lividans* TK24 that expressed pTGO or pTGOm was inoculated into 30 mL of medium (which contained 50 μg/mL apramycine) and cultured at 30 °C and 200 rpm for 3 days

analysis, the upstream sequence of the ORF contained a putative promoter [10]. However, the efficiency and the exact site of this endogenous promoter were not clear. Expression of the TGase ORF with the upstream sequence obtained extracellular TGase activity in *S. lividans*, confirming the existence of the endogenous promoter (Fig. 1). Based on deletion analysis, the complete sequence of the TGase endogenous promoter is restricted to the region from −693 to −48, and a negative element (−198 to −148) was identified (Fig. 2). Finally, the TGase production in *S. lividans* was increased by 81.3 % through the deletion of this element (Fig. 2b). Further investigation should be focused on the action mode of the negative element, which may serve to understand the physiological function of the TGase in *Streptomyces*.

Codon optimization was also used to improve the TGase expression in *S. lividans*. There is evidence of improved expression in the host strain when certain rare codons are replaced with preferred codons [24–26]. This phenomenon is thought to be related to the relative levels of the intracellular pool of charged transfer RNA molecules, which are low for rare codons and high for abundant codons [27]. As indicated by our previous study [10], *S. hygroscopicus* TGase ORF contains a rare codon TTA (leucine, codon usage 0.2 %) in *Streptomyces* [28]. Thus, it could prevent TGase expression because of the low level of transfer RNA molecules. In this study, the gene sequence optimization of TGase ORF according

to the codon bias of *Streptomyces* resulted in 73.6 % enhanced TGase production in *S. lividans*. To be noted, the codon optimization reduced the fermentation period for the highest TGase activity by 6 h (Fig. 4). After the sequence optimization, TTA that encoded Leu in the TGase ORF was mutated to the *Streptomyces* preferred codon CTC (Fig. 3). Because *bldA*, which encodes tRNA(Leu) (UUA), has been reported to be expressed only during the late stage of growth [28], the replacement of TTA by the *Streptomyces* preferred codon could account for the reduced fermentation period of the cells that expressed optimized TGase gene.

To further increase the production level of recombinant TGase at a large scale, the optimization medium and culture conditions will be performed in fermentors [29, 30].

Conclusions

In conclusion, we constructed an active TGase-producing strain with a high yield and productivity, which could be a good candidate strain for industrial production of this enzyme. Moreover, the optimized TGase promoter and site-directed of rare codon TTA may also useful for improving other protein expression in *S. lividans*.

Methods
Bacterial strains, plasmids, and culture conditions
S. hygroscopicus WSH03-13 that produces TGase was stored in our lab [10]. *E. coli* JM109 was used for gene cloning. *S. lividans* TK24 (*Str*-6, tipAp induced, SLP2-, SLP3-) and pIJ86 (*Streptomyces* complementation plasmid; oriColE1 SCP2* *aac(3)IV ermE**p) were used as the expression host and plasmid, respectively. *Streptomyces* cultures were grown on R2YE agar [31] or in liquid that contained glycerol 20 g/L, peptone 20 g/L, yeast extract 5 g/L, MgSO$_4$ 2 g/L, K$_2$HPO$_4$ 2 g/L, KH$_2$PO$_4$ 2 g/L, and CaCl$_2$ 1 g/L. A loop of fresh spore suspension of *S. hygroscopicus* WSH03-13 or *S. lividans* TK24 was inoculated into 30 mL of medium and cultured at 30 °C and 200 rpm for 2–3 days. *E. coli* JM109 was grown in Luria-Bertani medium at 37 °C.

Construction of plasmids that express TGase with their endogenous promoter
To obtain a plasmid that expresses TGase with its endogenous promoter, a 2.6-kb DNA fragment (*tg1*) that contained the TGase gene with a flanking sequence from *S. hygroscopicus* WSH03-13 was amplified by PCR using the primer pairs TGUF/TGDR (Table 1), and the fragment was then inserted into the *Kpn* I-*Bgl* II sites of pIJ86, which resulted in the plasmids pIJ86/*tg1* (Fig. 1a).

Table 1 Primers used in this study

Primer	Sequence (5'–3')
TGUF	CGGGGTACCCCGTAGCGGGTGGCGAAGAT
TGDR	GGAAGATCTCACGAGGACACCGAACGACTG
TG100F	TTCGAGCTCGGTACCCACCCCGCTGAATGGGACTCTTCGT
TG200F	TTCGAGCTCGGTACCCCAGGAGCAGGGGAACGCTGC
TG300F	TTCGAGCTCGGTACCGACGTTGCCGGGGAGTTGGCGC
TG400F	TTCGAGCTCGGTACCCTCTCCCTGCGGTCGCCGTGACAG
TGQ1R	ACATGCATGCGACCTCAGCCGCGCTGTCCTGGGTC
TGQ2R	ACATGCATGCCCGCCACGAGGCGGAAGGAGATGC
TGQ3R	ACATGCATGCGCGTGGCCGTCGCCGGTCATGACCTGGTG
TGQ4R	ACATGCATGCGGCGGCACCGGTGCCTCGCTACATC
TGQ5R	ACATGCATGCGGGCCCGGTCCGGGGGCCGAGG
TGQ6R	ACATGCATGCGGGAGTGCATGAAGTCGGTGTC
TGQ7R	ACATGCATGCACAGCGGCGGTCGCCGGGGCGACGG
TGQ8R	ACATGCATGCGCCTCGCCGCGAACCGCACGCCAGG
TGQ9R	ACATGCATGCGGCAGGTCGGGAGCGCCTGTC

Construction of plasmids that express TGase with partially deleted endogenous promoters

To partially delete the 5'-end of the promoter region, each gene fragment of *tg1* with a 5'-end deletion at the promoter region (Fig. 2a) was amplified from pIJ86/*tg1* by PCR using a specific forward primer and a constant reverse primer (TGDR) (Table 1). For the deletion of the first 100 bp nucleotides at the 5'-end of the promoter region, TG100F (Table 1) was used as a forward primer. For further deletions at the 5'-end of the promoter region, TG200F, TG300F, and TG400F were in turn used as the forward primer (Table 1). The resulting PCR products were inserted into the *Kpn* I-*Bgl* II sites of pIJ86 to produce pTGU2, pTGU3, pTGU4, and pTGU5, respectively (Fig. 2a).

To partially delete the 3'-end of the promoter region, the gene fragment that contained the complete open reading frame (ORF) of TGase and the fragments that encoded the promoter region with 3'-end deletions were amplified from pIJ86/*tg1* by PCR, separately (Fig. 2b). Each gene fragment that encoded the promoter region with the 3'-end deletion was obtained by using a constant forward primer (TGUF) and a specific reverse primer (Table 1). For the deletion of the first 50 bp nucleotides at the 3'-end of the promoter region (Fig. 2b), TGQ1R (Table 1) was used as a reverse primer. For further 3'-end deletion of the promoter, TGQ2R, TGQ3R, TGQ4R, TGQ5R, TGQ6R, TGQ7R, TGQ8R, and TGQ9R were in turn used as the forward primer (Table 1). The resulting PCR products were inserted into the *Kpn* I-*Sph* I sites of pIJ86/*tg1* to produce pTGD2, pTGD3, pTGD4, pTGD5, pTGD6, pTGD7, pTGD8, pTGD9, and pTGD10, respectively (Fig. 2b).

The negative element (−197 to −149) was removed from the TGase promoter (−693 and −48) by chemical synthesis, and the resulting gene fragment was cloned into the *Kpn* I-*Sph* I sites of pIJ86/*tg1* to produce pTGO (Fig. 2b).

Codon optimization of the TGase gene in *S. lividans*

According to the codon preference of *S. lividans*, the *S. hygroscopicus* TGase ORF was optimized and synthesized by Genscript (Nanjing, China). The codon-optimized TGase ORF with an intact upstream (−48 to −1) and downstream (1258 to 1715) was then cloned into the *Sph* I-*Bgl* II sites of pTGO, which yielded pTGOm.

Expression of the TGase gene in *S. lividans*

Molecular methods for *Stretomyces* were used as described by Hopwood et al. [31]. Plasmids that expressed TGase with an endogenous promoter or its partially deleted versions were transformed into *S. lividans* TK24. The *S. lividans* transformants were selected on a plate that contained 50 μg/mL apramycine. When the transformants were grown in liquid medium, 50 μg/mL apramycine were added. The recombinant *S. lividans* TK24 were inoculated into 30 mL of medium and cultured at 30 °C and 200 rpm for 2–3 days.

Assay of TGase activity

TGase activity was measured using a colorimetric procedure in which N-α-carbobenzoxyl-glutaminyl-glycine (N-CBZ-Gln-Gly) (Sigma, Shanghai, China) was used as the substrate [8]. Forty microliters of substrate solution (30 mmol/L N-CBZ-Gln-Gly, 100 mmol/L hydroxylamine, 10 mmol/L glutathione, 200 mmol/L Tris-HCl buffer, pH6) was added to 100 μL of TGase solution to initiate the enzymatic reaction. After 10 min, the reaction was stopped by the addition of a 40-μL terminator (1 mol/L HCl, 4 % (v/v) trichloroacetic acid, 2 % (m/v) $FeCl_3 \cdot 6H_2O$), and the reaction solution was subjected to spectrophotometry analysis at 525 nm. A calibration curve was obtained using L-glutamic acid γ-monohydroxamate (Sigma, Shanghai, China). One unit of TGase was defined as that required to generate 1 μmol of L-glutamic acid γ-monohydroxamate per min at 37 °C.

Protein analysis

SDS-PAGE was performed on a 12 % running gel, and the resolved proteins were visualized by staining with Coomassie Brilliant Blue R250. Protein concentrations were measured using the Bradford method, with bovine serum albumin as the standard.

Abbreviations

N-CBZ-Gln-Gly: N-α-carbobenzoxyl-glutaminyl-glycine; ORF: complete open reading frame; TGase: transglutaminase

Acknowledgments
We would like to thank Professor Nam Sun Wang for critically reading and providing suggestions for the manuscript.

Funding
This work was financially supported by the National Natural Science Foundation of China (No. 31401638), the National High Technology Research and Development Program of China (No. 2015AA021003), the Key Technologies R & D Program of Jiangsu Province (No. BE2016629), and the Natural Science Foundation of Jiangsu Province (No. BK20130132).

Authors' contributions
SL conducted the molecular genetic studies and drafted the manuscript. MW participated in the design of the study. GD and JC conceived the study and participated in its design and coordination and helped to draft the manuscript. All of the authors read and approved the final manuscript.

Competing interests
All authors declare that they have no competing interests.

Author details
[1]Key Laboratory of Industrial Biotechnology, Ministry of Education, School of Biotechnology, Jiangnan University, Wuxi, China. [2]National Engineering Laboratory for Cereal Fermentation Technology, Jiangnan University, Wuxi, China. [3]Key Laboratory of Carbohydrate Chemistry and Biotechnology, Ministry of Education, School of Biotechnology, Jiangnan University, Wuxi, China. [4]School of Food Science and Technology, Jiangnan University, Wuxi 214122, China.

References
1. Yokoyama K, Nio N, Kikuchi Y. Properties and applications of microbial transglutaminase. Appl Microbiol Biotechnol. 2004;64(4):447–54.
2. Martins IM, Matos M, Costa R, Silva F, Pascoal A, Estevinho LM, Choupina AB. Transglutaminases: recent achievements and new sources. Appl Microbiol Biotechnol. 2014;98(16):6957–64.
3. Serafini-Fracassini D, Del Duca S. Transglutaminases: widespread cross-linking enzymes in plants. Ann Bot. 2008;102(2):145–52.
4. Nemes Z, Marekov LN, Fesus L, Steinert PM. A novel function for transglutaminase 1: attachment of long-chain omega-hydroxyceramides to involucrin by ester bond formation. Proc Natl Acad Sci U S A. 1999;96(15): 8402–7.
5. Zhang D, Zhu Y, Chen J. Microbial transglutaminase production: understanding the mechanism. Biotechnol Genet Eng Rev. 2010;26:205–22.
6. Yang MT, Chang CH, Wang JM, Wu TK, Wang YK, Chang CY, Li TT. Crystal structure and inhibition studies of transglutaminase from Streptomyces mobaraense. J Biol Chem. 2011;286(9):7301–7.
7. Kashiwagi T, Yokoyama K, Ishikawa K, Ono K, Ejima D, Matsui H, Suzuki E. Crystal structure of microbial transglutaminase from Streptoverticillium mobaraense. J Biol Chem. 2002;277(46):44252–60.
8. Liu S, Zhang DX, Wang M, Cui WJ, Chen KK, Du GC, Chen J, Zhou ZM. The order of expression is a key factor in the production of active transglutaminase in Escherichia coli by co-expression with its pro-peptide. Microb Cell Fact. 2011;10:112.
9. Yurimoto H, Yamane M, Kikuchi Y, Matsui H, Kato N, Sakai Y. The pro-peptide of Streptomyces mobaraensis transglutaminase functions in cis and in trans to mediate efficient secretion of active enzyme from methylotropic yeasts. Biosci Biotechnol Biochem. 2004;68(10):2058–69.
10. Liu S, Zhang DX, Wang M, Cui WJ, Chen KK, Liu Y, Du GC, Chen J, Zhou ZM. The pro-region of Streptomyces hygroscopicus transglutaminase affects its secretion by Escherichia coli. FEMS Microbiol Lett. 2011;324(2):98–105.
11. Yu YJ, Wu SC, Chan HH, Chen YC, Chen ZY, Yang MT. Overproduction of soluble recombinant transglutaminase from Streptomyces netropsis in Escherichia coli. Appl Microbiol Biotechnol. 2008;81(3):523–32.
12. Kikuchi Y, Date M, Yokoyama K, Umezawa Y, Matsui H. Secretion of active-form Streptoverticillium mobaraense transglutaminase by Corynebacterium glutamicum: Processing of the pro-transglutaminase by a cosecreted subtilisin-like protease from Streptomyces albogriseolus. Appl Environ Microbiol. 2003;69(1):358–66.
13. Washizu K, Ando K, Koikeda S, Hirose S, Matsuura A, Takagi H, Motoki M, Takeuchi K. Molecular-cloning of the gene for microbial transglutaminase from Streptoverticillium and its expression in Streptomyces-lividans. Biosci Biotechnol Biochem. 1994;58(1):82–7.
14. Lin YS, Chao ML, Liu CH, Chu WS. Cloning and expression of the transglutaminase gene from Streptoverticillium ladakanum in Streptomyces lividans. Process Biochem. 2004;39(5):591–8.
15. Lin YS, Chao ML, Liu CH, Tseng M, Chu WS. Cloning of the gene coding for transglutaminase from Streptomyces platensis and its expression in Streptomyces lividans. Process Biochem. 2006;41(3):519–24.
16. Ali N, Herron PR, Evans MC, Dyson PJ. Osmotic regulation of the Streptomyces lividans thiostrepton-inducible promoter, ptipA. Microbiol-Sgm. 2002;148:381–90.
17. Liu X, Yang X, Xie F, Qian S. Cloning of transglutaminase gene from Streptomyces fradiae and its enhanced expression in the original strain. Biotechnol Lett. 2006;28:1319–25.
18. Ueda Y, Taguchi S, Nishiyama K, Kumagai I, Miura K. Effect of a rare leucine codon, TTA, on expression of a foreign gene in Streptomyces lividans. Biochim Biophys Acta. 1993;1172(3):262–6.
19. Chen K, Zhang D, Liu S, Wang NS, Wang M, Du G, Chen J. Improvement of transglutaminase production by extending differentiation phase of Streptomyces hygroscopicus: mechanism and application. Appl Microbiol Biotechnol. 2013;97(17):7711–9.
20. Cui L, Du GC, Zhang DX, Liu H, Chen J. Purification and characterization of transglutaminase from a newly isolated Streptomyces hygroscopicus. Food Chem. 2007;105(2):612–8.
21. Noda S, Miyazaki T, Tanaka T, Chiaki O, Kondo A. High-level production of mature active-form Streptomyces mobaraensis transglutaminase via pro-transglutaminase processing using Streptomyces lividans as a host. Biochem Eng J. 2013;74:76–80.
22. Marx CK, Hertel TC, Pietzsch M. Purification and activation of a recombinant histidine-tagged pro-transglutaminase after soluble expression in Escherichia coli and partial characterization of the active enzyme. Enzyme Microb Technol. 2008;42(7):568–75.
23. Kim DW, Kang SG, Kim IS, Lee BK, Rho YT, Lee KJ. Proteases and protease inhibitors produced in streptomycetes and their roles in morphological differentiation. J Microbiol Biotechn. 2006;16(1):5–14.
24. McVey JH, Ward NJ, Buckley SMK, Waddington SN, VandenDriessche T, Chuah MKL, Nathwani AC, McIntosh J, Tuddenham EGD, Kinnon C, et al. Codon optimization of human factor VIII cDNAs leads to high-level expression. Blood. 2011;117(3):798–807.
25. Fang BS, Li W, Ng IS, Yu JC, Zhang GY. Codon optimization of 1,3-propanediol oxidoreductase expression in Escherichia coli and enzymatic properties. Electron J Biotechnol. 2011;14(4):7.
26. Hou Y, Wang H, Wang QL, Zhang FF, Huang YH, Ji YL. Protein expression and purification of human Zbtb7A in Pichia pastoris via gene codon optimization and synthesis. Protein Expr Purif. 2008;60(2):97–102.
27. Binnie C, Cossar JD, Stewart DI. Heterologous biopharmaceutical protein expression in Streptomyces. Trends Biotechnol. 1997;15(8):315–20.
28. Liu L, Yang H, Shin HD, Li J, Du G, Chen J. Recent advances in recombinant protein expression by Corynebacterium, Brevibacterium, and Streptomyces: from transcription and translation regulation to secretion pathway selection. Appl Microbiol Biotechnol. 2013;97(22): 9597–608.
29. Lin SJ, Hsieh YF, Lai LA, Chao ML, Chu WS. Characterization and large-scale production of recombinant Streptoverticillium platensis transglutaminase. J Ind Microbiol Biotechnol. 2008;35(9):981–90.
30. Noda S, Miyazaki T, Tanaka T, Ogino C, Kondo A. Production of Streptoverticillium cinnamoneum transglutaminase and cinnamic acid by recombinant Streptomyces lividans cultured on biomass-derived carbon sources. Bioresour Technol. 2012;104:648–51.
31. Bibb MJ TK, MJ B, KF C, DA H. Practical Streptomyces Genetics. 2nd ed. Norwich: The John Innes Foundation; 2000.

11

Reliable handling of highly A/T-rich genomic DNA for efficient generation of knockin strains of *Dictyostelium discoideum*

Asuka Mukai, Aya Ichiraku and Kazuki Horikawa[*]

Abstract

Background: Social amoeba, *Dictyostelium discoideum*, is a well-established model organism for studying cellular physiology and developmental pattern formation. Its haploid genome facilitates functional analysis of genes by a single round of mutagenesis including targeted disruption. Although the efficient generation of knockout strains based on an intrinsically high homologous recombination rate has been demonstrated, successful reports for knockin strains have been limited. As social amoeba has an exceptionally high adenine and thymine (A/T)-content, conventional plasmid-based vector construction has been constrained due to deleterious deletion in *E. coli*.

Results: We describe here a simple and efficient strategy to construct *GFP*-knockin cassettes by using a linear DNA cloning vector derived from N15 bacteriophage. This allows reliable handling of DNA fragments whose A/T-content may be as high as 85 %, and which cannot be cloned into a circular plasmid. By optimizing the length of recombination arms, we successfully generate *GFP*-knockin strains for five genes involved in cAMP signalling, including a triple-colour knockin strain.

Conclusions: This robust strategy would be useful in handling DNA fragments with biased A/T-contents such as the genome of lower organisms and the promoter/terminator regions of higher organisms.

Keywords: A/T-rich genome, Linear DNA cloning, Knockin, *Dictyostelium discoideum*

Background

The eukaryote *Dictyostelium discoideum* (*D. discoideum*), also called social amoeba, is an excellent model to study the principles of unicellular physiology and multicellular development. In addition to the small genome size (34 Mb encoding ~13,500 genes) [1], its haploidy is highly compatible with the functional analysis of genes by mutational approaches. Genome-wide mutations have been randomly introduced by chemical mutagenesis and semi-randomly by restriction enzyme mediated integration mutagenesis (REMI) [2, 3]. Homologous recombination has also been effective for introducing site-specific mutations, or for the targeted disruption of genes of interest (GOI). As with other model organisms, gene targeting has been performed by introducing a DNA cassette consisting of a selection marker flanked by 5′ and 3′ recombination arms. For *D. discoideum*, up to a few kb of homology arm was sufficient to obtain knockout strains with high efficiency (>20 %) [4, 5].

Knockin, while also utilizing homologous recombination, would be a powerful strategy. 5′- or 3′-terminal tagging of genes with fluorescent proteins (FPs) at the endogenous locus is highly useful for quantitative analysis of protein abundance and cellular localization [6, 7]. Replacing the endogenous promoter with a synthetic one such as the tetracycline-inducible system (Tet-system) is another application which allows efficient control of the gene expression level [8–10]. In spite of potential applicability, successful generation of knockin strains has been limited to a few loci, while knockout strains have been routinely generated [6–12].

One possible explanation for this constraint is the technical difficulty in constructing knockin DNA cassettes compared with knockout DNA cassettes. The DNA cassette for gene tagging and promoter replacement must harbour

* Correspondence: horikawa.kazuki@tokushima-u.ac.jp
Division of Bioimaging, Institute of Biomedical Sciences, Tokushima University Graduate School, 3-18-15 Kuramoto-cho, Tokushima City, Tokushima 770-8503, Japan

terminator or promoter sequences from the endogenous targeted gene. The genome of *D. discoideum* has an intrinsically high A/T-content, with the promoter, terminator and intron sequences often showing >85 % A/T-content while that of exons is more moderate (~70 % of A/T) [1]. As has been widely accepted, A/T-rich or repetitive DNA fragments are notoriously unstable in circular plasmids due to secondary structures and the functional interference with transcription and replication systems of the host *E. coli* [13–16].

Cloning of unstable DNA has been enhanced by the use of modified plasmid vectors and host *E. coli* [14]. Low-copy number and transcription-free plasmids have been utilized for cloning small A/T-rich fragments. Recombinase-deficient *E. coli* strains have been developed to increase the stability of cloned DNAs containing repetitive sequences. However, construction of tagging vectors by using a circular plasmid was unsuccessful in our trial. We routinely experienced difficulties in the cloning of promoter or 3′ UTR/terminator sequences whose A/T-contents exceeded 75 %. Once a knockin cassette was successfully constructed, we often experienced serious instability during its maintenance in *E. coli* even when we utilized the improved materials described above.

To circumvent these issues, we employed a recently developed linear DNA cloning system that allows robust handling of large and A/T-rich DNA fragments [17–19]. We first demonstrated the efficient construction of 3′-tagging vectors by a simple restriction-ligation method. We then optimized the size of recombination arms for efficient knockin, and identified that the critical arm length differs depending on the target locus. Robustness of our strategy was finally demonstrated by multiple labelling of a gene with three different fluorescent proteins. These results suggest that our simple strategy would facilitate genomic manipulations which had previously been hampered by the inability to clone DNA fragments with biased A/T-contents for a variety of model organisms.

Methods
General molecular biology
For the preparation of genomic DNA, amoeba cells were suspended in quick lysis solution (50 mM KCl, 10 mM Tris pH 8.3, 2.5 mM MgCl$_2$, 0.45 % NP40, 0.45 % Tween 20 and 1 μg/μl of proteinase K) to ~50 cells/μl. Cells were lysed at 55 °C for 10 min, then heat denatured at 95 °C for 5 min. 1 μl of cell lysate was analysed by PCR. PCR for DNA cloning and genome analysis was performed by using KODplus (TOYOBO) and EmeraldAmp (TAKARA), respectively. PCR was performed according to the touch-down protocol as follows: (94°C_1 min, 60 °C_30 s, 60 °C_1 min/kb) × 5 cycles; followed

by (94°C_1 min, 53°C_30 s, 60 °C_1 min/kb) × 5 cycles; (94 °C_1 min, 46°C_30 s, 60 °C_1 min/kb) × 5 cycles; and (94°C_1 min, 40°C_30 s, 60 °C_1 min/kb) × 25 cycles. Transcripts from knockin loci were RT-PCR amplified using an oligo-dT primer. Sequence verified cDNAs were subcloned into *pDM304*, the extrachromosomal plasmid vector for *D. discoideum*, to establish cells over-expressing tagged proteins [20]. Southern blotting was performed as described in Additional file 1.

Handling A/T-rich DNA
Extension of PCR was carried out at 60 °C as A/T-rich fragments are often poorly amplified at higher temperatures [21]. DNA fragments separated by gel electrophoresis were stained with Gel-Red (Biotium) then were visualized by a blue light illuminator instead of UV to minimise photo-induced DNA damage. DNA fragments were collected by mechanical filtration by using a GenElute Agarose spin column kit (SigmaAldrich: G2291-70EA). Note that the chemical elution kit was not appropriate as A/T-rich DNAs are irreversibly denatured in the presence of chaotropic agents such as guanidinium and iodide ions [22].

Universal knockin module
To create universal knockin modules encoding fusion-ready fluorescent proteins followed by the excisable selection marker (Additional file 2: Figure S1), three amplicon units were sequentially cloned between the BamHI and ApaI site of pBluescript SK. Unit 1: cDNA of FPs sandwiched between a fusion linker (encoding 26 amino acids, Additional file 2: Figure S1) and the 1st LoxP site (BamHI > HindIII). Unit 2: Terminator of *myosin heavy chain A* (HindIII > SalI). Unit 3: *BsR*-selection marker with the 2nd LoxP site amplified from pLPBLP [23] (SalI > ApaI). To allow efficient expression of fluorescent proteins in *D.d* cells, cDNA from *Aequorea victoria* (*a.v.*) was utilized. To generate *a.v._mEGFP*, F64L, S65T, Y100F, S108T, M141L, A206K and I219V mutations were introduced into the original *GFP* by site directed mutagenesis. To generate *a.v._Turquoise2* [24], F64L, Y66W, S72A, Y100F, S108T, M141L, N146F, H148D, M153T, V163A, S175G, I219V and H231L were introduced to the original GFP. *mRFPmars* whose codon usage was optimised for *D. discoideum*, was PCR amplified from *pmRFPmars-LimΔcoil* (Dicty Stock Center ID_475) [25]. Resulting plasmids, *pUniv_CKI_mEGFP*, *_Turq2* and *_mRFPmars* whose GenBank Accession numbers are KU163138, KU163139 and KU163140, respectively, were deposited in the Dicty stock center.

Construction of knockin vectors by using linear cloning system
DNA fragments for the knockin vector were assembled by using the pJAZZ linear DNA cloning system

(Lucigen, #43036 and 43042). We first PCR amplified the 3′ recombination arm (i.e., 3′ UTR/terminator) and cloned it into the pJAZZ-OK_blunt vector according to the manufacturer's instructions. 10–30 μg of pJAZZ vector harbouring the 3′ recombination arm in the correct orientation was digested with ApaI and the resulting shorter fragment was separated by gel electrophoresis. The fragment was filter-eluted and concentrated to >100 ng/μl after desalting by ethanol precipitation. Similarly, a long arm of NotI digested pJAZZ control vector, NotI/BamHI digested 5′ recombination arm of GOI and BamHI/ApaI digested knockin module were prepared. 100 ng each of these four fragments were directionally ligated by single tube reaction and electroporated into TSA *E. coli*. Colonies were isolated from a YT-agar plate containing 30 μg/ml of kanamycin and screened by PCR and restriction enzyme digestion. For a large-scale preparation of the knockin vector, transformed TSA *E. coli* were cultured in 100 ml of TB medium. Approximately 100 μg of knockin vector was routinely obtained using the NucleoBond Xtra Midi kit (MACHEREY-NAGEL). To prepare the transformation-ready knockin cassette, 100 μg of knockin vectors were digested with NotI for 10 h then were cleaned-up by phenol-chloroform extraction and ethanol precipitation without DNA-size fractionation. The knockin DNA cassettes were suspended in sterile water (2 μg/μl) and were stored at -30 °C until use.

Cell culture

Axenic strain Ax2 was cultured and transformed as described elsewhere [4, 23, 26]. For knockin transformation, cells were washed and suspended with ice-cold EP buffer (6.6 mM KH_2PO_4, 2.8 mM Na_2HPO_4, 50 mM sucrose, pH 6.4) to 1×10^7 cells/ml. 800 μl of cell suspension mixed with 10–40 μg of NotI digested knockin vector in a 4-mm width cuvette was subjected to electroporation (5 s separated two pluses with 1.0 kV and 1.0 ms of the time constant) by using a MicroPulser (Bio-Rad). These cells were plated on 4–6 pieces of 10 cm-petri dishes with HL5 medium and incubated at 22 °C for 18 h under the non-selective condition, then were cultured in the presence of 12.5 μg/ml Blasticidin S (Wako). After 4–7 days, colonies of candidate recombinants were manually picked and were transferred into 96-well plates, then subjected for the genome check by PCR and Southern blotting analysis. Cre-recombinase mediated removal of the selection cassette was performed as described previously [23]. Briefly, NLS-Cre was transiently expressed by introducing *pDEX_NLS-Cre*. Candidate clones were selected in the presence of 5 μg/ml of G418 for 2–5 days followed by an additional few days' culture in the absence of G418 (LifeTechnologies). Optionally, single cell sorting of isolated clones

was performed by a JSAN cell sorter equipped with a CloneMate module (Bay Bioscience).

Immunoblotting

Cells starved for 8 h were lysed with SDS sample buffer. Proteins (3.6×10^5 cells/lane) were separated by SDS-polyacrylamide gel electrophoresis (5–15 % gradient gel, BIO CRAFT) and blotted on the PVDF membrane. *GFP/Turquoise2-* or *mRFPmars*-tagged proteins were detected with a rabbit monoclonal antibody to *GFP* (Abcam, ab183735) and a rat monoclonal antibody to *RFP* (ChromoTek, 5 F8), respectively.

Imaging

Triple-colour knockin cells on a 10-cm petri dish (8×10^6 cells/dish) were starved for 8–10 h in Development Buffer (5 mM Na_2HPO_4, 5 mM KH_2PO_4, 1 mM $CaCl_2$, 2 mM $MgCl_2$). These cells were re-plated on a 35 mm-glass bottomed dish (Iwaki) and were left at 22 °C for 30 min allowing them to establish the aggregation stream. Live cell images were captured on an inverted confocal microscope (Nikon A1R, Nikon) equipped with PlanApoVC 60xWI (NA 1.2), multi Argon gas laser (457, 488 nm, Melles Griot) and 561 nm DPSS laser (Melles Griot).

Results

Construction of A/T-rich knockin vector by a linear cloning system

To construct >4 kb of A/T-rich *GFP* knockin cassette consisting of 5′ and 3′ recombination arms (~1.0 kb each) centred by the knockin module encoding the fusion ready *GFP* and the selection marker (*BsR*; blasticidin resistance), we designed a simple experimental scheme that utilizes a linear DNA cloning system which allows robust handling of large and A/T-rich DNA (Fig. 1a). To facilitate vector construction, a universal knockin module was independently prepared as *pUniv_CKI_mEGFP* (Fig. 1a and Additional file 2: Figure S1). In brief, it contains cDNA encoding *mEGFP* followed by a generic terminator from *myosin heavy chain A* (*MHC*), which allows the expression of a *GFP*-fusion protein just after the knockin event. Two LoxP sites were introduced for Cre-mediated excision of the *MHC* terminator and *BsR* expression unit, which is needed for recycling the *BsR* marker and for restoring appropriate genomic organization to allow normal gene expression under the control of the endogenous 3′ UTR/terminator of the GOI.

To validate this set-up, we created a *GFP*-knockin cassette for *carA-1*(DDB_G0273397), encoding cAMP receptor during aggregation (Fig. 1b). We cloned PCR amplified 3′ UTR/terminator (1.0 kb) of *carA-1* into pJAZZ-OK_blunt vector [17]. Although it was not possible to clone this 1.0 kb fragment with 85 % A/T

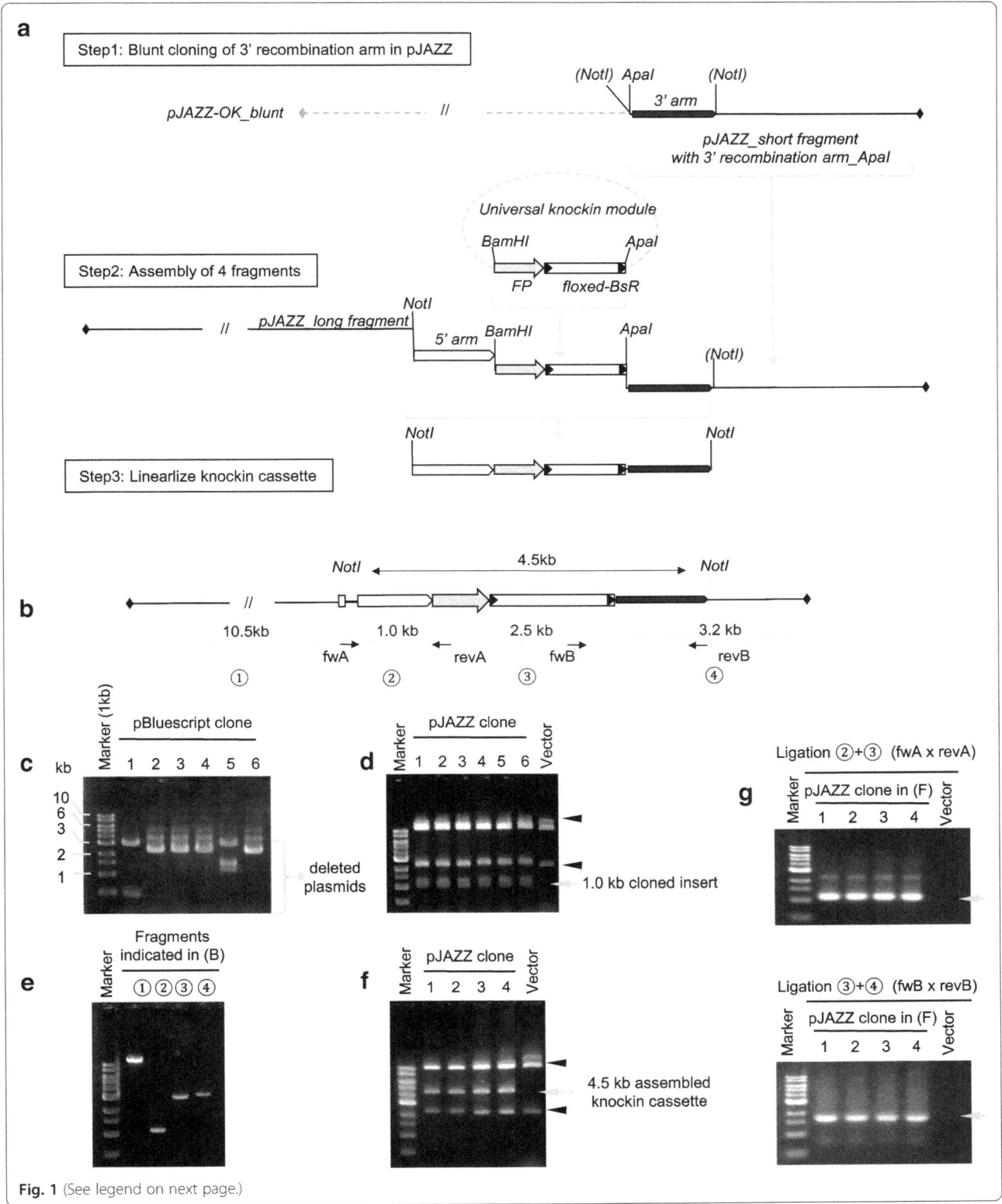

Fig. 1 (See legend on next page.)

(See figure on previous page.)
Fig. 1 Knockin vector construction by a linear DNA cloning system. **a** 3-step construction of the 3'-tagging vector. Step 1: Preparation of pJAZZ vector harbouring A/T-rich 3' recombination. Step 2: Assembly of the knockin vector by 4-piece ligation. Step 3: Release of knockin cassette by NotI digestion. **b** Design of *GFP* knockin vector for DDB_G0273397/*carA-1* harbouring 1.0 kb each of 5' and 3' recombination arms. **c, d** Stable cloning of A/T-rich 3' UTR/terminator of *carA-1* by linear cloning system. 1 kb of 3' UTR/terminator of *carA-1* were blunt cloned into pBluescript (**c**) and pJAZZ vector (**d**). Release of the insert in randomly selected 6 DNA clones was checked by restriction enzyme digestion with XhoI and SpeI for pBluescript and with NotI for pJAZZ (these are the multiple cloning sites on each vector). Variable size of released fragments in C indicates deletions of circular plasmids. Appropriate size of inserts (*arrow* in **d**) were released from all the 6 clones of pJAZZ vector. The lane for negative control (Vector) was loaded with NotI-digested pJAZZ vector carrying no insert. Arrow heads represent the long and short arm of NotI-digested pJAZZ vector. **e** Four DNA fragments as depicted in B were subjected to directional ligation. **f** DNAs from randomly selected TSA *E. coli* clones were digested with NotI to excise the 4.5 kb of assembled knockin cassette (*arrow*). **g** Appropriate DNA assembly in 4 clones (same as in **f**) was detected by PCR for fragment ligation between 2 and 3 (*upper column*, 2 + 3) or fragment 3 and 4 (*bottom column*, 3 + 4). Primer position was depicted in (**b**). All the molecular marker was1 kb DNA ladder

content into a circular plasmid due to the occurrence of deletions (Fig. 1c), pJAZZ vector allowed robust cloning without any signs of deletions and rearrangements (Fig. 1d) which was separately confirmed by nucleotide sequencing (data not shown). The unique ApaI site at the cloning site was utilized to prepare the shorter arm of pJAZZ vector harbouring a 3' recombination arm of *carA-1* and the kanamyicin resistance unit. Other DNA fragments corresponding to the longer arm of the pJAZZ vector (a NotI-digested 10.5 kb fragment including the replication origin), PCR amplified and NotI/BamHI digested 5' recombination arm (72 % A/T), and BamHI/ApaI digested universal knockin module (72 % of A/T) were similarly prepared (Fig. 1e and Table 1). These four DNA fragments were directionally ligated in vitro and were transformed into TSA competent cells, which is the optimized host *E. coli* for pJAZZ vector [17]. Assembly of these fragments was efficient as 75 % of checked clones were identified as carrying an appropriately ligated 4.5 kb knockin cassette for the *carA-1* gene (Fig. 1f-g). We also successfully constructed knockin vectors for other genes involved in cAMP signalling, containing a range of different recombination arm lengths (100–50 % efficiency, Table 1), demonstrating

the robustness and general applicability of this strategy to construct the large and A/T-rich knockin DNA cassette.

Optimized conditions for *GFP* knockin
The amount of targeting cassette and the length of recombination arms have been shown to control the homologous recombination rate. To determine the minimum amount of DNA vector for 3'-terminal knockin, 10, 20 or 40 μg of NotI-digested knockin vector for *carA-1* was electroporated into wildtype cells (Ax2). While none or few blasticidin resistant (*BsR*) colonies were obtained from cells transformed with 10 and 20 μg of linearized vector, more than 500 candidate clones grew under the selective culture conditions when 40 μg of total DNA was introduced. PCR analysis specifically detecting the homologous recombination event at both the 5' and 3' arms identified that 28 % of cells transformed with 40 μg of vector were positive clones whose *carA-1* locus was replaced by the knockin cassette (Fig. 2a-b and Table 2). Southern blotting (Additional file 3: Figure S2), genomic sequencing (data not shown) and immunoblotting analysis (Fig. 2c) confirmed the specific expression of *cARA-1-GFP* protein that is not the result of random

Table 1 Efficiency of knockin vector construction by using a linear DNA cloning system

Gene	Vector name	Arm length		A/T-content			Positive clone	Efficiency
		5' arm	3' arm	5' arm	3' arm	Knockin cassette Total		
*carA-1*DDB_G0273397	pJ_carA1_GFP	1.0 kb	1.0 kb	72 %	85 %	75 % (4.6 kb)	12/16	75 %
*acaA*DDB_G0281545	pJ_acaA_GFP	0.6 kb	0.5 kb	74 %	83 %	72 % (3.7 kb)	8/8	100 %
*regA*DDB_G0284331	pJ_regA_GFP_0.3	0.3 kb	1.0 kb	69 %	79 %	73 % (3.8 kb)	8/8	100 %
	pJ_regA_GFP_0.5	0.5 kb	1.0 kb	69 %	79 %	73 % (4.0 kb)	7/8	88 %
	pJ_regA_GFP_1.3	1.3 kb	1.0 kb	76 %	79 %	74 % (4.8 kb)	7/8	88 %
*dagA/crac*DDB_G0285161	pJ_crac_GFP_0.3	0.3 kb	0.6 kb	69 %	80 %	72 % (3.7 kb)	8/8	100 %
	pJ_crac_GFP_1.9	1.9 kb	0.6 kb	70 %	80 %	72 % (5.0 kb)	11/16	69 %
*erkB*DDB_G0283903	pJ_erkB_GFP_0.5/0.6	0.5 kb	0.6 kb	64 %	66 %	70 % (3.6 kb)	8/8	100 %
	pJ_erkB_GFP_1.0/0.6	1.0 kb	0.6 kb	72 %	66 %	71 % (4.1 kb)	7/8	88 %
	pJ_erkB_GFP_1.0/1.7	1.0 kb	1.7 kb	72 %	77 %	74 % (5.2 kb)	4/8	50 %
	pJ_erkB_GFP_0.5/1.7	0.5 kb	1.7 kb	64 %	77 %	73 % (4.7 kb)	4/8	50 %

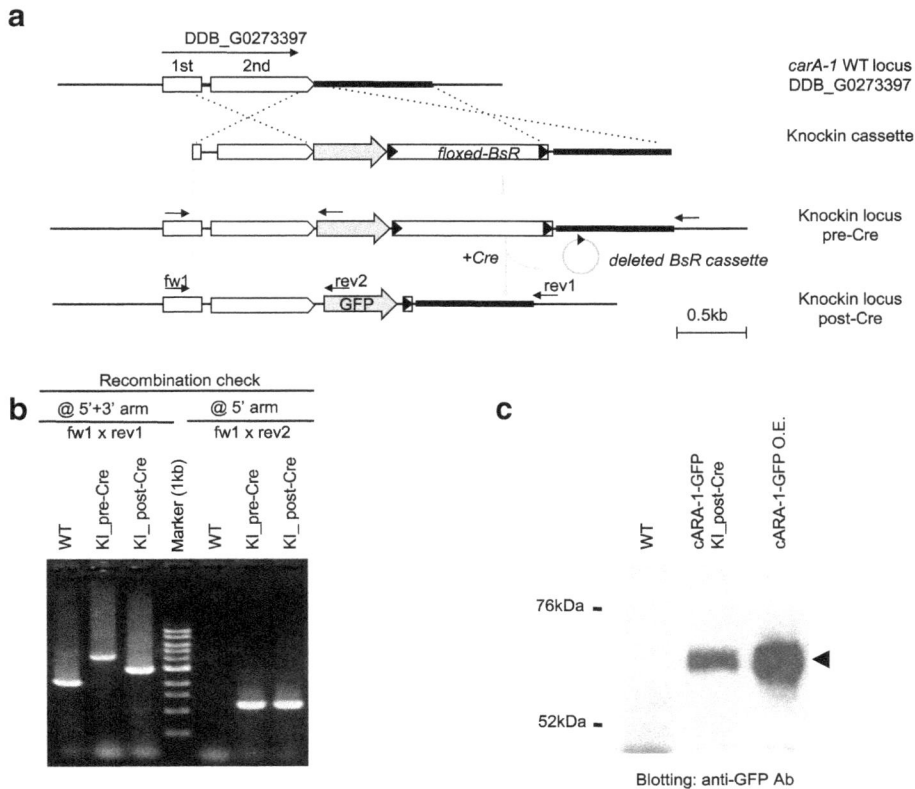

Fig. 2 Generation of *GFP* knockin strain for *carA-1*. **a** Genomic organization of wild-type (WT) and *GFP* knockin locus for DDB_G0273397/*carA-1*. **b** WT and knockin locus before and after the removal of *BsR* cassette was detected by PCR. The primer set fw1/rev1, both located outside the homology arms, detects WT and knockin locus (pre_Cre) as 2.0 and 4.5 kb, respectively. *BsR*-removal was detected as the decrease in the size of target locus from 4.5 kb (pre-Cre) to 2.7 kb (post-Cre). Primer combination of rev2 for *GFP* and fw1 confirms specific recombination at the 5′ arm by yielding a 1.0 kb fragment. **c** Expression of *GFP*-tagged *cARA-1* protein of the knockin strain detected by western blotting. Lysate of cells over-expressing *cARA-1-GFP* from extrachromosomal plasmid were loaded as the detection control (1/10 volume, *cARA-1-GFP* O.E)

integration of knockin vector. We concluded that 40 μg of total DNA, which corresponds to ~10 μg of knockin cassette, is optimal to obtain knockin cells.

We next investigated how the length of the recombination arm affects knockin efficiency. It has been demonstrated that the minimum length of homology arms differs significantly among eukaryotic species. For knockout, as short a length as 20 nucleotides is sufficient for budding yeast, *S. cerevisiae* [27]. Much longer arms up to 10 kb are needed for mice and malaria parasites [18, 28], while less than a few kb are sufficient for *D. discoideum* [23]. By using knockin vectors with distinct lengths of recombination arms, we tried to generate knockin strains for an additional four genes involved in cAMP signalling, as the recombination efficacy was expected to differ depending on the genomic locus. When cells were transformed with a knockin vector harbouring short recombination arms (0.5 kb each for 5′ and 3′) for the *acaA* gene, encoding adenylate cyclase during aggregation, appropriate recombinants were efficiently obtained (46 %), suggesting that homology arms as short as

a few hundred base-pairs would be sufficient. For the *regA* gene, encoding cAMP specific intracellular phosphodiesterase, three constructs with 0.3, 0.5 and 1.3 kb of 5′ arm in combination with 1.0 kb of 3′ arm were examined. Although the vector with 0.3 or 0.5 kb of 5′ arm yielded no recombinants, that with 1.3 kb of 5′ arm yielded 58 % of positive colonies, indicating that a longer 5′ arm is needed for *regA* locus. This is also the case for the *crac* gene (DDB_G0285161); encoding cytoplasmic regulator of *acaA*, such that 1.9 kb, but not 0.3 kb of 5′ recombination arm was needed for successful knockin. In the case of MAP kinase gene, *erkB*, a longer 3′ arm rather than 5′ arm was required. Vectors with 0.5 kb of each of homology arm yielded no recombinants from 96 screened colonies. Extending the 5′ arm to 1.0 kb was not effective, although one clone with a single cross-over at the 5′ arm (i.e., integration) was obtained from 288 *BsR*-clones. We then tested a knockin cassette whose 3′ arm was extended to 1.7 kb, reaching into the 2nd exon of the inversely located neighbouring gene, *pigA* (DDB_G0283965), while the 5′ arm was kept

Table 2 Effect of the arm length variation on knockin efficiency

Gene	Chromosome	Vector name	5′ arm length	3′ arm length	Total amount of knockin vector	[a]BsR clone	[b]Positive clone	Efficiency
carA-1DDB_G0273397	2	pJ_cAR1_GFP	1.0 kb	1.0 kb	10 μg	0	-	0 %
					20 μg	3	0/3	0 %
							0/0	-
					40 μg	>500	27/96	28 %
							3/16	19 %
acaADDB_G0281545	3	pJ_acaA_GFP	0.6 kb	0.5 kb	40 μg	>200	5/11	46 %
							3/8	38 %
regADDB_G0284331	4	pJ_regA_GFP_0.3	0.3 kb	1.0 kb	40 μg	124	0/96	0 %
							0/28	0 %
		pJ_regA_GFP_0.5	0.5 kb	1.0 kb	40 μg	138	0/96	0 %
							0/48	0 %
		pJ_regA_GFP_1.3	1.3 kb	1.0 kb	40 μg	180	56/96	58 %
							9/16	56 %
dagA/cracDDB_G0285161	4	pJ_crac_GFP_0.3	0.3 kb	0.6 kb	40 μg	>500	0/96	0 %
							0/96	0 %
		pJ_crac_GFP_1.9	1.9 kb	0.6 kb	40 μg	>500	10/16	63 %
							8/51	16 %
erkBDDB_G0283903	4	pJ_erkB_GFP_0.5/0.6	0.5 kb	0.6 kb	40 μg	>500	0/96	0 %
							0/96	0 %
		pJ_erkB_GFP_1.0/0.6	1.0 kb	0.6 kb	40 μg	>500	0/288[c]	0 %
							0/96	0 %
		pJ_erkB_GFP_1.0/1.7	1.0 kb	1.7 kb	40 μg	>500	16/30	53 %
							4/16	25 %
		pJ_erkB_GFP_0.5/1.7	0.5 kb	1.7 kb	40 μg	>500	14/30	47 %
							5/16	31 %

[a]Total number of BsR clones from two experiments by using independently prepared materials
[b]The number of homologous recombinants obtained from two independent experiments was separately provided
[c]One clone with single cross-over at 5′ arm obtained from 288 of BsR colonies

short (0.5 kb). In this case, 48 % of screened clones were identified to be knockin cells (14/30). The efficiency was comparable to the targeting vector harbouring long homology arms both for 5′ and 3′ arms (1.0 and 1.7 kb, respectively, 53 %). Although we did not examine much longer recombination arm, these results suggest that the efficiency of homologous recombination in *D. discoideum* was not linearly enhanced by longer arm, rather it was simply controlled in all-or-none fashion (Table 2). All together, these results demonstrate that the length of recombination arm up to 3 kb would be sufficient for 3′-tagging, in which downsizing was possible depending on the genomic loci.

Multicolour labelling by iterative knockin

To assess the applicability of the developed strategy, we tried to generate a knockin strain in which multiple genes were labelled with three different coloured fluorescent proteins. For this, we constructed universal knockin modules carrying fusion ready cyan and red fluorescent proteins in addition to *GFP* (Additional file 2: Figure S1B). To overcome the limited availability of the selectable marker in *dicty* cells, *Cre*-mediated recycling of *BsR* gene was utilized [23]. *BsR*-selection marker of *acaA-GFP* strain was removed by the transient expression of *NLS-Cre* (Fig. 3a left column). Restored sensitivity to blasticidin allowed the selection of secondary knockin of *regA-mRFPmars* (Fig. 3a middle column). High knockin efficiency of *regA-mRFPmars* (48 %) was reproduced as in the case for *regA-GFP*. Similarly, removal of *BsR* gene was repeated before and after introduction of *carA-1-Turq2* (Fig. 3a right column). PCR and immunoblotting analysis confirmed successful labelling of *regA* and *carA-1* with red and cyan fluorescent protein, respectively, yielding a triple-colour knockin strain (Fig. 3b). Live imaging of starved cells allowed simultaneous detection of subcellular localization of endogenously expressed *ACAA*, *REGA* and *cARA-1* (Fig. 3c). As expected, *REGA-mRFPmars* was distributed

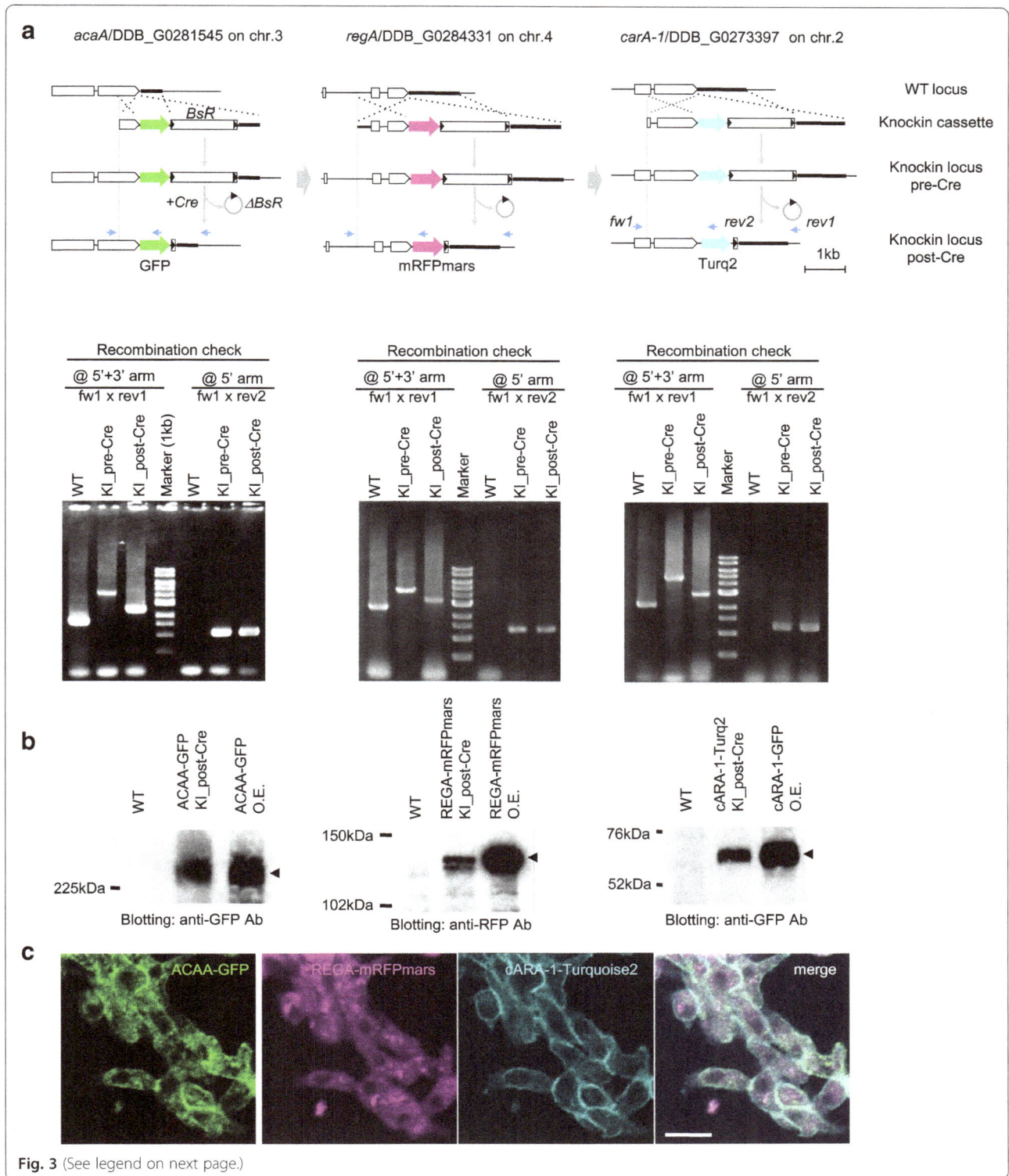

a

acaA/DDB_G0281545 on chr.3   regA/DDB_G0284331 on chr.4   carA-1/DDB_G0273397 on chr.2

WT locus
Knockin cassette
Knockin locus pre-Cre
Knockin locus post-Cre

BsR   +Cre   ΔBsR   GFP   mRFPmars   fw1   rev2   rev1   Turq2   1kb

Recombination check

@ 5'+3' arm   @ 5' arm
fw1 x rev1   fw1 x rev2

WT   KI_pre-Cre   KI_post-Cre   Marker (1kb)   WT   KI_pre-Cre   KI_post-Cre

b

WT   ACAA-GFP KI_post-Cre   ACAA-GFP O.E.
225kDa

Blotting: anti-GFP Ab

WT   REGA-mRFPmars KI_post-Cre   REGA-mRFPmars O.E.
150kDa
102kDa

Blotting: anti-RFP Ab

WT   cARA-1-Turq2 KI_post-Cre   cARA-1-GFP O.E.
76kDa
52kDa

Blotting: anti-GFP Ab

c

ACAA-GFP   REGA-mRFPmars   cARA-1-Turquoise2   merge

Fig. 3 (See legend on next page.)

(See figure on previous page.)
Fig. 3 Triple-colour knockin for *ACAA-GFP*, *REGA_mRFPmars* and *cARA-1-Turq2*. **a** Genomic organization of wild-type (WT) and knockin locus (KI) for DDB_G0281545/*acaA* (*left column*), DDB_G0284331/*regA* (*middle column*) and DDB_G0273397/*carA-1* (*right column*) tagged with green, red and cyan fluorescent proteins, respectively. Serial knockin in this order was performed, each followed by *Cre*-mediated *BsR*-recycling. Specific recombination was detected by PCR as in Fig. 2b. Primer fw1/rev1 detects a knockin event (WT to knockin_pre-Cre) and removal of *BsR* cassette (knockin_pre-Cre to post-Cre) as a 2.5 kb increase and 1.7 kb decrease of the amplified band, respectively. The fw1/rev2 set detects specific recombination at the 5′ recombination arm. **b** Protein expression of triple-colour knockin strain. Lysate from over-expressing cells (1/10 volume) with *ACAA-GFP*, *REGA-mRFPmars* and *cARA-1-GFP* were loaded as the positive control. **c** Live confocal images of the triple-colour knockin strain developed for 8 h. *ACAA-GFP* and *cARA-1-Turq2* were localized at the cell surface. *REGA-mRFPmars* was detected in the cytoplasm excluded from the nucleus. Bar: 10 μm

throughout the cytoplasm, except for the nucleus. *cARA-1-Turq2* was evenly presented at the cell membrane [29, 30]. *ACA-GFP* was also detected at the cell periphery like cARA-1, and there were no signs for the polarized accumulation at the rear end of chemotacting cells as reported for over-expressed *ACAA-YFP* [30, 31]. It is not clear the reasons for this, but the different experimental conditions such as the expression level, genetic backgrounds, linkers connecting ACA with FP or imaging conditions might explain mechanisms which should be carefully analyzed in future studies.

Discussion

We demonstrate here a simple and reliable strategy to establish knockin strains of *Dictyostelium discoidium*, which had been hampered by the difficulty in cloning A/T-rich DNA into circular plasmids. To our knowledge, the pJAZZ system is the unique solution allowing robust handling of unstable DNAs including genomic sequences of *D. discoideum* [17]. As has been recently reported, this enabled development of high quality genomic libraries for the malaria parasite *Plasmodium berghei* (*P. berghei*) and *D. discoideum*; both are characterized by exceptionally high A/T-contents [18, 19]. The unbiased coverage and increased stability with no limitation for the insert size suggest promising applicability for cloning A/T-rich sequences, including promoters and introns of *D. discoideum* which are required for 5′-tagging or promoter replacement.

The linear DNA cloning platform also allows the secondary modification of the cloned insert. As demonstrated here, simple restriction-ligation is feasible to assemble multiple pieces of A/T-rich DNA fragments without any special requisites. Compatibility with other technologies such as recombineering, utilizing *lambda* Red/ET-recombination, and the Gateway system yielded a scalable pipeline that can convert the pJAZZ-based genomic library to thousands of reverse genetic vectors for *P. berghei*). [18]. Its utility has been further strengthened by locus specific barcoding providing a versatile community resource named *Plasmo*GEM [32, 33]. Building the analogous pipeline based on the pJAZZ vector would no doubt accelerate the research activity of *dicty* community.

Another emphasis of our demonstration is in providing practical guidelines in designing a knockin vector for *D. discoideum*. While increasing the length of the homology arm up to 10 kb linearly boosts the recombination efficiency in *P. berghei*). and mice [18, 28], this is not the case for *D. discoideum*. Rather, the critical arm length estimated to be ~3 kb simply controls the knockin event in an almost all-or-none fashion (Table 2). Previously, poor targeting frequency was reported for some loci in which the single cross-over dominated over the double recombination event [34, 35]. *erkB* locus is one such example, as one insertion clone at the 5′ arm was obtained even after extensive screening. In our trial, this was simply overcome by a slightly longer recombination arm, which had not been testable under the constraints of conventional circular plasmid. As the critical length differs depending on the genomic locus, it would be cost and time saving to start with a total of 3 kb of homology arms for the manual construction of reverse genetic vectors.

Conclusion

The linear DNA cloning system is milestone technology allowing reliable handling of A/T-rich genomic DNAs needed for the reverse genetic analysis of *D. discoideum*. Promoter exchange, site-directed mutagenesis and functional tagging of endogenous proteins will bring new insights into the molecular mechanisms of intercellular communication, self-recognition and kin-discrimination, all of which are involved in social behaviour, one of the most popular research targets of *D. discoideum*. [11, 36, 37].

Abbreviations
A/T: adenine and thymine; GFP: green fluorescent protein; RFP: red fluorescent protein; cAMP: 3′,5′-cyclic adenosine monophosphate.

Competing interests
The authors declare that they have no competing interests.

Authors' contributions
AM and AI performed cellular and molecular biology work. AM and KH designed the experiment, analysed the data and wrote the manuscript. All authors read and approved the final version of manuscript.

Acknowledgements

We thank H. Urushibara at Tsukuba University for helpful discussion, Y. Okamura and M. Kitamura at the Support Center for Advanced Medical Sciences, Tokushima University for cell-sorting. We also acknowledge the Dicty Stock Center for providing *pLPBLP, pDEX_NLS-Cre, pDM304* and *pmRFPmars-LimΔcoil*. This work was supported in part by a Grant-in-Aid for Challenging Exploratory Research (15 K14491), by a Grant-in-Aid for Scientific Research on Innovative Areas "Spying minority in biological phenomena (No. 3306)" from The Ministry of Education, Culture, Sports, Science and Technology (MEXT) (23115003) to KH. Funding for open access charge: [MEXT/23115003].

References

1. Eichinger L, Pachebat JA, Glockner G, Rajandream MA, Sucgang R, Berriman M, Song J, Olsen R, Szafranski K, Xu Q et al. The genome of the social amoeba Dictyostelium discoideum. Nature. 2005;435(7038):43–57.

2. Kuspa A, Loomis WF. Tagging developmental genes in Dictyostelium by restriction enzyme-mediated integration of plasmid DNA. Proc Natl Acad Sci U S A. 1992;89(18):8803–7.

3. Yanagisawa K, Loomis Jr WF, Sussman M. Developmental regulation of the enzyme UDP-galactose polysaccharide transferase. Exp Cell Res. 1967;46(2):328–34.

4. Gaudet P, Pilcher KE, Fey P, Chisholm RL. Transformation of Dictyostelium discoideum with plasmid DNA. Nat Protoc. 2007;2(6):1317–24.

5. Wiegand S, Kruse J, Gronemann S, Hammann C. Efficient generation of gene knockout plasmids for Dictyostelium discoideum using one-step cloning. Genomics. 2011;97(5):321–5.

6. Bukharova T, Weijer G, Bosgraaf L, Dormann D, van Haastert PJ, Weijer CJ. Paxillin is required for cell-substrate adhesion, cell sorting and slug migration during Dictyostelium development. J Cell Sci. 2005;118(Pt 18):4295–310.

7. Zhang XY, Langenick J, Traynor D, Babu MM, Kay RR, Patel KJ. Xpf and not the Fanconi anaemia proteins or Rev3 accounts for the extreme resistance to cisplatin in Dictyostelium discoideum. PLoS Genet. 2009;5(9):e1000645.

8. Blaauw M, Linskens MH, van Haastert PJ. Efficient control of gene expression by a tetracycline-dependent transactivator in single Dictyostelium discoideum cells. Gene. 2000;252(1–2):71–82.

9. Funamoto S, Meili R, Lee S, Parry L, Firtel RA. Spatial and temporal regulation of 3-phosphoinositides by PI 3-kinase and PTEN mediates chemotaxis. Cell. 2002;109(5):611–23.

10. Myers SA, Han JW, Lee Y, Firtel RA, Chung CY. A Dictyostelium homologue of WASP is required for polarized F-actin assembly during chemotaxis. Mol Biol Cell. 2005;16(5):2191–206.

11. Hirose S, Benabentos R, Ho HI, Kuspa A, Shaulsky G. Self-recognition in social amoebae is mediated by allelic pairs of tiger genes. Science. 2011;333(6041):467–70.

12. Thompson CR, Bretscher MS. Cell polarity and locomotion, as well as endocytosis, depend on NSF. Development. 2002;129(18):4185–92.

13. Pan W, Ravot E, Tolle R, Frank R, Mosbach R, Turbachova I, Bujard H. Vaccine candidate MSP-1 from Plasmodium falciparum: a redesigned 4917 bp polynucleotide enables synthesis and isolation of full-length protein from Escherichia coli and mammalian cells. Nucleic Acids Res. 1999;27(4):1094–103.

14. Godiska RPM, Schoenfeld T, Mead DA. Beyond pUC: vectors for cloning unstable DNA. In: Kieleczawa J, editor. Optimization of the DNA Sequencing Process. vol. 36. Sudbury, Massachusetts: Jones and Bartlett Publishers; 2005. p. 55–75.

15. Inagaki H, Ohye T, Kogo H, Yamada K, Kowa H, Shaikh TH, Emanuel BS, Kurahashi H. Palindromic AT-rich repeat in the NF1 gene is hypervariable in humans and evolutionarily conserved in primates. Hum Mutat. 2005;26(4):332–42.

16. Burrow AA, Marullo A, Holder LR, Wang YH. Secondary structure formation and DNA instability at fragile site FRA16B. Nucleic Acids Res. 2010;38(9):2865–77.

17. Godiska R, Mead D, Dhodda V, Wu C, Hochstein R, Karsi A, Usdin K, Entezam A, Ravin N. Linear plasmid vector for cloning of repetitive or unstable sequences in Escherichia coli. Nucleic Acids Res. 2010;38(6):e88.

18. Pfander C, Anar B, Schwach F, Otto TD, Brochet M, Volkmann K, Quail MA, Pain A, Rosen B, Skarnes W et al. A scalable pipeline for highly effective genetic modification of a malaria parasite. Nat Methods. 2011;8(12):1078–82.

19. Rosengarten RD, Beltran PR, Shaulsky G. A deep coverage Dictyostelium discoideum genomic DNA library replicates stably in Escherichia coli. Genomics. 2015;106(4):249–55.

20. Veltman DM, Akar G, Bosgraaf L, Van Haastert PJ. A new set of small, extrachromosomal expression vectors for Dictyostelium discoideum. Plasmid. 2009;61(2):110–8.

21. Su XZ, Wu Y, Sifri CD, Wellems TE. Reduced extension temperatures required for PCR amplification of extremely A + T-rich DNA. Nucleic Acids Res. 1996;24(8):1574–5.

22. Prevorovsky M, Puta F. A/T-rich inverted DNA repeats are destabilized by chaotrope-containing buffer during purification using silica gel membrane technology. Biotechniques. 2003;35(4):698–700. 702.

23. Faix J, Kreppel L, Shaulsky G, Schleicher M, Kimmel AR. A rapid and efficient method to generate multiple gene disruptions in Dictyostelium discoideum using a single selectable marker and the Cre-loxP system. Nucleic Acids Res. 2004;32(19):e143.

24. Goedhart J, von Stetten D, Noirclerc-Savoye M, Lelimousin M, Joosen L, Hink MA, van Weeren L, Gadella TW, Jr., Royant A. Structure-guided evolution of cyan fluorescent proteins towards a quantum yield of 93 %. Nat Commun. 2012;3:751.

25. Fischer M, Haase I, Simmeth E, Gerisch G, Muller-Taubenberger A. A brilliant monomeric red fluorescent protein to visualize cytoskeleton dynamics in Dictyostelium. FEBS Lett. 2004;577(1–2):227–32.

26. Fey P, Kowal AS, Gaudet P, Pilcher KE, Chisholm RL. Protocols for growth and development of Dictyostelium discoideum. Nat Protoc. 2007;2(6):1307–16.

27. Eason RG, Pourmand N, Tongprasit W, Herman ZS, Anthony K, Jejelowo O, Davis RW, Stolc V. Characterization of synthetic DNA bar codes in Saccharomyces cerevisiae gene-deletion strains. Proc Natl Acad Sci U S A. 2004;101(30):11046–51.

28. Manis JP. Knock out, knock in, knock down–genetically manipulated mice and the Nobel Prize. N Engl J Med. 2007;357(24):2426–9.

29. Xiao Z, Zhang N, Murphy DB, Devreotes PN. Dynamic distribution of chemoattractant receptors in living cells during chemotaxis and persistent stimulation. J Cell Biol. 1997;139(2):365–74.

30. Kriebel PW, Barr VA, Rericha EC, Zhang G, Parent CA. Collective cell migration requires vesicular trafficking for chemoattractant delivery at the trailing edge. J Cell Biol. 2008;183(5):949–61.

31. Kriebel PW, Barr VA, Parent CA. Adenylyl cyclase localization regulates streaming during chemotaxis. Cell. 2003;112(4):549–60.

32. Gomes AR, Bushell E, Schwach F, Girling G, Anar B, Quail MA, Herd C, Pfander C, Modrzynska K, Rayner JC et al. A genome-scale vector resource enables high-throughput reverse genetic screening in a malaria parasite. Cell Host Microbe. 2015;17(3):404–13.

33. Schwach F, Bushell E, Gomes AR, Anar B, Girling G, Herd C, Rayner JC, Billker O. PlasmoGEM, a database supporting a community resource for large-scale experimental genetics in malaria parasites. Nucleic Acids Res. 2015;43(Database issue):D1176–82.

34. Charette SJ, Cornillon S, Cosson P. Identification of low frequency knockout mutants in Dictyostelium discoideum created by single or double homologous recombination. J Biotechnol. 2006;122(1):1–4.

35. Calvo-Garrido J, Carilla-Latorre S, Lazaro-Dieguez F, Egea G, Escalante R. Vacuole membrane protein 1 is an endoplasmic reticulum protein required for organelle biogenesis, protein secretion, and development. Mol Biol Cell. 2008;19(8):3442–53.

36. Li SI, Purugganan MD. The cooperative amoeba: Dictyostelium as a model for social evolution. Trends Genet. 2011;27(2):48–54.

37. Strassmann JE, Queller DC. How social evolution theory impacts our understanding of development in the social amoeba Dictyostelium. Dev Growth Differ. 2011;53(4):597–607.

12

Identification of residues important for the activity of aldehyde-deformylating oxygenase through investigation into the structure-activity relationship

Qing Wang[1,2†], Luyao Bao[1,2†], Chenjun Jia[2,3], Mei Li[3], Jian-Jun Li[1,4*] and Xuefeng Lu[1*]

Abstract

Background: Aldehyde-deformylating oxygenase (ADO) is a key enzyme involved in the biosynthetic pathway of fatty alk(a/e)nes in cyanobacteria. However, cADO (cyanobacterial ADO) showed extreme low activity with the k_{cat} value below 1 min^{-1}, which would limit its application in biofuel production. To identify the activity related key residues of cADO is urgently required.

Results: The amino acid residues which might affect cADO activity were identified based on the crystal structures and sequence alignment of cADOs, including the residues close to the di-iron center (Tyr39, Arg62, Gln110, Tyr122, Asp143 of cADO-1593), the protein surface (Trp 178 of cADO-1593), and those involved in two important hydrogen bonds (Gln49, Asn123 of cADO-1593, and Asp49, Asn123 of cADO-sll0208) and in the oligopeptide whose conformation changed in the absence of the di-iron center (Leu146, Asn149, Phe150 of cADO-1593, and Thr146, Leu148, Tyr150 of cADO-sll0208). The variants of cADO-1593 from *Synechococcus elongatus* PCC7942 and cADO-sll0208 from *Synechocystis* sp. PCC6803 were constructed, overexpressed, purified and kinetically characterized. The k_{cat} values of L146T, Q49H/N123H/F150Y and W178R of cADO-1593 and L148R of cADO-sll0208 were increased by more than two-fold, whereas that of R62A dropped by 91.1%. N123H, Y39F and D143A of cADO-1593, and Y150F of cADO-sll0208 reduced activities by ≤ 20%.

Conclusions: Some important amino acids, which exerted some effects on cADO activity, were identified. Several enzyme variants exhibited greatly reduced activity, while the k_{cat} values of several mutants are more than two-fold higher than the wild type. This study presents the report on the relationship between amino acid residues and enzyme activity of cADOs, and the information will provide a guide for enhancement of cADO activity through protein engineering.

Keywords: Aldehyde-deformylating oxygenase, Site-directed mutagenesis, Structure-activity relationship, Fatty alk(a/e)ne, *Synechococcus elongatus* PCC7942, *Synechocystis* sp. PCC6803

Background

Fatty alk(a/e)nes, which can be produced by plants, insects, birds, green algae, and cyanobacteria, are the main components of conventional fuels, and have been considered as the ideal replacement for fossil-based fuels [1–5]. It has been accepted that a two-step pathway for fatty alk(a/e)ne biosynthesis exists, involving reduction of fatty acyl-ACP or -CoA into corresponding aldehyde by acyl-ACP reductase and conversion of fatty aldehyde into alk(a/e)ne by aldehyde decarbonylase. In 2010, Schirmer et al. identified two genes involved in fatty alk(a/e)ne biosynthesis in cyanobacteria: acyl-ACP reductase and aldehyde decarbonylase (renamed as aldehyde-deformylating oxygenase, ADO) [1, 6]. Since then cADO (cyanobacterial ADO) has attracted particular interest due to the difficult and unusual reaction it catalyses [7].

* Correspondence: jjli@ipe.ac.cn; lvxf@qibebt.ac.cn
†Equal contributors
[1]Key Laboratory of Biofuels, Shandong Provincial Key Laboratory of Energy Genetics, Qingdao Institute of Bioenergy and Bioprocess Technology, Chinese Academy of Sciences, Qingdao 266101, China
Full list of author information is available at the end of the article

The crystal structures of cADO revealed that cADO belongs to the non-heme dinuclear iron oxygenase family of enzymes exemplified by methane monooxygenase, type I ribonucleotide reductase, and ferritin. The di-iron center is contained within an antiparallel four-α-helix bundle, where two histidines and four carboxylates (aspartate or glutamate) supply the protein ligands to the metal ions [1, 8–11]. The C1-derived co-product of the cADO-catalyzed reaction is formate (Fig. 1) [12]. Oxygen is needed, and one O-atom is incorporated into formate [13]. The auxiliary reducing system (biological or chemical) providing four electrons is required, and the endogenous electron transfer system worked more effectively than the heterologous and chemical ones in supporting cADO activity [1, 12, 14–16]. Self-sufficient cADOs fused to homogenous ferredoxin and ferredoxin-NADP+ reductase could efficiently catalyze the conversion of aldehydes into alk(a/e)nes [17]. It has been found that cADO also produces n-1 aldehydes and alcohols in addition to alk(a/e)ne [18]. Mechanistic studies have demonstrated that a radical intermediate is involved in the cADO-catalyzed reaction, and a possible catalytic process has been proposed based on the crystal structures of cADO from *Synechococcus elongates* strain PCC7942 [9, 19–21]. Moreover, cADO was engineered to improve specificity for short- to medium-chain aldehydes [22]. Recently, Hayashi et al. investigated the role of three cysteine residues of cADO in the structure, stability and alk(a/e)ne production [23].

However, cADO showed extreme low activity with the k_{cat} value below 1 min^{-1}, which would present a major barrier to its application in biofuel production [8, 14–17, 22]. In order to address this issue, protein engineering of cADO for improved activity including rational design and/or directed evolution is urgently needed. The knowledge about structure-function relationship of cADOs is the prerequisite for rational protein engineering. Until now, no detailed studies towards structure-function relationship of cADOs have been carried out. Some crystal structures of cADOs from *P. marinus* strain MIT9313 and *Synechococcus elongates* strain PCC7942 have been resolved, which have provided a base for identification of the residues important for cADO activity [8, 9].

In the current study, some amino acids which might affect cADO activity were identified through analysis of the cADO crystal structures and sequence alignment of some cADOs. The corresponding enzyme variants were made and characterized. We have found some essential residues for the cADO-catalyzed reaction, which will lay the foundation for improvement of cADO activity through protein engineering.

Results

Identification of the target residues for mutagenesis

The following residues were identified and investigated in the current study. Since the variants including mutations of the residues involved in coordinating the di-iron center negatively affected cADO activity, they were not included in the current study [9].

Residues whose conformations have changed in the absence of the di-iron center

Based on the structures of 1593 (cADO from *Synechococcus elongates* PCC7942) (PDB code: 4RC5) and sll0208 (cADO from *Synechocystis* sp. PCC6803) (PDB code: 4Z5S), the overall structures are similar with or without the di-iron center, except the conformation change of the helix H5 in the structure of metal-free cADO [9]. The switch of the helix H5 from helix to loop resulted in conformational changes of a number of amino acids (residues 144 to 150), including the two iron-coordinating residues Glu144 and His147. The observation suggests that the residues involved in the oligopeptide may impact cADO activity. The residues (144 to 150 for 1593) comprising of that oligopeptide are conservable among cADOs (sequence alignment of more than 150 cADOs, which were found by subjecting to a BLAST search of the sequence of 1593) (Additional files 1 and 2) [24]. For example, Glu144 and His147 (1593 used as a reference), two ligands of the di-iron center, are completely conserved. Tyr145 is highly conserved, and is Ser/Ala in several cADOs (Additional files 1 and 2), and is also involved in a hydrogen bond with Asn123 (described in detail in the following part). Residue 146 is variable, and is Leu/Thr/Ser/Glu, etc. Residue 148 is Leu among more than around 140 cADOs, and is Arg in 4 cADOs. Residue Asn149 is highly conservable, and is Asp/Lys in a few cADOs.

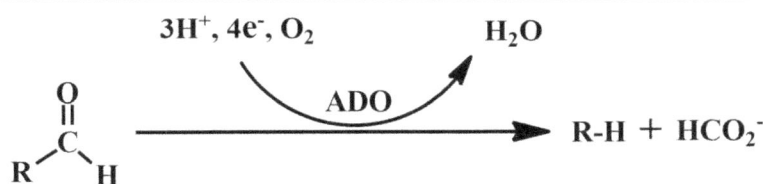

Fig. 1 cADO-catalyzed reaction [12–16]

Tyr150 shows high conservativeness among ~100 cADOs, and is Phe in ~50 cADOs. As mentioned above, the ligands of the di-iron center such as Glu144 and His147 were not investigated. The residues Leu146, Asn149 and Phe150 of 1593, and Thr146, Leu148 and Tyr150 of sll0208, which might have some influence on cADO activity due to the different properties of their side chains, were chosen for mutagenesis (Figs. 2 and 3).

Residues involved in two hydrogen bonds

According to the crystal structures of sll0208 and PMT1231 (ADO from *Prochlorococcus marinus* MIT9 313) (PDB code: 2OC5), it was observed that residues Tyr145 and Tyr150 which are equivalent to Tyr158 and Tyr163 of PMT1231 respectively were involved in two hydrogen bonds (between sll0208^{Asp49}/PMT1231^{His62} and sll0208^{Tyr150}/PMT1231^{Tyr163}, and between sll0208^{Asn123}/ PMT1231^{His136} and sll0208^{Tyr145}/PMT1231^{Tyr158}) (Fig. 4). However, there is only one hydrogen bond between Asn123 and Tyr145 in 1593, and the corresponding residues forming the other hydrogen bond in sll0208 and PMT1231 are Gln49 and Phe150 in 1593 (Figs. 2 and 3). Residue 49 is variable, and is Gln/Asn/His/ Asp/Ser/Glu, etc., all containing the polar side chains (Additional files 1 and 2). Residue Asn123 is conservable among ~120 cADOs, and is His in 32 cADOs. Considering that these three cADOs showed different activities against *n*-hexadecanal (unpublished results),

these two hydrogen bonds may have some effects on cADO activity. The residues Gln49, Asn123 of 1593 and Asp49, Asn123 of sll0208 were investigated.

Residues close to the di-iron center

The residues around the di-iron center provide conformational constraints that control the geometry of the di-iron center. According to the structure of 1593, residues Tyr39, Arg62, Gln110, Tyr122 and Asp143 are close to the di-iron center, which might interact with the nearby residues including the ligands of di-iron: Tyr39 with Glu60 (ligand), His147 (ligand), Gln110 and Ser111; Arg62 with His147 (ligand) and Asp143; Gln110 with Glu115 (ligand), Asn38 and Tyr39; Tyr122 with Glu32 (ligand), Glu144 (ligand) and Val28 (Fig. 2); Asp143 with Arg62 and His147 (ligand). Residue Tyr39 is highly conserved, and is Phe in 8 cADOs (Additional files 1 and 2). Arg62, Gln110, Tyr122 and Asp143 show high conservativeness among ~150 cADO. Likewise, the residues coordinating the di-iron center like Glu32, Glu44, Glu60, Glu115, Glu144 and His147 were excluded in this study. Residues Tyr39, Arg62, Gln110, Tyr122 and Asp143 of 1593 were selected for investigation.

Residues located on the protein surface

Based on sequence alignment of 150 cADOs, Trp178 of 1593 is highly conserved among ~120 cADOs, and is Arg/Lys in ~30 cADOs. The side-chains of Trp and Arg/

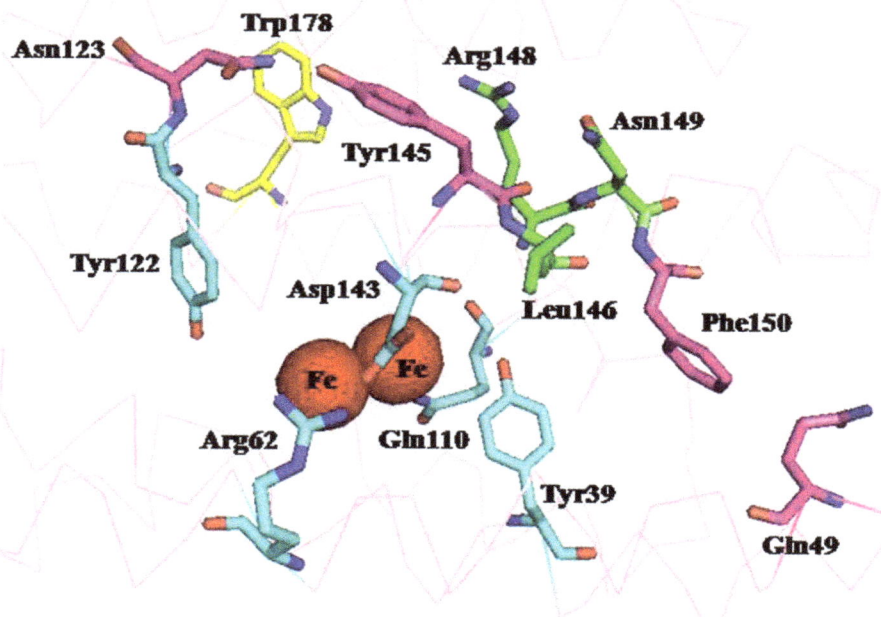

Fig. 2 Identified residues based on the crystal structure of ADO from *Synechococcuselongatus* PCC7942 (1593; PDB code:4RC5). The identified residues include those close to the di-iron center (Tyr39, Gln110, Tyr122), the protein surface (Trp178), and involved in the hydrogen-bonding network (Arg62, Asp143) and the oligopeptide whose conformation changed (Leu/Thr146, Leu148, Asn149 and Tyr/Phe150) in the absence of the diiron center

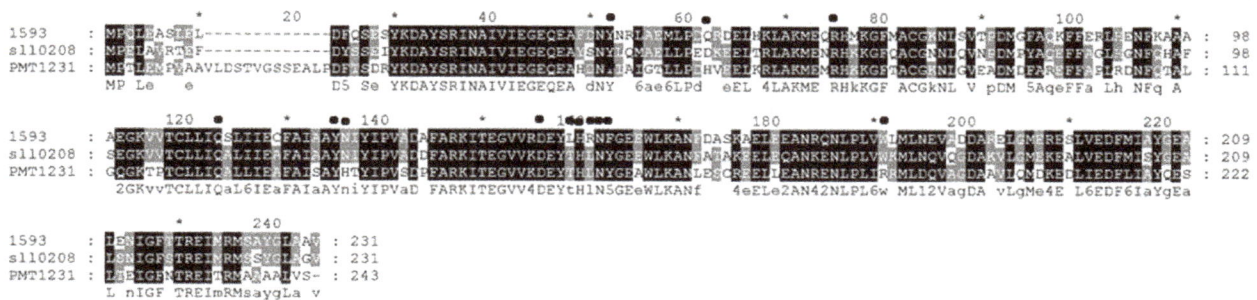

Fig. 3 Sequence alignment of 1593, sll0208 and PMT1231. The residues investigated in this paper are labelled with *black dots* above the sequence

Lys exhibited completely different properties: hydrophobicity versus hydrophilicity. Moreover, according to the crystal structure of 1593, Trp178 is positioned at Helix 6 and exposed to the protein surface (Fig. 2). These facts prompted us to presume that hydrophobicity or hydrophilicity of this residue might have some influence on cADO activity. Therefore, this residue was studied.

Site-directed mutagenesis, overexpression and purification

We have observed that 1593 is more active than sll0208 against *n*-hexadecanal, so Thr146, Leu148 and Tyr150 of sll0208 were respectively mutated into the counterparts of 1593 - Leu, Arg and Phe (Fig. 3). For comparison, Leu146 and Phe150 of 1593 were also mutated into the corresponding ones of sll0208 - Thr and Tyr respectively. Since PMT1231 is more active than sll0208 towards *n*-hexadecanal under our assay conditions (unpublished results), Asp49 and Asn123 of sll0208 were mutated into the corresponding residues of PMT1231 - His and His respectively (Fig. 3). In order to investigate

the effects of the polar side chains of some residues on cADO activity, the single site-directed mutants Y39F, R62A, Q110L, Y122F, D143A and N149A were constructed for 1593. In addition, N123H and W178R of 1593 were made based on the conservativeness.

Enzymatic activities of wild-type cADOs and variants

n-Hexadecanal and *n*-heptanal were used as the substrates to investigate the effects of mutations on enzymatic activities of 1593 and sll0208. When *n*-hexadecanal was used as a substrate, the yields of *n*-pentadecane were quantified. While *n*-heptanal was used as a substrate, the apparent k_{cat} values were measured.

Compared with WT (the wild type) 1593, the apparent k_{cat} values of 1593^{W178R}, $1593^{Q49H/N123H/F150Y}$, 1593^{L146T}, 1593^{F150Y} and $1593^{Q49H/F150Y}$ were enhanced by 226.7 ± 10.3%, 93.3 ± 8.6%, 93.3 ± 8.1%, 68.9 ± 6.2%, and 60 ± 5% respectively (Table 1). 1593^{R62A} exhibited significantly reduced k_{cat}^{app} value, and dropped by 91.1 ± 7.4% (Table 1). 1593^{Y122F} and 1593^{Q110L} displayed moderate activity,

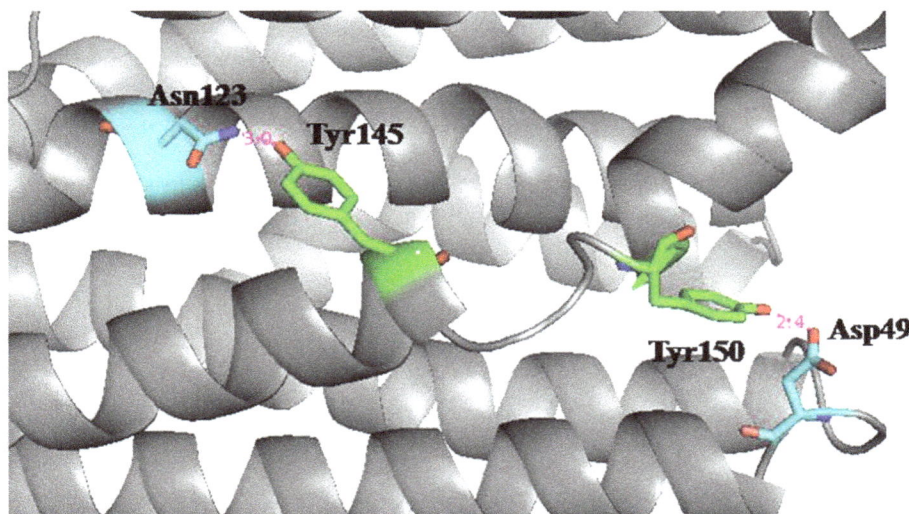

Fig. 4 Two hydrogen bonds in sll0208. The hydrogen-bond lengths between Asn123 and Tyr145 and between Tyr150 and Asp49 are 3.0 and 2.4 Å respectively

Table 1 Apparent k_{cat} values and yields of n-pentadecane of WT 1593, WT sll0208 and variants. The apparent k_{cat} values were determined using 2 mM n-heptanal as the substrate

		k_{cat}^{app} (Min^{-1})	Yield of n-pentadecane (μM)
1593	WT	0.45 ± 0.06	12.1 ± 0.4
	L146T	0.87 ± 0.1	14.2 ± 1.2
	N149A	0.31 ± 0.04	8.7 ± 0.7
	F150Y	0.76 ± 0.08	12.2 ± 0.6
	N123H	0.43 ± 0.05	10.4 ± 0.9
	Q49H/F150Y	0.72 ± 0.07	14.4 ± 0.6
	Q49H/N123H/F150Y	0.87 ± 0.09	15.0 ± 0.1
	Y39F	0.37 ± 0.05	10.4 ± 0.8
	Q110L	0.23 ± 0.03	5.8 ± 0.2
	Y122F	0.15 ± 0.02	2.1 ± 0.04
	R62A	0.04 ± 0.001	0.4 ± 0.06
	D143A	0.36 ± 0.05	9.2 ± 0.2
	W178R	1.47 ± 0.1	17.4 ± 0.2
sll0208	WT	0.44 ± 0.05	1.4 ± 0.1
	T146L	0.17 ± 0.02	1.4 ± 0.1
	L148R	0.75 ± 0.08	5.1 ± 0.2
	Y150F	0.42 ± 0.04	2.2 ± 0.1
	D49H	0.22 ± 0.03	1.8 ± 0.1
	N123H	0.14 ± 0.02	1.0 ± 0.08
	D49H/N123H	0.73 ± 0.08	3.4 ± 0.3

The yield of n-pentadecane was determined using 150 μM using n-hxadecanal as the substrate

Table 2 Kinetic parameters of WT 1593, WT sll0208 and some variants

		K_m (mM)	k_{cat} (min^{-1})	k_{cat}/K_m (min^{-1}mM^{-1})
1593	WT [17	0.30 ± 0.02	0.48 ± 0.01	1.6 ± 0.2
	L146T	0.34 ± 0.08	0.94 ± 0.08	2.76 ± 0.3
	F150Y	0.20 ± 0.04	0.75 ± 0.04	3.75 ± 0.4
	Q49H/F150Y	0.33 ± 0.05	0.87 ± 0.04	2.64 ± 0.2
	Q49H/N123H/F150Y	0.35 ± 0.02	1.04 ± 0.02	2.97 ± 0.3
	W178R	0.34 ± 0.08	1.81 ± 0.15	5.32 ± 0.4
sll0208	WT	0.35 ± 0.07	0.59 ± 0.04	1.69 ± 0.2
	L148R	0.32 ± 0.08	0.91 ± 0.08	2.84 ± 0.3
	D49H/N123H	0.38 ± 0.08	0.83 ± 0.08	2.18 ± 0.2

The kinetic parameters against n-heptanal were determined

whereas the catalytic activities of 1593^{Y39F}, 1593^{N123H}, 1593^{N149A} and 1593^{D143A} were reduced by ≤ 31.1 ± 2.6%. The yields of n-pentadecane of the enzyme variants showed the same trend as the apparent k_{cat} values (Table 1).

Compared with WT sll0208, the apparent k_{cat} values of sll0208^{L148R} and sll0208$^{D49H/N123H}$ were increased by 70.5 ± 5.3% and 65.9 ± 4.4% respectively, whereas those of sll0208^{D49H}, sll0208^{N123H} and sll0208^{T146L} were significantly reduced (Table 1). The activity of sll0208^{Y150F} was not almost affected. The yields of n-pentadecane of the enzyme variants demonstrated similar trend to the apparent k_{cat} values (Table 1).

The kinetic parameters towards n-heptanal were also determined for some variants showing higher activity than WT. The variants L148R and D49H/N123H of sll0208, and W178R, Q49H/N123H/F150Y, L146T and Q49H/F150Y of 1593 showed higher k_{cat} values than WT and comparable K_m values to WT, indicating that these mutations had significant effects on activity, but no big impact on substrate binding (Table 2). The observation that the k_{cat} value of F150Y was higher than that of WT 1593 and its K_m value was much lower than that of WT 1593 suggested that replacement of

phenylalanine 150 with tyrosine not only affected activity but also substrate binding (Table 2).

Interestingly, the k_{cat} values of 1593^{W178R}, 1593$^{Q49H/N123H/F150Y}$ and 1593^{L146T}, close to or above 1 min^{-1}, were obtained. Especially, 1593^{W178R} showed 3.8-fold improved k_{cat} value than WT 1593, and exhibited the highest catalytic efficiency (k_{cat}/K_m) among all variants, 3.3-fold higher than WT (Table 2).

Discussion

Although the conditions of the cADO-catalyzed reaction have been optimized through different attempts, the turnover numbers are still below ~1 min^{-1} [8, 14–17, 22]. Therefore, the sluggish kinetics of cADO has become a bottle-neck for biofuel production in cyanobacteria, and to identify the activity related key residues for cADO is definitely required. Since it was observed that cADO-1593 showed the highest activity under our assay conditions among cADOs tested (unpublished results), it was selected as the test sequence for BLAST. In the current paper, we have identified some amino acid residues which impact the enzyme activity of cADO through structure and sequence analysis of some cADOs.

Firstly, the results of 1593^{L146T}, 1593^{F150Y}, sll0208^{T146L} and sll0208^{L148R} have clearly demonstrated that the side chains of the residues consisting of the oligopeptide (residues 144 to 150 for 1593) whose conformation changed in the absence of the di-iron center had important effects on cADO activity. Crystal structures reveal that these polar residues form important hydrogen bond interactions with residues from other helices, which may contribute to the local structural stability of the oligopeptide. In addition, the polar side chains of these residues might interact with the nearby residues, for example, in 1593 (Protein code: 4RC5) Arg148 with Asn149 and Glu152, Asn149 with Arg148 and Glu152; in sll0208 (Protein code: 4Z5S) Thr146 with Asp143 and Asn149, Asn149 with Thr146 and Glu152, Tyr150 with

Gln49; in PMT1231 (Protein code: 2OC5) Thr159 with Asp156 and Asn162, Asn162 with Thr159 and Glu165, Tyr163 with His62, which are equivalent to the residues 146, 143, 149, 152, 150 and 49 of 1593 and sll0208 respectively. Thr159 of PMT1231 and the equivalent one (Thr146) of sll0208 are also involved in the hydrogen-bonding network close to the di-iron center. Replacement of the residues containing the polar side chains with ones containing the nonpolar side chains led to removal of the possible interaction with nearby residues, which might have severe influence on protein folding, stability, even activity. In contrast, substitution of amino acids with the nonpolar side chains by ones with the polar side chains gave the reverse impact, as observed for 1593^{L146T}, sll0208^{L148R} and 1593^{F150Y}. The enhanced activities of these three variants are presumably due to local stabilization caused by newly established interaction of three residues with the nearby amino acids.

Secondly, based on the structures of sll0208 and PMT1231, a hydrogen bond is possibly established between Tyr150 and Gln49 in 1593 while Phe150 of 1593 was substituted by Tyr. Activities of both F150Y and Q49H/F150Y of 1593 were enhanced, suggesting that the hydrogen bond between these two residues is important for 1593. The triple mutant $1593^{Q49H/F150Y/N123H}$ showed higher activity than WT 1593, $1593^{Q49H/F150Y}$ and 1593^{F150Y}, and the double mutant sll0208$^{D49H/N123H}$ exhibited improved activity than WT sll0208, sll0208^{D49H} and sll0208^{N123H}. Thus, some synergistic effects were observed for $1593^{Q49H/F150Y/N123H}$ and sll0208$^{D49H/N123H}$, implying that these two hydrogen bonds are beneficial for cADO activity and the coexistence of two His in both hydrogen bonds is more important than the presence of only one His. Unexpectedly,

the Y150F substitution of sll0208, in which the hydrogen bond between Tyr150 and Asp49 is removed, had negligible effect on the catalytic activity.

Thirdly, replacement of the residues Arg62, Gln110 and, Tyr122 of 1593 having the polar side chains by those containing the nonpolar ones resulted in significantly decreased activity, which could be possibly due to removal of the interaction with the nearby residues. These results confirmed that the residues close to the di-iron center impacted cADO activity and their polar side chains are important for cADO.

Finally, when the highly conserved residue Trp178 of 1593 was mutated into Arg, the W178Rvariant showed the highest cADO activity by far. Trp178 of 1593 and the equivalent one (Arg191) of PMT1231 are rightly located on the protein surface according to the crystal structures of cADOs (PDB codes: 2OC5, 4KVQ, 4KVR and 4KVS). The enhanced activity of W178R could be possibly due to increased protein stability resulting from mutation of exposed hydrophobic amino acids into hydrophilic ones such as arginine or the change of hydrophobicity into hydrophilicity of Trp178 exposed to the solvent at the protein surface contributes to cADO activity [25]. Moreover, according to the crystal structure of the mutant L194A of PMT1231, the substrate might enter the protein at Leu194. This mode of substrate binding is different from previously determined cADO structures with long-chain fatty acids bound [8]. In this structure, the side-chain of Arg191, equivalent to Trp178 of 1593, points away from the substrate, which may be beneficial for substrate entry (Fig. 5). However, according to the structure of 1593 (PDB code: 4QUW), the side-chain of Trp178 points

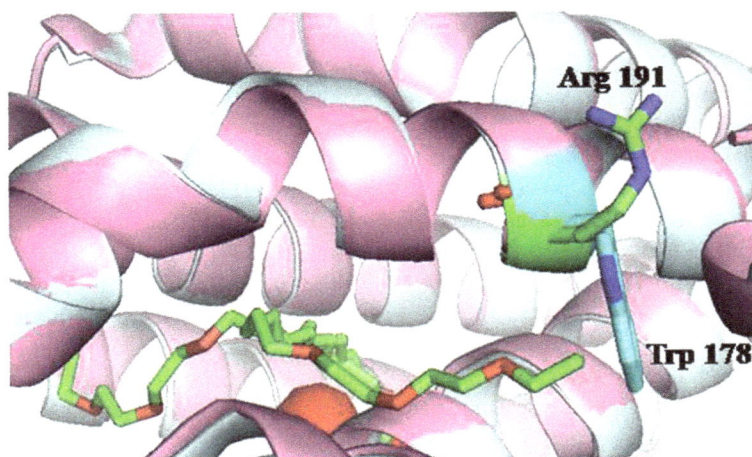

Fig. 5 Structural superimposition of L194A of PMT1231 (palecyan, PDB code: 4PGI) and 1593 (*light pink*, PDB code: 4QUW). Arg191 of PMT1231, Trp178 of 1593 and two substrate-binding modes were shown

towards the substrate, which might interfere with substrate entry (Fig. 5). Thus, mutation of Trp178 into Arg might contribute to substrate entry and binding, leading to enhanced activity observed for W178R.

Conclusions

Some amino acids which could affect cADO activity were identified based on the crystal structures and sequence alignment of cADOs. The residues close to the di-iron center, the protein surface, and those involved in the hydrogen-bonding network and the oligopeptide whose conformation changed, exerted great influence on cADO activity. Several mutations led to significantly decreased activity. Some enzyme variants showed improved activity than the wild type, and the k_{cat} values of several of them close to or above 1 min^{-1} were achieved for the first time in particular. We identified some important residues for the catalytic activity of cADO. The study would be helpful for establishing an efficient cell factory of biofuel production in cyanobacteria.

Methods

Materials

Chemicals were from Sigma, Merck or Ameresco. Oligonucleotides and the gene encoding ADO (aldehyde-deformyla tingoxygenase) Synpcc7942_1593 from *Synechococcus elongates* PCC7942 with codon optimization were synthesized by Shanghai Sangon Biotech Co. Ltd (China) [16]. *Pfu* DNA polymerase, restriction endonucleases *EcoRI* and *NotI* were from Fermentas or Takara Biotechnology. *DpnI* was from New England BioLabs. The kits used for molecular cloning were from Omega Bio-tek or Takara Biotechnology. Nickel column and expression vector pET-28a(+) were from Novagen. Amicon YM10 membrane was from Millipore.

Bacterial strains, plasmids, and media

E. coli DH5α was used for routine DNA transformation and plasmid isolation. *E. coli* BL21(DE3) was utilized for overexpression of cADO. *E. coli* strains were routinely grown in Luria-Bertani broth at 37 °C with aeration or on LB supplemented with 1.5% (w/v) agar. 100 μg/ml Kanamycin was added when required.

DNA manipulations

General molecular biology techniques were carried out by standard procedures [26]. DNA fragments were purified from agarose gels using the DNA gel extraction kit. Plasmid DNA was isolated using the plasmid miniprep kit.

The gene *sll0208* encoding ADO from *Synechocystis* sp. PCC6803 (sll0208) was amplified with the forward primer (5'- GCCTTACATATGATGCCCGAGCTTGC TG-3', the *Nde*I I restriction site underlined) and the reverse primer (5'- CAACTACTCGAGCTAGACTCC GGCCAAACC-3', the *Xho*I restriction site underlined) using genomic DNA as a template. The PCR products were recovered, and digested with restriction enzymes *Nde*I and *Xho*I respectively, and re-cloned into the vector pET-28a(+) digested with *Nde*I and *Xho*I, respectively.

Construction of site-directed mutants

Site-directed mutants were constructed according to the standard QuikChange Site-Directed Mutagenesis protocol (Stratagene Ltd) using wild-type (WT) 1593 or sll0208 as templates and the primers listed in Table S1 (Additional file 3).

For construction of double and triple mutants, WT 1593 or sll0208 harboring single or double mutation(s) was used as a template respectively following the same protocol as above.

Protein overexpression and purification

WT cADOs and variants were overexpressed in *E. coli* BL21(DE3) and purified on Nickel column as reported [16]. The purity of protein was checked by SDS-PAGE (Additional file 4). Apo-ADO was prepared according to the published procedure, and the diferrous form of ADO was reconstituted by the addition of the stoichiometric amounts of ferrous ammonium sulfate to the apo-ADO prior to assay [14–16]. Proteins were concentrated with Amicon YM10 membrane (10 kDa cut-off). The protein concentration was determined by the Bradford method using bovine serum albumin as a standard.

Enzyme activity assay

All enzymatic assays were carried out in triplicate. According to the published procedure [16, 17], assays were carried out in HEPES buffer, containing 100 mM KCl and 100 mM HEPES, pH 7.2. The reaction mixtures contain NADH (750 μM), catalase (1 mg/ml), ferrous ammonium sulfate (80 μM), PMS (Phenazine methosulfate) (75 μM), appropriate amount of aldehydes (150 μM for *n*-hxadecanal, 2 mM for *n*-heptanal), cADO (20 μM for *n*-hxadecanal, 5 μM for *n*-heptanal). When *n*-hxadecanal was used as the substrate, the yields of *n*-pentadecane were quantified by GC-MS. While *n*-heptanal was used as substrate, the apparent k_{cat} values (2 mM of *n*-heptanal was utilized) were measured. The control using *E. coli* containing the vector pET-28a only didn't show any ADO activity (data not shown).

K_m and V_{max} values of WT 1593, WT sll0208 and some variants against n-heptanal were determined according to the Michaelis-Menten equation of Graph-Pad Prism 5 (Additional file 5). The k_{cat} values were calculated from V_{max} on the basis of the molecular weight of enzymes.

Additional files

Additional file 1: Sequence alignment of cADOs (1–100).

Additional file 2: Sequence alignment of cADOs (101–150).

Additional file 3: Table S2. Primers used for construction of site-directed mutants.

Additional file 4: SDS PAGE analysis of WT 1593, WT sll0208 and all variants.

Additional file 5: Original data for determination of the kinetic parameters.

Abbreviations
ACP: Acyl carrier protein; ADO: Aldehyde-deformylating oxygenase; cADO: Cyanobacterial aldehyde-deformylating oxygenase; Fd: Ferredoxin; FNR: Ferredoxin-NADP$^+$ reductase; PMS: Phenazine methosulfate

Acknowledgements
Not applicable.

Funding
This work was supported by grants from the National Basic Research Program of China (973: 2011CBA00907 and 2011CBA00902), National Science Foundation of China (31170765 and 31370799), CAS Cross and Cooperation Team for Scientific Innovation (Y31102110A), the Excellent Youth Award of the Shandong Natural Science Foundation (JQ201306 to X. Lu) and the Shandong Taishan Scholarship (X. Lu).

Authors' contributions
JJL and XL designed the experiments. QW and LB performed the experiments, including gene cloning, construction of site-directed mutants, overexpression, purification, characterization, and enzymatic assays. CJ and ML performed structural analysis of cADOs, and assisted in manuscript preparation. JJL, XL, QW, and LB drafted the manuscript. All authors read and approved the final manuscript.

Competing interests
The authors declare that they have no competing interests.

Author details
[1]Key Laboratory of Biofuels, Shandong Provincial Key Laboratory of Energy Genetics, Qingdao Institute of Bioenergy and Bioprocess Technology, Chinese Academy of Sciences, Qingdao 266101, China. [2]University of Chinese Academy of Sciences, Beijing 100049, China. [3]National Laboratory of Biomacromolecules, Institute of Biophysics, Chinese Academy of Sciences, Beijing 100101, China. [4]Present Address: National Key Laboratory of Biochemical Engineering, Institute of Process Engineering, Chinese Academy of Sciences, Beijing 100190, China.

References
1. Schirmer A, Rude MA, Li X, Popova E, del Cardayre SB. Microbial biosynthesis of alkanes. Science. 2010;329:559–62.
2. Lu X. A perspective: Photosynthetic production of fatty acid-based biofuels in genetically engineered cyanobacteria. Biotechnol Adv. 2010;28:742–6.
3. Zhang F, Rodriguez S, Keasling JD. Metabolic engineering of microbial pathways for advanced biofuels production. Curr Opin Biotechnol. 2011;22:1–9.
4. Gronenberg LS, Marcheschi RJ, Liao JC. Next generation biofuel engineering in prokaryotes. Curr Opin Chem Biol. 2013;17:462–71.
5. Wen M, Bond-Watts BB, Chang MC. Production of advanced biofuels in engineered E. coli. Curr Opin Chem Biol. 2013;17:472–9.
6. Li N, Chang W, Warui DM, Booker SJ, Krebs C, Bollinger Jr JM. Evidence for Only Oxygenative Cleavage of Aldehydes to Alk(a/e)nes and Formate by Cyanobacterial Aldehyde Decarbonylase. Biochemistry. 2012;51:7908–16.
7. Marsh ENG, Waugh MW. Aldehyde Decarbonylases: Enigmatic Enzymes of Hydrocarbon Biosynthesis. ACS Catal. 2013;3:2515–21.
8. Buer BC, Paul B, Das D, Stuckey JA, Marsh ENG. Insights into Substrate and Metal Binding from the Crystal Structure of Cyanobacterial Aldehyde Deformylating Oxygenase with Substrate Bound. ACS Chem Biol. 2014;9:2584–93.
9. Jia C, Li M, Li J, Zhang J, Zhang H, Cao P, Pan X, Lu X, Chang W. Structural insights into the catalytic mechanism of aldehyde-deformylating oxygenases. Protein Cell. 2014;6:55–67.
10. Shanklin J, Guy JE, Mishra G, Lindqvist Y. Desaturases: emerging models for understanding functional diversification of diiron-containing enzymes. J Biol Chem. 2009;284:18559–63.
11. Sazinsky MH, Lippard SJ. Correlating structure with function in bacterial multicomponent monooxygenases and related diiron proteins. Acc Chem Res. 2006;39:558–66.
12. Warui DM, Li N, Nørgaard H, Krebs C, Bollinger Jr JM, Booker SJ. Detection of formate, rather than carbon monoxide, as the stoichiometric co-product in conversion of fatty aldehydes to alkanes by a cyanobacterial aldehyde decarbonylase. J Am Chem Soc. 2011;133:3316–9.
13. Li N, Nørgaard H, Warui DM, Booker SJ, Krebs C, Bollinger Jr JM. Conversion of Fatty Aldehydes to Alka(e)nes and Formate by a Cyanobacterial Aldehye Decarbonylase: Cryptic Redox by an Unusual Dimetal Oxygenase. J Am Chem Soc. 2011;133:6158–61.
14. Das D, Eser BE, Han J, Sciore A, Marsh ENG. Oxygen-independent decarbonylation of aldehydes by cyanobacterial aldehyde decarbonylase: a new reaction of di-iron enzymes. Angew Chem Int Ed. 2011;50:7148–52.
15. Eser BE, Das D, Han J, Jones PR, Marsh ENG. Oxygen-Independent Alkane Formation by Non-Heme Iron-Dependent Cyanobacterial Aldehyde Decarbonylase: Investigation of Kinetics and Requirement for an External Electron Donor. Biochemistry. 2012;50:10743–50.
16. Zhang J, Lu X, Li J-J. Conversion of fatty aldehydes into alk (a/e)nes by in vitro reconstituted cyanobacterial aldehyde-deformylating oxygenase with the cognate electron transfer system. Biotechnol Biofuels. 2013;6:86.
17. Wang Q, Huang X, Zhang J, Lu X, Li S, Li J-J. Engineering self-sufficient aldehyde deformylating oxygenase fused to alternative electron transfer systems for efficient conversion of aldehydes into alkanes. Chem Commun. 2014;50:4299–301.
18. Aukema KG, Makris TM, Stoian SA, Richman JE, Münck E, Lipscomb JD, Wackett LP. Cyanobacterial aldehyde deformylase oxygenation of aldehydes yields n-1 aldehydes and alcohols in addition to alkanes. ACS Catal. 2013;3:2228–38.
19. Paul B, Das D, Ellington B, Marsh EN. Probing the mechanism of cyanobacterial aldehyde decarbonylase using a cyclopropyl aldehyde. J Am Chem Soc. 2013;135:5234–7.
20. Das D, Ellington B, Paul B, Marsh EN. Mechanistic insights from reaction of α-oxiranyl-aldehydes with cyanobacterial aldehyde deformylating oxygenase. ACS Chem Biol. 2014;9:570–7.
21. Waugh MW, Marsh ENG. Solvent Isotope Effects on Alkane Formation by Cyanobacterial Aldehyde Deformylating Oxygenase and Their Mechanistic Implications. Biochemistry. 2014;53:5537–43.

22. Khara B, Menon N, Levy C, Mansell D, Das D, Marsh EN, Leys D, Scrutton NS. Production of propane and other short-chain alkanes by structure-based engineering of ligand specificity in aldehyde-deformylatingoxygenase. Chembiochem. 2013;14:1204–8.

23. Hayashi Y, Yasugi F, Arai M. Role of Cysteine Residues in the Structure, Stability, and Alkane Producing Activity of Cyanobacterial Aldehyde Deformylating Oxygenase. PLoS One. 2015;10:e0122217.

24. Altschul SF, Gish W, Miller W, Myers EW, Lipman DJ. Basic local alignment search tool. J Mol Biol. 1990;215:403–10.

25. Strub C, Alies C, Lougarre A, Ladurantie C, Czaplicki J, Fournier D. Mutation of exposed hydrophobic amino acids to arginine to increase protein stability. BMC Biochem. 2004;5:9.

26. Sambrook J, Fitsch EF, Maniatis T. A Laboratory Manual. Cold Spring Harbor: Cold Spring Harbor Press, Molecular Cloning; 1989.

13

Identification and characterization of two new 5-keto-4-deoxy-D-Glucarate Dehydratases/Decarboxylases

André Pick[1], Barbara Beer[1], Risa Hemmi[2], Rena Momma[2], Jochen Schmid[1], Kenji Miyamoto[2] and Volker Sieber[1*]

Abstract

Background: Hexuronic acids such as D-galacturonic acid and D-glucuronic acid can be utilized via different pathways within the metabolism of microorganisms. One representative, the oxidative pathway, generates α-keto-glutarate as the direct link entering towards the citric acid cycle. The penultimate enzyme, keto-deoxy glucarate dehydratase/decarboxylase, catalyses the dehydration and decarboxylation of keto-deoxy glucarate to α-keto-glutarate semialdehyde. This enzymatic reaction can be tracked continuously by applying a pH-shift assay.

Results: Two new keto-deoxy glucarate dehydratases/decarboxylases (EC 4.2.1.41) from *Comamonas testosteroni* KF-1 and *Polaromonas naphthalenivorans* CJ2 were identified and expressed in an active form using *Escherichia coli* ArcticExpress(DE3). Subsequent characterization concerning K_m, k_{cat} and thermal stability was conducted in comparison with the known keto-deoxy glucarate dehydratase/decarboxylase from *Acinetobacter baylyi* ADP1. The kinetic constants determined for *A. baylyi* were K_m 1.0 mM, k_{cat} 4.5 s^{-1}, for *C. testosteroni* K_m 1.1 mM, k_{cat} 3.1 s^{-1}, and for *P. naphthalenivorans* K_m 1.1 mM, k_{cat} 1.7 s^{-1}. The two new enzymes had a slightly lower catalytic activity (increased K_m and a decreased k_{cat}) but showed a higher thermal stability than that of *A. baylyi*. The developed pH-shift assay, using potassium phosphate and bromothymol blue as the pH indicator, enables a direct measurement. The use of crude extracts did not interfere with the assay and was tested for wild-type landscapes for all three enzymes.

Conclusions: By establishing a pH-shift assay, an easy measurement method for keto-deoxy glucarate dehydratase/decarboxylase could be developed. It can be used for measurements of the purified enzymes or using crude extracts. Therefore, it is especially suitable as the method of choice within an engineering approach for further optimization of these enzymes.

Keywords: Keto-deoxy-D-Glucarate, *Acinetobacter baylyi*, *Comamonas testosteroni*, *Polaromonas naphthalenivorans*, Dehydratase

Background

Renewable biogenic resources such as lignocellulosic hydrolysates, often referred to as second-generation feedstock, represent an increasingly important raw material for chemicals production. Complete exploitation of these substrates is still a challenging task due to their heterogeneous composition. Besides various hexoses and pentoses, which constitute the main fraction of the hydrolysates, sugar derivatives such as sugar acids accumulate. The latter include hexuronic acids such as D-galacturonic acid and D-glucuronic acid, which are mainly present when pectin-rich waste streams or plant xylans are utilized [1, 2]. Both acids are abundantly available in agricultural waste or forestry residues. In particular, plant pathogenic bacteria such as *Pseudomonas syringae*, *Agrobacterium tumefaciens* or *Erwinia carotovora* as well as *Escherichia coli* or *Thermotoga maritima* possess metabolic pathways for hexuronic acid utilization [3–7]. Up to now, three pathways have been identified for the utilization of hexuronic acids via isomerization, reduction or oxidation [8].

* Correspondence: sieber@tum.de
[1]Technical University of Munich, Straubing Center of Science, Chair of Chemistry of Biogenic Resources, Schulgasse 16, 94315 Straubing, Germany
Full list of author information is available at the end of the article

The oxidative pathway comprises four enzymatic steps (Fig. 1), generating α-keto-glutarate as the direct link entering the citric acid cycle [8]. The first oxidative step is catalysed by uronate dehydrogenase, which produces an aldaric acid lactone that hydrolyses spontaneously [9, 10]. Several uronate dehydrogenases of different origins have been described [11–14].

The subsequent two steps are catalysed by the enzymes glucarate dehydratase and keto-deoxy glucarate dehydratase/decarboxylase (KdgD). Both enzymes are responsible for the defunctionalisation of glucarate. First, glucarate dehydratase removes water, leading to keto-deoxy glucarate, which is the substrate for KdgD; this in turn catalyses the dehydration and decarboxylation into α-keto-glutarate semialdehyde [15]. In the final step, α-keto-glutarate semialdehyde dehydrogenase oxidizes the semialdehyde to α-keto-glutarate [16]. The glucarate dehydratase belongs to the mechanistically diverse enolase superfamily, which is known to catalyse at least 14 different reactions [17]. Within this superfamily, glucarate dehydratase is assigned to the mandelate racemase subgroup [18]. The reaction mechanism and protein structure of several members have been studied in detail [19, 20]. The bifunctional enzyme KdgD belongs to the class I aldolase family and is further sub-grouped into the N-acetylneuraminate lyase superfamily [21]. Only little attention has been devoted towards this enzyme even though it catalyses a very interesting reaction. Just recently, the crystal structure for KdgD from A. tumefaciens was solved [22] in parallel with investigations to gain a deeper understanding of the catalytic mechanism, which led to the identification of catalytically relevant amino acids [23].

For thorough characterization, easy monitoring of the enzymatic reaction is one of the main challenges. Neither the substrate nor the product can be detected photometrically; moreover, no cofactor is involved in the catalytic reaction. Therefore, all studies performed up to now have used a coupled enzyme assay with α-keto-glutarate semialdehyde dehydrogenase, following the formation of NAD(P)H at 340 nm [15]. However, the reaction catalysed by KdgD is well suited to establish a direct method for measuring enzymatic activity.

The release of CO_2 from a carboxylate leads to the consumption of protons and an increase in pH, which in theory can be monitored by a pH indicator and no additional enzyme is necessary to detect the reaction. Colorimetric assays based on a pH indicator system have been successfully used to monitor several enzymatic reactions, e.g. hydrolysis of esters, transfer of sugars, phosphate or nucleotides, as well as decarboxylation of amino acids [24–30].

Here, we report the identification and characterization of two novel KdgDs from Comamonas testosteroni KF-1 (Ct) and Polaromonas naphthalenivorans CJ2 (Pn). For better evaluation and validation, an already known KdgD from Acinetobacter baylyi ADP1 (Ab) was used as the reference. A first characterization and comparison was done by developing an easy and direct measurement method based on a pH indicator system using bromothymol blue (BTB) as the indicator and potassium phosphate as the buffer. The assay could be easily adopted to allow measurements in crude cell extracts and therefore will be very useful for screening approaches.

Methods
Reagents

D-Saccharic acid potassium salt (glucarate), magnesium chloride heptahydrate and BTB sodium salt were purchased from Sigma Aldrich (Seelze, Germany). Restriction enzyme BsaI, alkaline phosphatase, Phusion™ polymerase, T4 ligase and T4 polynucleotide kinase were purchased from New England Biolabs (Frankfurt, Germany). Taq polymerase was obtained from Rapidozym (Berlin, Germany). Oligonucleotides were synthesized by Thermo Fisher Scientific (Waltham, MA, USA). DNaseI was obtained from Applichem (Darmstadt, Germany). All other chemicals were purchased from Carl Roth (Karlsruhe, Germany) or Merck (Darmstadt, Germany) and were used without further purification. All columns used for protein purification were from GE Healthcare (Munich, Germany).

Sequence selection and comparison

The publicly available protein sequence of the 5-keto-4-D-deoxyglucarate dehydratase/decarboxylase of A. baylyi

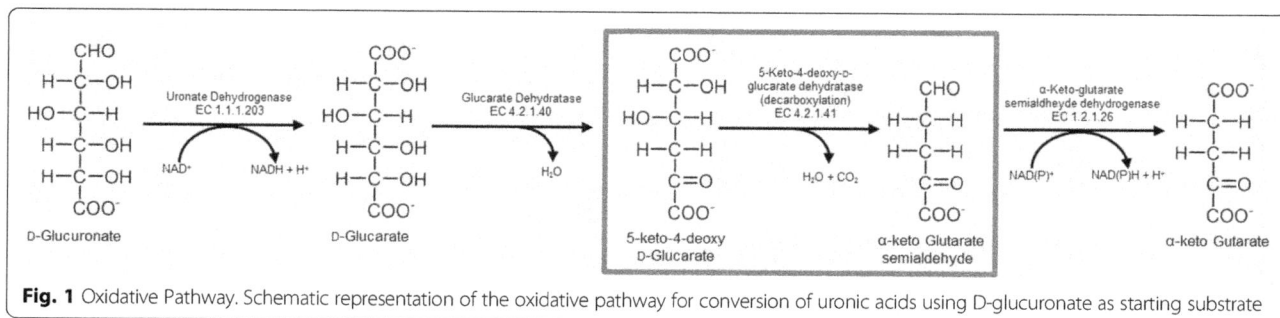

Fig. 1 Oxidative Pathway. Schematic representation of the oxidative pathway for conversion of uronic acids using D-glucuronate as starting substrate

ADP1 was used as the query sequence for BLAST analysis (blastp) for the identification of potential candidates [31]. Candidate proteins belonging to another species with a maximum identity of 70 % were chosen. In the next step, to verify the possible occurrence of the oxidative pathway for D-glucuronate and D-glucarate conversion, the occurrence of the upstream enzyme D-glucarate dehydratase was confirmed by screening the genome sequence of *P. naphthalenivorans* CJ2 (NC_008781.1) and *C. testosteroni* KF-1 (AAUJ02000001.1). Identification of both enzymes was the final criterion for selection.

Four protein sequences encoding for KdgD were aligned using the web-based program T-COFFEE. 5-keto-4-D-deoxyglucarate dehydratase/decarboxylase of *A. baylyi* ADP1, the enzyme from *A. tumefaciens* C58, whose structure was recently determined, were chosen as references [23]. Based on the BLAST results, the two enzyme candidates from *P. naphthalenivorans* CJ2 (WP_011800997.1) and *C. testosteroni* KF-1 (WP_003059546.1) were chosen.

Strains and plasmids

The following strains were used in this study: *E. coli* XL1-Blue (Stratagene), *E. coli* BL21(DE3) (Novagen) and *E. coli* ArcticExpress(DE3) (Agilent Technologies). The DNA sequences for the corresponding genes of keto-deoxy-D-glucarate dehydratase/decarboxylase (kdgD) from *A. baylyi* (ADP1) (protein sequence GenBank™ WP_004930673.1), which is identical to *A. baylyi* DSM 14961 (protein sequence GenBank™ ENV53020.1), from *C. testosteroni* KF-1 (protein sequence GenBank™ WP_003059546.1) and from *P. naphthalenivorans* (protein sequence GenBank™ WP_011800997.1) were synthesized with optimized codon usage for expression in *E. coli* (Additional file 1: Figures S1-S3) (Life Technologies, Regensburg, Germany). The following primers were used for amplification: F-kdgD-A.b.- CAGCAA**GGT CTCA**CATATGGATGCCCTGGAACTG, R-kdgD-A.b.- CTGCGG-ACCCAGGGTTG, F-kdgD-C.t.-CAGCAA**GG TCTCA**CATATGACACCGCAGG-ATCTGAAAG, R-kdg D-C.t.-xCTGCGGACCCAGTTTATCAATC, F-kdgD-P.n.- CAGCAA**GGTCTCA**CATATGAATCCGCAGGATCTGA AAAC, R-kdgD-P.n.- CTGCGGACCCAGGCTTTTAATC. The restriction enzyme recognition site for BsaI is underlined and the start codon is marked in bold. The reverse primers were phosphorylated using T4 polynucleotide kinase according to the supplier's manual. PCR products were digested with BsaI and cloned into pCBR, pCBRHisN and pCBRHisC, which are derivatives of pET28a (Novagen). The cloning strategy of all pET28 derivatives is described by Guterl et al. [32]. Ligation of the PCR products and the following transformation led to the plasmids pCBR-kdgD-A.b., pCBR-kdgD-C.t., pCBR-kdgD-P.n., pCBRHisN-kdgD-A.b., pCBRHisN-kdgD-C.t., pCBRHisN-kdgD-P.n., pCBRHisC-kdgD-A.b.,

pCBRHisC-kdgD-C.t., and pCBRHisC-kdgD-P.n. Multiplication of the plasmids was performed by *E. coli* XL1 Blue (Stratagene) in Luria–Bertani medium containing 30 µg/ml kanamycin. *E. coli* BL21(DE3) or *E. coli* ArcticExpress(DE3) were used for expression.

Overexpression and FPLC purification

Protein expression was performed with two different *E. coli* expression strains depending on the target enzyme. *E. coli* BL21(DE3) [pET28a-NH-kdgD-A.b.] was cultivated with a slightly modified protocol described by Aghaie et al. [15]. The preculture was incubated in 4 ml of Terrific broth medium supplemented with 1 M sorbitol and 5 mM betaine with 100 µg/ml kanamycin at 37 °C overnight on a rotary shaker (180 rpm). The expression culture consisted of the same media and was inoculated with a 1:100 dilution of the preculture. Incubation was performed at 37 °C until an OD_{600} of 2 was reached. Protein expression was induced with the addition of IPTG to a final concentration of 0.5 mM followed by incubation for 21 h at 16 °C. For *E. coli* ArcticExpress(DE3) [pET28a-NH-kdgD-C.t. or pET28a-CH-kdgD-P.n.] the preculture was cultivated in Luria–Bertani media with 100 µg/ml kanamycin and 15 µg/ml gentamycin over night at 37 °C. The expression culture consisted of autoinduction media with both antibiotics and was inoculated with a 1:100 dilution of the preculture [33]. Incubation was performed for 3 h at 37 °C followed by the second step at 12 °C for 45 h. Afterwards, cells were harvested by centrifugation and washed one time with 50 mM sodium phosphate buffer (pH 8.0) and frozen at –20 °C or resuspended in a binding buffer (50 mM potassium phosphate, pH 8.0, 20 mM imidazol, 500 mM NaCl and 10 % glycerol). Crude extracts were prepared using a Basic-Z cell disrupter (IUL Constant Systems) and the subsequent addition of $MgCl_2$ to a final concentration of 2.5 mM in combination with DNaseI (10 µg/ml), followed by an incubation for 20 min at room temperature for successful DNA degradation. The insoluble fraction of the lysate was removed by centrifugation at 20,000 rpm for 20 min at 4 °C. The supernatant was filtered through a 0.45 µm syringe filter and applied to an IMAC affinity resin column, 5 ml HisTrapTM FF, equilibrated with the binding buffer using the ÄKTA purifier system. The enzyme was washed with 20 ml of binding buffer and eluted with 50 mM potassium phosphate buffer (pH 8.0, 500 mM imidazol, 500 mM NaCl and 10 % glycerol). Aliquots of each eluted fraction were subjected to 12 % SDS-PAGE. The fractions containing the eluted protein were pooled and the protein was desalted using a HiPrepTM 26/10 desalting column, which was preliminary equilibrated with 50 mM Tris-HCl (pH 7.5) or 5 mM sodium phosphate buffer (pH 7.0).

Protein concentrations were determined using a Bradford assay (Roti®-nanoquant, Carl Roth).

Enzyme expression in 96-deep well scale

For all three enzymes, an expression in the 96-deep well scale was performed. Therefore, electrocompetent *E. coli* ArcticExpress(DE3) cells were transformed with the corresponding plasmid. Single clones were picked using the CP7200 Colony Picker (Norgren Systems) and transferred to 96-deep well plates filled with 1.2 ml autoinduction media [33] by a MicroFlo Select dispenser (Bio-Tek Instruments). After incubation (36 h, 37 °C at 1,000 rpm), further processing was done manually. First, 100 µl of cell culture was transferred into a 96-well plate (U-shaped bottom) and harvested by centrifugation (4,570 rpm, 10 min at RT) while the supernatant was discarded. The frozen pellets (1 h at −20 °C) were thawed at room temperature for one hour to improve cell lysis. Lysis was continued by the addition of 30 µl lysis buffer (3 h, 1,000 rpm, 37 °C) containing 2 mM KP_i, pH 7.0, 2 mM $MgCl_2$, 10 µg/ml DNaseI, 100 µg/ml lysozyme. Next, 120 µl buffer (2 mM KP_i, pH 7.0) was added followed by centrifugation (3,000 rpm, 15 min at RT). For the photometric measurement, 20 µl of the crude extract was transferred to a 96-well plate (F-shaped bottom) and the reaction was started by adding 180 µl master mix to give a final volume of 200 µl (2.5 mM KP_i, pH 7.0, 2 mM $MgCl_2$, 25 µg/ml BTB and 5 mM keto-deoxy-D-glucarate). The measurements were carried out for 60 min at 2-min intervals. Depending on the enzyme, different time windows were used for the activity calculation.

Substrate preparation

5-keto-4-deoxy-D-glucarate is not commercially available. Therefore, it was prepared using an enzymatic conversion of D-glucarate. For that, a 250–500 mM solution of D-glucarate containing 2 mM $MgCl_2$ was prepared. Potassium glucarate is not completely soluble at this concentration, and the pH value was around 4.5. The pH was shifted to 8.0 by adding potassium hydroxide. A sample was taken as zero-point control and the reaction was started by the addition of D-glucarate dehydratase (Beer et al., manuscript in preparation). Using HPLC, a sample was analysed at regular intervals. After full conversion, the reaction was stopped by removing the enzyme by filtration (spin filters, 10 kDa MWCO, modified PES; VWR, Darmstadt, Germany). In the last step, the pH was adjusted to 7.0 using HCl.

HPLC analysis

D-glucarate, 5-keto-4-D-deoxyglucarate and α-ketoglutarate semialdehyde were separated by HPLC, using an Ultimate-3000 HPLC system (Dionex, Idstein, Germany), equipped with an autosampler (WPS 3000TRS), a column compartment (TCC3000RS) and a diode array detector (DAD 3000RS). The column Metrosep A Supp10−250/40 column (250 mm, particle size 4.6 mm; Metrohm, Filderstadt, Germany) at 65 °C was used for separation by isocratic elution with 30 mM ammonium bicarbonate (pH 10.4) as the mobile phase at 0.2 mL min^{-1}. Samples were diluted in water, filtered (10 kDa MWCO, modified PES; VWR, Darmstadt, Germany) and 10 µL of the samples was applied on the column. Data was analysed with Dionex Chromelion software.

Determination of $\Delta\varepsilon_{617}$ for BTB and Q factor calculation

The extinction coefficient difference $\Delta\varepsilon_{617}$ was determined experimentally. The protonated form of BTB (0–100 µg/mL) was measured in different potassium phosphate buffer concentrations (2.5–10 mM) at pH 5.5. In addition, the deprotonated form was measured at pH 8.0 under identical conditions. For both measurements, the concentration multiplied by the pathlength ((mol/L) (cm)) was plotted against the absorbance and the slope was determined. $\Delta\varepsilon_{617}$ was calculated by subtracting the value of the protonated BTB from the deprotonated BTB.

After determination of $\Delta\varepsilon_{617}$, the buffer factor Q factor, a constant relating absorbance change and reaction rate for a given buffer/indicator system [30, 34, 35], was calculated for different buffer and indicator concentrations by using equation (1). C_B and C_{In} are the total molar concentrations of the buffer and the indicator, respectively, and l represents the path length.

$$Q = \frac{C_B/C_{In}}{\Delta_{\varepsilon_{617} \times l}} \tag{1}$$

Colorimetric assay

For direct detection of KdgD activity, a colorimetric assay in a 96-well microplate format was developed in a Multiskan® spectrum spectrophotometer (Thermo Fisher Scientific). The total reaction volume was 200 µl and consisted of 2.5 mM potassium phosphate buffer (pH 7.0), 2 mM $MgCl_2$, 25 µg/ml BTB and the substrate at 37 °C. Every measurement was conducted at least three times. Addition of the enzyme solution initiated the measurement. Enzyme concentration for each KdgD varied and corresponded to a suitable signal over time. One unit of enzyme activity was defined as the amount of protein that converts 1 µmol of substrate/min at 37 °C. Calculation of the enzyme velocity was performed using equation (2), where dA/dt is the rate of absorbance change, V_R and V_E represent the reaction volume and the enzyme volume, c_E is the enzyme concentration and D is a measure of the dilution factor for the enzyme solution. The enzyme concentration allowed use of an 8–10 min time window for a linear slope. Substrate conversion was always below 10 % for each concentration during the kinetic measurements.

$$U/mg = \frac{\frac{dA}{dt} \times Q \times V_R \times D}{V_E \times C_E} \tag{2}$$

Enzyme characterization

Each enzyme was investigated concerning K_m and k_{cat}. The substrate concentration for the kinetic measurements was in the range 0.05–20 mM. The other conditions remained the same as was described in the previous section for the colorimetric assay. Calculation of the Michaelis-Menten kinetic parameters was done by fitting the data to the Michaelis-Menten equation (3)

$$v = \frac{v_{max} \times [s]}{K_m \times [s]} \tag{3}$$

using Sigma–Plot 11.0 (Systat Software). The Michaelis-Menten equation consists of the following terms: v is the

reaction rate (µmol/min/mg), V_{max} is the maximum reaction rate (µmol/min/mg), *[S]* is the varying substrate concentration (mM), and K_m is the Michaelis-Menten constant (mM).

The enzyme stability of the variants was investigated for storage at 8 °C (refrigerator) and in the context of cryo-conservation and reuse. Enzyme stock solutions in 50 mM Tris-HCl (pH 7.5) were stored without additional cryo-protectants, such as glycerol, at –20 °C.

Enzyme stability was investigated using two different temperatures: 37 °C and 65 °C. Therefore, aliquots of each enzyme for each measuring point with a volume of 100 µl were incubated in a water bath. The enzymes were incubated using a 5 mM KPi buffer (pH 7.0) at a concentration of 0.13 mg/ml. Therefore, the storage buffer 50 mM Tris-HCl (pH 7.5) was exchanged with the ÄKTA purifier system using a HiPrepTM 26/10

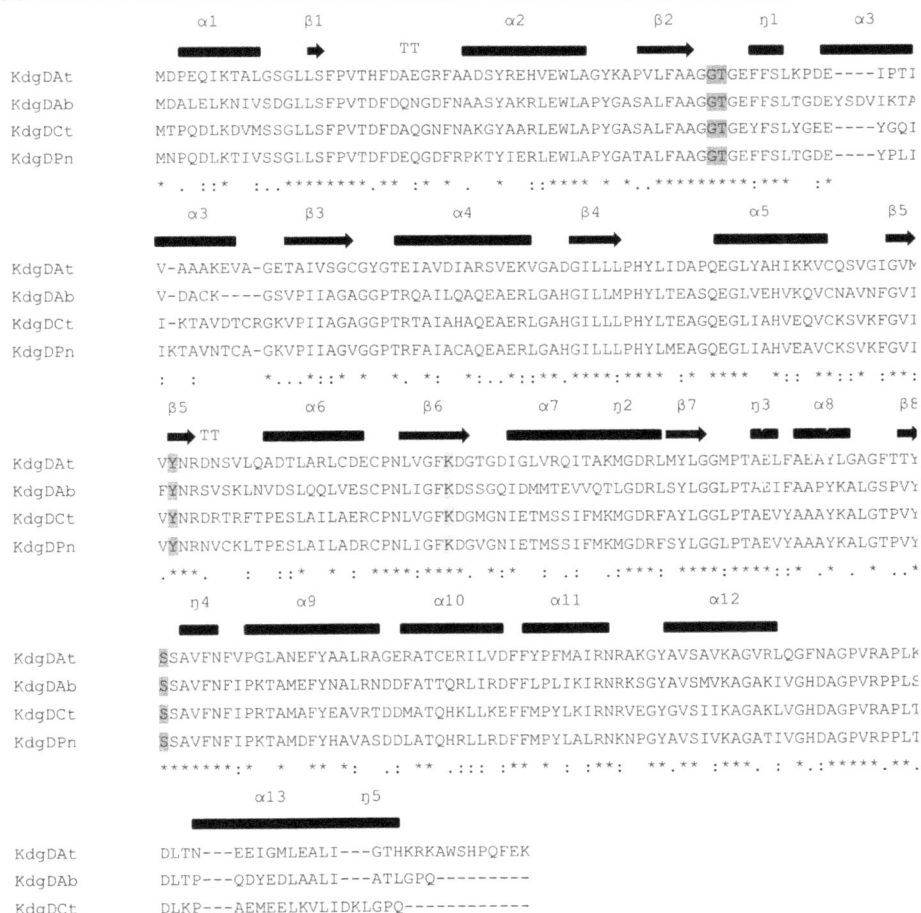

Fig. 2 Sequence alignment. Multiple sequence alignment of known 5-keto-4-deoxy-D-glucarate dehydratases/decarboxylases using clustal omega [46]. The secondary structures are shown above with thick bars representing α-helices and arrows representing β-sheets. Coloured residues represent conserved residues of the active centre involved in the specific substrate recognition. The lysine residue (light grey) forms the intermediate Schiff base with the substrate. The sequence identity for all three dehydratases/decarboxylases obtained by referring to KdgDAt as the standard and performing pairwise alignment with EMBOSS Needle [40] are as follows: KdgDAb = 47 %, KdgDCt = 49 %, KdgDPn = 52 %

desalting column. In case of KdgDAb, an additional buffer system of 10 mM NH_4HCO_3 (pH 7.9) was used. For the measurements, 5 mM substrate was used. The half-life for each enzyme at the incubation temperature of 65 °C was calculated according to Rogers and Bommarius [36].

Results and discussion
Selection of KdgDs

Until now, only a few KdgD enzymes are described in the literature [37–39]. Recently, the complete oxidative pathway for *A. baylyi* ADP1 was elucidated and the enzymes involved were recombinantly expressed in *E. coli* [15]. Therefore, the selection of the novel KdgD candidates was performed by a BLAST analysis based on the amino acid sequence of the known enzyme derived from *A. baylyi* ADP1 identified by Aghaie et al. [15]. Two candidates were chosen, showing less than or equal to 70 % identity toward *A. baylyi* ADP1 based on a pairwise alignment performed by EMBOSS Needle [40], *P. naphthalenivorans* CJ2 (69.6 %) and *C. testosteroni* KF-1 (67.7 %) (Fig. 2). With an eightfold (βα) barrel structure, these enzymes share a ubiquitously found motif that is capable of catalysing many different reactions [41]. As a member of the *N*-acetylneuraminate lyase superfamily, KdgD exhibits a conserved lysine residue (Fig. 2, light grey) that forms the Schiff base essential for the enzymatic reaction at the end of the sixth β-sheet. Furthermore, a tyrosine residue located at the end of the fifth β-sheet (Fig. 3, dark grey) is also conserved and catalyses the deprotonation of the β-carbon after the Schiff base had been formed. In the next step, the hydroxyl group at the fourth carbon atom is protonated and subsequently

released as a water molecule; this is mediated by a serine positioned at the C-terminus of the eighth β-sheet. A glycine and threonine (Fig. 3, dark grey) at the end of the second β-sheet coordinate the C_6-carboxylate group. The conserved residues described by Taberman et al. can be found in all investigated enzymes [23].

Heterologous expression

The codon-optimized kdgD genes of *A. baylyi*, *C. testosteroni* and *P. naphthalenivorans* were heterologously expressed in different *E. coli* expression strains. Untagged versions as well as His-tagged versions (oH = without His-Tag, CH = C-terminal His-Tag and NH = N-terminal His-Tag) were constructed for all genes of interest. Using *E. coli* BL21(DE3) with autoinduction media resulted in inclusion bodies for all three enzymes in every His-tag as well as without His-tag version. Therefore, the expression using slightly modified conditions that was described by Aghaie et al. was used [15]. Terrific broth medium containing 1 M sorbitol and 5 mM betaine was used for the expression starting at 37 °C until A600 reached ≥1.5. After the addition of IPTG to a final concentration of 1 mM, cultivation was continued at 16 °C for 20 h. Under these conditions, a soluble expression of the NHKdgDAb (hereafter, referred as KdgDAb), oHKdgDAb and oHKdgDCt was possible. Using an ArcticExpress(DE3) *E. coli* expression strain in combination with autoinduction media enabled the expression of NHKdgDCt (KdgDCt) and CHKdgDPn (KdgDPn) in a soluble form. The distribution between the insoluble and the soluble proteins varied from almost complete solubility of NHKdgDAb to an 80:20 ratio for

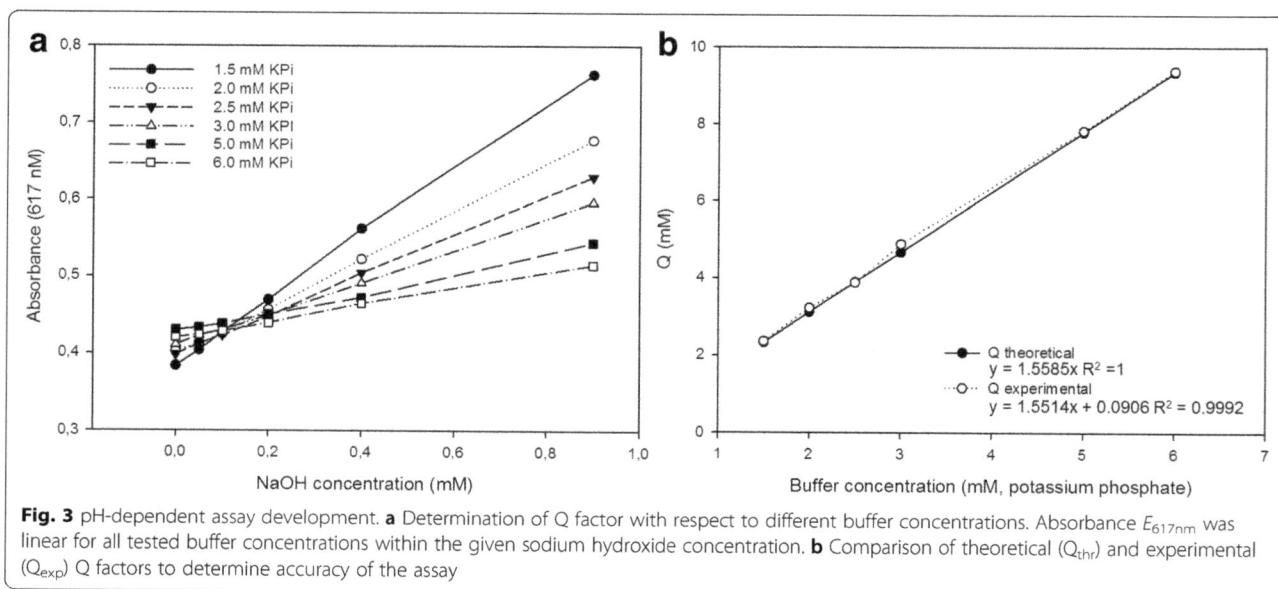

Fig. 3 pH-dependent assay development. **a** Determination of Q factor with respect to different buffer concentrations. Absorbance E_{617nm} was linear for all tested buffer concentrations within the given sodium hydroxide concentration. **b** Comparison of theoretical (Q_{thr}) and experimental (Q_{exp}) Q factors to determine accuracy of the assay

Table 1 Kinetic parameters determined for KdgDAb, KdgDCt and KdgDPn at 37 °C using the pH-shift assay with 2.5 mM KP_i buffer (pH 7.0)

Enzyme	K_m [mM]	k_{cat} [1/s]	k_{cat}/K_m [1/mM s]	v_{max} [U/mg]
KdgDAb	1.0 ± 0.1	4.5 ± 0.02	4.5	8.0 ± 0.27
KdgDCt	1.1 ± 0.1	3.1 ± 0.02	2.8	5.3 ± 0.11
KdgDPn	1.1 ± 0.1	1.7 ± 0.01	1.5	3.0 ± 0.06
KdgDAt[a]	0.5	8.83 ± 0.16	17.0	
KdgDAb[b]	0.2	3.9 ± 0.04	19.2	

[a] Assay conditions: 50 mM NaP_i pH 7.5, 100 mM NaCl, 4 mM $NADP^+$, 22 °C [23]
[b] Assay conditions: 50 mM Hepes/NaOH pH 7.5, 100 mM NaCl, 5 mM $MgCl_2$, 4 mM $NADP^+$, 10 % glycerol, 22 °C [15]

NHKdgDCt and CHKdgDPn (data not shown). Eluted proteins appeared as a single band on SDS polyacrylamide gels and no further bands indicating co-elution of chaperonins of *E. coli* ArcticExpress(DE3) expression were detected. This is notwithstanding the publication of several reports that mention co-elution as a problem for successful enzyme purification [42, 43]. Enzyme activities of the different KdgDs were stable after storage at 8 °C in desalting buffer (50 mM Tris-HCl, pH 7.5) for at least 14 days. Independent stable long-term storage, preserving the catalytic activity of the enzymes, (four month) at −20 °C and −80 °C was realized without cryo-protectants.

Assay validation

A reliable and sensitive pH-shift assay requires the pKa of the indicator and the buffer to be very similar to allow a direct correlation between the colour change and the changing concentration of hydrogen ions within the assay solution. The pKa of the phosphate buffer is 7.2 and that of the BTB lies between 7.1 and 7.3, resulting in a suitable combination.

The absorbance spectra of protonated and deprotonated BTB were determined and the maximum difference in extinction coefficient was observed at 617 nm, which compares well with the wavelength reported in the literature [44]. The large difference in the extinction coefficient

between protonated and deprotonated BTB ($\Delta\varepsilon_{617}$ 28101 M^{-1} cm^{-1}) results in a low Q value (Eq. 1), and this in turn ensures a high sensitivity (dA/dt) of the assay. For a final validation, the buffer factor (Q) was calculated and determined experimentally to guarantee the reliability of the assay. Q was experimentally determined (Q_{exp}) by testing several buffer concentrations and titrations of sodium hydroxide in the range of 0–1 mM (Fig. 3a). The reciprocal of the slopes directly correlates to Q, and the theoretical Q (Q_{thr}) value for each buffer concentration was calculated using equation (1). Buffer concentration was plotted against Q_{thr} and Q_{exp} to check the extent of the correlation (Fig. 3b). Although a concentration of 1.5 mM resulted in almost the same Q_{thr} and Q_{exp}, a higher buffer concentration of 2.5 mM was chosen. The reasons for this are an increased robustness of the assay and better pH stability in the initial phase of the measurement. The Q value at this buffer concentration was still sufficiently low to guarantee high sensitivity.

Enzyme characterization

The purified enzymes were used to determine their kinetic parameters k_{cat}, K_m and v_{max} for the substrate keto-deoxy glucarate (Table 1). The enzyme KdgDAb characterized by Aghaie et al. using a coupled enzyme method [15] was used to allow comparison of the same parameters obtained here using the pH-shift assay and to compare with the activity of KdgDCt and KdgDPn. The K_m for keto-deoxy glucarate was almost similar for all three enzymes (1.0–1.1 mM). The graphs for determination of the Michaelis−Menten constants are shown in Fig. 4. There were differences in k_{cat}, and v_{max} among the three enzymes. KdgDAb showed the highest k_{cat} with 4.5 s^{-1} followed by KdgDCt (3.1 s^{-1}) and KdgDPn (1.7 s^{-1}). The kinetic parameters for KdgDAb differ slightly from those reported by Aghaie et al., especially the higher K_m, but also k_{cat}. The latter appear to be similar, however Aghaie et al. and Taberman et al. measured the enzymatic activity at a lower temperature (22 °C) compared to the pH-shift assay (37 °C) indication a 2-fold lower k_{cat} value obtained within this study [15, 23]. This difference

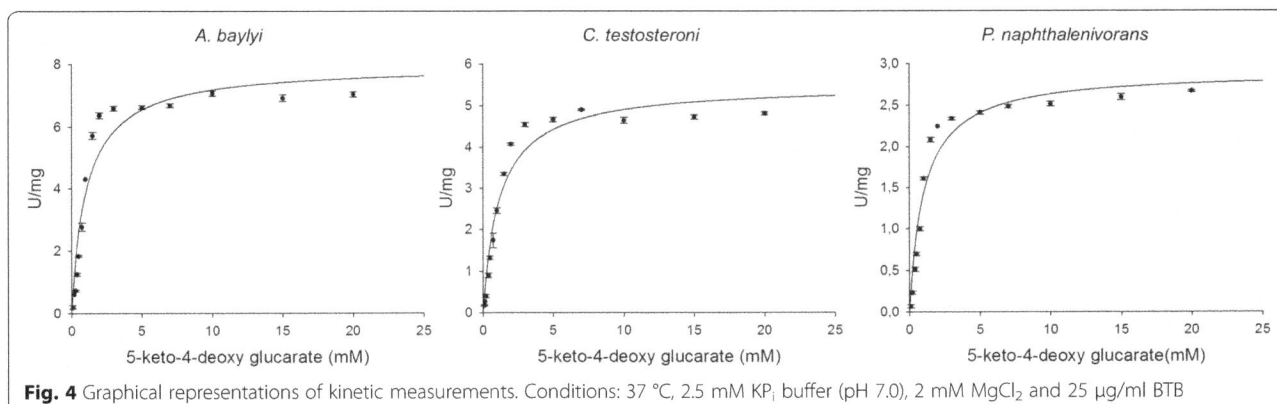

Fig. 4 Graphical representations of kinetic measurements. Conditions: 37 °C, 2.5 mM KP_i buffer (pH 7.0), 2 mM $MgCl_2$ and 25 µg/ml BTB

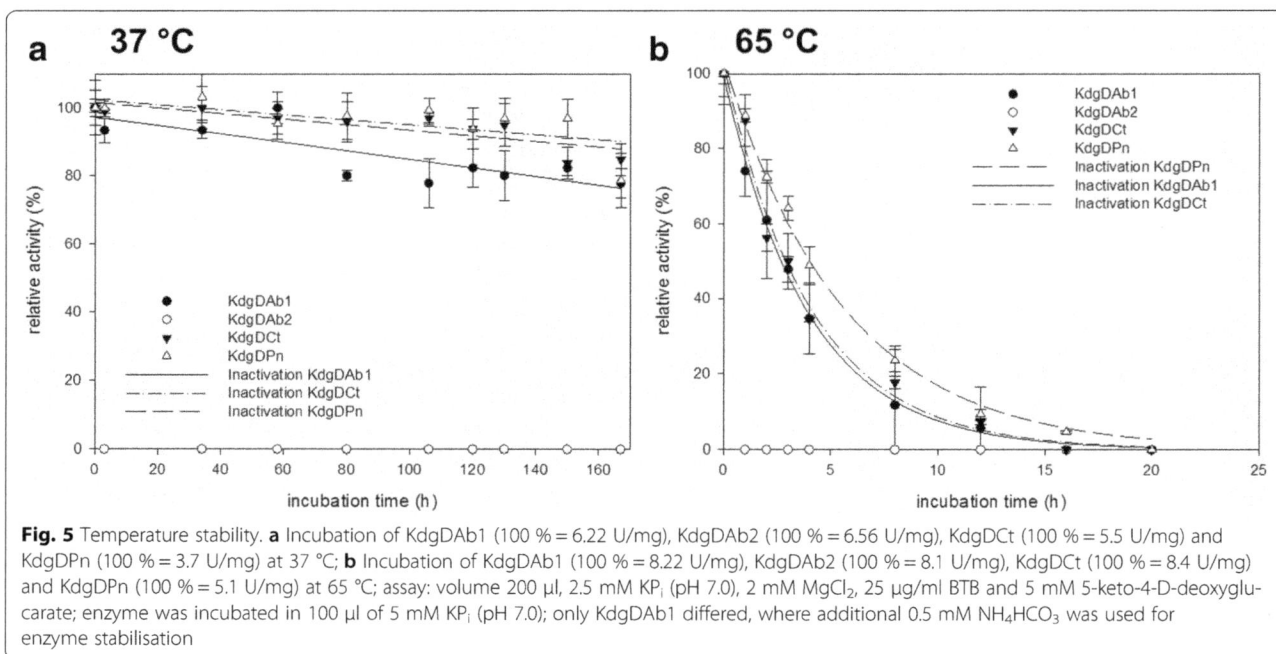

Fig. 5 Temperature stability. **a** Incubation of KdgDAb1 (100 % = 6.22 U/mg), KdgDAb2 (100 % = 6.56 U/mg), KdgDCt (100 % = 5.5 U/mg) and KdgDPn (100 % = 3.7 U/mg) at 37 °C; **b** Incubation of KdgDAb1 (100 % = 8.22 U/mg), KdgDAb2 (100 % = 8.1 U/mg), KdgDCt (100 % = 8.4 U/mg) and KdgDPn (100 % = 5.1 U/mg) at 65 °C; assay: volume 200 µl, 2.5 mM KP_i (pH 7.0), 2 mM $MgCl_2$, 25 µg/ml BTB and 5 mM 5-keto-4-D-deoxyglucarate; enzyme was incubated in 100 µl of 5 mM KP_i (pH 7.0); only KdgDAb1 differed, where additional 0.5 mM NH_4HCO_3 was used for enzyme stabilisation

might be explained by the differing buffer conditions during the kinetic parameter measurement. The different buffer system as well as the reduced buffer concentration without any stabilizers (glycerol, DTT or NaCl) in combination with a slight decrease in the pH value might be responsible for this observation. Taberman et al. described the KdgD of *A. tumefaciens* and identified the pH optimum to be in the range 7.5–8.0 [23]. For this measurement, a phosphate buffer (pH 7.5) with additional NaCl was used and a K_m of 0.5 mM was reported. The enzyme showed only 65 % activity at pH 7.0 compared with the activity at pH 8.0 (12 U/mg) using a phosphate buffer. Application of the pH-shift assay system using BTB/phosphate buffer at a higher pH is not suitable.

In addition, the stability as a function of temperature was investigated. Activities of KdgDAb, KdgDCt and KdgDPn were determined at two different temperatures: 37 and 65 °C (Fig. 5a and b). KdgDCt and KdgDPn showed a similar behaviour at both temperatures. For both enzymes, activity was stable over 150 h at 37 °C, whereas after 4 h at 65 °C the initial activity decreased below 50 %. The calculated $t_{1/2}$ was 2.79 h (KdgDCt) and 3.83 h (KdgDPn). KdgDAb (Fig. 5a, KdgDAb2) showed a different behaviour. A fast decrease in enzyme activity was observed already at 37 °C. It appeared that the KP_i buffer system had a negative influence on the stability of this enzyme. Changing the incubation conditions and adding an additional buffer at a low concentration, in this case, NH_4HCO_3 at a concentration of 0.5 mM, maintained stability of KdgDAb (Fig. 5a, KdgDAb1) at a similar level as seen for KdgDCt and KdgDPn. The calculated $t_{1/2}$ at 65 °C for KdgDAb was 2.71 h and this is almost

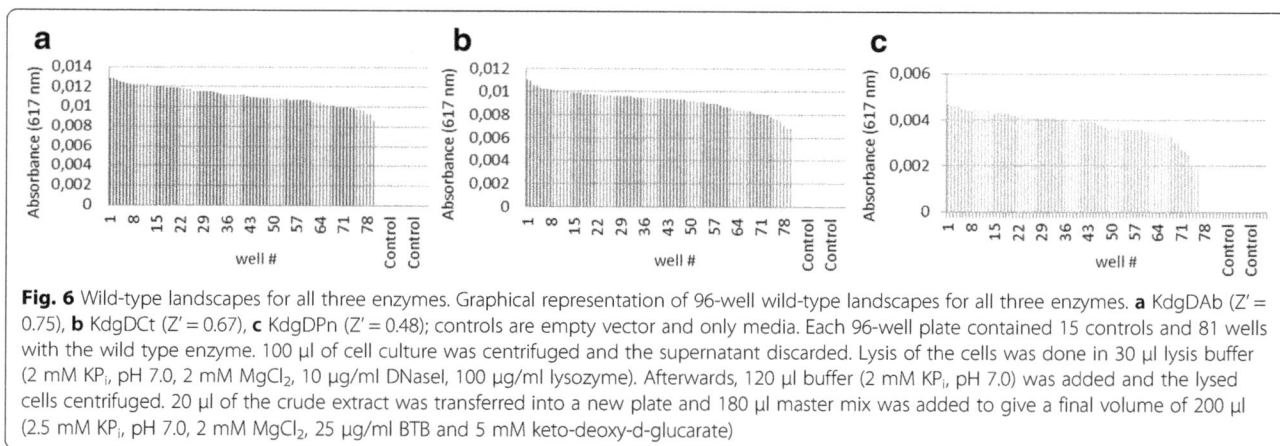

Fig. 6 Wild-type landscapes for all three enzymes. Graphical representation of 96-well wild-type landscapes for all three enzymes. **a** KdgDAb (Z' = 0.75), **b** KdgDCt (Z' = 0.67), **c** KdgDPn (Z' = 0.48); controls are empty vector and only media. Each 96-well plate contained 15 controls and 81 wells with the wild type enzyme. 100 µl of cell culture was centrifuged and the supernatant discarded. Lysis of the cells was done in 30 µl lysis buffer (2 mM KP_i, pH 7.0, 2 mM $MgCl_2$, 10 µg/ml DNaseI, 100 µg/ml lysozyme). Afterwards, 120 µl buffer (2 mM KP_i, pH 7.0) was added and the lysed cells centrifuged. 20 µl of the crude extract was transferred into a new plate and 180 µl master mix was added to give a final volume of 200 µl (2.5 mM KP_i, pH 7.0, 2 mM $MgCl_2$, 25 µg/ml BTB and 5 mM keto-deoxy-d-glucarate)

comparable to that of the other two enzymes. Additional, KdgDCt and KdgDPn were tested with NH_4HCO_3 at a concentration of 0.5 mM but no difference in stability or activity was detectable for both enzymes.

High-throughput decarboxylase screening assay

After the verification of the pH-shift assay for the measurement of purified enzymes, its transferability to parallelization and HTS was tested by measuring Z-factor of the wild-type landscapes for all three KdgDs [45]. Again, *E. coli* ArcticExpress(DE3) was used as the expression strain for KdgDAb because of the positive results obtained for the expression of the other two KdgDs and in order to use the same standardized 96-well expression protocol for all three enzymes. Despite the fact that *E. coli* ArcticExpress(DE3) was developed to allow the expression of target proteins at a low temperature, the temperature was maintained constant at 37 °C. The reason for this was the inhomogeneous growth behaviour due to temperature switches during cultivation in 96-well scale observed for several different enzymes using autoinduction media. The cells were cultivated for 36 h to ensure that every expression culture reached their stationary phase. KdgDCt and KdgDPn showed a similar cell density at the end, whereas cell density of KdgDAb was lower compared with the other two strains (data not shown). The use of a commercial protein extraction reagent (B-PER, Thermo Fisher) did not improve cell lysis. Therefore, only DNaseI and lysozyme were used in the lysis step. After the lysis step, the lysate volume was increased to 150 µl to guarantee that no cell debris was transferred into the assay plate since the remaining cell debris showed a negative impact on the assay which resulted in a shift towards acidic pH conditions.

Based on the pH-shift measurements, the Z'-factor was calculated for all three wild-type landscapes (Fig. 6). Z' was 0.75 for KdgDAb, 0.67 for KdgDCt and 0.48 for KdgDPn. The linear slope was measured in a suitable range with an overall measurement time of 60 min. These findings are consistent with the activity of the enzymes discussed above. The Z'-values correspond well to the relative activities of the three enzymes. For screening purposes, a Z'-value of ca. 0.5 or above is desired. The developed method therefore was shown to be suitable for application in the directed evolution of all three dehydratase/decarboxylase enzymes.

Conclusions

In conclusion, we identified two KdgD enzymes from *C. testosteroni* and *P. naphthalenivorans* and compared them with the already known enzyme from *A. baylyi* ADP1 concerning catalytic activity and stability. Therefore, we developed a pH-shift assay on the basis of BTB as the pH indicator and potassium phosphate as the buffer component. We were able to reduce the buffer concentration to 2.5 mM while maintaining reliability, reproducibility and sensitivity at a high level. KdgD enzymes from *C. testosteroni* (K_m 1.1 mM and k_{cat} 3.1 s^{-1}) and *P. naphthalenivorans* (K_m 1.1 mM and k_{cat} 1.7 s^{-1}) were characterized using this assay system. The calculated $t_{1/2}$ was 2.79 h (KdgDCt) and 3.83 h (KdgDPn) at 65 °C. In addition, the assay system was successfully tested with crude extracts and a high reliability. Therefore, the 96-well based screening system enables further optimization of KdgD enzymes.

Abbreviations

BTB: Bromothymol blue; IMAC: Immobilized metal ion affinity chromatography; IPTG: Isopropyl-β-D-thiogalactopyranosid; KdgD: Keto-deoxy glucarate dehydratase/decarboxylase; MWCO: Molecular weight cut-off

Acknowledgements

Financial support for travelling was obtained from the German academic exchange service (DAAD) through project no. 54365152, "Discovery of novel biocatalysts from bacterial consortia". This work was supported by the German Research Foundation (DFG) and the Technical University of Munich within the funding programme Open Access Publishing.

Authors' contributions

AP participated in the design of the study, performed all the experiments, analysed the data and wrote the manuscript. BB participated in the research and performed the HPLC measurements. RH participated in the evaluation of the pH assay. RM participated in the evaluation of the pH assay. JS participated in the design of the study and commented on the manuscript. KM commented on the manuscript. VS participated in the design of the study and commented on the manuscript. All authors have read and approved the final manuscript.

Competing interests

The authors declare that they have no competing interests.

Author details

[1]Technical University of Munich, Straubing Center of Science, Chair of Chemistry of Biogenic Resources, Schulgasse 16, 94315 Straubing, Germany. [2]Department of Biosciences and Informatics, Keio University, 3-14-1 Hiyoshi, 2238522 Yokohama, Japan.

References

1. Willats WG, McCartney L, Mackie W, Knox JP. Pectin: cell biology and prospects for functional analysis. In: Plant Cell Walls. Berlin: Springer; 2001. p. 9–27.

2.   Ebringerova A, Heinze T. Xylan and xylan derivatives–biopolymers with valuable properties, 1. Naturally occurring xylans structures, isolation procedures and properties. Macromol Rapid Commun. 2000;21(9):542–56.

3.   Adams E, Rosso G. α-ketoglutaric semialdehyde dehydrogenase of pseudomonas properties of the purified enzyme induced by hydroxyproline and of the glucarate-induced and constitutive enzymes. J Biol Chem. 1967;242(8):1802–14.

4.   Chang YF, Feingold DS. D-Glucaric Acid and Galactaric Acid Catabolism by Agrobacterium Tumefaciens. J Bacteriol. 1970;102(1):85–96.

5.   Ashwell G, Wahba AJ, Hickman J. A new pathway of uronic acid metabolism. Biochim Biophys Acta. 1958;30(1):186–7.

6.   Rodionova IA, Scott DA, Grishin NV, Osterman AL, Rodionov DA. Tagaturonate–fructuronate epimerase UxaE, a novel enzyme in the hexuronate catabolic network in Thermotoga maritima. Environ Microbiol. 2012;14(11):2920–34.

7.   Van Gijsegem F, Toussaint A. In vivo cloning of Erwinia carotovora genes involved in the catabolism of hexuronates. J Bacteriol. 1983;154(3):1227–35.

8.   Richard P, Hilditch S. D-galacturonic acid catabolism in microorganisms and its biotechnological relevance. Appl Microbiol Biotechnol. 2009;82(4):597–604.

9.   Boer H, Maaheimo H, Koivula A, Penttilä M, Richard P. Identification in Agrobacterium tumefaciens of the D-galacturonic acid dehydrogenase gene. Appl Microbiol Biotechnol. 2010;86(3):901–9.

10. Bouvier JT, Groninger-Poe FP, Vetting M, Almo SC, Gerlt JA. Galactaro δ-lactone isomerase: lactone isomerization by a member of the amidohydrolase superfamily. Biochemistry. 2014;53(4):614–6.

11. Yoon S-H, Moon TS, Iranpour P, Lanza AM, Prather KJ. Cloning and characterization of uronate dehydrogenases from two Pseudomonads and Agrobacterium tumefaciens strain C58. J Bacteriol. 2009;191(5):1565–73.

12. Wagschal K, Jordan DB, Lee CC, Younger A, Braker JD, Chan VJ. Biochemical characterization of uronate dehydrogenases from three Pseudomonads, Chromohalobacter salixigens, and Polaromonas naphthalenivorans. Enzyme Microb Technol. 2015;69:62–8.

13. Parkkinen T, Boer H, Janis J, Andberg M, Penttila M, Koivula A, Rouvinen J. Crystal structure of uronate dehydrogenase from Agrobacterium tumefaciens. J Biol Chem. 2011;286(31):27294–300.

14. Pick A, Schmid J, Sieber V. Characterization of uronate dehydrogenases catalysing the initial step in an oxidative pathway. J Microbial Biotechnol. 2015;8(4):633–43.

15. Aghaie A, Lechaplais C, Sirven P, Tricot S, Besnard-Gonnet M, Muselet D, de Berardinis V, Kreimeyer A, Gyapay G, Salanoubat M, et al. New insights into the alternative D-glucarate degradation pathway. J Biol Chem. 2008;283(23):15638–46.

16. Watanabe S, Kodaki T, Makino K. A Novel α-ketoglutaric semialdehyde dehydrogenase: evolutionary insight into an alternative pathway of bacterial l-arabinose metabolism. J Biol Chem. 2006;281(39):28876–88.

17. Glasner ME, Gerlt JA, Babbitt PC. Evolution of enzyme superfamilies. Curr Opin Chem Biol. 2006;10(5):492–7.

18. Babbitt PC, Hasson MS, Wedekind JE, Palmer DR, Barrett WC, Reed GH, Rayment I, Ringe D, Kenyon GL, Gerlt JA. The enolase superfamily: a general strategy for enzyme-catalyzed abstraction of the α-protons of carboxylic acids. Biochemistry. 1996;35(51):16489–501.

19. Gulick AM, Hubbard BK, Gerlt JA, Rayment I. Evolution of enzymatic activities in the enolase superfamily: crystallographic and mutagenesis studies of the reaction catalyzed by d-glucarate dehydratase from Escherichia coli. Biochemistry. 2000;39(16):4590–602.

20. Gulick AM, Palmer DR, Babbitt PC, Gerlt JA, Rayment I. Evolution of enzymatic activities in the enolase superfamily: crystal structure of (D)-glucarate dehydratase from Pseudomonas putida. Biochemistry. 1998;37(41):14358–68.

21. Babbitt PC, Gerlt JA. Understanding enzyme superfamilies Chemistry as the fundamental determination in the evolution of new catalytic activities. J Biol Chem. 1997;272(49):30591–4.

22. Taberman H, Andberg M, Parkkinen T, Richard P, Hakulinen N, Koivula A, Rouvinen J. Purification, crystallization and preliminary X-ray diffraction analysis of a novel keto-deoxy-d-galactarate (KDG) dehydratase from Agrobacterium tumefaciens. Acta Crystallogr Sect F. 2014;70(1):49–52.

23. Taberman H, Andberg M, Parkkinen T, Jänis J, Penttilä M, Hakulinen N, Koivula A, Rouvinen J. Structure and function of a decarboxylating Agrobacterium tumefaciens Keto-deoxy-D-galactarate dehydratase. Biochemistry. 2014;53(51):8052–60.

24. Tang L, Li Y, Wang X. A high-throughput colorimetric assay for screening halohydrin dehalogenase saturation mutagenesis libraries. J Biotechnol. 2010;147(3):164–8.

25. Persson M, Palcic MM. A high-throughput pH indicator assay for screening glycosyltransferase saturation mutagenesis libraries. Anal Biochem. 2008;378(1):1–7.

26. Chapman E, Wong C-H. A pH sensitive colorometric assay for the high-Throughput screening of enzyme inhibitors and substrates: a case study using kinases. Bioorg Med Chem. 2002;10(3):551–5.

27. Rosenberg RM, Herreid RM, Piazza GJ, O'Leary MH. Indicator assay for amino acid decarboxylases. Anal Biochem. 1989;181(1):59–65.

28. Yu K, Hu S, Huang J, Mei L-H. A high-throughput colorimetric assay to measure the activity of glutamate decarboxylase. Enzyme Microb Technol. 2011;49(3):272–6.

29. He N, Yi D, Fessner W-D. Flexibility of substrate binding of Cytosine-5'-Monophosphate-N-Acetylneuraminate Synthetase (CMP-Sialate Synthetase) from Neisseria meningitidis: an enabling catalyst for the synthesis of neo-sialoconjugates. Adv Synthesis Catalysis. 2011;353(13):2384–98.

30. Janes LE, Löwendahl AC, Kazlauskas RJ. Quantitative screening of hydrolase libraries using pH indicators: identifying active and enantioselective hydrolases. Chem Eur J. 1998;4(11):2324–31.

31. Altschul SF, Madden TL, Schäffer AA, Zhang J, Zhang Z, Miller W, Lipman DJ. Gapped BLAST and PSI-BLAST: a new generation of protein database search programs. Nucleic Acids Res. 1997;25(17):3389–402.

32. Guterl J-K, Garbe D, Carsten J, Steffler F, Sommer B, Reiße S, Philipp A, Haack M, Rühmann B, Koltermann A, et al. Cell-free metabolic engineering: production of chemicals by minimized reaction cascades. ChemSusChem. 2012;5(11):2165–72.

33. Studier FW. Protein production by auto-induction in high-density shaking cultures. Protein Expr Purif. 2005;41(1):207–34.

34. Gibbons BH, Edsall JT. Rate of hydration of carbon dioxide and dehydration of carbonic acid at 25. J Biol Chem. 1963;238(10):3502–7.

35. Khalifah RG. The carbon dioxide hydration activity of carbonic anhydrase I. Stop-flow kinetic studies on the native human isoenzymes B and C. J Biol Chem. 1971;246(8):2561–73.

36. Rogers TA, Bommarius AS. Utilizing simple biochemical measurements to predict lifetime output of biocatalysts in continuous isothermal processes. Chem Eng Sci. 2010;65(6):2118–24.

37. Jeffcoat R, Hassall H, Dagley S. Purification and properties of D-4-deoxy-5-oxoglucarate hydro-lyase (decarboxylating). Biochem J. 1969;115(5):977–83.

38. Jeffcoat R, Hassall H, Dagley S. The metabolism of D-glucarate by Pseudomonas acidovorans. Biochem J. 1969;115(5):969–76.

39. Sharma BS, Blumenthal HJ. Catabolism of d-Glucaric Acid to α-Ketoglutarate in Bacillus megaterium. J Bacteriol. 1973;116(3):1346–54.

40. Rice P, Longden I, Bleasby A. EMBOSS: the European molecular biology open software suite. Trends Genet. 2000;16(6):276–7.

41. Nagano N, Orengo CA, Thornton JM. One fold with many functions: the evolutionary relationships between TIM Barrel families based on their sequences, structures and functions. J Mol Biol. 2002;321(5):741–65.

42. Lee K-H, Kim H-S, Jeong H-S, Lee Y-S. Chaperonin GroESL mediates the protein folding of human liver mitochondrial aldehyde dehydrogenase in Escherichia coli. Biochem Biophys Res Commun. 2002;298(2):216–24.

43. Hartinger D, Heinl S, Schwartz HE, Grabherr R, Schatzmayr G, Haltrich D, Moll W-D. Enhancement of solubility in Escherichia coli and purification of an aminotransferase from Sphingopyxis sp. MTA 144 for deamination of hydrolyzed fumonisin B 1. Microb Cell Fact. 2010;9:62.

44. Banerjee A, Kaul P, Sharma R, Banerjee UC. A high-throughput amenable colorimetric assay for enantioselective screening of nitrilase-producing microorganisms using pH sensitive indicators. J Biomol Screen. 2003;8(5):559–65.

45. Zhang J-H, Chung TD, Oldenburg KR. A simple statistical parameter for use in evaluation and validation of high throughput screening assays. J Biomol Screen. 1999;4(2):67–73.

46. Sievers F, Wilm A, Dineen D, Gibson TJ, Karplus K, Li W, Lopez R, McWilliam H, Remmert M, Soding J, et al. Fast, scalable generation of high-quality protein multiple sequence alignments using Clustal Omega. Mol Syst Biol. 2011;7(1):539.

14

Genomic variation and DNA repair associated with soybean transgenesis: a comparison to cultivars and mutagenized plants

Justin E. Anderson, Jean-Michel Michno, Thomas J. Y. Kono, Adrian O. Stec, Benjamin W. Campbell, Shaun J. Curtin and Robert M. Stupar[*]

Abstract

Background: The safety of mutagenized and genetically transformed plants remains a subject of scrutiny. Data gathered and communicated on the phenotypic and molecular variation induced by gene transfer technologies will provide a scientific-based means to rationally address such concerns. In this study, genomic structural variation (e.g. large deletions and duplications) and single nucleotide polymorphism rates were assessed among a sample of soybean cultivars, fast neutron-derived mutants, and five genetically transformed plants developed through *Agrobacterium* based transformation methods.

Results: On average, the number of genes affected by structural variations in transgenic plants was one order of magnitude less than that of fast neutron mutants and two orders of magnitude less than the rates observed between cultivars. Structural variants in transgenic plants, while rare, occurred adjacent to the transgenes, and at unlinked loci on different chromosomes. DNA repair junctions at both transgenic and unlinked sites were consistent with sequence microhomology across breakpoints. The single nucleotide substitution rates were modest in both fast neutron and transformed plants, exhibiting fewer than 100 substitutions genome-wide, while inter-cultivar comparisons identified over one-million single nucleotide polymorphisms.

Conclusions: Overall, these patterns provide a fresh perspective on the genomic variation associated with high-energy induced mutagenesis and genetically transformed plants. The genetic transformation process infrequently results in novel genetic variation and these rare events are analogous to genetic variants occurring spontaneously, already present in the existing germplasm, or induced through other types of mutagenesis. It remains unclear how broadly these results can be applied to other crops or transformation methods.

Keywords: Somaclonal variation, Structural variation, Genetic engineering, Biotechnology, Transgenic crops, Soybean

Background

Plant breeders use genetic variation from elite and diverse lines as the primary source for cultivar development and trait improvement. In some cases, traits of interest cannot be found within this "standing" variation in the current germplasm. However, mutagenesis or genetic transformation can provide a means to introduce such traits. Standard mutagenesis treatments, such as Fast Neutron (FN) irradiation, alter DNA sequences at random loci throughout the genome in an attempt to generate novel trait variation [1]. Genetic transformation, alternatively, attempts to insert one or few transgenes to confer a novel trait or disrupt the activity of an endogenous gene.

The genetic transformation of most crop species requires plant tissue culture methods, which can introduce heritable phenotypes caused by unintended genetic and epigenetic changes [2]. These unintended changes, known as somaclonal variation, may theoretically compromise

---

* Correspondence: stup0004@umn.edu
Department of Agronomy & Plant Genetics, University of Minnesota, 1991 Upper Buford Circle, 411 Borlaug Hall, St. Paul MN 55108, USA

the safety of transgenic plants [3]. Therefore, it is important to understand the coupled effects of genetic transformation and tissue culture [4] and how these compare to standing and other types of induced variation.

Naturally occurring variation is a well-established source of novel phenotypes in many vegetatively propagated fruits and vegetables, where they are commonly known as 'sports' [5]. Somaclonal variation induced through tissue culture, first observed in sugarcane (*Saccharum*) [6], has been reported in many other plant species [2]. Desirable agronomic traits and released cultivars have even been derived from this type of induced variation [7]. The molecular underpinnings of somaclonal variation can include DNA sequence changes, chromosome rearrangements, aneuploidy, activation of transposable elements, and epigenetic restructuring [2]. Genome-wide single nucleotide changes resulting from tissue culture have been recently observed using next-generation sequencing (NGS) in Arabidopsis [8] and rice [9–13]. These studies suggest tissue culture might increase the single nucleotide mutation rate and may activate transposons [14].

The insertion of a transgene is also known to create localized or dispersed genomic changes. Recent studies found that transformation can result in DNA inserted at multiple loci, multiple transgenes per locus, fragmented T-DNA, and chromosome rearrangements [15–19], though such complex events are rare and often discarded rather than commercialized. In Arabidopsis, transgene insertion is generally random across chromosomes, in both genic and non-genic sequences, and frequently associated with a deletion ranging from 11 to 100 bp in size [20]. For soybean (*Glycine max*), *Agrobacterium* based transformation methods occasionally result in multiple insertion sites, tandem insertions, and integration of plasmid backbone sequences [21]. Recently, resequencing methods have been used to accurately localize and resolve transgene insertions in different plant species [19, 22–24]. While advanced technologies have helped detect the local and dispersed effects of

tissue culture and transformation, limitations still exist due to sequencing errors, genetic heterogeneity of plant accessions, and reference bias [25].

Separating the changes induced by transformation from pre-existing genetic variation can be a challenge [26]. Plant genomes can vary dramatically between cultivars. A large portion of this variation occurs as genomic structural variants (SV), such as large deletions and duplications [27]. These SV are associated with a number of biologically and agriculturally important traits [27]. Previous studies in soybean have used array-based comparative genomic hybridization (CGH) or resequencing approaches to observe levels of standing SV among accessions [28, 29], or SV induced through FN mutagenesis [1]. However, no comparable studies have addressed the incidence of tissue culture and transformation on rates of genome-wide SV in soybean.

This study investigates five transgenic (T$_1$ generation) soybean plants derived from standard *Agrobacterium*-mediated transformation. SV in these five plants was assessed by CGH and two of these plants were resequenced to ascertain the frequency of nucleotide substitutions. These data allow for comparisons of genomic variation in transgenic plants to the genomic variation observed in mutagenized and standing accessions. These analyses provide new insight towards understanding somaclonal variation, the effects of transgene insertion, the inheritance of SV, and the genomic consequences of developing mutant and transgenic stocks as compared to the standing variation already present in the soybean germplasm.

Results

Genome-wide structural variation

A CGH tiling microarray with 1.4 million features was used to estimate the genomic locations and sizes of SV events in the genomes of three classes of germplasm. The first class consisted of five transgenic plants each derived from a unique *Agrobacterium*-based transformation event. Each transgenic plant contained a different

Table 1 Results from CGH, breakpoint sequencing, TAIL-PCR, and resequencing of transgenic plants

Transgenic Genotype	Construct	Data Types	Background	CGH-detected SV	T-DNA Location[a]	T-DNA Direction	SV adjacent to T-DNA	No. Transgenes
WPT_384-1-1	TALEN	CGH, TAIL-PCR	'Bert-MN-01'	23,406 bp; Gm01 deletion	Gm07:35,729,562	+	Untested	Likely 1
WPT_389-2-2	mPing-Pong Transposon	CGH, NGS, TAIL-PCR, Southern Blot	'Bert-MN-01'	125,228 bp; Gm11 deletion	Gm13:35,614,273	+	1,533 bp deletion + 37 bp deletion	1
WPT_301-3-13	GFP + RNAi Hairpin	CGH, TAIL-PCR, Southern Blot	'Wm82-ISU-01'	6,869 bp; Gm13 duplication	Gm04:2,695,263	-	Untested	1
WPT_391-1-6	Magnesium Chelatase RNAi Hairpin	CGH, NGS	'Bert-MN-01'	7,854 bp; Gm19 deletion	Gm05:38,834,281	+	~1,200 bp deletion	1
WPT_312-5-126	Zinc Finger Nuclease	CGH, Southern Blot	'Bert-MN-01'	None	Untested	NA	Untested	1

[a]Genome coordinates adjacent to left border according to the soybean genome assembly version 1.0 (Wm82.a1.v1.1)

transgene (Table 1), and each event was specified by a unique Whole Plant Transformation (WPT) identifier. A range of different transgene types were represented among the five plants, including a green fluorescence protein (GFP) transgene, an RNAi hairpin, a zinc-finger nuclease (ZFN), a transcription activator-like effector nuclease (TALEN), and an mPing-Pong transposon. Genotyping was done on the T_1 generation. Genome-wide CGH screens for deletions and duplications revealed single, unique SV in four of the five genotypes. These consisted of three deletions and one duplication (Table 1). The plant WPT_312-5-126 (ZFN transgene) did not exhibit any SV.

The second class, sampling FN-induced variation, consisted of a sub-set of plants from a larger mutant population developed in the genotype 'M92-220' [1]. This subset included ten plants with an associated mutant phenotype (Additional file 1: Table S1) and 35 plants that exhibited no obvious mutant phenotypes, and were thus referred to as "no-phenotype". The final class, representing inter-cultivar variation, came from a previous study of genic SV [29], and consisted of 41 parental lines from a soybean Nested Association Mapping (SoyNAM) population (www.Soybase.org/SoyNAM).

All three datasets (transgenic, FN, and inter-cultivar) were designed to detect SV in each individual genotype as compared to an appropriate reference (Additional file 1: Table S2). The transgenic plants were compared to the transformation parent line ('Bert' [30] for four of the

plants and 'Williams 82' for one plant; see Additional file 1: Table S2), the FN plants compared to the mutagenesis parent line ('M92-220'), and the SoyNAM parents were compared to the reference genotype 'Williams 82'. The Methods section includes analysis details and information on how extant heterogeneity within the background cultivars was addressed.

As shown in Fig. 1, CGH results varied by chromosome and by class. In this figure each black dot represents a single probe's log_2 ratio score. Clusters of dots above or below zero are putative duplications or deletions, respectively. Inter-cultivar variation, shown as the comparison of SoyNAM parent LD02-9050 to 'Williams 82' (Fig. 1a), occurs frequently and on nearly every chromosome. The amount of inter-cultivar variation is strikingly high when compared to a FN or transgenic plant (Fig. 1b and c, respectively). The SV observed in FN or transformed plants was easier to detect, but occurred much less frequently.

The number of genes putatively deleted or duplicated varied widely among the classes. Among the inter-cultivar comparisons, the total number of genes overlapping with duplications ranged from 45 to 124 per pairwise cultivar comparison, while the number of genes overlapping with deletions varied from 156 to 362 per comparison (Fig. 2). The FN class had a lower median genic SV per plant (Table 2) but was highly variable, as the number of genes overlapping with duplications

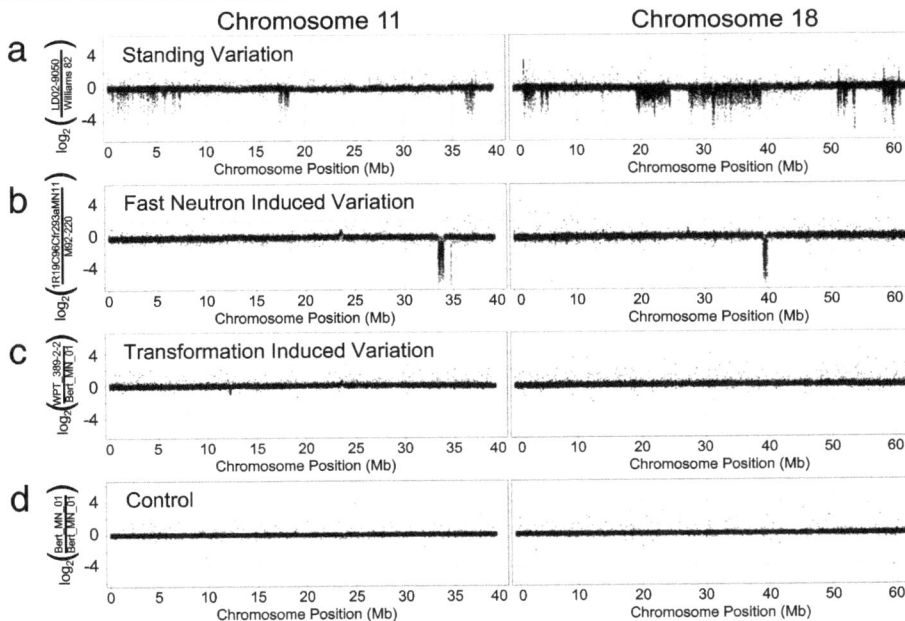

Fig. 1 Visual comparison of CGH data for individuals from the three germplasm classes and control. Each *black dot* represents a single probe and its log_2 ratio score. Data are shown from chromosome 11 on the left and chromosome 18 on the right for all samples. **a** The standing inter-cultivar variation between lines LD02-9050 and 'Williams 82' is shown. **b** Fast neutron (No-phenotype) plant 1R19C96Cfr293aMN11 is compared to the FN parent line 'M92-220'. **c** Transgenic plant WPT_389-2-2 is compared to the parent line 'Bert'; it shows relatively little noise and one true SV on chromosome 11. **d** The control CGH compared 'Bert-MN-01' to itself

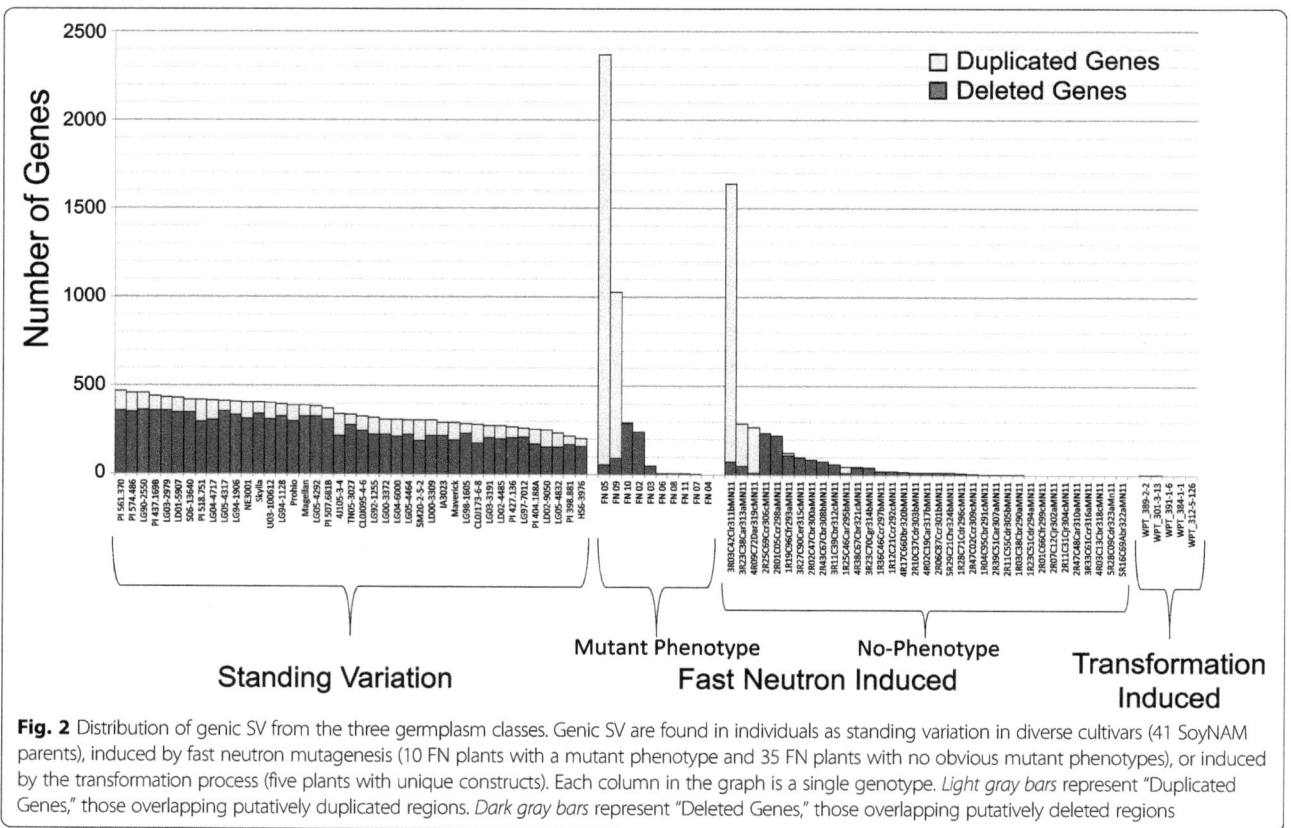

Fig. 2 Distribution of genic SV from the three germplasm classes. Genic SV are found in individuals as standing variation in diverse cultivars (41 SoyNAM parents), induced by fast neutron mutagenesis (10 FN plants with a mutant phenotype and 35 FN plants with no obvious mutant phenotypes), or induced by the transformation process (five plants with unique constructs). Each column in the graph is a single genotype. *Light gray bars* represent "Duplicated Genes," those overlapping putatively duplicated regions. *Dark gray bars* represent "Deleted Genes," those overlapping putatively deleted regions

ranged from 0 to 2,312 per plant, and the number of genes overlapping with deletions ranged from 0 to 290 per plant. The average size of the SV in the FN plants was over 500,000 bp, which is inflated by a small number of exceptionally large SV. Nevertheless, this value is substantially larger than those observed in the inter-cultivar class, where the average was less than 15,000 bp (Table 2). Of the four SV events in the transgenic plants, only two affected gene space. This included one deletion in plant WPT_389-2-2 that affected four genes on

Table 2 Summary of SV frequency in Inter-cultivar, Fast Neutron (FN), and Transgenic genotypic classes

		Inter-Cultivar	FN Mutant Phenotype	FN No-Phenotype	Transgenic
Unique Up CNV Genes	Total genes in class	223	3253	2118	2
	Maximum among genotypes	124	2312	1568	2
	Median among genotypes	83	0	0	0
	Minimum among genotypes	45	0	0	0
Unique Down CNV Genes (homozygous or heterozygous deletions)	Total genes in class	1126	987	1231	4
	Maximum among genotypes	362	290	236	4
	Median among genotypes	244	48	12	0
	Minimum among genotypes	156	0	0	0
Up SV (homozygous duplications)	Total genic segments in class	117	11	9	1
	Mean Size	13,580 bp	4,671,937 bp	2,447,335 bp	6,434 bp
	Median Size	3,182 bp	2,802,275 bp	747,592 bp	6,434 bp
Down SV (homozygous or heterozygous deletion)	Total genic segments in class	547	23	49	1
	Mean Size	14,958 bp	1,276,033 bp	515,051 bp	125,228 bp
	Median Size	2,775 bp	110,656 bp	131,036 bp	125,228 bp

chromosome 11 (Fig. 3) and one duplication in plant WPT_301-3-13 that encompassed two genes on chromosome 13 (Fig. 4). Overall, the average number of genes affected by CGH-detectable SV in transgenic plants was estimated to be one order of magnitude less than that induced by FNs and two orders less than that observed among soybean varieties.

Validation of SV in the transgenic plants

The four incidences of SV detected with CGH in the transgenic plants were confirmed using PCR. Two SV events overlapped with genes, including a 125,228 bp deletion on chromosome 11 in WPT_389-2-2 (Fig. 3) and a 6,869 bp duplication on chromosome 13 in WPT_301-3-13 (Fig. 4). The two non-genic deletions included 23,406 bp on chromosome 1 in WPT_384-1-1 (Additional file 2: Figure S1) and 7,854 bp on chromosome 19 in WPT_391-1-6 (Additional file 2: Figure S2). Sequence data from all four SV junctions showed evidence of microhomology-mediated DNA repair (Figs. 3c and 4c, and Additional file 2: Figure S1c and S2d).

Screening a subset of these SV by PCR confirmed they were not intra-cultivar variation in the 'Bert' or 'Williams 82' backgrounds, as is known to exist at some loci [31] (Additional file 2: Figure S3), or derived from contamination or outcrossing from other lines (Additional file 2: Figure S4). The deletions on chromosome 1 and chromosome 11 were stably inherited in T_1 siblings and T_2 offspring (Additional file 2: Figure S1 and S5),

indicating these events were both present in their respective T_0 generations. The deletion on chromosome 19 was homozygous and therefore present in the T_0 generation assuming SV is induced on a single chromosome and then becomes a homozygous deletion through genetic segregation. These data indicate these SV were derived *de novo*. The duplication on chromosome 13, however, is not found in any individual other than the T_1 transgenic genotype, WPT_301-3-13. The offspring ($T_{1:2}$), siblings (T_1), and parent (T_0) of this individual were all tested and showed no evidence of the duplication on chromosome 13 (Additional file 2: Figure S6). This evidence suggests the duplication arose in a post transformation generation and may not be directly attributable to the transformation process.

Transgene insertion sites

Transgenic plants were analyzed for number of transgene insertions and location of transgene(s). Southern blots of siblings or parents of WPT_301-3-13, WPT_312-5-126, and WPT_389-2-2 each showed evidence for single locus integration (Additional file 2: Figure S7). Thermal Asymmetric Interlaced PCR (TAIL-PCR) mapped the single insertion sites in WPT_389-2-2, WPT_384-1-1, and WPT_301-3-13. Resequencing data were also used to localize the T-DNA insertion site in WPT_389-2-2 and WPT_391-1-6. Transgene results are summarized in Table 1. Transgenes were all found to occur on different chromosomes than the aforementioned

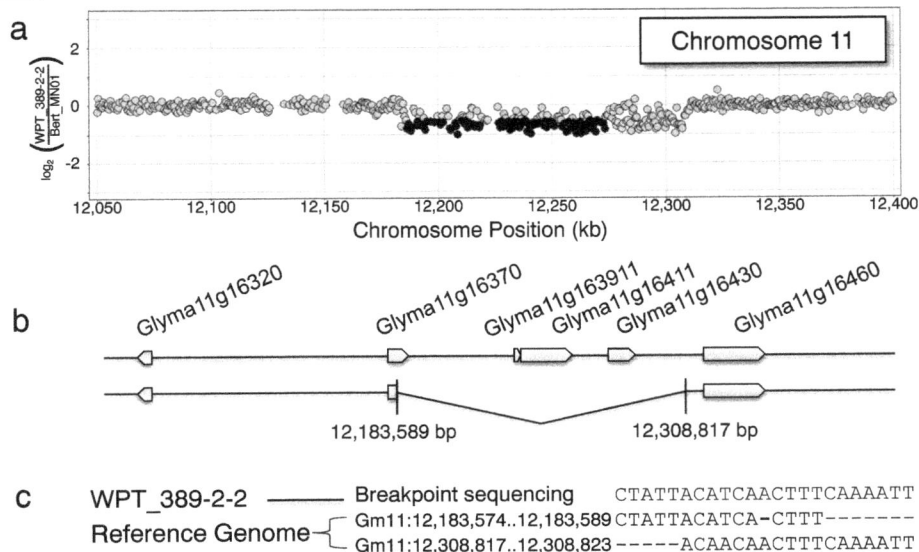

Fig. 3 A novel deletion on chromosome 11 in transgenic plant WPT_389-2-2. **a** A plot of CGH data for the transgenic plant versus 'Bert' is shown, zoomed in on the chromosome 11 deletion seen in Fig. 1c. Probes are plotted as *dots* corresponding to the $\log_2$ ratio from the CGH array. *Dark gray dots* represent probes within significant SV segments that exceed the empirical threshold. Even with the extremely low detection threshold, part of this deletion could not be verified via CGH alone, necessitating visual inspection and sequencing of the deletion breakpoint. **b** Graphical interpretation of the hemizigous deletion found in WPT_389-2-2 is shown. **c** Sequence data from the breakpoint junction shows moderate homology on either end of the breakpoint

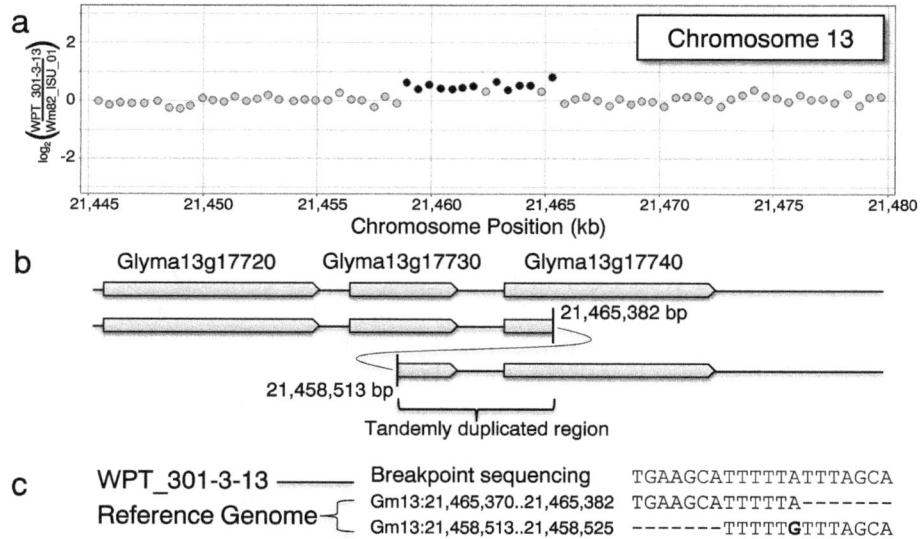

Fig. 4 A novel duplication on chromosome 13 in transgenic plant WPT_301-3-13. **a** A plot of CGH data for the transgenic plant versus 'Williams 82' is shown, zoomed in on the chromosome 13 duplication. Probes are plotted as *dots* corresponding to the $\log_2$ ratio from the CGH array. *Dark gray dots* represent probes within significant SV segments that exceed the empirical threshold. **b** A graphical interpretation of the heterozygous duplication found in WPT_301-3-13, which includes a portion of Glyma13g17730 and a portion of Glyma13g17740, is shown. **c** The sequence data from the breakpoint junction shows five base pairs of homology on either end of the breakpoint

SV (Table 1). Transgene insertion and repair was observed to coincide with microhomology between the genome and the left border (Fig. 5 and Additional file 2: Figure S8).

According to the whole genome resequencing data from two transgenic plants, transgene insertions in WPT_389-2-2 and WPT_391-1-6 both induced adjacent deletions too small for CGH detection (the other three transgenic plants were not analyzed by whole genome resequencing). The deletion induced by the transgene insertion in WPT_391-1-6 was ~1,200 bp and occurred adjacent to the transgene (Additional file 2: Figure S9).

The transgene locus from WPT_389-2-2 was more complex. As outlined in Fig. 5a, the transgene (an mPing-Pong transposon construct) induced two deletions and a 6-bp insertion of filler sequence in the T-DNA integration process. This transgene integration and associated mutations occurred in the promoter region and 5'UTR of Glyma13g33960. The WPT_389-2-2 T-DNA and adjacent mutations were homozygous in this T_1 plant. The resequencing data aligned to the transgene found nine read-pairs that spanned the mPing-Pong portion of the construct (Additional file 2: Figure S10a) suggesting one of the homologous chromosomes has a transgene where this mPing-Pong portion was deleted or jumped out (Additional file 2: Figure S10b), as has been demonstrated with this element [32]. Had this transposon reintegrated in the genome, the methodology used for transgene mapping should have detected it.

Genome-wide single nucleotide substitutions

Resequencing data were used to assess the frequency of nucleotide substitutions within the inter-cultivar, FN, and transgenic classes. Based on earlier studies, it has been established that pairwise comparisons of soybean cultivars typically identify over one-million single base substitutions [33, 34]. We tested our substitution identification pipeline by resequencing cultivars 'Archer' and 'Noir 1'. These data corroborated earlier studies, as 'Archer' and 'Noir 1' respectively exhibited 1,110,325 and 1,904,061 homozygous substitutions compared to the soybean reference genome 'Williams 82'.

Resequencing data were then used to assess the frequency of nucleotide substitutions in the previously sequenced ten mutant phenotype FN plants and the FN parent 'M92-220' [1] (Additional file 1: Table S1). Substitutions were detected and filtered so only those homozygous and novel to one plant were included. This filtering method was based on previous mutation accumulation studies [8, 35, 36]. The FN mutagenized plants had on the order of tens of unique homozygous substitutions per individual (Additional file 1: Table S3), with the highest individual exhibiting 73 substitutions. However, most of these substitutions may be attributed to spontaneous processes [36] rather than the FN treatment, as the nonmutagenized 'M92-220' control also exhibited 41 unique substitutions relative to the ten FN plants. As shown in Fig. 6a, substitutions in the FN plants were distributed across many more chromosomes than SV.

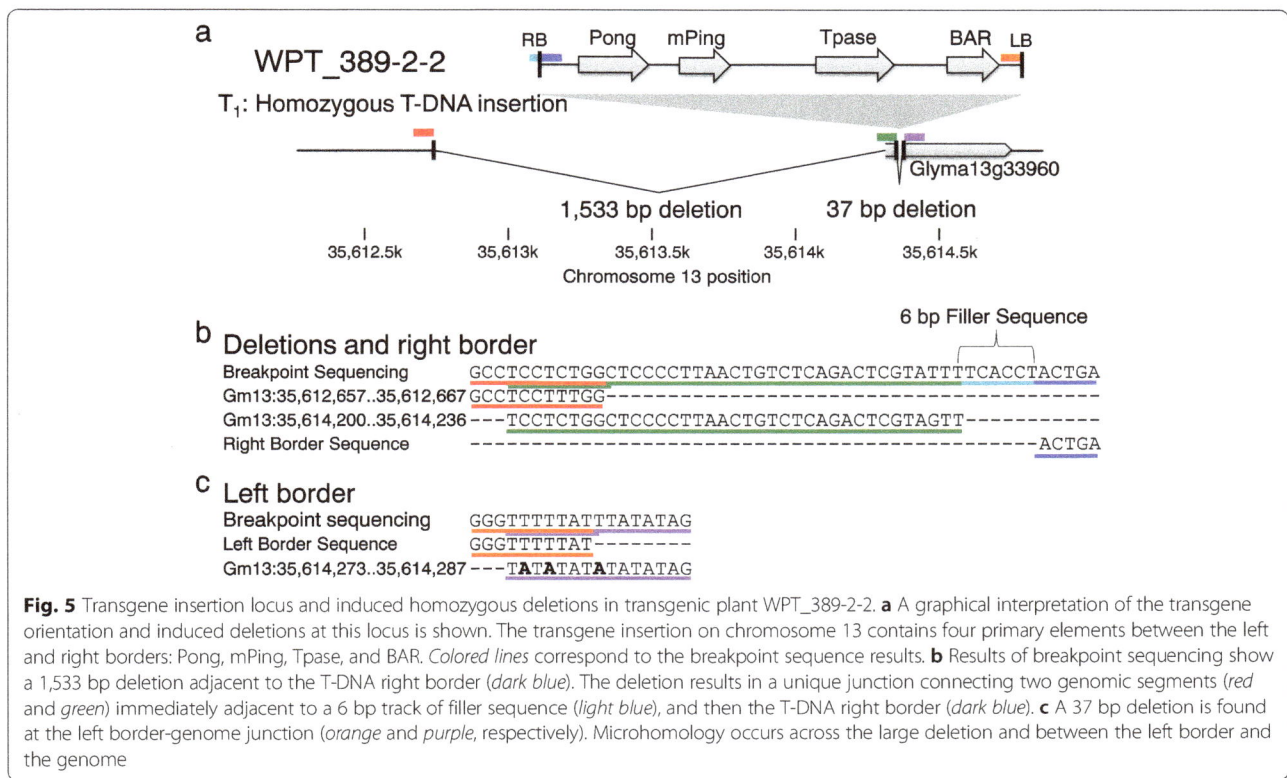

Fig. 5 Transgene insertion locus and induced homozygous deletions in transgenic plant WPT_389-2-2. **a** A graphical interpretation of the transgene orientation and induced deletions at this locus is shown. The transgene insertion on chromosome 13 contains four primary elements between the left and right borders: Pong, mPing, Tpase, and BAR. *Colored lines* correspond to the breakpoint sequence results. **b** Results of breakpoint sequencing show a 1,533 bp deletion adjacent to the T-DNA right border (*dark blue*). The deletion results in a unique junction connecting two genomic segments (*red* and *green*) immediately adjacent to a 6 bp track of filler sequence (*light blue*), and then the T-DNA right border (*dark blue*). **c** A 37 bp deletion is found at the left border-genome junction (*orange* and *purple*, respectively). Microhomology occurs across the large deletion and between the left border and the genome

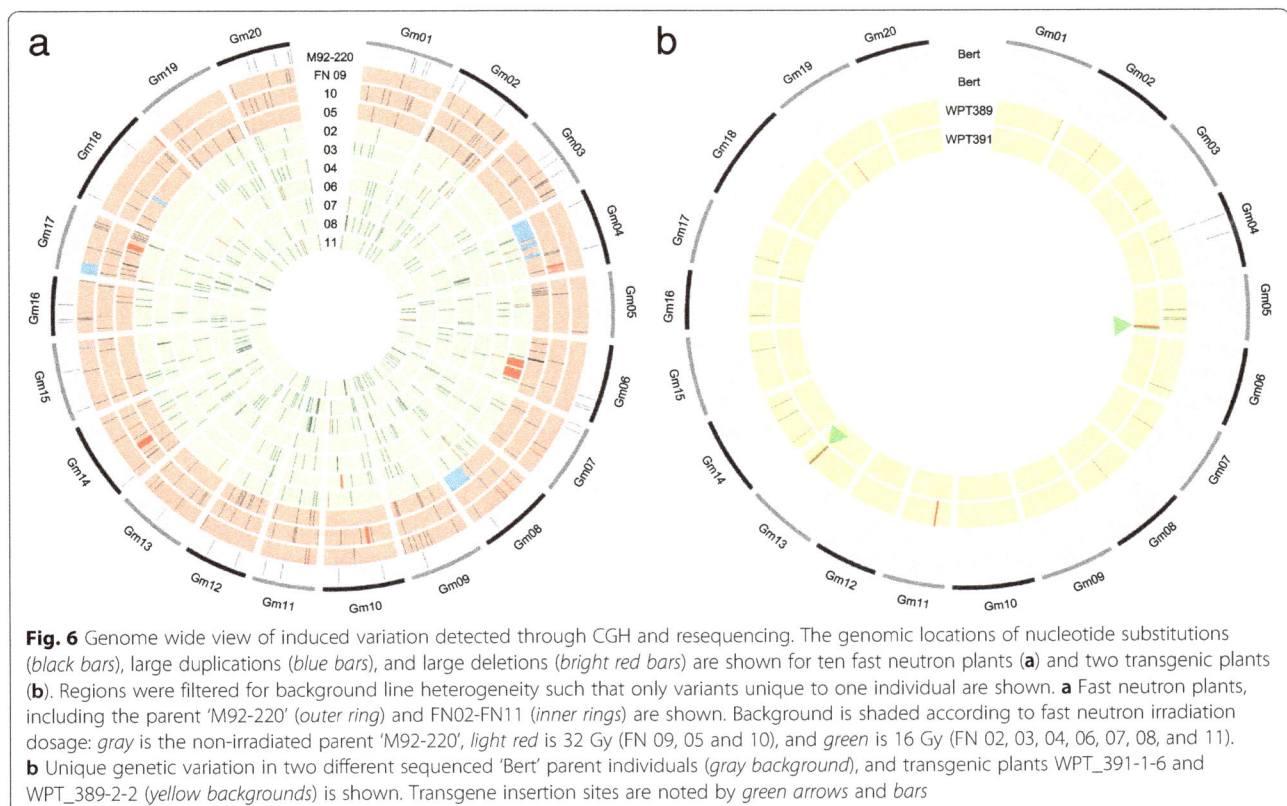

Fig. 6 Genome wide view of induced variation detected through CGH and resequencing. The genomic locations of nucleotide substitutions (*black bars*), large duplications (*blue bars*), and large deletions (*bright red bars*) are shown for ten fast neutron plants (**a**) and two transgenic plants (**b**). Regions were filtered for background line heterogeneity such that only variants unique to one individual are shown. **a** Fast neutron plants, including the parent 'M92-220' (*outer ring*) and FN02-FN11 (*inner rings*) are shown. Background is shaded according to fast neutron irradiation dosage: *gray* is the non-irradiated parent 'M92-220', *light red* is 32 Gy (FN 09, 05 and 10), and *green* is 16 Gy (FN 02, 03, 04, 06, 07, 08, and 11). **b** Unique genetic variation in two different sequenced 'Bert' parent individuals (*gray background*), and transgenic plants WPT_391-1-6 and WPT_389-2-2 (*yellow backgrounds*) is shown. Transgene insertion sites are noted by *green arrows* and *bars*

The two resequenced transgenic plants were also analyzed for homozygous and novel substitutions, along with two non-transgenic 'Bert' control plants (Additional file 1: Table S3). The number of novel homozygous base-pair substitutions per individual were as follows: two in plant WPT_391-1-6, eighteen in plant WPT_389-2-2, one in the first 'Bert' control plant, and two in the second 'Bert' control plant. The location of the substitutions in the transgenic plants appeared unrelated to the location of the transgene insertions or the induced SV (Fig. 6b) and did not occur in coding regions (Additional file 1: Table S3).

Discussion

In this study, we observed the rates of SV and single nucleotide substitutions in transgenic and FN plants and explored the genetic nature of the unintended consequences of these breeding practices. The primary safety concern relating to these genomic changes is that novel genetic variants might disrupt genes or pathways leading to an unforeseen harmful byproduct [3]. Therefore, we focused our comparisons specifically on the protein-encoding gene space, with less emphasis on intergenic space and heterochromatin. Furthermore, we focused on the number of genes affected by new mutations rather than on the risk associated with a specific mutation or disruption of a specific gene. While the latter is of critical importance, it is impossible to estimate the specific effects of mutating each of the over 40,000 soybean genes. Therefore, for this discussion, differences in the number of total genes disrupted serves as a proxy for the amount of risk associated with each of these tools for genetic variation.

The SV observed in the inter-cultivar comparison was widespread throughout the genome, including many events that were repeatedly found in multiple lines and several events that encompassed only a single gene. This diversity has presumably accumulated through ongoing spontaneous mutation over numerous generations. Each of the genetic variants seen in this class would not be perceived to pose a new risk to consumers, as they likely already exist in the current marketplace. Furthermore, the genetic variation currently segregating in these lines represents only a subset of the total genetic diversity found in *Glycine max* or the wild progenitor *Glycine soja* [33, 34]. Genetic variation arising spontaneously, or introgressed from diverse lines into elite cultivars, is a process by which even cultivars developed through traditional breeding methodology unintentionally introduce novel variants to the marketplace.

The SV observed in the FN plants contrasts with the patterns of SV in the inter-cultivar class. SV induced through FN mutagenesis are oftentimes large and highly variable from plant to plant in terms of the number of genes affected. The large sizes of some of the SV observed in the no-phenotype FN plants were unexpected, as multigene deletions and duplications would be expected to cause noticeable phenotypic changes.

The transgenic class had so few SV that it is difficult to compare with the other classes. The events observed through CGH are moderate in size, impacting a combined total of only six genes among the five plants. While this likely represents a single generation increase in the SV mutation rate compared to the spontaneous SV mutation rate in soybeans, the total amount of genetic disturbance is substantially less than that observed in the standing soybean collection or the FN-induced plants. Working under the aforementioned assumption that each gene deleted or duplicated may pose a safety risk, one would conclude that the transgenic plants in this study are of lower risk than the vast majority of the FN plants analyzed, in terms of background genome disruption. Furthermore, while some induced variation may occur at the transgene locus, extensive backcrossing to introgress transgenes into elite backgrounds (which is the current common practice in many crop species) makes any new SV event(s) unlinked to the transgene inconsequential to the final cultivar.

While these transformation-induced events seem inconsequential when compared to those induced through FNs or found as standing variation, the novel SV identified in these plants exhibited several interesting properties. The discovery of locally induced deletions, the addition of filler sequence, and microhomology between the left border and the insertion site, corroborate previous patterns of T-DNA insertion in Arabidopsis [20]. The ~1 kb deletions found at transgene insertion sites in both of the resequenced soybean plants are substantially larger than the deletions previously found in Arabidopsis, but the sampling of only two plants is not sufficient to infer a general pattern. Short sequence homology was observed at the T-DNA insertion sites and also at the breakpoints of the four SV observed at non-transgene loci in these plants. These results imply that the microhomology-mediated end joining pathway [37] may be frequently involved in the DNA repair of these events.

The use of FN mutagenesis or tissue culture/transformation has been previously reported to result in a single generation increase in single nucleotide substitutions [8–11, 35]. A single nucleotide substitution disrupting a coding or regulatory region could similarly have an assumed safety risk associated with a novel byproduct. The FN and transgenic plants in this study accumulated a similar number of unique homozygous substitutions compared to a subset of previously published results. For example, a FN mutagenesis study in Arabidopsis detected between 5 and 18 novel homozygous substitutions per M_3 plant [35] and a similar study of Arabidopsis tissue culture reported between 9 and 65 novel homozygous substitutions per R_1 (the generation following tissue culture regeneration; analogous to T_1) plant [8]. In rice, a FN mutagenesis study observed between 28 and 78 mutations per line in an M_3 population, with the majority of mutations being single base substitutions [38], and a tissue culture study found no considerable difference in the number of variants in transgenic compared to control (wild type) plants [12]. In the present study, the number of unique homozygous substitutions observed in our control plants was similar to the number in the FN or transgenic plants, respectively. This implies that most of the identified substitutions were likely due to spontaneous mutation rather than a treatment effect of mutagenesis or transformation. In terms of single nucleotide substitutions, this result indicates that there is minimal difference in the safety risks associated with the three germplasm classes. This result stands in contrast to some of the previous studies of tissue culture in rice, where the authors concluded that there was a significantly higher number of induced homozygous substitutions and associated mutation rates [10, 11, 13]. A number of confounding factors might affect these incongruities, including differences in the species examined, SNP calling methods and thresholds, adjustments for intra-cultivar heterogeneity, FN dosage or tissue culture conditions and timeline, the inclusion of a control plant, and the number of plants sampled.

Based on data from the present study, it appears the use of FN mutagenesis can produce profound new SV events and may slightly increase the number of single nucleotide substitutions. Tissue culture/transformation methodologies can also produce new SV and possibly increase the nucleotide substitution rate. However, the number of SV and single nucleotide polymorphisms existing as standing variation in soybean cultivars dwarfs the induced variation observed in both FN and transformed plants. While these findings are noteworthy, it is unclear how broadly they can be applied. All of the transgenic plants in this study were obtained from *Agrobacterium*-mediated transformation; further work would test other transformation techniques such as biolistic-based methods. Similarly, FN irradiation was the only mutagenesis system tested; other mutagens (EMS, ENU, X-rays, etc.) would likely induce different mutational profiles. Furthermore, a deeper sampling of mutated and transformed plants, perhaps among different plant species, would be required to generalize the SV and nucleotide trends observed. Detailed sequence analysis of specific transgene loci did identify a small number of intermediate-sized deletions adjacent to transgenes, but

there was no systematic attempt to detect intermediate-sized (1–2,000 bp) deletions/duplications genome-wide. Additional variants have also been reported to exist in FN [1, 38] and transgenic plants [12, 17, 39–41] but were not assessed within this dataset, including insertions, inversions and translocations, as well as epigenetic or transcriptional perturbations. Lastly, soybean is a palaeopolyploid species. It is likely that a true polyploid (or true diploid) species may exhibit differential tolerance or lack of tolerance to the type of genetic perturbations associated with these technologies.

Conclusions

The total findings of this study help to inform the discussion currently surrounding the unintended consequences of genetic transformation in crop improvement [4, 42]. First, the frequency of induced SV events appears to be low, particularly in comparison to the frequency of those induced by FNs. Additionally, these rare SV events are likely indistinguishable from other spontaneously occurring SV or those already present in the existing germplasm. As demonstrated by the genetic variability in the no-phenotype FN plants, SV generated *de novo* are not necessarily associated with novel or noticeable phenotypic traits, even when these SV events are large. Therefore, the speculated risk of unintended genetic consequences in tissue culture/transformation may only merit as much consideration as given to variation arising spontaneously, through traditional breeding practices, or other genetic variation induction methods.

Methods
Plant materials and genetic transformation
The plant materials comprising the inter-cultivar and FN classes included in this study have been previously described [1, 29]. Briefly, the inter-cultivar group consists of 41 soybean accessions used as parents in developing the SoyNAM population. The FN population was developed in the background of the variety 'M92-220' [43] derived from the 2006 Crop Improvement Association seed stock of variety 'MN1302' [44]. Two types of FN plants were studied, including ten with detectable mutant phenotypes and 35 with no detectable phenotype. All FN plants were descendants of unique M_1 individuals that were treated with either 4, 16, or 32 Gy of FN radiation [1].

Genetic transformation using *Agrobacterium rhizogenes* followed published methods [45, 46]. Each plant was confirmed to be transgenic based on PCR analysis and survival on selective (herbicide-treated) medium. The five T_1 soybean individuals were from unique transformation events. The constructs for these transformations included a zinc finger nuclease [47], transcription activator-like effector nuclease, GFP and RNAi hairpin, mPing-Pong transposon [32], and a magnesium chelatase [48] RNAi hairpin. These transformations were in a 'Bert' cultivar [30] background (subline 'Bert-MN-01') or 'Williams 82' (subline 'Wm82-ISU-01') [31, 49]. The 'Bert-MN-01' subline (referred to as 'Bert' throughout this study) was derived from a single 'Bert' individual to reduce heterogeneity between transformed plants. The 'Wm82-ISU-01' subline (referred to as 'Williams 82' throughout this study) was derived from a single 'Williams 82' individual and is the nearest known match to the soybean reference genome assembly version 1.0 (Wm82.a1.v1.1) [31, 50].

Comparative genome hybridization

The CGH data for all comparisons used in this study have been deposited in the National Center for Biotechnology Information Gene Expression Omnibus (http://www.ncbi.nlm.nih.gov/geo). The data for the inter-cultivar, FN, and transgenic plant comparisons can be found as accession numbers GSE56351, GSE58172, and GSE73596, respectively.

As with previous CGH analyses [1, 29], the DEVA software algorithm SegMt was used to generate raw data and identify segments in the transgenic plants. DNA samples from transgenic plants were labeled with Cy3 and the appropriate reference individual ('Bert' or 'Williams 82') was labeled with Cy5. Program parameters were: minimum segment difference = 0.1, minimum segment length (number of probes) = 2, acceptance percentile = 0.99, number of permutations = 10. Spatial correction and qspline normalization were applied. The resulting segments were processed based on their $\log_2$ ratio mean. Segments that exceeded the upper threshold were considered "UpCNV". Segments that were less than the lower threshold were considered "DownCNV". The upper threshold of 0.3484 and lower threshold of −0.5257 were based on empirical data from hemizygous deletions and duplications in eight previously characterized FN plants (Additional file 1: Table S4) [1]. A custom Perl script calculated the number of genes overlapping these significant segments. Minimum segment length was adjusted to three probes to account for noise seen in control arrays. Structural variants in the transgenic plants were further investigated through visual inspection, to identify any obvious SVs that were not detected by the threshold based pipeline.

SV attributable to intra-cultivar heterogeneity were removed, as has been done in the previous studies [1, 29]. Intra-cultivar heterogeneity was seen as significant segments of the exact same location occurring in multiple plants. By overlaying the raw CGH data of the four transgenic plants in the 'Bert' background, heterogeneous SV in the 'Bert' cultivar were removed. A similar method was used to filter out heterogeneity in the

transformed 'Williams 82' background. The comparison array in this case was 'Williams' (the backcross parent in 'Williams 82' [49]) also hybridized to 'Williams 82'. Any identical SV event discovered in both 'Williams' and transformed 'Williams 82' was considered heterogeneity and removed.

The CGH platform, methods, and filtering steps of the inter-cultivar and FN data have been previously described [1, 29]. The SV detected in the inter-cultivar variation study were all cross validated with resequencing data and conservative thresholds. For all CGH arrays, test genotypes were labeled with Cy3 and the appropriate reference individual was labeled with Cy5 in all hybridizations (Additional file 1: Table S2).

Visual displays of the CGH data were created using Spotfire DecisionSite software. Additional file 1: Table S5 provides a list of soybean plants chosen for analysis, corresponding publication, and hybridization reference. Our previous study [29] of inter-cultivar variation assessed CNV on a gene-by-gene cross-validated basis across all 41 SoyNAM genotypes, concluding that SV affected 1528 genes. We conservatively converted this to SV genes per genotype using the CGH thresholds from the study and probe-based $\log_2$ ratio score for each of the 1528 genes. FN data came from the "no-phenotype" class of 35 plants as described above, and ten "mutant phenotype" lines described in Additional file 1: Table S1 [1]. Only SV overlapping with genes were included in segment size summaries in all three genotypic classes.

Confirming novel SV

PCR was used to confirm structural variants found with CGH in the transgenic plants. PCR and Sanger sequencing across breakpoints was used to confirm the four CGH observed events. Confirmed events and internal primers were used to genotype the structural variants in additional plants. Primer sequences are provided in Additional file 1: Table S6. In three of these lineages, siblings and offspring of the transgenic plants were genotyped to test if the SV were heritable. The events were confirmed not to be intra-cultivar heterogeneity by PCR-genotyping 47 untransformed individuals (either in the corresponding 'Bert' or 'Williams 82' background) at these three loci. Furthermore, the SoyNAM parents as well as cultivars 'Archer', 'Minsoy', and 'Noir1' were also PCR-genotyped with the breakpoint and internal primers to test for novelty of the SV events.

Analyzing transgene insertion sites

Transgene integrations were analyzed using TAIL-PCR [51], Southern blot, and resequencing data. Southern blots used a BAR gene probe to detect the number of T-DNA insertions in the plants tested. TAIL-PCR was used to detect T-DNA locations in WPT_384-1-1, WPT_389-

2-2 and WPT_301-3-13. Transgene insertion sites and counts were also determined by resequencing according to steps one through six outlined by Srivastava et al. [52]. Briefly, raw paired-end reads were aligned using Bowtie2 to the transgene sequence between the left and right border and the orphaned mapped reads were then aligned to the host soybean genome. The resulting putative transgene integration locations were filtered on prior knowledge of homology between components of the transgene (i.e. GmUbi promoter, RNAi hairpin targets, and their paralogs) and the genome. The location of the mapped orphaned reads, read depth coverage, and paired-end read spacing were further used to detect SV induced locally to transgene insertions. Integrated Genome Viewer (IGV) version 2.3.52 was used to visualize alignment results [53].

Sequence handling, alignment, and calling of nucleotide substitutions

The sequence read data from the ten "mutant phenotype" fast neutron plants analyzed in this study, along with the parent line of the population (cv. 'M92-220'), are deposited in the Sequence Read Archive (http://www.ncbi.nlm.nih.gov/sra/) under accession number SRP036841. The sequence read data from the two transgenic plants, along with two individuals of the parent line (cv. 'Bert'), and the cultivars 'Archer' and 'Noir 1' are deposited in the Sequence Read Archive under accession number SRP063738.

To determine the relative rates of base substitution due to FN mutagenesis, we used resequencing data from the aforementioned ten FN plants that had associated mutant phenotypes as reported in [1] (see Additional file 1: Table S1). We sequenced two transgenic plants and two controls to estimate the base substitution rate and localize T-DNA insertion sites. See Additional file 2: Figure S11 for the transgenic resequencing data analysis pipeline. All individuals were sequenced with Illumina 100 bp paired end reads.

FastQC version 0.11.2 was used on initial read data (and after any modifications to sequence data) to ensure that tools were used properly and the data was of acceptable quality for downstream applications [54]. Forward and reverse reads were treated separately, and then resynchronized for alignment using resync.pl (Riss util version 1.0, http://msi-riss.readthedocs.org/en/latest/software/riss_util.html). Cutadapt version 1.6 was used to remove adapter sequences using −b to specify both adapter sequences (GATCGGAAGAGCACACGTCTGA ACTCCAGTCAC-NNNNNN-ATCTCGT-ATGCCGTC TTCTGCTTG, AATGATACGGCGACCACCGAGAT CTACACTCTTTCCC-TACACGACGCTCTTCC-GAT CT) where NNNNNN specifies the unique 6 bp sequence attached to samples when multiplexing.

Sequence artifacts (low-complexity reads) were removed using fastx artifacts filter (Fastx toolkit version 0.0.14). Read quality was further filtered using fastq quality trimmer in the fastxtoolkit. Bases with phred quality of less than 20 were removed, and reads that were shorter than 30 bp after trimming were discarded.

We chose to align reads to the reference with two different read mapping programs, BWA mem (v. 0.7.10) [55], and Bowtie2 (v. 2.2.4) [56]. BWA mem alignments allowed for more accurate single base substitution calls, and Bowtie2 produces alignments more suitable for confirming CGH-identified SV. For BWA mem, the mismatch penalty was set to 6 (−B 6), which allows for approximately seven high-quality mismatches per read. Bowtie2 alignments were produced with default parameters. In both cases, reads were mapped to the *Glycine max* assembly version 1.0 (Wm82.a1.v1.1) [50]. Read cleaning and post-alignment filtering resulted in a realized mean coverage of 35x for the FN mutagenized plants, and 20x for WPT_389-2-2, and 21x for WPT_391-1-6.

Genotype calls for all sites were generated with the UnifiedGenotyper in the Genome Analysis Tool Kit (GATK) version 3.3 [57]. Pairwise comparisons of soybean varieties typically identify over one-million single base substitutions [33, 34]. This BWA mem resequencing and SNP detection pathway identified 1,110,325 substitutions between genotype 'Archer' and the 'Williams 82' reference genome sequence, and 1,904,061 substitutions between genotype 'Noir 1' and 'Williams 82'. These findings served as a control to demonstrate our analysis pipeline identified similar polymorphism counts as have been previously reported in soybean studies.

We then applied a set of filtering criteria to look at only substitutions that are private to a single individual (termed "unique" or "novel" throughout the paper) across the most confidently called portions of the genome. This excluded sites with less than five reads per sample, sites that were monomorphic for the reference base, sites with heterozygous or missing calls, and sites with a homozygous alternate base call in more than one individual. Applied together, these filtering criteria produced variant calls that were homozygous private differences from reference. The filtering criteria assumed *de novo* mutations at a single base position will only be observed once. A large section in FN plant 07 on Chromosome 12 between 10 and 23 Mb was found to contain a disproportionate number of substitutions. CGH results from other FN individuals [1], not included in this sample, suggest this region is heterogeneous in the 'M92-220' cultivar. We therefore excluded this region of 183 substitutions when analyzing FN plant 07. The observed transition:transversion ratios were too variable between individuals to compare to previously reported ratios in FN mutagenesis [35].

Circos plots [58] were generated using 2d tile data tracks, plotting unique substitutions detected, previously published FN-induced SV [1], detected transformation-induced SV, and T-DNA mapping results. Scripts to perform data handling and analysis are available at https://github.com/TomJKono/Unintended_Consequences.

Additional files

Additional file 1: Table S1. Resequenced fast neutron genotypes, all from the forward screen family, Bolon et al. [1]. **Table S2.** Summary of data type, CGH design, and analysis method for Inter-cultivar, Fast Neutron, and Transgenic genotypic classes. **Table S3.** Summary of SNP frequencies in a subsample fast neutron and transgenic plants. **Table S4.** Genotypes and regions used to develop CGH log$_2$ ratio empirical thresholds. **Table S5.** Genotypes examined by CGH. **Table S6.** Sequences of PCR primers used for genotyping.

Additional file 2: Figure S1. A novel deletion detected on chromosome 01 in transgenic plant WPT_384-1-1. **Figure S2.** A novel deletion on chromosome 19 in transgenic plant WPT_391-1-6. **Figure S3.** Test for intracultivar variation in the parental lines by genotyping 47 individuals taken from GRIN stocks of the varieties 'Bert' and 'Williams 82'. **Figure S4.** Genotyping diverse lines including the 41 SoyNAM parents, cultivars 'Archer', 'Minsoy', and 'Noir1', 'Bert-MN-01', and 'Wm82-ISU-01,' for previous evidence of SV found in transformed plants. **Figure S5.** Novel deletion on chromosome 11 in transgenic plant WPT_389-2-2. **Figure S6.** Novel duplication on chromosome 13 in transgenic plant WPT_301-3-13. **Figure S7.** Southern blot analysis of HindIII digested genomic DNA. **Figure S8.** Microhomology of sequences at the T-DNA left border and the sites of genomic integration for three transgenic plants. **Figure S9.** Structure of the heterozygous transgene insertion on chromosome 05 in transgenic plant WPT_391-1-6. **Figure S10.** Transgene insertion on chromosome 13 in transgenic plant WPT_389-2-2. **Figure S11.** Pipeline for utilizing resequencing data in this study.

Abbreviations

CGH: array-based comparative genomic hybridization; FN: fast neutron; GFP: green fluorescent protein; NGS: next-generation sequencing; SoyNAM: soybean nested association mapping; SV: structural variants; TAIL-PCR: thermal asymmetric interlaced PCR; TALEN: transcription activator-like effector nuclease; WPT: whole plant transformation; ZFN: zinc-finger nuclease.

Competing interests

The authors declare that they have no competing interests.

Authors' contributions

JEA, and RMS designed the experiments. AOS, JM, TJYK, and JEA, performed the experiments. JEA, TJYK, and JM performed the data analysis. BWC and SJC developed the T-DNA constructs. The article was written by JEA, JM, TJYK, and RMS. All listed authors improved the manuscript. All authors read and approved the final manuscript.

Acknowledgments

The authors thank Peter Morrell, Michael Kantar, and Wayne Parrot for reviewing the manuscript. We are grateful to Carroll Vance and Gary Muehlbauer for contributing towards the resequencing of the FN plants. This work was supported by the United Soybean Board (Projects #1520-532-5601 and #1520-532-5603), the Minnesota Soybean Research and Promotion Council (Project #18-15C), the National Science Foundation (Project IOS-1127083), the United States Department of Agriculture (Biotechnology Risk Assessment Project #2015-33522-24096) and the MnDRIVE 2014 Global Food Ventures Fellowship program in support of T.J.Y.K. This work was carried out in part using hardware and software provided by the University of Minnesota Supercomputing Institute.

References

1. Bolon YT, Stec AO, Michno JM, Roessler J, Bhaskar PB, Ries L, Dobbels AA, Campbell BW, Young NP, Anderson JE, et al. Genome resilience and prevalence of segmental duplications following fast neutron irradiation of soybean. Genetics. 2014;198(3):967–81.
2. Neelakandan AK, Wang K. Recent progress in the understanding of tissue culture-induced genome level changes in plants and potential applications. Plant Cell Rep. 2012;31(4):597–620.
3. Latham JR, Wilson AK, Steinbrecher RA. The mutational consequences of plant transformation. J Biomed Biotechnol. 2006;2006(2):25376.
4. Schnell J, Steele M, Bean J, Neuspiel M, Girard C, Dormann N, Pearson C, Savoie A, Bourbonniere L, Macdonald P. A comparative analysis of insertional effects in genetically engineered plants: considerations for pre-market assessments. Transgenic Res. 2015;24(1):1–17.
5. D'Amato F. Role of somatic mutations in the evolution of higher plants. Caryologia. 1997;50(1):1–15.
6. Heinz DJ, Mee GWP. Morphologic, cytogenetic, and enzymatic variation in saccharum species hybrid clones derived from callus tissue. Am J Bot. 1971; 58(3):257–62.
7. Jain SM. Tissue culture-derived variation in crop improvement. Euphytica. 2001;118(2):153–66.
8. Jiang C, Mithani A, Gan X, Belfield EJ, Klingler JP, Zhu JK, Ragoussis J, Mott R, Harberd NP. Regenerant Arabidopsis lineages display a distinct genome-wide spectrum of mutations conferring variant phenotypes. Curr Biol. 2011; 21(16):1385–90.
9. Endo M, Kumagai M, Motoyama R, Sasaki-Yamagata H, Mori-Hosokawa S, Hamada M, Kanamori H, Nagamura Y, Katayose Y, Itoh T, et al. Whole-genome analysis of herbicide-tolerant mutant rice generated by Agrobacterium-mediated gene targeting. Plant Cell Physiol. 2015;56(1):116–25.
10. Miyao A, Nakagome M, Ohnuma T, Yamagata H, Kanamori H, Katayose Y, Takahashi A, Matsumoto T, Hirochika H. Molecular spectrum of somaclonal variation in regenerated rice revealed by whole-genome sequencing. Plant Cell Physiol. 2012;53(1):256–64.
11. Zhang D, Wang Z, Wang N, Gao Y, Liu Y, Wu Y, Bai Y, Zhang Z, Lin X, Dong Y, et al. Tissue culture-induced heritable genomic variation in rice, and their phenotypic implications. PLoS One. 2014;9(5):e96879.
12. Kashima K, Mejima M, Kurokawa S, Kuroda M, Kiyono H, Yuki Y. Comparative whole-genome analyses of selection marker-free rice-based cholera toxin B-subunit vaccine lines and wild-type lines. BMC Genomics. 2015;16:48.
13. Kawakatsu T, Kawahara Y, Itoh T, Takaiwa F. A whole-genome analysis of a transgenic rice seed-based edible vaccine against cedar pollen allergy. DNA Res. 2013;20(6):623–31.
14. Sabot F, Picault N, El-Baidouri M, Llauro C, Chaparro C, Piegu B, Roulin A, Guiderdoni E, Delabastide M, McCombie R, et al. Transpositional landscape of the rice genome revealed by paired-end mapping of high-throughput re-sequencing data. Plant J. 2011;66(2):241–6.
15. Nacry P, Camilleri C, Courtial B, Caboche M, Bouchez D. Major chromosomal rearrangements induced by T-DNA transformation in Arabidopsis. Genetics. 1998;149(2):641–50.
16. Svitashev SK, Somers DA. Characterization of transgene loci in plants using FISH: A picture is worth a thousand words. Plant Cell Tissue Organ Cult. 2002;69(3):205–14.
17. Clark KA, Krysan PJ. Chromosomal translocations are a common phenomenon in Arabidopsis thaliana T-DNA insertion lines. Plant J. 2010;64(6):990–1001.

18. Muskens MWM, Vissers APA, Mol JNM, Kooter JM. Role of inverted DNA repeats in transcriptional and post-transcriptional gene silencing. Plant Mol Biol. 2000;43(2):243–60.

19. Ming R, Hou S, Feng Y, Yu Q, Dionne-Laporte A, Saw JH, Senin P, Wang W, Ly BV, Lewis KL, et al. The draft genome of the transgenic tropical fruit tree papaya (Carica papaya Linnaeus). Nature. 2008;452(7190):991–6.

20. Forsbach A, Schubert D, Lechtenberg B, Gils M, Schmidt R. A comprehensive characterization of single-copy T-DNA insertions in the Arabidopsis thaliana genome. Plant Mol Biol. 2003;52(1):161–76.

21. Olhoft PM, Flagel LE, Somers DA. T-DNA locus structure in a large population of soybean plants transformed using the Agrobacterium-mediated cotyledonary-node method. Plant Biotechnol J. 2004;2(4):289–300.

22. Guttikonda SK, Marri P, Mammadov J, Ye L, Soe K, Richey K, Cruse J, Zhuang M, Gao Z, Evans C, et al. Molecular characterization of transgenic events using next generation sequencing approach. PLoS One. 2016;11(2):e0149515.

23. Kovalic D, Garnaat C, Guo L, Yan Y, Groat J, Silvanovich A, Ralston L, Huang M, Tian Q, Christian A, et al. The use of next generation sequencing and junction sequence analysis bioinformatics to achieve molecular characterization of crops improved through modern biotechnology. Plant Genome. 2012;5(3):149–63.

24. Kanizay LB, Jacobs TB, Gillespie K, Newsome JA, Spaid BN, Parrott WA. HtStuf: High-throughput sequencing to locate unknown DNA junction fragments. Plant Genome. 2015;8(1): doi: 10.3835/plantgenome2014.10.0070.

25. Sims D, Sudbery I, Ilott NE, Heger A, Ponting CP. Sequencing depth and coverage: key considerations in genomic analyses. Nat Rev Genet. 2014; 15(2):121–32.

26. Ladics GS, Bartholomaeus A, Bregitzer P, Doerrer NG, Gray A, Holzhauser T, Jordan M, Keese P, Kok E, Macdonald P, et al. Genetic basis and detection of unintended effects in genetically modified crop plants. Transgenic Res. 2015;24(4):587–603.

27. Zmienko A, Samelak A, Kozlowski P, Figlerowicz M. Copy number polymorphism in plant genomes. Theor Appl Genet. 2014;127(1):1–18.

28. McHale LK, Haun WJ, Xu WW, Bhaskar PB, Anderson JE, Hyten DL, Gerhardt DJ, Jeddeloh JA, Stupar RM. Structural variants in the soybean genome localize to clusters of biotic stress-response genes. Plant Physiol. 2012;159(4): 1295–308.

29. Anderson JE, Kantar MB, Kono TY, Fu F, Stec AO, Song Q, Cregan PB, Specht JE, Diers BW, Cannon SB, et al. A roadmap for functional structural variants in the soybean genome. G3 (Bethesda). 2014;4(7):1307–18.

30. Orf JH, Kennedy BW. Registration of "Bert" soybean. Crop Sci. 1992;32(3):830.

31. Haun WJ, Hyten DL, Xu WW, Gerhardt DJ, Albert TJ, Richmond T, Jeddeloh JA, Jia G, Springer NM, Vance CP, et al. The composition and origins of genomic variation among individuals of the soybean reference cultivar Williams 82. Plant Physiol. 2011;155(2):645–55.

32. Hancock CN, Zhang F, Floyd K, Richardson AO, Lafayette P, Tucker D, Wessler SR, Parrott WA. The rice miniature inverted repeat transposable element mPing is an effective insertional mutagen in soybean. Plant Physiol. 2011;157(2):552–62.

33. Lam HM, Xu X, Liu X, Chen W, Yang G, Wong FL, Li MW, He W, Qin N, Wang B, et al. Resequencing of 31 wild and cultivated soybean genomes identifies patterns of genetic diversity and selection. Nat Genet. 2010;42(12): 1053–9.

34. Zhou Z, Jiang Y, Wang Z, Gou Z, Lyu J, Li W, Yu Y, Shu L, Zhao Y, Ma Y, et al. Resequencing 302 wild and cultivated accessions identifies genes related to domestication and improvement in soybean. Nat Biotechnol. 2015;33(4):408–14.

35. Belfield EJ, Gan X, Mithani A, Brown C, Jiang C, Franklin K, Alvey E, Wibowo A, Jung M, Bailey K, et al. Genome-wide analysis of mutations in mutant lineages selected following fast-neutron irradiation mutagenesis of Arabidopsis thaliana. Genome Res. 2012;22(7):1306–15.

36. Ossowski S, Schneeberger K, Lucas-Lledo JI, Warthmann N, Clark RM, Shaw RG, Weigel D, Lynch M. The rate and molecular spectrum of spontaneous mutations in Arabidopsis thaliana. Science. 2010;327(5961):92–4.

37. McVey M, Lee SE. MMEJ repair of double-strand breaks (director's cut): deleted sequences and alternative endings. Trends Genet. 2008;24(11):529–38.

38. Li G, Chern M, Jain R, Martin JA, Schackwitz WS, Jiang L, et al. Genome-wide sequencing of 41 rice (Oryza sativa L.) mutated lines reveals diverse mutations induced by fast-neutron irradiation. Mol Plant. 2016;(in press) doi: 10.1016/j.molp.2016.03.009.

39. Majhi BB, Shah JM, Veluthambi K. A novel T-DNA integration in rice involving two interchromosomal translocations. Plant Cell Rep. 2014;33(6):929–44.

40. Tax FE, Vernon DM. T-DNA-associated duplication/translocations in Arabidopsis. Implications for mutant analysis and functional genomics. Plant Physiol. 2001;126(4):1527–38.

41. Cheng KC, Beaulieu J, Iquira E, Belzile FJ, Fortin MG, Stromvik MV. Effect of transgenes on global gene expression in soybean is within the natural range of variation of conventional cultivars. J Agric Food Chem. 2008;56(9):3057–67.

42. Weber N, Halpin C, Hannah LC, Jez JM, Kough J, Parrott W. Editor's choice: Crop genome plasticity and its relevance to food and feed safety of genetically engineered breeding stacks. Plant Physiol. 2012;160(4):1842–53.

43. Bolon YT, Haun WJ, Xu WW, Grant D, Stacey MG, Nelson RT, Gerhardt DJ, Jeddeloh JA, Stacey G, Muehlbauer GJ, et al. Phenotypic and genomic analyses of a fast neutron mutant population resource in soybean. Plant Physiol. 2011;156(1):240–53.

44. Orf JH, Denny RL. Registration of "MN1302" soybean. Crop Sci. 2004;44(2):693.

45. Curtin SJ, Zhang F, Sander JD, Haun WJ, Starker C, Baltes NJ, Reyon D, Dahlborg EJ, Goodwin MJ, Coffman AP, et al. Targeted mutagenesis of duplicated genes in soybean with zinc-finger nucleases. Plant Physiol. 2011; 156(1):466–73.

46. Paz MM, Martinez JC, Kalvig AB, Fonger TM, Wang K. Improved cotyledonary node method using an alternative explant derived from mature seed for efficient Agrobacterium-mediated soybean transformation. Plant Cell Rep. 2006;25(3):206–13.

47. Curtin SJ, Michno JM, Campbell BW, Gil-Humanes J, Mathioni SM, Hammond R, Gutierrez-Gonzalez JJ, Donohue RC, Kantar MB, Eamens AL, et al. MicroRNA maturation and microRNA target gene expression regulation are severely disrupted in soybean dicer-like1 double mutants. G3 (Bethesda). 2015;6(2):423–33.

48. Campbell BW, Mani D, Curtin SJ, Slattery RA, Michno JM, Ort DR, Schaus PJ, Palmer RG, Orf JH, Stupar RM. Identical substitutions in magnesium chelatase paralogs result in chlorophyll-deficient soybean mutants. G3 (Bethesda). 2014;5(1):123–31.

49. Bernard RL, Cremeens CR. Registration of "Williams 82" soybean. Crop Sci. 1988;28(6):1027–8.

50. Schmutz J, Cannon SB, Schlueter J, Ma J, Mitros T, Nelson W, Hyten DL, Song Q, Thelen JJ, Cheng J, et al. Genome sequence of the palaeopolyploid soybean. Nature. 2010;463(7278):178–83.

51. Singer T, Burke E. High-throughput TAIL-PCR as a tool to identify DNA flanking insertions. Methods Mol Biol. 2003;236:241–72.

52. Srivastava A, Philip VM, Greenstein I, Rowe LB, Barter M, Lutz C, Reinholdt LG. Discovery of transgene insertion sites by high throughput sequencing of mate pair libraries. BMC Genomics. 2014;15:367.

53. Thorvaldsdottir H, Robinson JT, Mesirov JP. Integrative Genomics Viewer (IGV): high-performance genomics data visualization and exploration. Brief Bioinform. 2013;14(2):178–92.

54. FastQC: A quality control tool for high throughput sequence data [http://www.bioinformatics.babraham.ac.uk/projects/fastqc]

55. Li H, Durbin R. Fast and accurate short read alignment with Burrows-Wheeler transform. Bioinformatics. 2009;25(14):1754–60.

56. Langmead B, Salzberg SL. Fast gapped-read alignment with Bowtie 2. Nat Methods. 2012;9(4):357–9.

57. DePristo MA, Banks E, Poplin R, Garimella KV, Maguire JR, Hartl C, Philippakis AA, del Angel G, Rivas MA, Hanna M, et al. A framework for variation discovery and genotyping using next-generation DNA sequencing data. Nat Genet. 2011;43(5):491–8.

58. Krzywinski M, Schein J, Birol I, Connors J, Gascoyne R, Horsman D, Jones SJ, Marra MA. Circos: an information aesthetic for comparative genomics. Genome Res. 2009;19(9):1639–45.

15

Heterologous expression of *Aspergillus aculeatus* endo-polygalacturonase in *Pichia pastoris* by high cell density fermentation and its application in textile scouring

Dede Abdulrachman[1], Paweena Thongkred[2], Kanokarn Kocharin[2], Monthon Nakpathom[3], Buppha Somboon[3], Nootsara Narumol[3], Verawat Champreda[2], Lily Eurwilaichitr[2], Antonius Suwanto[1], Thidarat Nimchua[2*] and Duriya Chantasingh[2*]

Abstract

Background: Removal of non-cellulosic impurities from cotton fabric, known as scouring, by conventional alkaline treatment causes environmental problems and reduces physical strength of fabrics. In this study, an endo-polygalacturonase (EndoPG) from *Aspergillus aculeatus* produced in *Pichia pastoris* was evaluated for its efficiency as a bioscouring agent while most current bioscouring process has been performed using crude pectinase preparation.

Results: The recombinant EndoPG exhibited a specific activity of 1892.08 U/mg on citrus pectin under the optimal condition at 50 °C, pH 5.0 with a V_{max} and K_m of 65,451.35 μmol/min/mL and 15.14 mg/mL, respectively. A maximal activity of 2408.70 ± 26.50 U/mL in the culture supernatant was obtained by high cell density batch fermentation, equivalent to a 4.8 times greater yield than that from shake-flask culture. The recombinant enzyme was shown to be suitable for application as a bioscouring agent, in which the wettability of cotton fabric was increased by treatment with enzyme at 300 U/mL scouring solution at 40 °C, pH 5.0 for 1 h. The bio-scoured fabric has comparable wettability to that obtained by conventional chemical scouring, but has higher tensile strength.

Conclusion: The work has demonstrated for the first time functions of *A. aculeatus* EndoPG on bioscouring in eco-textile processing. EndoPG alone was shown to possess effective scouring activity. High expression level and homogeneity could be achieved in bench-scale bioreactor.

Keywords: Acidic pectinase, Bioscouring, Cotton fabric, Heterologous expression, High cell density fermentation

Background

Cotton is an important natural resource used in the textile industry. It is a highly pure natural cellulosic material, comprising approximately 90% cellulose that is organized into strong microfiber structures. Pectin is a major non-cellulosic impurity present in the cuticle and primary cell wall of cotton. It acts as a cementing material for the cellulosic network and as a hydrating agent that controls the movement of water and plant fluids [1], providing firmness and rigidity to the cotton fibers. The backbone of pectin is made of α-1,4-linked D-galacturonic acid residues, in which the carboxylic groups of galacturonic acid are largely esterified with methoxy groups, while the hydroxyl groups on the backbone can be partially acetylated. Xylose, galactose and arabinose are present as side chain sugars of pectin [2]. The hydrophobic nature of pectin is responsible for the non-wetting behavior of the native cotton, which creates difficulty for fabric dyeing in textile processing [3].

Conventionally, pectin and other waxy substances in cotton fabrics are removed by boiling alkali treatment in a process known as scouring. This process makes the fabric more hydrophilic and more accessible for

---

* Correspondence: thidarat.nim@biotec.or.th; duriya@biotec.or.th
[2]Microbial Biotechnology and Biochemicals Research Unit, National Center for Genetic Engineering and Biotechnology, 113 Thailand Science Park, Pahonyothin Rd, Khlong Luang, Patumthani 12120, Thailand
Full list of author information is available at the end of the article

subsequent textile processing. The scouring process carries an environmental burden owing to the large consumption of sodium hydroxide and water (the latter is required for an intensive rinsing step). In addition, scouring also causes non-specific degradation of cellulose, which decreases the tensile strength of fibers, and consequently leads to lower fabric quality [4]. In contrast to the harsh conditions employed in the conventional scouring process, enzymatic processing (bioscouring) can be performed under mild reaction conditions using individual or combinations of different enzymes including lipases, proteases, cellulases and pectinases [1–5]. Among them, pectinases are the most promising [6]. Pectinases release waxes and other non-cellulosic components of the cotton fibers with minimal non-specific damage to the cellulosic structure. Pectinase penetrates the cuticle layer of cotton fiber through cracks or micropores and then partially breaks down pectin in the primary wall matrix [1, 5]. The breakdown of pectin loosens linkages between the cuticle and the cellulose body leading to increased absorbency of the fabric.

According to the mode of action and substrate preference, pectinases are classified into two groups, namely (i) pectin esterases (EC 3.1.1.11) and (ii) pectin depolymerases, which are further divided into polygalacturonases (EC 3.2.1.15, EC 3.2.1.67 and EC 3.2.1.82) and lyases (EC 4.2.2.2 and EC 4.2.2.10) [5]. Polygalacturonases usually work optimally under an acidic range and require no co-factors for their catalysis. Conversely, lyases require divalent metal co-factors and are optimally active under alkaline conditions [7]. Crude pectinase preparations isolated from fungi, for example *Aspergillus* spp. and *Penicillium* spp. that have been tested for potential as bioscouring agents of cotton textiles, possess strong pectate lyase activity. This enzyme works optimally under alkaline conditions, which is compatible with the production of peroxide in the subsequent bleaching step [8, 9] while there has been no report on application of purified or recombinant pectate lyases or polygalacturonases in bio-scouring. In addition to bioscouring, enzymatic starch removal (desizing) using amylases prior to the bioscouring step has been shown to have potential for reducing operating time and energy consumption in textile processing [5, 10]. Combined desizing and bioscouring treatment using amylases and pectinases is hindered by the very different optimal pH of these enzymes. Searching for acidic pectinase working on bioscouring under acidic conditions optimal is thus challenging.

In this study, we aimed to express a recombinant acidic endo-polygalacturonase from *A. aculeatus* in methylotrophic yeast *P. pastoris* using high cell density fermentation. The recombinant enzyme was evaluated in eco-friendly textile processing in comparison to the

conventional alkaline scouring process. The developed biscouring process has demonstrated for the first time the action of polygalacturonase on textile scouring with advantages on improved physical properties of the fabrics and allows further optimization of the one-step desizing-scouring for textile processing.

Methods

Strains, culturing conditions, genes, and primers

The endo-polygalacturonase gene (*endoPG*), encoding the mature endo-polygalacturonase (EndoPG) without a signal peptide was synthesized based on the coding sequence of *A. aculeatus* ATCC16872. The codon usage was analyzed by Codon Usage Analyzer [11] and optimized for expression in *P. pastoris* using Gene Designer Software [12]. The optimized gene was synthesized by Genescript (Piscataway, NJ, USA). *Escherichia coli* DH5α was used as a host for plasmid propagation. *E. coli* was propagated in low salt Luria-Bertani (LB) medium (1% (*w/v*) tryptone, 0.5% (*w/v*) NaCl, and 0.5% (*w/v*) yeast extract) at 37 °C. *P. pastoris* KM71 (Invitrogen, Carlsbad, CA, USA) was used as a host for EndoPG expression and was cultured in YPD (1% (*w/v*) yeast extract, 2% (*w/v*) peptone, and 2% (*w/v*) glucose). *P. pastoris* transformants were selected on YPD agar containing 100 μg/mL Zeocin (Invitrogen, Carlsbad, CA, USA). pPICZαA vector (Invitrogen) was used as the expression vector. All synthetic oligonucleotides used in this study were purchased from 1st Base (Selangor, Malaysia).

Construction of expression plasmids

The expression plasmid carrying *endoPG* gene was constructed by amplification of the *endoPG* in pUC57-EndoPG plasmid with a primer pair, EndoPG-F21 (GCAT<u>GAATTC</u>GCACCTACAGACATCGAGAAGAG ATC; *Eco*RI site underlined) and EndoPG-R1 (TACA<u>TC TAGA</u>TTAGCAACTGGCACCGGAAG; *Xba*I underlined) for the native mature enzyme. The amplicon was digested with *Eco*RI and *Xba*I and ligated with pPICZαA-digested vector. The recombinant pPICZαA containing *endoPG* (designated as pPIC-PG1), was transformed into *E. coli* DH5α by the heat shock method according to Sambrook and Russel [13]. Transformants were cultivated and selected on LB plates supplemented with 25 μg/mL Zeocin. The DNA sequences of the recombinant plasmids were verified by Sanger sequencing (1st Base, Selangor, Malaysia).

Heterologous expression of EndoPG in *P. pastoris* KM71

The plasmid pPIC-PG1 was linearized with *Pme*I (Thermo Scientific, Waltham, MA, USA) and then transformed into *P. pastoris* KM71 by electroporation (Gene Pulser, Bio-Rad, Hercules, CA, USA) according to the Easy Select™ *Pichia* Expression Kit instructions

(Invitrogen, Carlsbad, CA, USA). Sixty colonies of putative recombinant clones were randomly selected from YPDS plates (1% (w/v) yeast extract, 2% (w/v) peptone, 2% (w/v) glucose, 18.2% (w/v) sorbitol, and 1.5% (w/v) bacto agar) supplemented with 100 μg/mL Zeocin. Integration of the gene into the *P. pastoris* genome was confirmed by colony PCR using 5′AOX1-F (5′-GACTGGTCCAATTGAC AAGC-3′) and 3′AOX1-R (5′-CGAAATGGCATTCT GACATGG-3′) primers according to Easy Select™ *Pichia* Expression Kit instructions (Invitrogen, Carlsbad, CA, USA).

Five mL of YPD medium was inoculated with a single colony of *P. pastoris* transformant and incubated at 30 °C for overnight with rotary shaking at 250 rpm until the OD_{600} reached 5–6. One milliliter of the seed culture was transferred to 25 mL of the buffered glycerol-complex medium (BMGY; 1% (w/v) yeast extract, 2% (w/v) peptone, 100 mM potassium phosphate buffer pH 6.0, 1.34% (w/v) YNB, 0.0004% (w/v) biotin, and 1% (v/v) glycerol) and the culture was grown under the same condition as described above for 24 h. The cell pellet was then harvested and resuspended in one-fifth of the original culture volume with methanol-minimal medium (BMMY; 1% (w/v) yeast extract, 2% (w/v) peptone, 100 mM potassium phosphate buffer pH 6.0, 1.34% (w/v) YNB, 0.0004% (w/v) biotin, and 3.0% (v/v) methanol). Absolute methanol was added to a final concentration of 3% (v/v) to induce expression of the heterologous gene. The cell-free supernatant was collected at 1, 2, 3, and 4 days after induction for monitoring protein expression by SDS-PAGE [13]. To remove residual culture media and contaminants, the enzyme used for biochemical characterization was further processed by using ultrafiltration (Amicon® Ultra-15, Millipore) with 50 mM potassium phosphate buffer pH 6. Protein concentration was analyzed with Bio-Rad Protein Assay Reagent based on Bradford's method (Bio-Rad, Hercules, CA, USA) using bovine serum albumin as the standard [14].

Enzyme activity assay

Pectinase activity was assayed based on hydrolysis of pectin. 0.34 mL assay reactions contained an appropriate enzyme dilution in 0.5% citrus pectin (w/v) in 100 mM sodium acetate buffer and incubated at 50 °C for 10 min. The amount of liberated reducing sugars was determined by the dinitrosalicylic acid (DNS) method using D-(+)-galacturonic acid as a standard [15]. The amylase, Avicelase, and CMCase activities were analyzed using 1% soluble starch, 1% microcrystalline cellulose (Avicel®) and 1% carboxymethylcellulose as the substrates, respectively, according to the reaction conditions described above using glucose as the standard. One unit of enzyme activity was defined as the amount of enzyme that catalyzes the formation of 1 μmol of reducing sugar

from substrate per min under its optimal conditions. The experiments were done in triplicate.

Characterization of recombinant EndoPG

The effects of pH on activity of EndoPG was determined at 50 °C for 10 min with a pH range from 2.0 to 9.0, using 100 mM citric acid-sodium citrate (pH 2.0-4.0), 100 mM acetic acid-sodium acetate (pH 4.0-6.0), 100 mM potassium phosphate (pH 6.0-8.0), and 100 mM Tris-Cl (pH 8-9). The optimum temperature was determined within the temperature range of 30–80 °C at pH 5 in 100 mM sodium acetate buffer. The pH stability of the enzyme was studied by incubation of the enzyme at 4 °C in buffers with varying pH for 24 h before determination of the residual pectinase activity under the standard conditions (50 °C, pH 5.0). Thermostability was determined by pre-incubating the enzymes in 100 mM acetate buffer, pH 5.0 at different temperatures (20–60 °C) for varying time intervals from 0.5 to 3 h before analyzing the residual pectinase activity relative to the enzyme activity without pre-incubation denoted as 100%.

The effect of salts (NaCl and KCl), divalent metal ions ($CaCl_2$, $CuSO_4$, $MgCl_2$, $MnSO_4$, $ZnSO_4$, $FeSO_4$, and $CoCl_2$) and divalent ion chelator (EDTA) on the enzyme activity was studied by pre-incubating an appropriate dilution of the enzyme in the presence of the compounds at 1 mM at 0 °C for 30 min before determining the pectinase activity under the standard assay conditions. The enzyme activity in the absence of supplemented chemicals was considered as 100%. The data were analyzed by Dunnett's multiple comparison test compared to the control.

The specificity of EndoPG was assayed on polygalacturonic acid with varying degrees of esterification. The assay reactions contained an appropriate enzyme dilution with 0.5% w/v of non-esterified polygalacturonic acid, [6.7%]- or [55–70%]-esterified-pectin from citrus peel (Sigma-Aldrich, USA) in 100 mM sodium acetate buffer, pH 5 and incubated at 50 °C for 10 min. The amount of released reducing sugars was determined using the DNS method as described above. The substrate specificity was expressed as relative percent activity compared with the activity on polygalacturonic acid.

Esterified pectin (6.7%) was used as substrate for the kinetic studies. The K_m and V_{max} were determined at different substrate concentrations ranging from 0.2 to 20 mg/mL under standard assay conditions (50 °C with an incubation time of 5 min). The kinetics data were calculated and defined by the SigmaPlot 7 (Systat Software, San Jose, CA).

High cell density fermentation

The recombinant *P. pastoris* expressing EndoPG was cultivated in high cell density fed-batch fermentation in a 5-L bioreactor (B. Braun Sartorius Ltd., Göttingen, Germany) with the method modified from Charoenrat et al. (2013) [16]. Briefly, the primary inoculum was prepared by inoculating a colony of the recombinant *P. pastoris* into a 125 mL flask containing 20 mL of YPD broth and incubated at 30 °C with rotary shaking at 250 rpm for 24 h before subcultivating 20 mL of the inoculum into 180 mL FM22 medium. The FM22 medium contained: KH_2PO_4, 42.9 g/L; $CaSO_4$, 0.93 g/L; K_2SO_4, 14.3 g/L; $MgSO_4 \cdot 7H_2O$, 11.7 g/L; $(NH_4)_2SO_4$, 5.0 g/L; glycerol, 20.0 g/L; histidine, 2.0 g/L and PTM1, 4.35 mL/L. The trace salt (PTM1) contained: $CuSO_4 \cdot 5H_2O$, 6.0 g/L; KI, 0.08 g/L; $MnSO_4 \cdot H_2O$, 3.0 g/L; $Na_2MoO_4 \cdot 2H_2O$, 0.2 g/L; H_3BO_3, 0.02 g/L; $ZnCl_2$, 20.0 g/L; $FeCl_3$, 13.7 g/L; $CoCl_2 \cdot 6H_2O$, 0.9 g/L; H_2SO_4, 5.0 mL/L; and biotin, 0.2 g/L. The culture was further incubated at 30 °C with rotary shaking at 250 rpm for 48 h. A 5-L bioreactor vessel containing 1.8 L basal salt medium (2.67% (*v/v*) phosphoric acid 85%, 0.093% (*w/v*) $CaSO_4$, 1.82% (*w/v* K_2SO_4, 1.49% (*w/v*) $MgSO_4.7H_2O$, 0.413% (*w/v*) KOH, 4% (*w/v*) glycerol) was inoculated with 200 mL of the secondary inoculum. The fermentation parameters were controlled at 30 °C, 500–700 rpm agitation, 30% dissolved oxygen (DO), with 1 vvm aeration, and pH 5.0, adjusted using NH_4OH. The fed-batch cultivation was divided into four stages: glycerol batch stage, glycerol fed-batch stage, methanol induction stage, and production stage as described previously by Jahic et al. [17]. The batch culture was grown until the glycerol was exhausted and the fed-batch mode was applied by constant glycerol feeding at 18.15 mL/h/L of initial fermentation volume for 4 h. The methanol induction feed was started after 3–5 h to avoid repression of AOX promoter [18]. Then, 100% methanol feed was added continuously into the fermentation culture, starting at a rate of 1.00 mL/h/L fermentation volume for the first 2 h and gradually increasing by 10% increments every 1 h to a target rate of 3.00 mL/h/L fermentation volume, which was maintained during the fermentation process. The pectinase activity, dry-cell weight, and protein concentration of the fermenter culture were monitored throughout the cultivation.

Enzymatic scouring of cotton fabrics

A 211 g/m^2 enzymatically desized plain-woven 100% cotton fabric (PYI) (Thanapaisal R.O.P, Samutprakarn, Thailand) with a weight of 3.0 g was placed in a 500 mL container with different concentrations of EndoPG (100, 200, 300 and 400 U/mL) at a liquor to fabric ratio of 20:1 in the presence of 0.2% (*w/v*) wetting agent (Hostapal® NIN liq c, Archroma, Singapore). The enzymatic scouring process was carried out in a DaeLim Starlet II DLS-8080 infrared lab dyeing machine at 40 °C for 1 h in 50 mM sodium acetate buffer, pH 5.0 with a heating rate of 2 °C/min and a shaking rate of 30 rpm. The enzyme was inactivated by boiling in distilled water for 10 min twice. The fabric samples were washed twice with distilled water at room temperature and then air-dried. A control treatment was carried out without enzyme addition. Conventional chemical scouring was performed in a 0.2 NaOH solution at 100 °C for 30 min [9]. All experiments were performed in triplicate.

The fabric weight loss was expressed as a percentage loss in weight of fabric after drying at 80 °C in an air-circulated oven water bath with respect to the initial dry weight of the fabric. The samples were weighed after cooling in a desiccator. Water absorbency was evaluated according to the AATCC test method 79-1995 [19]. Tensile strength test was performed according to the ASTM 5034-09 protocol using a bench top material testing machine (H5K-T UTM, Tinius Olsen, PA, USA) with a 75 mm gauge length at an extension rate of 300 mm/min. Three experiments in warp and weft directions were carried out for each sample.

Results

Expression of *A. aculeatus* EndoPG

In this study, mature *A. aculeatus* EndoPG was expressed in *P. pastoris* KM71 in a secreted form. The *endoPG* gene (1029 bp) was fused in frame to the α−factor secretion signal, under the control of the AOX1 promoter. Induction with methanol led to strong expression of EndoPG in the culture supernatant with high homogeneity (>90%) as shown by SDS-PAGE (Fig. 1). The apparent size of the secreted protein was concordant with the theoretical MW of 38.7 kDa. The highest enzyme expression was obtained after 48 h induction by methanol with the maximal polygalacturonase activity in the shake-flask culture of 503 ± 3 U/mL and a specific activity of 1892 U/mg.

Biochemical characterization and kinetic study of EndoPG

Effects of pH and temperature on the EndoPG activity and stability were studied in order to determine the enzyme's optimal working conditions (Additional file 1). The recombinant enzyme demonstrated optimal pH and temperature of 5.0 (Fig. 2a) and 50 °C (Fig. 2b), respectively. It retained more than 60% of its initial activity after incubation at 40 °C for 180 min (Fig. 3a). However, the enzyme rapidly lost activity at 50 °C and 60 °C. The pH stability test showed that more than 80% of enzyme activity was retained after incubation in pH 2.0 to 6.0 for 24 h (Fig. 3b). A marked decrease in enzyme activity was observed at pH 7.0 to 9.0.

The effects of metal ions and divalent ion chelator on the activity of EndoPG were examined by determining the enzyme activity in the presence of the additives. Among the various additives tested, Cu^{2+} and Fe^{2+} stimulated the activity of EndoPG ($p < 0.05$) (Table 1). In contrast, Mn^{2+}, Na^+ and Zn^{2+} inhibited the enzyme. K^+,

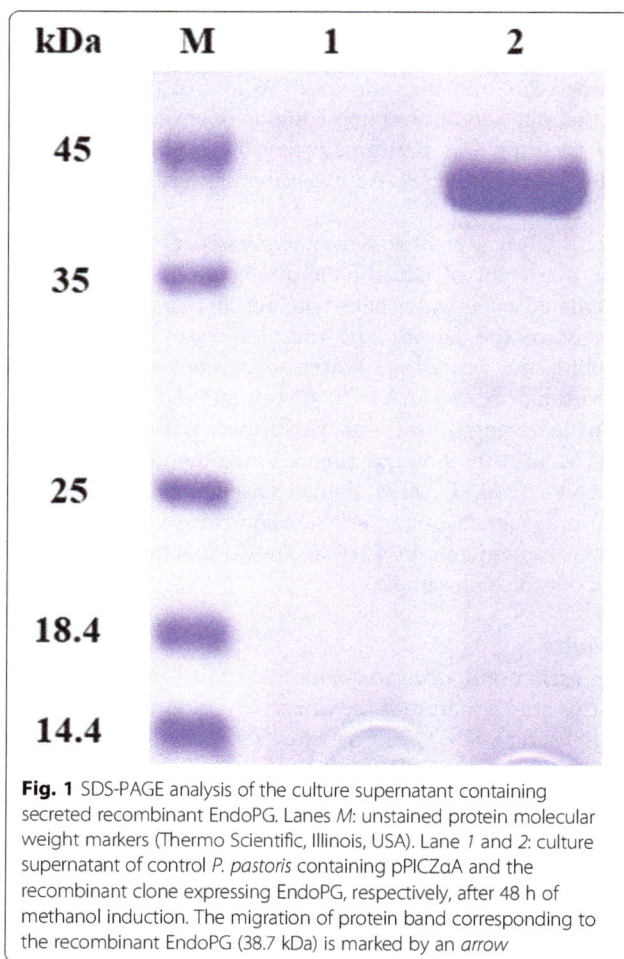

Fig. 1 SDS-PAGE analysis of the culture supernatant containing secreted recombinant EndoPG. Lanes *M*: unstained protein molecular weight markers (Thermo Scientific, Illinois, USA). Lane *1* and *2*: culture supernatant of control *P. pastoris* containing pPICZαA and the recombinant clone expressing EndoPG, respectively, after 48 h of methanol induction. The migration of protein band corresponding to the recombinant EndoPG (38.7 kDa) is marked by an *arrow*

Li^+, Mg^{2+} and the divalent ion chelator (EDTA) was found to have no significant effect on the enzyme activity.

The recombinant EndoPG showed the highest activity on polygalacturonic acid, followed by pectin with different degrees of esterification with the relative activity 22–25% of that obtained on the non-esterified substrate (Table 2). There was no non-specific hydrolysis activity against starch or cellulose. The estimated enzyme kinetic parameters of EndoPG for pectin digestion were K_m of 15.14 mg/mL and V_{max} of 65,451.35 µmol/min/mL.

High cell-density fermentation

Production of EndoPG was further studied using high cell-density fermentation approach in a laboratory-scale fermenter. The fed-batch fermentation process comprised three steps including glycerol batch, transition or glycerol fed-batch, and methanol induction or production phases (Fig. 4). The inoculum was prepared in synthetic FM22 medium in order to increase cell density and enzyme production. The glycerol batch stage was carried out for the first 30 h. At the end of glycerol batch stage, the dry cell weight reached 23 g/L. Subsequently, the glycerol fed-batch stage was initiated using constant feeding at the rate of 18.15 mL/h/L of the initial fermentation volume for 4 h. During the glycerol fed-batch stage, the cell dry-weight increased almost two fold from 23 to 41 g_{CDW}/L. The EndoPG activity reached the highest level of 2408.00 U/mL in the induction stage (111 h) with the specific production yield of 43,781.81 U/g biomass. The high cell density fermenter thus resulted in 4.8 times greater enzyme production yield compared with the shake-flask culture.

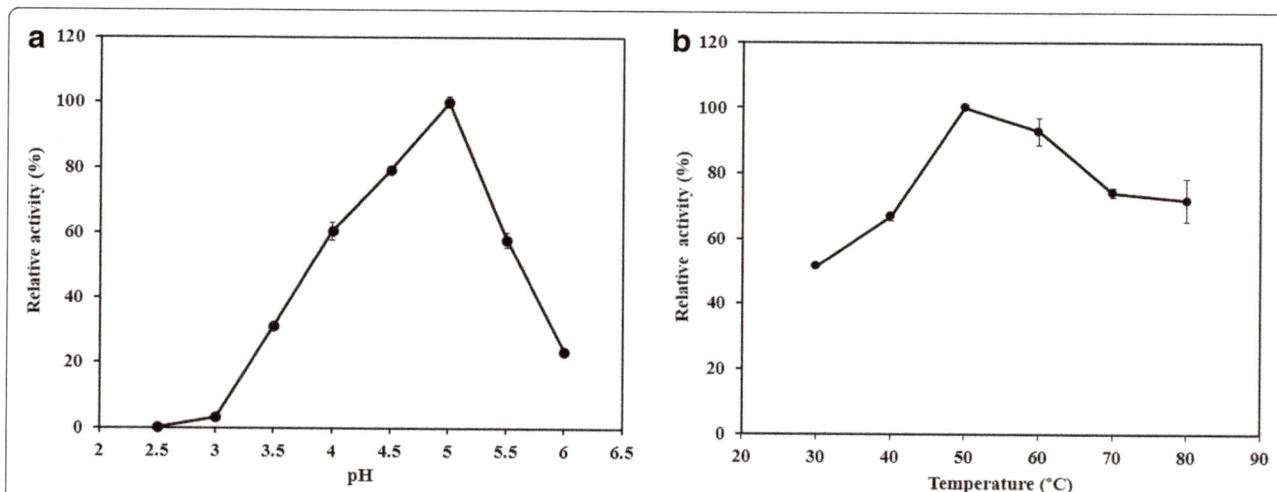

Fig. 2 Effects of (**a**) pH and (**b**) temperature on EndoPG activity. The highest activity under the optimal condition (in sodium acetate buffer pH 5 at 50 °C) was considered as 100% relative activity. *Data points* are the mean values and *error bars* represent standard deviations from triplicate experiments

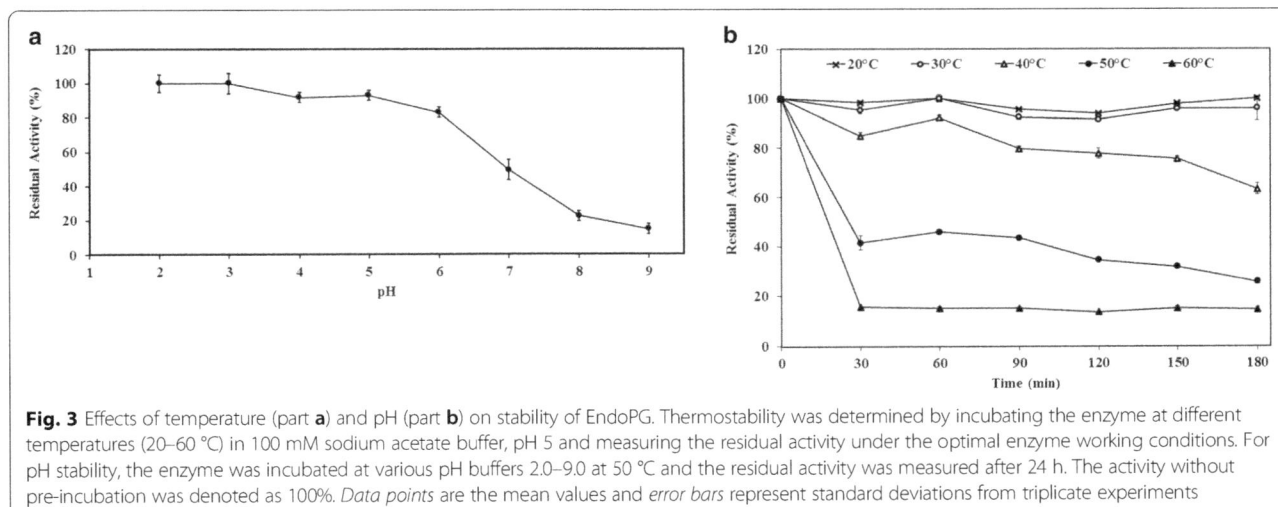

Fig. 3 Effects of temperature (part **a**) and pH (part **b**) on stability of EndoPG. Thermostability was determined by incubating the enzyme at different temperatures (20–60 °C) in 100 mM sodium acetate buffer, pH 5 and measuring the residual activity under the optimal enzyme working conditions. For pH stability, the enzyme was incubated at various pH buffers 2.0–9.0 at 50 °C and the residual activity was measured after 24 h. The activity without pre-incubation was denoted as 100%. *Data points* are the mean values and *error bars* represent standard deviations from triplicate experiments

Bioscouring by EndoPG

The performance of EndoPG on bioscouring of cotton fabric was tested based on weight loss and wettability assay (Table 3). Enzymatic treatment of the fabric with EndoPG led to a marked improvement on water absorbency of the fabric samples compared with the control of no enzyme treatment. The control fabric showed low water absorbency with a wettability time greater than 3 min. Increasing enzyme dosage led to increased fabric wettability (shorter wettability time) and corresponding increased weight loss of fabric. Maximum water absorbency of 16.5 s was observed at the highest enzyme dosage, which was comparable to that measured for fabric scoured by alkali treatment; however, with a lower weight loss of the textile. In addition, the fabrics treated with EndoPG showed a greater tensile strength for both the weft yarn (722.33 N) and warp yarn (986.67 N)

compared with that obtained using chemical scouring (Table 4). The results clearly indicated potential and advantages of the bioscouring process using EndoPG as a promising alternative to the conventional alkaline scouring method. Similar results on weight loss, wettability and tensile compared to those obtained using buffer alone were observed using the heat-inactivated enzyme (data not shown).

Discussion

Heterologous expression of composite enzymes attacking pectin including polygalacturonases, pectate lyases, and pectin esterases from various ascomycetes and basidiomycetes fungi has been reported in yeasts [20–24]. *P. pastoris* is considered an efficient expression system owning to its high production yield of heterologous proteins in secreted forms, allowing cost competitive enzyme production with minimal downstream processing [17, 18]. Recombinant endo-polygalacturonases from *A. niger*, *Bispora sp.*, *Phytophthora parasitica* and *Penicillium sp.* produced in *P. pastoris* expression system have been reported with varying biochemical characteristics [20–23]. *A. aculeatus* is a promising producer of multi-component starch and cell wall degrading enzymes with potent application in various biotechnological processes [25, 26]. In

Table 1 Effects of additives (metal ions and chelator) on the activity of EndoPG

Additive (1 mM final)	Relative activity[a] (%)
Control	100
Ca^{2+}	99.75 ± 2.68
Cu^{2+}	112.70 ± 5.02*
Fe^{2+}	112.03 ± 0.61*
K^+	99.90 ± 1.19
Li^+	100.62 ± 2.74
Mg^{2+}	95.55 ± 3.58
Mn^{2+}	73.76 ± 2.56*
Na^+	90.88 ± 2.29*
Zn^{2+}	88.81 ± 3.43*
EDTA	98.39 ± 2.73

[a]Mean and standard deviations from triplicate experiments
*Indicates significantly differently from control; $p < 0.05$

Table 2 Substrate specificity of EndoPG

Substrate	Relative activity[a] (%)
Polygalacturonic acid	100.00
Pectin (6–7% esterified)	24.99 ± 0.56
Pectin (55–70% esterified)	22.09 ± 0.15
Soluble Starch	nd
Avicel	nd
Carboxymethyl cellulose	nd

[a]Mean and standard deviations from triplicate experiments; *nd* not detected

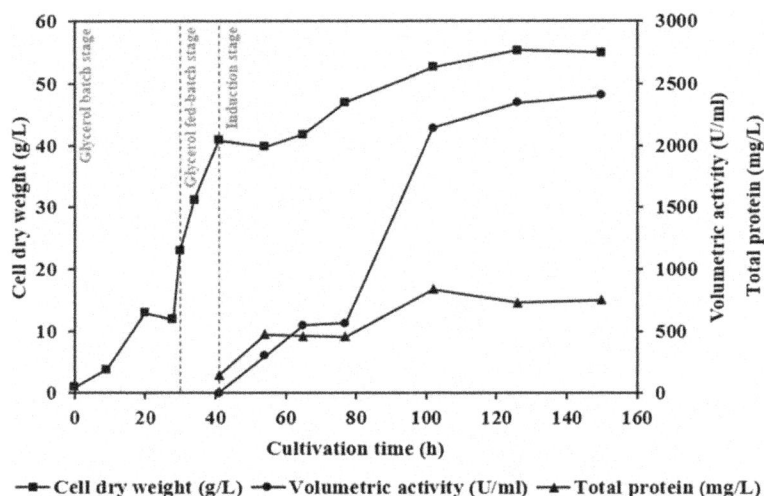

Fig. 4 High density fermentation profile of EndoPG production in *P. pastoris*. The cultivation condition during the production stage was maintained at 30 °C and pH 5.0

this study, an EndoPG from *A. aculeatus* ATCC16872 was expressed as a major protein in the culture supernatant with >90% homogeneity, allowing production of the enzyme with no need for subsequent downstream purification. The enzyme showed a specific activity of 1892 U/mg total protein (50 °C, pH 5) which is comparable to the equivalent recombinant enzyme from *A. niger* [20]. The optimal activity of the enzyme at 50 °C was relatively higher than that of the commercial acid fungal polygalacturonase, which works optimally at 40 °C [9].

Production of EndoPG by high cell density fermentation was achieved yielding maximal activity of 2408 U/mL. The greater yield of enzyme from fermentation compared with batch cultivation is due to the higher cell density achievable in the former, as shown by OD_{600} of 81.60, equivalent to a cell concentration of 40.83 g/L at the exponential glycerol feeding stage before induction by methanol. The benefit of improved yield of enzyme using high cell density fermentation has also been reported for other types of recombinant enzymes expressed in *P. pastoris* e.g., xylanase and cellulase [18, 27]. To our knowledge, our work represents the first report on heterologous expression and high cell density fermentation of polygalacturonase in *P.*

pastoris with the highest production yield, 43,781.81 U/g biomass, resulting in the higher enzyme productivity compared to that obtained using batch cultivation using wild type or recombinant microorganisms [22–26].

Bioscouring of cotton fabrics using pectinases has been shown as an interesting approach for reducing chemical and energy consumption in textile processing. Compared to conventional chemical scouring using alkalis such as sodium hydroxide and soaps at high temperature, the bioscouring alternative can result in complete replacement of the scouring chemicals and as the enzymatic process is performed under lower temperature, which leads to substantial saving of cost for energy and waste water treatment [5, 9, 28]. Although EndoPG showed the optimal temperature at 50 °C, the bioscouring process was carried out at 40 °C at which the enzyme showed high catalytic activity with high operational stability (>80% of its maximal activity was retained after 60 min). The lower operating temperature in the scouring step also has advantage on a lower energy cost. Crude pectinases from various bacterial and fungal origins have been used for bioscouring of cottons and other types of textiles e.g., linen and ramie [2]. The use of purified or recombinant composite pectinases on bioscouring has not been previously reported. The recombinant pectinase reported here

Table 3 Comparison of weight loss and wettability of the fabrics scoured by enzyme (EndoPG) and alkali treatment (sodium hydroxide)

Sample	Weight loss (%)	Wettability (s)
Buffer	0.16	181.20
EndoPG (100 U/mL$_{scoured\ solution}$)	2.91	19.92
EndoPG (200 U/mL$_{scoured\ solution}$)	4.22	17.88
EndoPG (300 U/mL$_{scoured\ solution}$)	5.27	16.50
Sodium hydroxide	7.64	15.84

Table 4 Comparison of tensile strength between fabrics scoured by enzyme (EndoPG) and alkali treatment (sodium hydroxide)

Sample	Tensile strength (N)	
	Weft Yarn	Warp Yarn
Buffer	741.00	1042.33
EndoPG (300 U/mL$_{scoured\ solution}$)	722.33	986.67
Sodium hydroxide	713.33	918.67

was shown to be capable of enzymatic bioscouring with a maximal water adsorbancy of 16.5 s, which is in the accepted range for industrial processing (<30 s). However, application of either crude enzymes rich in polygalacturonase or pectate lyase activity alone has been reported to show incomplete scouring on the fabric in some cases [9, 29, 30]. Pectate lyases catalyze trans-elimination of a-1,4-glycosidic linkage in pectic acid and form products with 4,5-unsaturated residues at the non-reducing end. These structural changes increase more polar functional groups and hence the hydrophilicity on the surface of cotton fibers, which could be an important factor for the improved water absorption and dyeability of the treated fabrics. The results thus suggest further formulation of the EndoPG with pectate lyases or other composite pectinolytic enzymes as well as other enzymes e.g., cellulases and cutinases [6, 31] to improve its performance. The EndoPG catalyzed process resulted in fabrics with higher tensile strength compared with the chemical scouring method. This reflected the higher specificity of the enzyme on attacking the polygalacturonic acid backbone of pectin with no decomposition of the cellulose fibers, which is an undesirable side activity of alkali in chemical scouring [9, 28]. This resulted in fabrics with higher strength with lower weight loss from the enzymatic process.

Conclusion

The applicability of a recombinant cellulase-free polygalacturonase (*A. aculeatus* EndoPG) on bioscouring has been demonstrated. The enzyme can be produced efficiently by high-cell density fermentation in the *P. pastoris* system and showed its potential on completely replacing the use of toxic scouring chemical with substantial energy saving. The EndoPG characterized in this study represents a promising alternative for efficient bioscouring towards establishment of an eco-friendly textile processing industry.

Abbreviations
AOX: Alcohol oxidase; CDW: Cell dry weight; EndoPG: Endo-polygalacturonase

Acknowledgements
Manuscript proofreading by Dr. Philip James Shaw is appreciated.

Funding
This research was supported by the grant (P-13-50429) from the National Center for Genetic Engineering and Biotechnology, National Science and Technology Development Agency, Thailand.

Authors' contributions
DA and DC carried out plasmid construction, enzyme production and characterization. DA and KK participated in the high cell density fermentation. DA, PT, MN, BS, NN, and TN participated in enzymatic scouring experiment. AS, LE, TN, and VC conceived and designed of the study. DA, VC, TN and DC contributed to the analysis of the results and manuscript writing. All authors read and approved the final manuscript.

Competing interests
The authors declare that they have no competing interests.

Author details
[1]Faculty of Biotechnology, Atmajaya Catholic University, Jl. Jend. Sudirman 51, Jakarta 12930, Indonesia. [2]Microbial Biotechnology and Biochemicals Research Unit, National Center for Genetic Engineering and Biotechnology, 113 Thailand Science Park, Pahonyothin Rd, Khlong Luang, Patumthani 12120, Thailand. [3]Textile Laboratory, Polymers Research Unit, National Metal and Materials Technology Center, 114 Thailand Science Park, Pahonyothin Rd, Khlong Luang, Patumthani 12120, Thailand.

References
1. Li Y, Hardin IR. Enzymatic scouring of cotton-surfactants, agitation and selection of enzymes. Text Chem Color. 1998;30:23–9.
2. Kashyap DR, Vohra PK, Chopra S, Tewari R. Applications of pectinases in the commercial sector: a review. Bioresour Technol. 2001;77:215–27.
3. Lin CH, Hsieh YL. Direct scouring of greige cotton fabrics with proteases. Text Res J. 2001;71:425–34.
4. Buschle-Diller G, El Mohgahz Y, Inglesby MK, Zeronian SH. Effects of scouring with enzymes, organic solvents, and caustic soda on the properties of hydrogen peroxide bleached cotton yarn. Text Res J. 1998;68:920–9.
5. Etters JN. Cotton preparation with alkaline pectinase: an environmental advance. Text Chem Color Am D. 1999;1:33–6.
6. Karapinar E, Sariisik MO. Scouring of cotton with cellulases, pectinases and proteases. Fibres Text East Eur. 2004;12:79–82.
7. Collmer A, Reid J, Mount M. Assay methods for pectic enzymes. Method Enzymol. 1988;161:329–35.
8. Miller CA, Jorgensen SS, Otto EW, Lange MK, Condon B, J L. Alkaline enzyme scouring of cotton textiles. US Patent. US 5,912,407.
9. Tzanov T, Calafell M, Guebitz GM, Cavaco-Paulo A. Bio-preparation of cotton fabrics. Enzyme Microb Tech. 2001;29:357–62.
10. Nimchua T, Rattanaphan N, Champreda V, Pinmanee P, Kittikhun S, Eurwilaichitr L, et al. One-step process for desizing and scouring of cotton fabric by multienzyme. Thai Patent 2012. TH 1203000885.
11. Fuhrmann M, Hausherr A, Ferbitz L, Schödl T, Heitzer M, Hegemann P. Monitoring dynamic expression of nuclear genes in Chlamydomonas reinhardtii by using a synthetic luciferase reporter gene. Plant Mol Biol. 2004;55:869–81.
12. Villalobos A. Gene designer: a synthetic biology tool for constructing artificial DNA segments. BMC Bioinformatics. 2006;7:285.
13. Sambrook J, Russel D. Molecular cloning: A laboratory manual. 3rd ed. New York: Cold Spring Harbor Laboratory Press; 2007. p. 1.116–8.
14. Bradford MM. A rapid and sensitive method for the quantitation of microgram quantities of protein utilizing the principle of protein-dye binding. Anal Biochem. 1976;72:248–54.
15. Miller GL. Use of dinitrosalicylic acid reagent for determination of reducing sugars. Anal Chem. 1959;31:426–8.
16. Charoenrat T, Khumruaengsri N, Promdonkoy P, Rattanaphan N, Eurwilaichitr L, Tanapongpipat S, et al. Improvement of recombinant endoglucanase produced in *Pichia pastoris* KM71 through the use of synthetic medium for inoculum and pH control of proteolysis. J Biosci Bioeng. 2013;116:193–8.
17. Jahic M, Rotticci-Mulder J, Martinelle M, Hult K, Enfors SO. Modeling of growth and energy metabolism of *Pichia pastoris* producing a fusion protein. Bioprocess Biosyst Eng. 2002;24:385–93.
18. Ruanglek V, Sriprang R, Ratanaphan N, Tirawongsaroj P, Chantasigh D, Tanapongpipat S, et al. Cloning, expression, characterization, and high cell-density production of recombinant endo-1,4-β-xylanase from *Aspergillus niger* in *Pichia pastoris*. Enzyme Microbial Tech. 2007;41:19–25.
19. AATCC test Method 79: Absorbency of Bleached Textiles. AATCC Technical Manual. American Association of Textile Chemists and Colorists; 1995.
20. Liu MQ, Dai XJ, Bai LF, Xu X. Cloning, expression of *Aspergillus niger* JL-15 endo-polygalacturonase A gene in Pichia pastoris and oligo-galacturonates production. Protein Expres Purif. 2014;94:53–9.

21. Yang J, Luo H, Li J, Wang K, Cheng H, Bai Y, et al. Cloning, expression and characterization of an acidic endo-polygalacturonase from *Bispora* sp. MEY-1 and its potential application in juice clarification. Process Biochem. 2011;46:272–7.

22. Yan HZ, Liou RF. Cloning and analysis of *pppg1*, an inducible endo-polygalacturonase gene from the oomycete plant pathogen *Phytophthora parasitica*. Fungal Genet Biol. 2005;42:339–50.

23. Yuan P, Meng K, Huang H, Shi P, Luo H, Yang P, et al. A novel acidic and low-temperature-active endo-polygalacturonase from *Penicillium* sp. CGMCC 1669 with potential for application in apple juice clarification. Food Chem. 2011;129:1369–75.

24. Zhou H, Li X, Guo M, Xu Q, Cao Y, Qiao D, et al. Secretory expression and characterization of an acidic endo-polygalacturonase gene from *Aspergillus niger* SC323 in *Saccharomyces cerevisiae*. J Microbiol Biotechnol. 2015;25:999–1006.

25. Rattanachomsri U, Tanapongpipat S, Eurwilaichitr L, Champreda V. Simultaneous non-thermal saccharification of cassava pulp by multi-enzyme activity and ethanol fermentation by *Candida tropicalis*. J Biosci Bioeng. 2009;107:488–93.

26. Poonsrisawat A, Wanlapatit S, Paemanee A, Eurwilaichitr L, Piyachomkwan K, Champreda V. Viscosity reduction of cassava for very high gravity ethanol fermentation using cell wall degrading enzymes from *Aspergillus aculeatus*. Process Biochem. 2014;49:1950–7.

27. Thongekkaew J, Ikeda H, Masaki K, Iefujii H. An acidic and thermostable carboxymethyl cellulase from yeast Cryptococcus sp. S-2: purification, characterization and improvement of its recombinant enzyme production by high cell-density fermentation of Pichia pastoris. Protein Expres Purif. 2008;60:140–6.

28. Anis P, Eren HA. Comparison of alkaline scouring of cotton vs. alkaline pectinase preparation. AATCC Review. 2002;2:22–6.

29. Calafell M, Klung-Santner B, Guebitz G, Garria P. Dyeing behaviour of cotton fabric bioscoured with pectate lyase and polygalacturonase: a review. Color Tech. 2005;121:291–7.

30. Calafell M, Garriga P. Effect of some process parameters in the enzymatic scouring of cotton using an acid pectinase. Enzym Microb Technol. 2004;34:326–31.

31. Agrawal PB, Nierstrasz VA, Bouwhuis GH, Warmoeskerken MMCG. Cutinase and pectinase in cotton bioscouring: an innovative and fast bioscouring process. Biocatal and Biotransfor. 2008;26:412–21.

16

Selection of oleaginous yeasts for fatty acid production

Dennis Lamers[1,2*†], Nick van Biezen[1†], Dirk Martens[2], Linda Peters[1], Eric van de Zilver[1], Nicole Jacobs-van Dreumel[1], René H. Wijffels[2,3] and Christien Lokman[1]

Abstract

Background: Oleaginous yeast species are an alternative for the production of lipids or triacylglycerides (TAGs). These yeasts are usually non-pathogenic and able to store TAGs ranging from 20 % to 70 % of their cell mass depending on culture conditions. TAGs originating from oleaginous yeasts can be used as the so-called second generation biofuels, which are based on non-food competing "waste carbon sources".

Results: In this study the selection of potentially new interesting oleaginous yeast strains is described. Important selection criteria were: a broad maximum temperature and pH range for growth (robustness of the strain), a broad spectrum of carbon sources that can be metabolized (preferably including C-5 sugars), a high total fatty acid content in combination with a low glycogen content and genetic accessibility.

Conclusions: Based on these selection criteria, among 24 screened species, *Schwanniomyces occidentalis* (*Debaromyces occidentalis*) CBS2864 was selected as a promising strain for the production of high amounts of lipids.

Keywords: Oleaginous yeast, *Schwanniomyces occidentalis*, Lipid production, TAG

Background

Mineral and vegetable oil is a crucial resource for the modern human civilization, but the worldwide amount is depleting rapidly and alternatives need to be explored. An interesting alternative to decrease the dependency of western societies on these fossil and vegetable sources might be the use of oleaginous micro-organisms as described by various authors [1–3]. Lipids isolated from oleaginous micro-organisms can be used as components in coatings, paints, personal care products, production of fine chemicals and biodiesel thereby decreasing the dependency on vegetable and mineral oil [4–7]. Fatty Acid Methyl Esters originating from the lipids of oleaginous micro-organisms (e.g. algae, yeast and fungi) show identical fuelling properties compared to conventional diesel and could be used in modern cars without major adaptations [8]. At this moment the majority of biodiesel is produced from lipids, which are also used in the food chain and thus compete with food for agricultural land [7]. Therefore, oleaginous micro-organisms growing on agricultural waste residues are an attractive class of micro-organisms for lipid production.

An interesting class of oleaginous micro-organisms are yeasts. Oleaginous yeasts are able to store large quantities of TAGs in the form of lipid bodies in the cells. Typical lipid contents range from 20 % to 76 % depending species and culture conditions. Oleaginous yeasts strains studied today are e.g. *Yarrowia lipolytica*, *Candida 107*, *Rhodotorula glutinis*, *Rhodosporidium toruloides*, *Cryptococcus curvatus*, *Trichosporon pullulan* and *Lipomyces lipofer*. Screening studies are still performed, leading to the identification of several new oleaginous yeast species [1, 9–11]. Lipid accumulation is triggered by a nutrient limitation combined with an excess of carbon. Mostly nitrogen limitation is used to trigger lipid accumulation, but also other nutrients as phosphorus and sulphur have been shown to induce lipid accumulation [12–15]. Oleaginous yeasts should preferably be able to grow to high cell densities combined with a high fatty acid content, have good growth characteristics at low pH and a broad temperature range (robust

* Correspondence: dennis.lamers@han.nl

†Equal contributors

[1]HAN BioCentre, University of Applied Sciences, P.O. Box 6960, 6503 GL Nijmegen, The Netherlands

[2]Bioprocess Engineering, Wageningen University and Research Centre, P.O. Box 8129, 6700 EV Wageningen, The Netherlands

Full list of author information is available at the end of the article

process conditions), which facilitate the process development for future industrial applications. Furthermore, the ability to grow on a broad spectrum of carbon sources make oleaginous yeasts economically interesting.

The aim of this study is to find new yeasts that meet the aforementioned criteria and are potentially suited for fatty acid production for industrial applications. To this extent 24 non-*Saccharomyces* yeast species were selected and tested for the above mentioned criteria. Some of these selected strains have been described as having an oleaginous character [10, 16–20].

After selection for growth rate, lipid accumulation capacity, ability to use different carbon sources, pH and temperature optimum, *Schwanniomyces occidentalis* was selected as the most promising strain.

Results and discussion
Selection of strains by TLC analysis

From a private collection 24 yeast strains were selected to investigate their possible oleaginous character, where for 4 of these strains 2 variants were included, resulting in a total of 28 yeasts tested (Table 1). Generally, it is considered that lipid accumulation is induced at a molar C/N ratio greater than 20 [20]. Previously, it was shown that lipid accumulation in *R. toruloides* is observed at a C/N ratio of 30 and increases with an C/N ratio up to 120 using glucose as carbon source [21]. When growing *Y. lipolytica* on glucose at a C/N ratio of 50 a lipid content of 36 % is reached [22]. In *T. cutaneum* a slight increase in lipid content was reached when increasing the C/N ratio from 60 to 180, followed by a sharp decrease when the C/N ratio was further increased to 200 [23]. Furthermore, for *C. freyschussi* similar lipid content was reached at a C/N ratios of 52 and 100 whilst an increase to C/N 200 had a negative effect on lipid content [24]. Not only the C/N ratio but also type of the carbon and nitrogen sources used can have an impact on lipid production [22, 25]. Therefore, in this study screening for novel oleaginous yeasts was performed using medium with a C/N ratio of 75, without optimizing growth conditions for each individual strain, using glucose as carbon and ammonium chloride as nitrogen source. The strains listed in Table 1 were cultivated in C/N 75 medium for three days. Cell mass was harvested and dry weight content and triacylglyceride content was determined after saponification.

In Fig. 1 the fatty acid content after saponification of the different strains is visualised by thin layer chromatography (TLC) using oleic acid as a positive control. Since equal amounts of dry cell mass were used, the intensity of the spot represents the triacylglyceride content per gram dry weight. From the TLC analysis, 10 strains could be identified as strains with a high triacylglyceride content, viz.; *H. californica, P. anomala, T. delbrueckii, H. beijerinckii, C.*

Table 1 Strains used in this study

No.	Strain names	Culture collection
1	*Hansenula californica*	CBS 5760
2	*Candida glabrata*	CBS 2663
3	*Kluyveromyces phaffii*	CBS 4417
4	*Pichia angusta*	CBS 4732
5	*Torulopsis glabrata*	own collection/HBC14
6	*Pichia anomala*	CBS 5759
7	*Candida glabrata*	CBS 2192
8	*Candida lipolytica*	own collection/HBC08
9	*Torulopsis glabrata*	CBS 858
10	*Torulaspora delbrueckii*	own collection/HBC36
11	*Hansenula beijerinckii*	CBS 2564
12	*Candida tropicalis*	own collection/HBC07
13	*Yarrowia lipolytica*	CBS 6124
14	*Candida lusitaniae*	IFFI 01461
15	*Pichia silvicola*	CBS 1706
16	*Schwanniomyces occidentalis*	CBS 2864
17	*Sporobolomyces roseus*	CBS 2841
18	*Metschinikowia pulcherrima*	CBS 5534
19	*Candida bombicola*	ATCC 22214
20	*Candida intermedia*	CBS 572
21	*Candida tropicalis*	CBS 94
22	*Kloeckera africana*	CBS 277
23	*Pichia petersonii*	CBS 5556
24	*Lodderomyces elongisporus*	CBS 2605
25	*Kloeckera apiculata*	CBS 104
26	*Waltomyces lipofer*	CBS 5841
27	*Yarrowia lipolytica*	CBS 2073
28	*Cryptococcus curvatus*	CBS 570

tropicalis (12), S. occidentalis, L. elongisporus, W. lipofer, Y. lipolytica (27) and *C. curvatus.*

In Fig. 2 the final cell mass concentrations of the various strains are shown. Based on the results of Figs. 1 and 2, five strains were selected for further research, viz.; *P. anomala, T. delbrueckii, H. beijerinckii, S. occidentalis* and *W. lipofer.* Selection was based on a high content of triacylglycerides combined with a high cell mass concentration. The strain *L. elongisporus* met the criteria, but was not selected due to its suspected potential pathogenic character [26]. In addition based on literature *S. cerevisiae* was taken along as negative control.

Growth of selected strains at various temperatures

Strains used in large scale production processes should preferably be robust. Robustness of a strain is defined as the possibility to withstand process disturbances (e.g. temperature and pH variations), without having a large

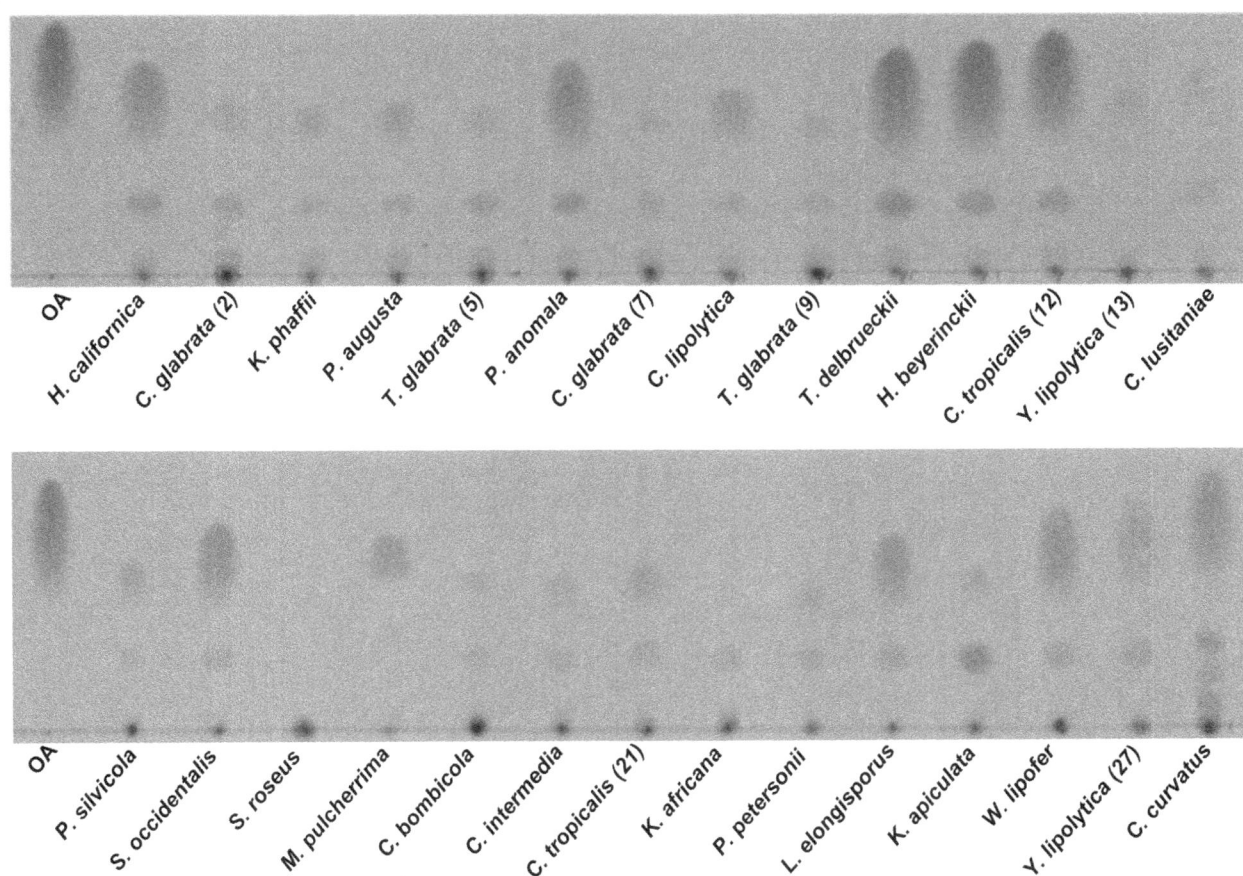

Fig. 1 TLC chromatogram of fatty acids isolated after saponification of 27 screened strains; Strains were grown in shake flasks containing 30 ml medium of a C/N ratio of 75 at 30 °C. After 72 h total lipids were extracted followed by TLC analysis. The black spots indicate the fatty acid content per sample. Oleic acid (OA) is used as a positive control ($C_{18:1}$). Each lane represents an equal amount of dry weight. When multiple variants of strains are used the number between brackets refers to the position of the strain variant in Table 1

influence on the productivity of the process. The effect of temperature (T) on cell growth was investigated for the selected yeast strains. A relatively broad T range at which the strains are still capable to grow without major changes in growth characteristics is desired in a large scale production process. In other words a shift in the temperature process control should have little effect on the final process. Furthermore, the ability to grow at higher temperatures is preferred to decrease the amount of cooling needed for cultivation [27].

Temperature profiles (Fig. 3) were obtained as described in the methods section. From these graphs two types of strains could be identified. Strains with a relatively narrow temperature range and strains having a broad temperature range in which growth is marginally influenced. Selecting the temperature area in which the growth rate is 80 % of the maximum value, it can be seen that strains *H. beijerinckii* and *S. cerevisiae* have a narrow optimum, with approximately a 5 °C bandwidth around the maximum growth. For all the other strains

this bandwidth ranges from 7 °C to 9 °C, which is almost twice as broad (see dashed rectangles in Fig. 3). Based on the temperature profiles *S. occidentalis*, *P. anomala*, *W. lipofer* and *T. delbrueckii* are the most robust of the strains tested.

Growth on different carbon sources

Production of bio-based materials should preferably be performed using renewable and low cost carbon sources (often harbouring C5-sugars e.g. xylose and arabinose). The results of the growth potential of all strains on various carbon sources is displayed in Table 2.

From Table 2 it can be concluded that from the strains tested *S. occidentalis* can grow on all carbon sources used in this study except cellulose and hemi-cellulose, which requires the action of multiple enzymes: exo,1,4-β-D-glucanase; endo,1,4-β-D-glucanase; cellobiohydolase and β-D-glucosidase, which are not commonly found in yeast [28, 29]. However, also some typical fungal characteristics can be observed in this strain like growth on

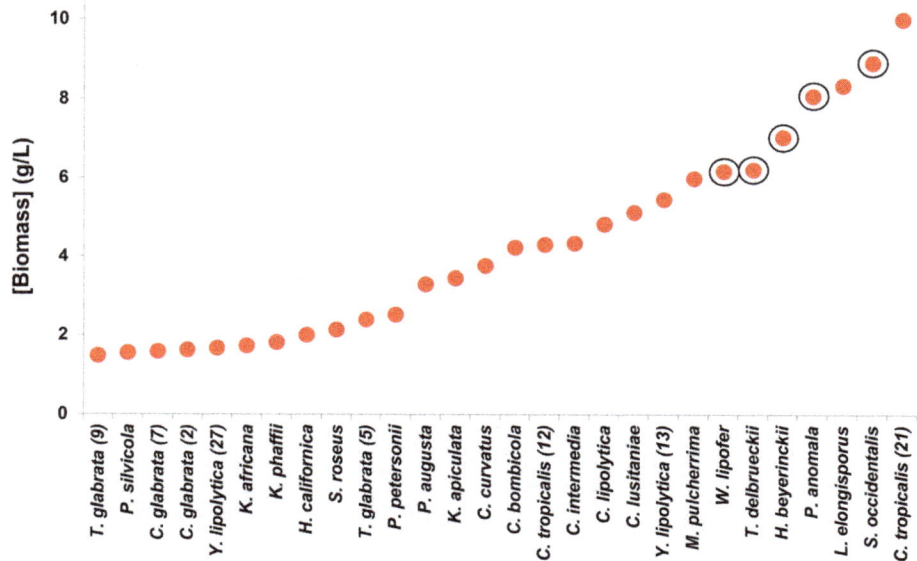

Fig. 2 Final biomass mass concentrations after three days of growth; Strains were grown in shake flasks containing 30 ml medium of a C/N ratio of 75 at 30°. After 72 h of growth the biomass concentrations were determined. Strains are ordered in increasing amounts of biomass concentration. When multiple variants of strains are used the number between brackets refers to the position of the strain variant in Table 1. Five promising fatty acid producing strains are indicated with a circle

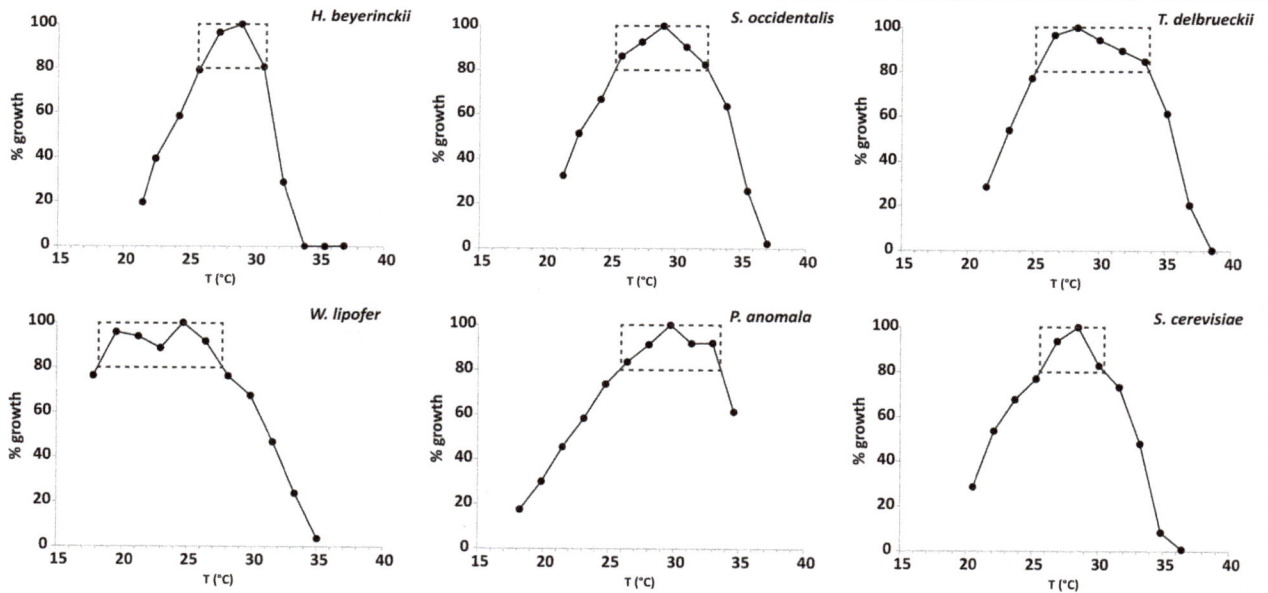

Fig. 3 Growth profiles at different temperatures; Strains were grown in static 2 ml cultures using medium with a C/N ratio of 5 at varying temperatures. Growth was determined by measuring biomass concentrations. The highest biomass concentration was defined as 100 %. The dashed rectangle indicates the temperature range in which the specific growth rate is higher than 80 % of the maximum value. Depicted strains are *H. beijerinckii, S. occidentalis, W. lipofer, P. anomala, T. delbrueckii* and *S. cerevisiae*

Table 2 Growth on agar plates containing different carbon sources

Carbon source	S. occidentalis	W. lipofer	H. beijerinckii	P. anomala	T. delbrueckii	S. cerevisiae
Glucose	+	+	+	+	+	+
Fructose	+	+	+	+	+	+
Galactose	+	+	+	+	+	+
Arabinose	+	+	-	-	-	-
Xylose	+	+	+	-	-	-
Glycerol	+	+	+	+	+	-
Sucrose	+	+	+	+	+	+
Lactose	+	-	-	-	-	-
Maltose	+	+	+	+	+	+
Cellobiose	+	+	+	+	-	-
Starch	+	-	-	-	-	-
Inulin	+	+	+	+	+	-
Xylan	-	-	-	-	-	-
Cellulose	-	-	-	-	-	-
No C source	-	-	-	-	-	-

starch that can only be achieved in the presence of glucoa-mylase activity, which is normally present in filamentous fungi and absent in yeasts. These findings are in line with the presence of glucoamylase activity which was already confirmed in *S. occidentalis* [30]. Striking was also the growth on cellobiose requiring β-glucosidase activity, which is part of the cellulolytic enzyme activity of fungi but not of that of yeasts. The genome of *S. occidentalis* was sequenced and assembled and the presence of a β-glucosidase gene having an identity of 77 % with the β-glucosidase sequence of *Scheffersomyces stipitis* CBS 6054 (XP_001387646) was confirmed using tblastn. Growth on glycerol was tested, since it is an abundantly available side product obtained from biodiesel production. All strains except *S. cerevisiae* were able to grow on gly-cerol. The results of carbon source utilization by *S. occi-dentalis* correspond with a recent study in which different oleaginous yeast species were screened for carbon source utilization and inhibitory tolerance in order to select yeasts suitable for specific industrial applications [31]. In the aforementioned study of Sitepu et al., *S. occidentalis* was found to be resistant to inhibitors at concentrations that are common in lignocellulosic hydrolysates (e.g. 2 g/l HMF, 1 g/l furfural and 2,5 g/l acetic acid) thereby indi-cating its potential to utilize these carbon sources for lipid production.

Genetic engineering of oleaginous yeasts is frequently used to expand substrate utilization and further increase lipid content and productivity. The oleaginous yeast *Y. lipolytica* is unable to utilize starch and by the combined expression of alpha-amylase and glucoamylase growth on starch led to a fatty acid accumulation of 21 % which was further increased to 27 % after media optimizations

[32]. Furthermore, Tai and Stephanopoulus report that co-expression of ACC1 and DGA1 increases fatty acid content from 8.77 % to 41.4 %% in *Y. lipolytica* [33]. A similar co-expression of ACC1 and DGA1 was per-formed in *R. toruloides* which increased lipid content from 31.3 % to 61.1 % [34]. Of the 5 strains tested only *S. occidentalis* is genetically accessible thereby indicating the potential to increase its fatty acid content, yield and carbon utilization [35, 36].

Analysis of yeast cell mass

The five selected strains and *S. cerevisiae* (negative control) were cultivated in media with different C/N ratios for three days. Cell mass was harvested and quan-titatively analysed for triacylglycerides content using gas chromatography (see Additional file 1 for fatty acid com-position of these strains). Results are shown as a func-tion of the C/N ratio of the medium in Fig. 4. Fatty acid per dry weight content of all strains, except *S. cerevisae*, increased with an increasing C/N ratio. The maximum lipid accumulation was 319 g/kg reached by *T. delbrueckii* at a C/N ratio of 75 whereas the minimum lipid accumu-lation was 152 g/kg reached by *P. anomala* at a C/N ratio of 90. Typical fatty acid content reached by oleaginous yeasts is higher and ranges from 360 g/kg for *Y. lipolytica*, 580 g/kg for *C. curvatus* to 720 g/kg for *R. glutinis* [37]. The lower fatty acid content in this study could be attrib-uted to the fact that cultivation conditions are not opti-mized. As previously demonstrated by Calvey et al. in *L. starkeyi* both the initial C/N ratio and the agitation rate can have an impact on lipid accumulation and by lowering the agitation rate from 300 rpm to 200 rpm lipid accumu-lation was increased from 28.43 % to 54.85 % [38].

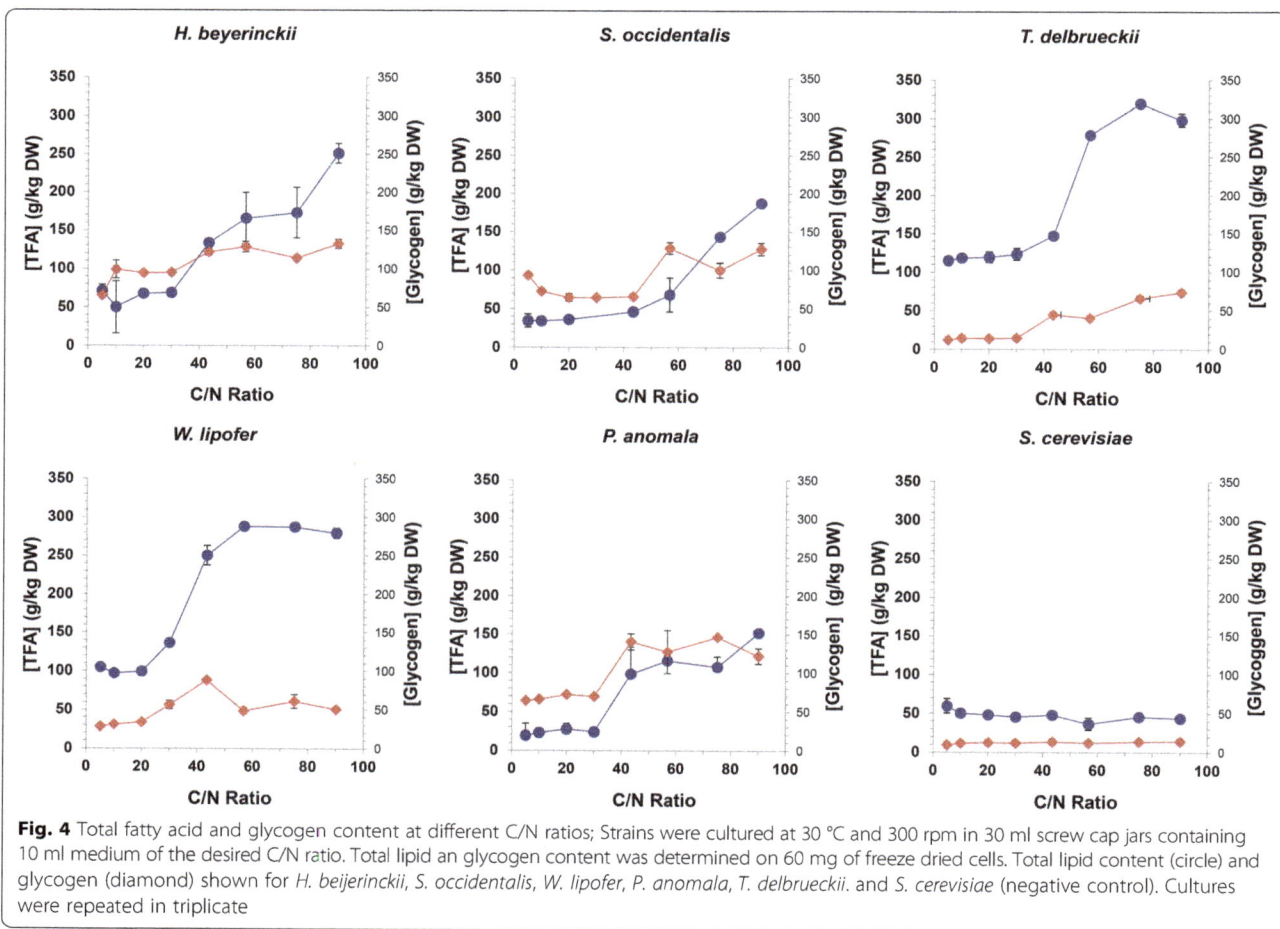

Fig. 4 Total fatty acid and glycogen content at different C/N ratios; Strains were cultured at 30 °C and 300 rpm in 30 ml screw cap jars containing 10 ml medium of the desired C/N ratio. Total lipid an glycogen content was determined on 60 mg of freeze dried cells. Total lipid content (circle) and glycogen (diamond) shown for *H. beijerinckii, S. occidentalis, W. lipofer, P. anomala, T. delbrueckii.* and *S. cerevisiae* (negative control). Cultures were repeated in triplicate

Furthermore, in *L. starkeyi* medium optimization while utilizing a mixture of glucose and xylose increased lipid content from 38.8 % to 61.5 % indicating that optimal lipid production can be achieved by further optimizations [39]. However, it has to be noted that comparison of triacylglyceride content with other studies described in literature is difficult, since they are highly dependent on used culture conditions and C/N ratios.

To analyse the overall potential of triacylglyceride production the glycogen content per cell mass was also measured (Fig. 4). Cells undergoing nitrogen starvation are able to channel their carbon into several compounds such as lipids and glycogen [40]. Glycogen production under nitrogen starvation indicates that lipid concentrations could potentially be increased by channeling carbon from glycogen to lipid production via metabolic engineering or by changing culture condition [41]. All strains tested were able to accumulate glycogen at different C/N ratio's. High glycogen contents were observed in *P. anomala, H. beyerinckii* and *S. occidentalis*, ranging from 125 g/kg to 146 g/kg. The *S. cerevisiae* strain contained low amounts of glycogen which is in line with the fact that in *S. cerevisiae*

production of trehalose is favoured over glycogen under nitrogen depletion [42].

Taken together all of the aforementioned criteria e.g. lipid content, growth on carbon sources, temperature and genetic accessibility we selected *S. occidentalis* as the most versatile strain.

A fed-batch fermentation of *S. occidentalis* was performed starting with an initial C/N ratio of 75. Maximal lipid content reached in the fed-batch fermentation was 41.9 %, lipid productivity was 0,083 g/l.h while DCW produced was 19,01 g/l in 96 h (Table 3).

In comparison the maximal lipid content in *S. occidentalis* of 41.9 % is higher than that of *Y. lipolytica* 36.73 % when grown in a fed-batch fermentation using glucose, however it has to be noted that the initial C/N ratio for *Y. lipolytica* was 50 [22]. Furthermore, the total amount of lipids produced by *L. starkeyi* on glucose at an initial C/N ratio of 72 is slightly higher than for *S. occidentalis* 8.00 g/l versus 10.03 g/l [38]. Both *R. glutinis* and *R. toruloides* surpass *S.occidentalis* both in lipid content (64.00 % and 58.60 % respectively) and in lipid productivity (0.950 g/l.h and 0.360 g/l.h respectively). Via medium optimization and genetic engineering both

Table 3 Productivity of fed-batch fermentations of oleaginous yeasts grown on glucose

Strain	Substrate	Initial C:N ratio	Culture time (h)	DCW produced (g/l)	Lipids produced (g/l)	Lipid %	Lipid yield (g/g sugar)	Lipid productivity (g/l/h)	DCW productivity (g/l/h)	Reference
S. occidentalis	Glucose	75	96	19.00	8.00	41.90	0.120	0.083	0.198	This study
L. starkeyi	Glucose	72		18.28	10.03	54.85	0.170	0.060	0.112	[39]
R. glutinis	Glucose		50	110.00	70.04	64.00		0.950	1.408	[41]
Y. lipolytica	Glucose	50	48	43.38	15.93	36.73	0.200	0.904	0.332	[22]
R. toruloides	Glucose				52.20	58.60	0.200	0.360		[45]

lipid content and productivity in *S. occidentalis* could be further increased [32, 39].

Growth characteristics at low pH

Strains growing at relatively low pH values are more interesting, since infection problems can be reduced significantly in a large scale production process as can be seen in the dairy industry [43]. For this reason we investigated if *S. occidentalis* was able to grow by performing batch fermentations at a pH ranging from 2.5 to 7.5. and determining biomass after 24 h of growth. The results demonstrate that *S. occidentalis* was able to grow in the pH range from 3.5 to 6.5. Optimal growth was observed at a pH range of 4.5 to 6.5 with a biomass concentration of 10.5 g/l to 12.7 g/l, whereas biomass concentrations at pH 3.5 where 8.0 g/l.

Conclusions

The aim of this study was to select strains with a high lipid production, broad temperature range for growth, the ability to use a wide variety of carbon sources and genetic accessibility. Selection based on these criteria resulted in the selection of *S. occidentalis* as the most promising strain for industrial applications due to its ability to grow at a broad temperature and pH range and the ability to utilize many different carbon sources. The fatty acid production was not optimized in this study and leaves room for further improvement by optimizing process conditions and via metabolic engineering.

Methods
Strains and media

Strains used in this study are described in Table 1. Strains were inoculated on YPD slants (1 % yeast extract, 2 peptone and 2 % glucose), grown at 30 °C and stored at 4 °C prior to use. A small amount of cells from the slants was resuspended in water, followed by centrifugation to remove any medium components. This cell suspension was used as an inoculum for growth experiments (typical seed rate was 0.01 %).

The composition of the C/N 75 medium (75 mol C/mol N) was: glucose.aq 33 g/l, NH_4Cl 0.139 g/l, yeast extract (Gistex LS from DSM, @ 10 % N) 1.5 g/l, KH_2PO_4 3.2 g/l, $MgSO_4.7H_2O$ 1.0 g/l. Glucose was sterilised

separately from the other medium ingredients (20 min, 121 °C). Filter sterilised biotin was added at a concentration of 0.02 mg/l. Adjustment of the C/N content of the medium was obtained by decreasing the NH_4Cl content. With this yeast extract medium a C/N ratio of 90 mol C/mol N could be obtained, without varying the yeast extract content.

Temperature gradient

Yeast strains were grown in static 2 ml cultures C/N 5 medium (addition of 9.7 g/l NH_4Cl), using a temperature gradient block, which applies a gradient from 20 to 40 °C. After two days of cultivation the OD_{600nm} was measured. OD_{600nm} values were corrected for the amount of evaporation, which was determined by weighing the culture tubes prior and after cultivation. The growth was expressed as relative value against the maximum obtained value.

Growth on various carbon sources

Agar plates were made from Yeast Nitrogen Base (YNB) medium (Roth art. no HP26.1) containing 2 % agar (Roth art. no 5210.1), which were supplemented with 2 % of the indicated carbon sources (see the results). Carbon sources were sterilised separately. Overnight cultures were diluted to an OD_{600nm} of 0.1 and 10 µl was spotted on the individual carbon source containing agar plates. Duplicate plates were incubated at 30 °C and checked daily for growth for 72 h. Growth in the spotted areas was analysed on plates with different carbon sources using plates lacking carbon source as a negative control.

C/N ratio experiments

10 ml of the desired C/N medium was inoculated with a washed suspension of cells (typical seed rate was 0.01 %) and grown in 30 ml plastic flat bottomed screw cap jars (VWR art. no. 216–2694, diameter 3 cm and height 7 cm) with a cotton wool stopper in the cap. The jars were incubated on a rotary shaker (type Innova, New Brunswick Scientific) set at 300 rpm. After three days, the cells were harvested, washed and freeze dried for further analysis.

Batch fermentations

Batch fermentations were performed in 7 l BioFlo115 fermenters using the C/N 5 medium. The pH was controlled with 2 M NaOH or 1 M H_3PO_4. During the fermentation process 4 ml samples were taken using an automatic sampler (Gilson art. no. F203B). Collected samples were cooled to 1 °C to quench the metabolism. Samples were used for dry weight determination and the presence of residual glucose was analysed on a Cobas Mira Plus using the Horiba ABX Pentra Glucose HK CP reagent (art. no. A11A01667).

Fed batch fermenation

A single colony was used to inoculate 100 ml of C/N 75 medium. After 24 h growth at 30 °C and 250 rpm agitation the inoculum was transferred into a 1,25 L fermenter.

Fed-batch fermentation was performed in 1 l BioFlo 115 fermenter using the C/N 75 medium. After depletion of the initial glucose concentration a glucose feed of 1,1 g/h was established. The pH was controlled with 2 M NaOH or 1 M H_3PO_4. During the fermentation process 4 ml samples were taken using an automatic sampler (Gilson art. no. F203B). Collected samples were cooled to 1 °C to quench the metabolism. Samples were used for dry weight determination and the presence of residual glucose was analysed on a Cobas Mira Plus using the Horiba ABX Pentra Glucose HK CP reagent (art. no. A11A01667).

Total lipids extraction and TLC analysis

500 µl of 50 % NaOH was added to a 2 ml cell suspension in capped glass tubes. Tubes were incubated overnight at 100 °C. After cooling down 1.1 ml of 37 % HCl was added to liquefy the soap. The mixture was extracted with 5 ml of ethyl acetate by vortexing, followed by centrifugation at 2000 g for 5 min. The top layer was transferred to another tube and the excess of ethyl acetate was evaporated by an air stream at 50 °C. After complete drying, 100 µl of ethyl acetate was added to dissolve the residue. To compare the strains, the residues were further diluted to obtain normalised concentrations according to their cell mass content. In total 30 mg of cells were used. Thin layer chromatography was used to separate and visualise the residues obtained from saponification. Diluted samples (10 µl) were loaded on TLC plates (MERCK art. no. code 1.05554.0001) and left at room temperature for drying. Plates were run in a pre-equilibrated TLC container with a mixture of hexane : ethyl acetate : acetic acid = 90 : 10 : 1 as the mobile phase. After 10 cm of front migration the plates were removed from the TLC container and air dried. To visualise the products the dried plates were sprayed with concentrated sulphuric acid : methanol = 1 : 1 and placed in an oven at 150 °C for approximately 30 min. Oleic acid was used as a control.

Fatty acid determination with gas chromatography

The total fatty acid content was measured according to a modified version of the method described by Kang and Wang (2005). To 30 mg of freeze dried cells in a capped test tube with screw cap (VWR art. no. SCERE5100160011G1 and SCERKSSR15415BY100) 1 ml BF_3/methanol reagent (Merck art. no. 8.01663.0500) and 1 ml heptane (Acros organics art. no. 120340025) was added. After overnight incubation at 70 °C 2 ml water was added and mixed. Subsequently, the tubes were centrifuged at 2000 g for 5 min, and the upper (heptane) layer was transferred to a GC vial and analysed on methylated fatty acids by using a Focus-GC (Interscience) equipped with FID. GC was equipped with a Stabilwax Column (Restek art. no. 10624) and uses hydrogen as carrier gas. The sum of the methylated fatty acids was quantified using methylheptadocanic acid as a standard.

Glycogen analysis

Glycogen was analysed according to a modified method as described by Aklujkar et al. [44]. To 30 mg of dry weight cells 500 µl of 2 M NaOH was added followed by boiling for 1 h. To neutralize this mixture, 60 µl of 9 M H_2SO_4 and 600 µl 1 M NaAc/HAc buffer (pH 4,5) were added. To 400 µl of sample, 50 units of glucoamylase (Novozymes art. no. NS22035) were added and incubated at 50 °C for 1 h. The final glucose concentration of the mixture was measured with and without amyloglucosidase treatment on a Cobas Mira Plus using the Horiba ABX Pentra Glucose HK CP reagent (art. no. A11A01667). Maltose was used as a positive control to check the activity of the amyloglucosidase.

Dry weight analysis

A culture sample of 10 ml was weighed on an analytical balance in a pre-weighed tube and centrifuged at 3500 g for 10 min. The cell pellet was washed with water, 20 % of the original volume, and centrifuged again (3500 g for 10 min.). The pellet was resuspended in 0.5 ml of water and frozen at −20 °C for 4 h. Cell pellets were dried by lyophilisation for 24 h on a Christ freeze dryer (type 2–4 LD) and dried pellets were weighed on an analytical balance. Biomass concentration is determined by dividing the weight of the dried biomass by the weight of the culture sample. After dry weight quantification, the freeze dried cells were used for total fatty acid and glycogen analysis.

Abbreviations

C/N ratio, molar ratio of carbon over nitrogen; FA, fatty acid; FID, flame ionization detector; GC, gas chromatography; T, temperature; TAGs, triacylglycerides; TLC, thin layer chromatography; YNB, yeast nitrogen base; YPD, yeast-petone-dextrose broth

Acknowledgements

The authors would like to thank Niek Klein for his technical assistance.

Funding

This research project was financially supported by HAN University of Applied Sciences and SIA.

Authors' contributions
DL, NvB, LP, EvdZ and NJvD performed all experiments. DL, NvB, LP, EvdZ, NJvD, DM drafted the manuscript. RW and CL revised the manuscript. All authors read and approved the final manuscript.

Competing interests
The authors declare that they have no competing interests.

Author details
[1]HAN BioCentre, University of Applied Sciences, P.O. Box 6960, 6503 GL Nijmegen, The Netherlands. [2]Bioprocess Engineering, Wageningen University and Research Centre, P.O. Box 8129, 6700 EV Wageningen, The Netherlands. [3]University of Nordland, Faculty of Biosciences and Aquaculture, N-8049 Bodø, Norway.

References
1. Ageitos JM, Vallejo JA, Veiga-Crespo P, Villa TG. Oily yeasts as oleaginous cell factories. Appl Microbiol Biotechnol. 2011;90:219–1227.
2. Ratledge C. Single Cell Oil. Enzyme Microb Technol. 1982;4:58–60.
3. Rattray JBM. Biotechnology and the fats and oils industry – an overview. JAOCS. 1984;61:1701–12.
4. Belgacem MN, Gandini A. Chapter 3 - Materials from Vegetable Oils: Major Sources, Properties and Applications. In Monomers, polymers and pomposites from renewable resources. First ed. Great Britain: Elsevier; 2008. pp. 39-66.
5. Beopoulos A, Verbeke J, Bordes F, Guicherd M, Bressy M, Marty A, Nicaud JM. Metabolic engineering for ricinoleic acid production in the oleaginous yeast Yarrowia lipolytica. Appl Microbiol Biotechnol. 2014;98:251–62.
6. Draaisma RB, Wijffels RH, Slegers PM, Brentner LB, Roy A, Barbosa MJ. Food commodities from microalgae. Curr Opin Biotechnol. 2013;24:167–77.
7. Luque R, Lovett JC, Datta B, Clancy J, Campelo JM, Romero AA. Biodiesel as feasible petrol fuel replacement: a multidisciplinary overview. Energy Enivron Sci. 2010;3:1706–21.
8. Yuan W, Hansen AC, Zhang Q. Vapor pressure and normal boiling point predictions for pure methyl esters and biodiesel fuels. Fuel. 2005;84:943–50.
9. Ratledge C. Regulation of lipid accumulation in oleaginous micro-organisms. Biochem Soc Trans. 2002;30:1047–50.
10. Ratledge C, Cohen Z. Microbial and algal oils: Do they have a future for biodiesel or as commodity oils ? Lipid Technol. 2008;20:155–61.
11. Sitepu IR, Sestric R, Ignatia L, Levin D, German JB, Gillies LA, Almada LA, Boundy-Mills KL. Manipulation of culture conditions alters lipid content and fatty acid profiles of a wide variety of known and new oleaginous yeast. Bioresour Technol. 2013;144:360–9.
12. Papanikolaou S, Aggelis G. Lipids of oleaginous yeasts. Part I:Biochemistry of single cell oil production. Eur J Lipid Sci Technol. 2011;113:1031–51.
13. Gill CO, Hall MJ, Ratledge C. Lipid accumulation in an oleaginous yeast (Candida 107) growing on glucose in single-stage continuous culture. Appl Environ Microbiol. 1977;2:231–9.
14. Wu S, Hu C, Jin G, Zhao X, Zhao ZK. Phosphate-limitation mediated lipid production by Rhodosporidium toruloides. Bioresour Technol. 2010;101:6124–9.
15. Wu S, Zhao X, Shen H, Wang Q, Zhao ZK. Microbial lipid production by Rhodosporidium toruloides under sulfate-limited conditions. Bioresour Technol. 2011;102:1803–7.
16. Beopoulos A, Nicaud JM, Gailardin C. An overview of lipid metabolism in yeasts and its impact on biotechnological processes. Appl Microbiol Biotechnol. 2011;90:1193–206.
17. Daniel HJ, Otto RT, Binder M, Reuss M, Syldatk C. Production of sophorolipids from whey: development of a two-stage process with Cryptococcus curvatus ATCC 20509 and Candida bombicola ATCC 22214 using deproteinized whey concentrates as substrates. Appl Microbiol Biotechnol. 1999;51:40–5.
18. Johnson EA. Biotechnology of non-Saccharomyces yeasts- the basidiomycetes. Appl Microbiol Biotechnol. 2013;97:7563–77.
19. Meesters P, Huijberts GNM, Eggink G. High cell density cultivation of the lipid accumulating yeast Cryptococcus curvatus using glycerol as a carbon source. Appl Microbiol Biotechnol. 1996;45:575–9.
20. Papanikolaou S, Aggelis G. Lipids of oleaginous yeasts. Part II:Biochemistry of single cell oil production. Eur J Lipid Sci Technol. 2011;113:1052–73.
21. Nicaud JM, Coq A, Rossignol T, Morin N. Protocols for monitoring growth and lipid accumulation in oleaginous yeasts. Berlin Heidelberg: Springer Protocols Handbooks. 2014; pp. 1–17.
22. Fontanille P, Kumar V, Christophe G, Nouaille R, Larroche C. Bioconversion of volatile fatty acids into lipids by the oleaginous yeast Yarrowia lipolytica. Bioresourc Techno. 2012;114:443–9.
23. Chen XF, Huang C, Yang XY, Xiong L, Chen XD, Ma LL. Evaluating the effect of medium composition and fermentation condition on the microbial oil production by Trichosporon cutaneum on corncob acid hydrolysate. Bioresourc Techno. 2013;143:18–24.
24. Raimondi S, Rossi M, Leonardi A, Bianchi M, Rinaldi T, Amaretti A. Getting lipids from glycerol: new perspectives on biotechnological exploitation of Candida freyschussii. Microbial Cell Factories. 2014;13:83.
25. Kolouchová I, Sigle K, Schreiberová O, Masák J, Řezanka T. New yeast-based approaches in production of palmitoleic acid. Bioresourc Techno. 2015;192: 726–34.
26. Bennet RJ. A Candida-based view of fungal sex and pathogenesis. Genome Biol. 2009;10:230.
27. Abdel-Fattah WR, Fadil M, Nigam P, Banat IM. Isolation of thermotolerant ethanologenic yeasts and use of selected strains in industr al scale fermentation in an Egyptian distillery. 2000;68:531–535.
28. Souza AC, Carvalho FP, Batista CF SE, Schwan RF, Dias DR. Sugarcane bagasse hydrolysis using yeast cellulolytic enzymes. J Microbiol Biotechnol. 2013;23:1403–12.
29. Wei H, Wang W, Alahuhta M, Vander Wall T, Baker JO, Taylor LE, Decker SR, Himmel ME, Zhang M. Engineering towards a complete heterologous cellulose secretome in Yarrowia lipolytica reveals its potential for consolidated bioprocessing. Biotechnol Biofuels. 2014;7:148.
30. Sato F, Okuyama M, Nakai H, Mori H, Kimura A, Chiba S. Glucoamylase originating from Schwanniomyces occidentalis is a typical alpha-glucosidase. Biosci Biotechnol Biochem. 2005;69:1905–13.
31. Sitepu I, Selby T, Lin T, Zhu S, Boundy-Mills K. Carbon source utilization and inhibitor tolerance of 45 oleaginous yeast species. J Ind Microbio Biotechnol. 2014;41:1061–70.
32. Ledesma-Amaro R, Dulermo T, Nicaud JM. Engineering Yarrowia lipolytica to produce lipidbase from raw starch. Biotechnol Biofuels. 2015;8:148.
33. Tai M, Stephanopoulos G. Engineering the push and pull of lipid biosynthesis in oleaginous yeast Yarrowia lipolytica for biofuel production. Metab Eng. 2013;15:1–9.
34. Zhang S, Skerker J, Rutter C, Maurer M, Arkin A, Rao C. Engineering Rhodosporidium toruloides for increased lipid production. Biotechnol Bioeng. 2015;112:1056–66.
35. Wang T, Lee C, Lee B. The molecular biology of Schwanniomyces occidentalis klocker. Crit Rev Biotechnol. 1999;19:113–43.
36. Alvaro-Benito M, Fernandez-Lobato M, Baronian K, Kunze G. Assessment of Schwanniomyces occidentalis as a host for protein production using the wide-range Xplor2 expression platform. Appl Microbiol Biotechnol. 2013;97:4443–56.
37. Ratledge C, Wynn J. The biochemistry and molecular biology of lipid accumulation in oleaginous microorganisms. Adv Appl Microbiol. 2002; 51:1–51.
38. Calvey C, Su Y, Willis L, McGee M, Jeffries T. Nitrogen limitation, oxygen limitation, and lipid accumulation in Lipomyces starkeyi. Bioresourc Technol. 2016;200:780–8.
39. Zhao X, Kong X, Hua Y, Feng B, Zhao Z. Medium optimization for lipid production through co-fermentation of glucose and xylose by the oleaginous yeast Lipomyces starkeyi. Eur J Lipid Sci Tech. 2008;110:405–12.
40. Cescut J, Fillaudeau L, Molina-Jouve C, Uribelarrea J. Carbon accumulation in Rhodotorula glutinis induced by nitrogen limitation. Biotechnol Biofuels. 2014;7:164.
41. Cupertino FB, Freitas FZ, de Paula RM, Bertolini MC. Ambient pH controls glycogen levels by regulation of glycogen synthase gene expression in Neurospora crassa. New insights into the pH signaling pathway. PLoS One. 2012;7:e44258.
42. Hazelwood LA, Walsh MC, Luttik MA, Daran-Lapujade P, Pronk JT, Daran JM. Identity of the growth-limiting nutrient strongly affects storage carbohydrate accumulation in anaerobic chemostat cultures of Saccharomyces cerevisiae. Appl Enivron Microbiol. 2009;75:6876–85.
43. Suriyarachchi VR, Fleet GH. Occurrence and growth of yeasts in yoghurts. Appl Environ Microbiol. 1981;42:574–9.
44. Aklujkar PP, Sank SN, Arvindekar AU. A simplified method for the isolation and estimation of cell wall bound glycogen in Saccharomyces cerevisiae. J Inst Brew. 2008;114:205–8.

17

Oligomerization triggered by foldon: a simple method to enhance the catalytic efficiency of lichenase and xylanase

Xinzhe Wang, Huihua Ge, Dandan Zhang, Shuyu Wu and Guangya Zhang[*] (iD)

Abstract

Background: Effective and simple methods that lead to higher enzymatic efficiencies are highly sough. Here we proposed a foldon-triggered trimerization of the target enzymes with significantly improved catalytic performances by fusing a foldon domain at the C-terminus of the enzymes via elastin-like polypeptides (ELPs). The foldon domain comprises 27 residues and can forms trimers with high stability.

Results: Lichenase and xylanase can hydrolyze lichenan and xylan to produce value added products and biofuels, and they have great potentials as biotechnological tools in various industrial applications. We took them as the examples and compared the kinetic parameters of the engineered trimeric enzymes to those of the monomeric and wild type ones. When compared with the monomeric ones, the catalytic efficiency (k_{cat}/K_m) of the trimeric lichenase and xylanase increased 4.2- and 3.9- fold. The catalytic constant (k_{cat}) of the trimeric lichenase and xylanase increased 1.8- fold and 5.0- fold than their corresponding wild-type counterparts. Also, the specific activities of trimeric lichenase and xylanase increased by 149% and 94% than those of the monomeric ones. Besides, the recovery of the lichenase and xylanase activities increased by 12.4% and 6.1% during the purification process using ELPs as the non-chromatographic tag. The possible reason is the foldon domain can reduce the transition temperature of the ELPs.

Conclusion: The trimeric lichenase and xylanase induced by foldon have advantages in the catalytic performances. Besides, they were easier to purify with increased purification fold and decreased the loss of activities compared to their corresponding monomeric ones. Trimerizing of the target enzymes triggered by the foldon domain could improve their activities and facilitate the purification, which represents a simple and effective enzyme-engineering tool. It should have exciting potentials both in industrial and laboratory scales.

Keywords: Foldon, Oligomerization, Non-chromatographic purification, Catalytic efficiency, Elastin-like polypeptides, Enzyme engineering

Background

Enzymatic catalysis played a significant role in industry and laboratory, especially in enzymatic hydrolysis of lignocellulose to produce fuel-grade ethanol. It was an attractive opportunity for producing renewable and environmentally friendly biofuels [1]. Within this context, since the limited catalytic performance in the reaction process, many studies on the functional characteristics such as substrate affinity, high catalytic properties received extensive attention [2]. Presently, researchers have proposed some effective approaches to improve the catalytic performance of the enzymes. They included the site-directed mutagenesis and directed evolution, which has successfully produced enzymes with optimized features such as activity, thermal stability and substrate specificity etc. Sometimes, immobilization and chemical modification of the target enzymes could also achieve the goal [3].

The scale of improved enzymes produced by mutants and others methods existed is large. For example, Zhang [4] and coworkers revealed a series of xylanases mutants which displayed 35–45% decrease in K_m and 75–105% increase in k_{cat} and leading to an approximately 200% increase in catalytic efficiencies by directed evolution.

* Correspondence: zhgyghh@hqu.edu.cn
Fujian Provincial Key Laboratory of Biochemical Technology, Huaqiao University, Xiamen, Fujian 361021, China

And, mutating Asn to Asp at position 35 adjacent to Glu172 enhanced the catalytic activity of xylanase from *Bacillus circulans* [5]. Indeed, those approaches achieve great success for improving the enzyme activities to varying degrees. However, there are still several drawbacks as described below: (1). The process of directed evolution must be iterated until the desired change is reached, or until no further change is elicited iteratively for at least 2- rounds. It needs a straightforward and efficient high-throughput screening method [6, 7]; (2). Site-directed mutagens through rational approaches should base on structural analysis. It could not be achieved without the well-known of the relationship between crystal structure and functional amino acid residues [8]; (3). Chemical modification is that of covalent attachment of special groups of modifiers to the side-chain group of certain residues in the enzyme. This method is often in severe reaction conditions and may cause unexpected loss of enzymatic activity by alteration of the active conformation or essential residues in the active sites [9].

Our purpose is to develop a convenient and efficient enzyme engineering method to improve the catalytic activities based on trimerizing the target enzymes. Oligomerization is a general way for many proteins who self-associate into oligomers to gain functional advantages [10]. The subunit assembly induced by the domains such as collagen triple helices and the obligatory oligomers like COMP and foldon usually results in improving thermostability [11, 12]. Foldon was a small 27-residue (GYIPEAPRDGQAYVRKDGEWVLLSTFL) β-propeller like trimer consisting of monomeric β-hairpin segments, which was originally identified at the C-terminus of bacteriophage T4 fibritin [13]. By gene fusion, this domain may be artificially linked to target enzymes to change their properties. Thermodynamic stability of several engineered proteins such as short collagen fibers [14, 15], HIV1 envelope glycoprotein has been enhanced by means of attachment of the foldon domain [16, 17].

Here, we fused foldon at the C-terminus of the elastin-like polypeptides-lichenase (defined as monomeric lichenase) and elastin-like polypeptides-xylanase (defined as monomeric xylanase) to induce these monomeric enzymes forming trimeric enzymes, respectively. The insertion of the two domain was expected to make purifying the recombinant proteins more effective and trigger some improvement of the catalytic properties. We found the trimeric lichenase and xylanase showed superior kinetic parameters and improved catalytic activities over their corresponding monomeric and wild-type counterparts. Meanwhile, the trimeric lichenase and xylanase could improve the activity recovery during the process of non-chromatographic purification by elastin-like polypeptides (ELPs). ELPs are stimulus-responsive polymers

consist of repeating pentapeptide (Gly-Xxx-Gly-Val-Pro), where X represents any amino acid except proline [18]. As a purification tag, ELPs were used to purify recombinant proteins and peptides without chromatography through undergoing an inverse transition cycling (ITC) within a narrow temperature range (2 ~ 3 °C) in aqueous solution [19]. The inverse phase transition can also be isothermally triggered by adding salt, which is a promising method both inexpensive and simple [20–22]. Also, the ELPs is in the random coil state when the target enzyme catalyzed the substrate, thus lessening the potentially unfavorable effects of the foldon domain on the active sites.

Results

Expression and purification of the recombinant lichenase and xylanase

We successfully expressed the monomeric and trimeric genes, purified the target enzymes and evaluated their purities by gel electrophoresis. For the monomeric ones, SDS-PAGE yield one band of 43 kDa and 39 kDa, denoting the monomeric lichenase (B-E) and xylanase (X-E) with ELPs tag, respectively (Fig. 1a,c). Quantity calculating results demonstrated the monomeric lichenase and xylanase comprised about 98% and 99% of the total soluble proteins after purification. Besides, the precise molecular weights (MWs) of purified B-E and X-E were further determined by MALDI-TOF mass spectrometry (MS). They presented with MWs of 42,696.3 Da and 39,562.6 Da respectively (Fig. 1b,d), which matched their theoretical values at 42262.5 Da and 39,509.3 Da calculated by ProtParam (http://web.expasy.org/protparam/). As for the trimeric lichenase, SDS-PAGE yield one band of 137 kDa, it comprised about 95% of the total proteins (Fig. 1e). The precise MW of its monomeric constituent determined by MALDI-TOF MS was 45,628.5 Da, which was 2932.2 Da (about 3 kDa, the MW of the foldon) more than the MW of the B-E. It was exactly the MW of one subunit of foldon (Fig. 1f). When it comes to the trimeric xylanases, SDS-PAGE yield three close bands ranging from 110 kDa to 130 kDa, they comprised about 98% of the total proteins (Fig. 1g). The precise MW of it was 42,664.0 Da, which was also 3.1 kDa more than X-E, standing for one subunit of the foldon (Fig. 1h). The reason for three close bands in the SDS-PAGE might be the sample was not heated enough before loading on the gel and some of them refolded. However, we are sure they were all trimeric xylanases. Because we only detected one molecule weight (42,664.0 Da) by the MALDI-TOF MS, which was exactly the MW of the monomer of the X-E-F, indicating there were no covalent bonds between monomers.

To illustrate the thermal stability of the trimeric enzymes, we pretreated them with different temperatures

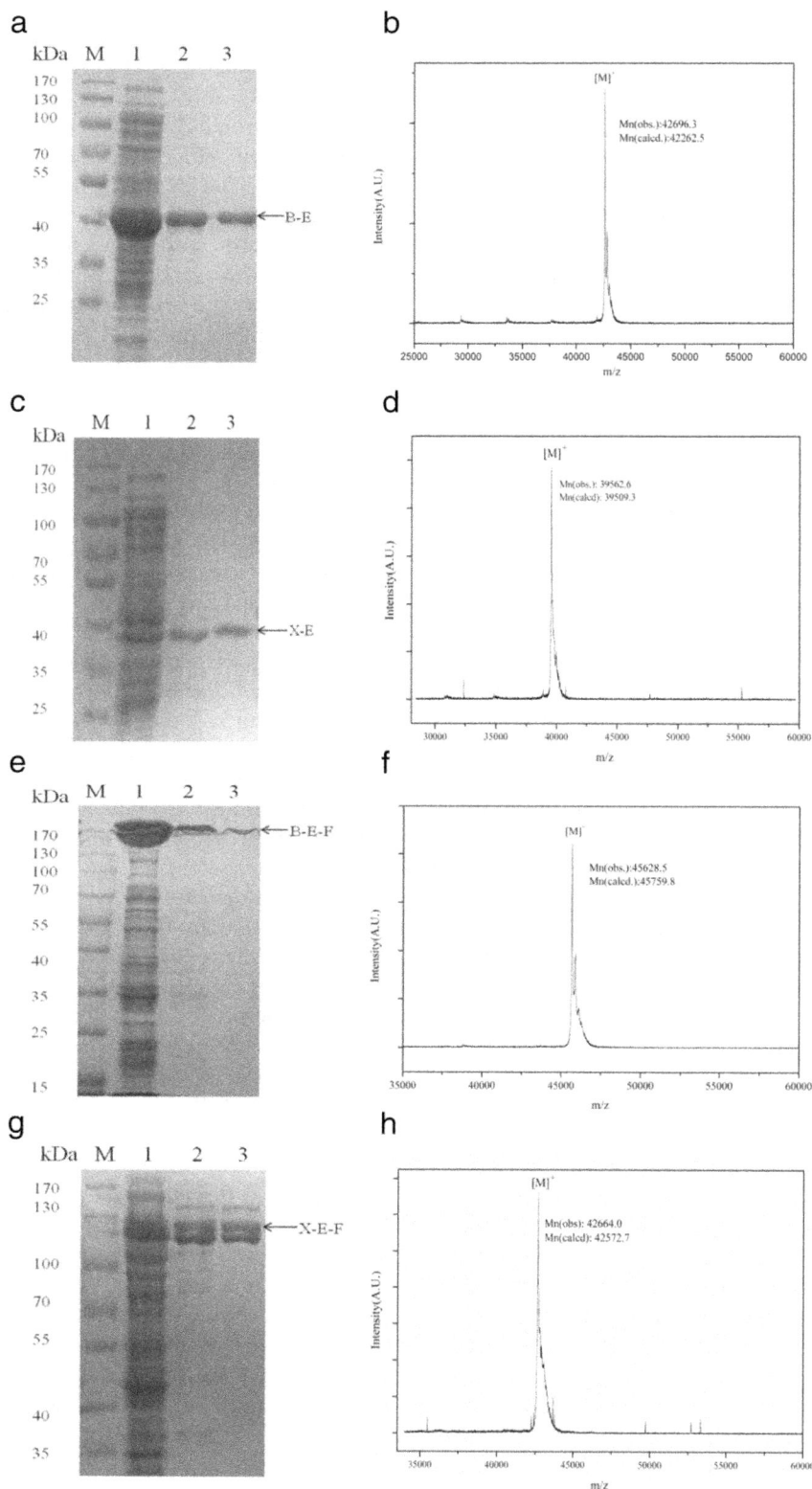

Fig. 1 Purification and analyzed the target enzymes by SDS-PAGE and MALDI-TOF. Monomeric enzymes B-E **a**, **b** and X-E **c**, **d**; trimeric enzymes B-E-F **e**, **f** and X-E-F **g**, **h** Lane M: marker; lane1 the lysate; lane 2 supernatant of first-round ITC, lane 3 supernatant of second-round ITC

(ranging from 25 °C to 100 °C) in loading buffer as reported [23] and then evaluated by gel electrophoresis. As Fig. 2a shows, when we pretreated the trimeric lichenase at the temperature of 25 °C, SDS-PAGE yielded one band of 137 kDa. It was about 3 times of the MWs of the monomer (46 kDa), indicating the trimeric lichenase remains folded. However, as the temperature changes from 40 °C to 70 °C, the trimeric lichenase began to unfolded partially, and thus the SDS-PAGE yielded two bands. Of them, one was the trimeric one and the other was the monomeric one with the MWs about 46 kDa. When the temperature is 70 °C or above, the SDS-PAGE yielded one band of 46 kDa, indicating the trimeric lichenase unfolded to the monomers. Besides, we also observed the unfolding monomers refolded into trimer after a rapid cooling in spite of withstanding the high temperature of 100 °C. Similar trends also existed in the trimeric xylanase as shown in Fig. 2b. However, there was one thing needed to point out. The SDS-PAGE yields three close bands when the trimeric xylanase was pretreated under the temperature of 40 °C, it yields two close bands under the temperature ranging from 50 °C to 65 °C, and it finally yields 1 bands when we pretreated the trimeric xylanase above 70 °C. These results proved the three bands in lane 1 and 2 were both the trimeric xylanase, which agreed with the results mentioned above.

We also compared the specific activity, purification fold and the activity recovery of the trimeric and monomeric enzymes during the ITC purification process and listed the results in Table 1. From it, we can see the trimeric enzymes are easier to purify with increased purification folds and activity recovery. For example, the purification fold of B-E-F and X-E-F was 7.4 and 12.5, which is 1.7 and 2.1 times of the monomeric ones. Meanwhile, the activity recovery increased by 12.5% and 6.2%, respectively. As shown in Table 1, the specific activity of the trimeric lichenase and

Table 1 Purification performance of the monomeric and trimeric enzymes

ITC purification		Specific activity(U/mg)	Purification fold	Recovery of activity (%)
B-E-F	Crude	11.9 ± 0.5	7.4	68.7 ± 2.2
	Purified	88.2 ± 3.5		
B-E	Crude	8.1 ± 0.8	4.4	56.2 ± 1.4
	Purified	35.4 ± 0.9		
X-E-F	Crude	41.4 ± 0.1	12.5	66.3 ± 1.4
	Purified	516.0 ± 10.3		
X-E	Crude	45.4 ± 0.1	5.8	60.2 ± 2.8
	purified	265.2 ± 3.9		

xylanase was 88.2 ± 3.5 and 516.0 ± 10.3 U/mg, which enhanced 149% and 94% than their corresponding monomer ones. These results demonstrated the foldon domain was beneficial to the purification of the target enzyme when it fused at the C-terminus of the ELPs tag. Considering the ELPs system can facilitate purifying enzymes (proteins) from the cell lysate and eliminates the need for expensive chromatography, our method may have great potentials in the recombinant enzyme (protein) bio-separation technology ranging from laboratory to manufacturing scales applications [24].

Optimal temperature and pH of the trimeric and monomeric enzymes

To evaluate the optimal temperature and pH profiles, we assayed the activities of the recombinant enzymes at various temperatures ranging from 30 °C to 65 °C, or pH ranging from 5.0 to 8.0. The relative activity was calculated by comparing to the highest enzyme activity (defined as 100%) to assure the optimal condition. For B-E-F and B-E, they have consistent optimal pH at 6.4–7.0 but the different optimal temperature of 40 °C and 55 °C, respectively (Fig. 3a,c), while the optimum pH

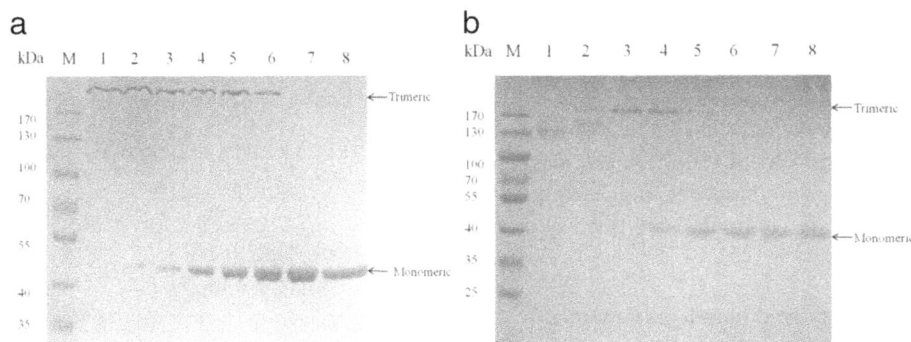

Fig. 2 The influences of temperature on the oligomeric state of lichenase and xylanase. The samples of trimeric lichenase **a** xylanase **b** were heated for 5 min at the designated temperatures prior to loading on the gel. The temperature was set to 25 °C (lane 1), 40 °C (lane 2), 50 °C (lane 3), 55 °C (lane 4), 60 °C (lane 5), 65 °C (lane 6), 70 °C (lane 7) and 100 °C (lane 8), respectively

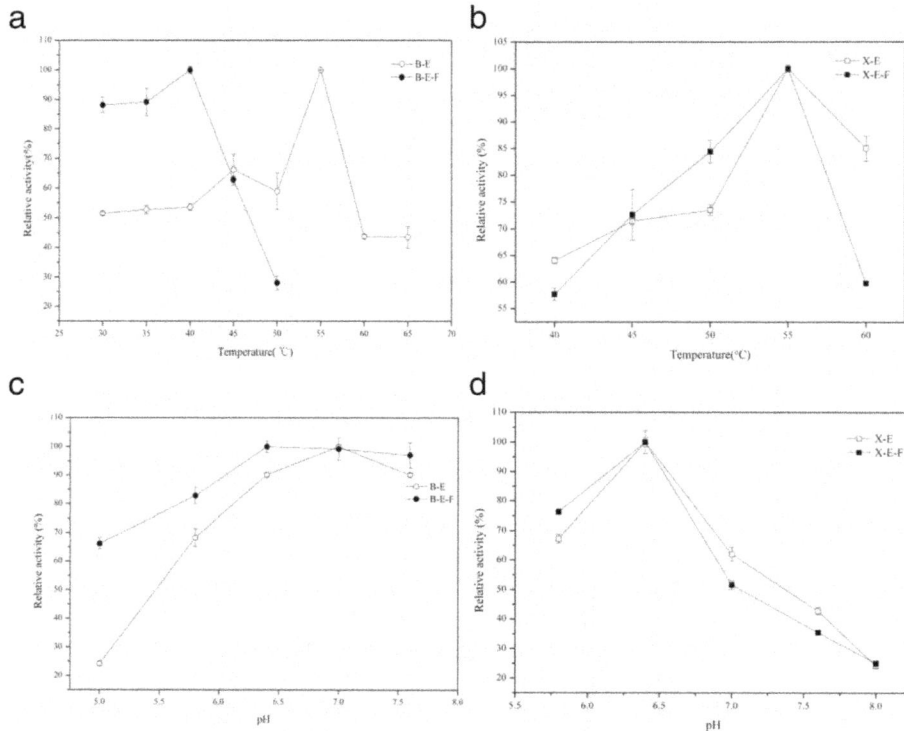

Fig. 3 The influences of temperature and pH on the relative activity of lichenase and xylanase. The influences of temperature on recombinant monomeric and trimeric lichenase **a** and xylanase **b** and the influences of pH on the relative activity of recombinant monomeric and trimeric lichenase **c** and xylanase **d** The error *bars* indicate the standard error of the mean (SEM)

and temperature of the wild type were 7.0 and 50 °C. For X-E-F and X-E, they show the same pH and temperature optimum of 6.4 and 55 °C (Fig. 3b,d), respectively. They were very similar with those of the wild-type xylanases, which had the optimal pH and temperature of 7.0 and 50 °C, respectively. For the thermostability, we assayed the residual activity of lichenase after incubated at the optimum condition for 4 h. The trimeric one remained about 65% of the relative activity, while the monomeric only 40%. In addition, the relative activity of monomeric xylanases decreased faster than the trimeric ones. Notably, the acid resistance of B-E-F was sharply improved. It remained 65% of the relative activity at pH 5.0, while B-E only remained 20%. Similarly, as previously reported, better pH resistance of GFP was obtained by formation of tunable dimerization, suggesting that protein conformations, in general, may be used to alter other protein properties [25].

Trimeric enzymes with improved kinetic parameters

We measured the kinetic parameters of the recombinant enzymes at their own optimal temperature and pH (shown in Table 2.) and compared them with the wild type ones previously reported [26]. As our sequence

came from it and we synthesized the gene coding the identical lichenase and xylanase with the reference. The K_m values of the B-E-F and X-E-F for glucan and birchwood xylan are 61.1 ± 2.8 mmol/L and 79.5 ± 4.0 mmol/L, which decreases 1.4- and 1.7- fold respectively when compared with their corresponding monomeric ones. This indicates the trimeric enzymes have the better affinity to their specific substrates than the monomeric ones. Besides, the catalytic constant (k_{cat}) of B-E-F is 361.2 ± 24.0 s^{-1}, which increases 3.0- and 1.8- fold according to their monomeric and wild-type lichenase. The k_{cat} of X-E-F is 870.1 ± 8.9 s^{-1}, which increases 2.3- and 5.0- fold according to their monomeric and wild-type xylanase, respectively. As mentioned above, the

Table 2 Kinetic parameter of trimeric, monomeric and wild-type enzymes

		K_{cat} (s^{-1})	K_m (mmol/L)	K_{cat}/K_m (L/mmol/s)
Lichenase	Wild type	197.2 ± 8.5	18.8 ± 1.8	10.5 ± 0.5
	Trimeric	361.2 ± 24.0	61.1 ± 2.8	5.9 ± 0.4
	Monomeric	119.7 ± 8.8	85.6 ± 4.0	1.4 ± 0.1
Xylanase	Wild type	173.7 ± 8.3	51.8 ± 4.6	3.4 ± 0.2
	Trimeric	870.1 ± 8.9	79.5 ± 4.0	10.9 ± 0.4
	Monomeric	379.8 ± 12.1	134.2 ± 14.7	2.8 ± 0.1

foldon domain in the trimeric lichenase and xylanase were partially unfolded at the temperature ranging from 45 °C to 60 °C. To compare the catalytic properties changed by oligomerization, we measured the catalytic efficiency (k_{cat}/K_m) over temperatures ranging from 30 to 60 °C. As Fig. 4 showed, the trimeric enzymes displayed distinguished catalytic efficiency (k_{cat}/K_m) than those of their corresponding monomeric ones in the measured temperature range. What needs to point out is the k_{cat} of the monomeric xylanase is higher (2.2 times) than the wild-type enzyme. Similar reports also existed in other researcher's results. For example, Yang and coworkers fused an oligopeptide to the N-terminus of an alkaline- amylase, the specific activity and catalytic constant (k_{cat}) of AmyK-p1 increased by 4.1- and 3.5-fold, respectively. They thought the main reason for the improved catalytic efficiency and the specific activity is the greater flexibility around the active site induced by the oligopeptide [27, 28]. Here, the catalytic efficiency (k_{cat}/K_m) of the monomeric xylanases is lower than the wild type although the values are quite similar, this is due to the increased K_m of the monomeric xylanases. For lichenase, the catalytic constant (k_{cat}) of the wild-type is higher than the monomeric although the values are quite similar. However, the catalytic efficiency of the monomeric lichenase is lower than the wild-type, this is also because of the increased K_m of the monomeric lichenase. The decrease of the substrate affinity of the engineered enzymes might relate to the fact the designed repeated pentapeptides were larger than the frequently used tag like histidine. It may have potential steric hindrance when the target enzyme binds the substrate.

Discussions

Enzyme engineering has become a common strategy to optimize the catalytic and biophysical properties of enzymes, and the improved performance makes biocatalysts attractive in both the laboratory and industrial processes. Directed evolution and site-directed mutagenesis techniques have been used with great success for diversifying gene sequences and optimizing enzyme phenotypes. Most of the engineered enzymes are those with improved biophysical properties, such as improved stability to high temperature or harsh pH, few of them with increased catalytic activities. For example, Zhang and coworkers conducted site-directed mutagenesis of β-1, 4-endoglucanase based on a rational design and obtained increased activity of 77% -87%. The V_{max} and K_m of the triple-sites mutant were 4.23 ± 0.15 µmol mg^{-1} min^{-1} and 1.97 ± 0.05 mM, respectively, about 2 times higher than those of the initial enzyme [29]. Chemical modification of the target enzyme is another effective way to increase the catalytic activities of some thermophilic enzymes. For instance, Lys modification of α-amylase from *Bacillus licheniformis* led to a 15-fold increase in activity of BLA at 15 °C with a 3-fold increase in k_{cat}/K_m at 37 °C [30]. Besides, Lin et al. obtained a mutant Q51H of glutamate decarboxylase with higher k_{cat} and k_{cat}/K_m of 47.67 ± 3.18 s^{-1} and 2.05 ± 0.09 s$^{-1} \cdot$ (mmol·L^{-1})$^{-1}$ compared to 18.06 ± 1.70 s^{-1} and 0.80 ± 0.06 s$^{-1} \cdot$(mmol· L^{-1})$^{-1}$ of the parental, respectively [31]. According to the previous results, our method is comparable with those conducted with enzyme mutagenesis or modification. However, our method is in mild condition, simple to operate and do not need to screen target enzymes from the mutant libraries. It should be a simple enzyme engineering method or at least be an effective complement to the existing methods.

Many researchers have illustrated subunit oligomerization motifs play an important role in protein function. Such as the parallel five-stranded coiled coil COMP, the obligatory dimerization POZ, the foldon domain and so on [11, 12]. For example, the thermal hysteresis activity of the antifreeze proteins (AFPs) was

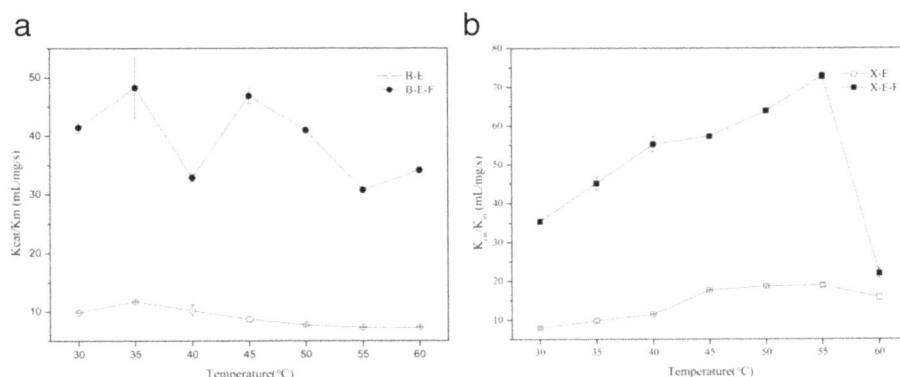

Fig. 4 The k_{cat}/K_m of lichenase and xylanase at different temperatures. K_{cat}/K_m of monomeric and trimeric lichenase **a** and xylanase **b** at various temperature from 30 to 60 °C, samples were performed at 5 °C intervals. The error bars indicate the standard error of the mean (SEM)

increased significantly, reaching a thermal hysteresis of >1.6 °C at the concentration of <1 mM [32]. Another example is the melting temperature (Tm) of a collagen-like peptide increased to around 75 °C from 24 °C when attaching a foldon domain to the C-terminus [33].

As the results displayed in this work, the trimeric lichenase and xylanase showed distinguished substrate affinity and superior catalytic efficiency likely accelerated owing to the high intrinsic concentration and diminished enthalpic interactions imposed by the trimeric foldon clamp [15]. Zanphorlin and coworkers have just recently shown such increased activity and k_{cat}. In their work, they showed that a tetramer leading to a 10-fold higher activity with higher k_{cat} compared to the disassembled monomers of the psychrophilic β-glucosidases [34]. However, the tetramer is the natural biological active unit of the β –glucosidase. While the monomers, which was just a small fraction of the predominant population of tetramers, were obtained during the size-exclusion chromatography. The monomer is about ten-fold less active than the tetramer, suggesting the quaternary structure is crucial for the proper function of this enzyme. Recently, Eshaghi and coworkers obtained dimeric fluorescent proteins based on a single-domain antibody, which led 7- to 8-fold increased pH resistance, and 3- fold higher brightness in vitro. The structural basis for improved brightness and acid resistance relies on stabilization of unfavorable protein conformations [25]. Besides, Yang and coworkers obtained trimeric nitrilase active inclusion bodies by fusing an amphipathic self-assembly peptide 18A at the C-terminus of nitrilase, which has higher specific activities and thermostability than the native nitrilase [35]. Our method is a little bit similar to their work and has a comparable improvement of activity but with different oligomerization trigger domain. The improved performance of the trimeric enzymes is likely accelerated owing to the increase in the local concentration of the enzymes. And it also relied on stabilizing the protein conformations triggered by the obligatory domain foldon, which accelerated the contact of active site and substrate [36]. These results are very encouraging, as the foldon-guided trimerization of the target enzymes will lead to improving their catalytic performances effectively and simply.

Thus, our results are the first report about increasing the activities of lichenase and xylanase by fusing a foldon domain at the C-terminus to trigger their trimerization. It should also be applicable to other enzymes by proper design. Thanks to the cheap and efficient gene synthesis technology, our methods without chemical modification and mutations was effective, simple and powerful. Perhaps, further research should pay more attention to understand why and how the oligomerization domains improve the functional properties of target enzymes. Besides, the relationship between the oligomeric states and the activity of the target enzymes is also an interesting area of concern and need more investigation.

It is well known that many factors could affect the thermal behaviors of ELPs, and the physical properties of the protein fused with ELPs was one of them. The foldon domain was expected to reduce the transition temperature of the ELPs because of increased chain length, molecular weight, intrinsic concentration and enhanced hydrophobic interactions [37]. We observed the transition temperatures of B-E-F and X-E-F decreased by 8 °C and 11 °C than those of B-E and X-E, respectively. This feature is beneficial for achieving effective purification at low salinity concentrations and temperature for the ELPs containing oligomerization domain.

The increased hydrophobic character, as well as increased concentration of ELPs, promoted chain folding and aggregation, and thus oligomerization of the target enzymes facilitated functional advantages such as superior catalysis properties, increased recovery of activity and specific activity. However, a latent problem is whether oligomerization of protein could be potentially adopted as a general strategy to obtain multiple functions in various enzymes. As the foldon domain in samples was unfolded when heated to 70 °C or above, it may not suitable for the thermophilic enzyme who have the optimal temperature above the unfolding temperature of foldon (70 °C). Therefore, enzymes with proper optimum temperature will be studied to verify the properties and mechanism changed by oligomerization domain in the future. Besides, introducing the disulfide bond in the foldon domain may improve its thermostability, thus the foldon-triggered oligomerization methods we proposed here may be useful for improving the catalytic performances of the thermophilic enzymes.

Conclusion

Our results presented here highlight two important properties of the foldon-triggered trimerization of the lichenase and xylanase. First, the trimeric enzymes induced by foldon have advantages in catalytic performances such as improved activity, catalytic efficiency and substrate affinity over the monomeric ones. Second, the foldon domain could increase the purification fold and decrease the loss of activity during the process of ITC purification. The foldon domain should be an effective tool to improve the catalysis performance and purification efficiency of the target enzymes, and it will have great potentials in enzyme engineering and purification.

Methods

Plasmid construction

The complete genetic sequence of lichenase (bglS, Gene ID: 937,470) and xylanase (xynA, Gene ID: 939,861) came from references [26]. They were synthesized and sequenced by Sangon Biotech Co., Ltd. (Shanghai, China) and then cloned into pET 22(b+) vector. $ELPs_{40}$ (VPGVG) was preserved in our lab and fused with the C-terminal of the target enzymes. Meanwhile, the gene encoding the monomeric ELPs-fused enzymes was ligated between the NedI and HindIII restriction sites of pET 22b(+), respectively (denoted as B-E and X-E). For comparison, we also constructed a DNA fragment that encodes the trimeric enzymes. The foldon gene directly fused with the C-terminal of ELPs, which was fused at the C-terminal of the enzymes using NedI and HindIII as the restriction sites as well (denoted as B-E-F and X-E-F). The gene encoding the foldon domain was synthesized by Sangon Biotech Co., Ltd. (Shanghai, China) by means of overlapping PCR amplification. The recombinant plasmids encoding the lichenase and xylanase from Bacillus subtilis 168 were transformed into E.coli BL21. Recombinant plasmids of xylanase were shown in Additional file 1: Figure S1 and Additional file 2: Figure S2.

Protein expression, extraction, and purification

Plasmids carrying the monomeric and trimeric enzymes were transformed and overexpressed in the E. coli strain BL21(DE3). Single colonies were incubated overnight (~12 h) at 37 °C in fresh Luria Bertani (LB) supplemented with 100 μg/ml ampicillin. And then inoculated into terrific broth (TB) medium for a continued incubation at 37 °C till optical density at 600 nm of the culture reached between 0.4 and 0.6. The culture was induced for 20 h at 20 °C with 0.5 mmol/L isopropyl-β-thiogalactopyranoside (IPTG). Cell pellets were collected by centrifugation at 4 °C (8000 g, 10 min), resuspended in cold 0.1 mol/L phosphate-buffered saline (PBS, pH 6.4) and lysed by ultrasonic disruption at 4 °C. We harvested the crude enzymes by centrifugation at 4 °C (8000 g, 10 min) to remove insoluble cell debris and obtained supernatant. The fusion proteins were purified by ITC, as described previously [17]. Briefly, the inverse phase transition of the ELPs-fused enzymes was triggered by adding crystalline NaCl (2.5 mol/L) and further incubated at 40 °C for 20 min. The aggregations were separated from the lysate by centrifugation at 40 °C (12,000 g, 10 min) immediately. After discarding the supernatant, the pellets containing the fusion enzymes were resuspended in cold PBS, and another centrifugation at 4 °C (12,000 g, 10 min) was performed to remove the insoluble portion and collect the supernatant from the spin. Two

steps constituted one round of ITC and the aggregation and resolubilization process were repeated twice [38]. The proteins purity was firstly determined with the 12% sodium dodecyl sulfate polyacrylamide gel electrophoresis (SDS-PAGE) with commassie brilliant blue staining, and then the UltrafleXtreme MALDI TOF mass spectrometer (Bruker Daltonics Inc., MA, USA) with sinapinic acid as the matrix. We removed the salts using the ultrafiltration centrifuge tube (Millpore). The protein concentration was measured by Coomassie brilliant blue method using bovine serum albumin serum (BSA) as a standard.

Enzyme activity assays

The activities of the lichenase and xylanase were determined by measuring the release of reducing sugars from the substrates by the dinitrosalicylic acid (DNS) procedure and monitored with the OD value at the wavelength of 540 nm [39]. The standard process was conducted at their respective optimum temperature for 10 min in 0.1 mol/L PBS, using 1% (w/v) lichenan (Megazyme, Wicklow, Ireland) or birchwood xylan (Sigma) as substrate respectively. One activity unit (U) was defined as the amount of enzyme releasing 1 mol of reducing sugar per minute under the assay condition. Purified lichenases and xylanases were used throughout the assays and all the assays were performed in triplicate.

To determine the optimal pH reaction condition, the enzymatic reaction was conducted at different pH ranging from 5.0 to 8.0 in PBS. The optimal temperature was determined by incubating the reaction mixture for 10 min at pH 6.4 at a temperature gradient ranging from 30 to 65 °C.

Enzyme kinetic parameters determination

We calculated the kinetic parameters by using substrate concentrations prepared at different concentrations (0, 1.1, 1.25, 1.43, 1.67, 2, 2.5, 3.33, 5, 10 g L^{-1}) in PBS with pH 6.4. A given volume of lichenase or xylanase was added to their corresponding substrate solutions and incubated for 10 min, respectively. All the enzymatic reactions were conducted at the optimal temperature and pH. The substrate saturation curves were fitted into Michaelis–Menten kinetics and the K_m value and V_{max} of lichenase and xylanase were calculated using Lineweaver–Burk plot.

Additional files

Additional file 1: Figure S1. The profiles of plasmid lichenase. The profiles of plasmid monomeric lichenase **a**, the gene was cloned between NedI and HindIII digestion sites in pET 22b(+); plasmid trimeric lichenase **b**, foldon was directly fused with the HindIII digestion sites in pET 22b(+).

> **Additional file 2: Figure S2.** The profiles of plasmid xylanase. The profiles of plasmid monomeric xylanase **a**, the gene was cloned between *Nde*I and *Hind*III digestion sites in pET 22b(+); plasmid trimeric xylanase **b**, foldon was directly fused with the *Hind*III digestion sites in pET 22b(+).

Abbreviations

BSA: Bovine serum albumin; DNS: 3, 5-dinitrosalicylic acid; ELPs: Elastin-like polypeptides; IPTG: Isopropyl-β-thiogalactopyranoside; ITC: Inverse transition cycling; LB: Luria-Bertani; MALDI-TOF: Matrix-assisted laser desorption ionization time of flight; MS: Mass spectrometer; MW: Molecular weight; PBS: Phosphate-buffered saline; SDS-PAGE: Sodium dodecyl sulfate polyacrylamide gel electrophoresis; TB: Terrific broth; Tm: Melting temperature; Tt: Transition temperature

Acknowledgements

We thank Dr. Hongchun Li at the University of Pittsburgh for his comments on an earlier version of the manuscript and revising the manuscript. We would also like to show our gratitude to the anonymous reviewers for their so-called insights.

Funding

This work was funded by the National Natural Science Foundation of China (21376103) and the Natural Science Foundation of Fujian Province (2017 J01065).

Authors' contributions

XW and GZ designed the research. XW, HG, DZ and SW performed the major experiments containing the expression of target enzymes, measurement of activity and experimental analysis. XW, and GZ analyzed the data and wrote the manuscript. All authors read and approved the final manuscript.

Competing interests

The authors declare that they have no competing interests.

References

1. Lin Y, Tanaka S. Ethanol fermentation from biomass resources: current state and prospects. Appl Microbiol Biot. 2006;69:627–42.
2. Cantone S, Hanefeld U, Basso A. Biocatalysis in non-conventional media-ionic liquids, supercritical fluids and the gas phase. Green Chem. 2007;9:954–71.
3. Zhang Y, Ge J, Liu Z. Enhanced activity of immobilized or chemically modified enzymes. ACS Catal. 2015;5:4503–13.
4. Zhang ZG, Yi ZL, Pei XQ, Wu ZL. Improving the thermostability of *Geobacillus stearothermophilus* xylanase XT6 by directed evolution and site-directed mutagenesis. Bioresour Technol. 2010;101:9272–8.
5. Li PP, Wang XJ, Yuan XF, Wang XF, Cao YZ, Cui ZJ. Screening of a composite microbial system and its characteristics of wheat straw degradation. Agric Sci China. 2011;10:1586–94.
6. Illanes A, Cauerhff A, Wilson L, Castro GR. Recent trends in biocatalysis engineering. Bioresour Technol. 2012;115:48–57.
7. Dalby PA. Strategy and success for the directed evolution of enzymes. Curr Opin Struc Biol. 2011;21:473–80.

8. Zhang W, Liu Y, Zheng H, Yang S, Jiang W. Improving the activity and stability of GL-7-ACA acylase CA130 by site-directed mutagenesis. Appl Environ Microb. 2005;71:5290–6.
9. Pešić M, Božić N, López C, Lončar N, Álvaro G, Vujčić Z. Chemical modification of chloroperoxidase for enhanced stability and activity. Process Biochem. 2014;49:1472–9.
10. Papanikolopoulou K, Forge V, Goeltz P, Mitraki A. Formation of highly stable chimeric trimers by fusion of an adenovirus fiber shaft fragment with the foldon domain of bacteriophage T4 fibritin. J Biol Chem. 2004;279:8991–8.
11. Muller SA, Sasaki T, Bork P, Wolpensinger B, Schulthess T, Timpl R, et al. Domain organization of Mac-2 binding protein and its oligomerization to linear and ring-like structures. J Mol Biol. 1999;291:801–13.
12. Malashkevich VN, Kammerer RA, Efimov VP, Schulthess T, Engel J. The crystal structure of a five-stranded coiled coil in COMP: a prototype ion channel? Science. 1996;274:761–5.
13. Tao Y, Strelkov SV, Mesyanzhinov VV, Rossmann MG. Structure of bacteriophage T4 fibritin: a segmented coiled coil and the role of the C-terminal domain. Structure. 1997;5:789–98.
14. Frank S, Kammerer RA, Mechling D, Schulthess T, Landwehr R, Bann J, et al. Stabilization of short collagen-like triple helices by protein engineering. J Mol Biol. 2001;308:1081–9.
15. Stetefeld J, Frank S, Jenny M, Schulthess T, Kammerer RA, Boudko S, et al. Collagen stabilization at atomic level: crystal structure of designed (GlyProPro)10 foldon. Structure. 2003;11:339–46.
16. Yang X, Lee J, Mahony EM, Kwong PD, Wyatt R, Sodroski J. Highly stable trimers formed by human immunodeficiency virus type 1 envelope glycoproteins fused with the oligomeric motif of T4 bacteriophage fibritin. J Virol. 2002;76:4634–42.
17. Sissoeff L, Mousli M, England P, Tuffereau C. Stable trimerization of recombinant rabies virus glycoprotein ectodomain is required for interaction with the p75NTR receptor. J Gen Virol. 2005;86:2543–52.
18. Li CC, Zhang GY. The fusions of elastin-like polypeptides and xylanase self-assembled into insoluble active xylanase particles. J Biotechnol. 2014;177:60–6.
19. Park JE, Won JI. Thermal behaviors of elastin-like polypeptides (ELPs) according to their physical properties and environmental conditions. Biotechnol Bioproc E. 2009;14:662–7.
20. Cho YH, Zhang YJ, Christensen T, Sagle LB, Chilkoti A, Cremer PS. Effects of Hofmeister anions on the phase transition temperature of elastin-like polypeptides. J Phys Chem B. 2008;112:13765–71.
21. Chu HS, Won JI. Improvement of the productivity of long elastin-like polypeptides (ELPs) in a cell-free protein synthesis system. J Biosci Bioeng. 2009;108:S68–S.
22. Lim DW, Trabbic-Carlson K, MacKay JA, Chilkoti A. Improved non-chromatographic purification of a recombinant protein by cationic elastin-like polypeptides. Biomacromolecules. 2007;8:1417–24.
23. Ali G, James TC, Nolan BH. Thermoreversible micelle formation using a three-armed star elastin-like polypeptide. Macromolecules. 2010;43:4340–5.
24. Mee C, Banki M, Wood D. Towards the elimination of chromatography in protein purification: expressing proteins engineered to purify themselves. Biochem Eng J. 2008;135:56–62.
25. Eshaghi M, Sun G, Grüter A, Lim CL, Chee YC, Jung G, et al. Rational structure-based design of bright GFP-based complexes with tunable dimerization. Angew Chem Int Edit. 2015;127:14158–62.
26. Cota J, Oliveira LC, Damásio ARL, Citadini AP, Hoffmam ZB, et al. Assembling a xylanase–lichenase chimera through all-atom molecular dynamics simulations. Biochim Biophys Acta. 2013;1834:1492–500.
27. Siddiqui KS, Parkin DM, Curmi PMG, Francisci DD, Poljak A, et al. A novel approach for enhancing the catalytic efficiency of a protease at low temperature: reduction in substrate inhibition by chemical modification. Biotechnol Bioeng. 2009;103:676–86.
28. Yang HQ, Lu XY, Liu L, Li JH, Shin HD, et al. Fusion of an oligopeptide to the N terminus of an alkaline- amylase from *Alkalimonas amylolytica* simultaneously improves the enzyme's catalytic efficiency, thermal stability, and resistance to oxidation. Appl Environ Microb. 2013;9:3049–58.
29. Zhang J, Shi H, Xu L, Zhu X, Li X. Site-directed mutagenesis of a hyperthermophilic endoglucanase Cel12B from *Thermotoga maritima* based on rational design. PLoS One. 2015;10:e0133824.
30. Khajeh K, Ranjbar B, Naderi-Manesh H, Habibi AE, Nemat-Gorgani M. Chemical modification of bacterial α-amylases: changes in tertiary structures and the effect of additional calcium. Bba-protein Struct M. 2001;1548:229–37.

31. Lin L, Hu S, Yu K, Huang J, Yao S, Lei Y, et al. Enhancing the activity of glutamate decarboxylase from *Lactobacillus brevis* by directed evolution. Chinese J Chem Eng. 2014;22:1322–7.

32. Can O, Holland NB. Utilizing avidity to improve antifreeze protein activity: a type III antifreeze protein trimer exhibits increased thermal hysteresis activity. Biochemistry-Us. 2013;52:8745–52.

33. Engel J, Kammerer RA. What are oligomerization domains good for? Matrix Biol. 2000;19:283–8.

34. Zanphorlin LM, de Giuseppe PO, Honorato RV, Tonoli CCC, Fattori J, Crespim E, et al. Oligomerization as a strategy for cold adaptation: structure and dynamics of the GH1 β-glucosidase from *Exiguobacterium* antarcticum B7. Sci Rep-UK. 2016;6:23776. doi:10.1038/srep23776.

35. Yang XF, Huang A, Peng JZ, Wang JF, Wang XN, et al. Self-assembly amphipathic peptides induce active enzyme aggregation that dramatically increases the operational stability of nitrilase. RSC Adv. 2014;4:60675–84.

36. Luo T, Kiick KL. Noncovalent modulation of the inverse temperature transition and self-assembly of elastin-b-collagen-like peptide bioconjugates. J Am Chem Soc. 2015;137:15362–5.

37. Holehouse AS, Pappu RV. Protein polymers: encoding phase transitions. Nat Mater. 2015;14:1083–4.

38. Christensen T, Hassouneh W, Trabbic-Carlson K, Chilkoti A. Predicting transition temperatures of elastin-like polypeptide fusion proteins. Biomacromolecules. 2013;14:1514–9.

39. Miller GL. Use of dinitrosalicylic acid reagent for determination of reducing sugar. Anal Chem. 1959;31:426–8.

18

Expression of *Vitreoscilla* hemoglobin enhances production of arachidonic acid and lipids in *Mortierella alpina*

Huidan Zhang[1,3,4], Yingang Feng[1,3], Qiu Cui[1,2,3]* and Xiaojin Song[1,3]* (iD)

Abstract

Background: Arachidonic acid (ARA, C20:4, n-6), which belongs to the omega-6 series of polyunsaturated fatty acids and has a variety of biological activities, is commercially produced in *Mortierella alpina*. Dissolved oxygen or oxygen utilization efficiency is a critical factor for *Mortierella alpina* growth and arachidonic acid production in large-scale fermentation. Overexpression of the *Vitreoscilla* hemoglobin gene is thought to significantly increase the oxygen utilization efficiency of the cells.

Results: An optimized *Vitreoscilla* hemoglobin (VHb) gene was introduced into *Mortierella alpina* via *Agrobacterium tumefaciens*-mediated transformation. Compared with the parent strain, the VHb-expressing strain, termed VHb-20, grew faster under both limiting and non-limiting oxygen conditions and exhibited dramatic changes in cell morphology. Furthermore, VHb-20 produced 4- and 8-fold higher total lipid and ARA yields than those of the wild-type strain under a microaerobic environment. Furthermore, ARA production of VHb-20 was also 1.6-fold higher than that of the wild type under normal conditions. The results demonstrated that DO utilization was significantly increased by expressing the VHb gene in *Mortierella alpina*.

Conclusions: The expression of VHb enhances ARA and lipid production under both lower and normal dissolved oxygen conditions. This study provides a novel strategy and an engineered strain for the cost-efficient production of ARA.

Keywords: Arachidonic acid, *Mortierella Alpina*, Hemoglobin, Dissolved oxygen, Fermentation

Background

Arachidonic acid (ARA, 5,8,11,14-cis-eicosatetraenoic acid), belonging to the omega-6 series of polyunsaturated fatty acids (PUFAs), is not only a structural component of the cell membrane but also the biogenetic precursor of prostaglandins, leukotrienes, thromboxanes, and other eicosanoid hormones [1, 2]. Furthermore, ARA is widely used in health food, pharmacology, agriculture, cosmetics and other industries [3, 4]. *Mortierella alpina*, which has a high ARA content, is considered one of the best ARA-producing strains and has been used in industrial applications [5].

The growth of *M. alpina* is closely related to the dissolved oxygen, especially in high cell density fermentation. The fast growth of the mycelium can result in an apparent increase in broth viscosity. The mycelium can agglomerate easily and further decrease the efficiency of oxygen mass transfer [6, 7]. Therefore, improving the concentration of dissolved oxygen in the fermentation process or increasing the utilization of dissolved oxygen is key to improving the yield of ARA.

At present, there are various methods to improve the oxygen supply, for example, by increasing the stirring speed or ventilation, or by adding surfactants to increase the concentration of dissolved oxygen in the fermentation. However, these methods require either special equipment or high energy consumption [8, 9]. It has been shown that the expression of the bacterial (*Vitreoscilla*) hemoglobin (VHb) is an effective method to solve the problem of oxygen supply in the fermentation process because VHb can

* Correspondence: cuiqiu@qibebt.ac.cn; songxj@qibebt.ac.cn
[1]Shandong Provincial Key Laboratory of Energy Genetics, Qingdao Institute of Bioenergy and Bioprocess Technology, Chinese Academy of Sciences, No.189 Songling Road, Laoshan District, Qingdao, Shandong Province 266101, China
Full list of author information is available at the end of the article

improve the efficiency of intracellular oxygen transport [10, 11]. VHb is an oxygen-binding protein produced by *Vitreoscilla*, which allows the *Vitreoscilla* (aerobic bacteria) to survive under limited oxygen conditions. Khosla and Bailey first cloned the *vgb* gene and successfully expressed it in *E. coli* [12].They demonstrated that the expression of VHb could promote cell growth, increase protein synthesis capacity under limited oxygen conditions [13, 14]. According to the characteristics of VHb, it has a good application, which has been successfully engineered into different bacteria, fungi, as well as some plants and animals, to improve cell growth and to increase the expression of exogenous protein and metabolite production [15, 16].

The metabolic engineering method also has been applied to improve the ARA production. For example, Hao et al. [17] studied overexpression of ME2 and G6PD2 in *M. alpina*, increasing the supply of intracellular NADPH, with a 1.7-fold increase in total fatty acid and a 1.5-fold increase in arachidonic acid (ARA) content. In 2005, Takeno et al. [18] improved of the fatty acid composition of *M. alpina* 1S-4, through RNA interference with Δ12 desaturase gene expression. But ARA biosynthesis is multi-step, subject to multiple factors. The modification of one or two genes is not easy to produce transformants with the desired trait. Overexpression of VHb could increase the intensity of respiratory chain and increase the supply of intracellular ATP, which may affect the expression of intracellular multi-gene expression and promote the formation of ARA.

To the best of our knowledge, VHb has not been used in *M. alpina* to date. In this study, we investigate the effect of VHb on growth and ARA production in *Mortierella alpina* and explored the mechanism of VHb in ARA synthesis.

Methods
Strains, plasmids and growth conditions
Wild-type *Mortierella alpina* ATCC 32222 was obtained from the American Type Culture Collection (ATCC, Manassas, VA, USA). The *Agrobacterium tumefaciens* strain C58C1 and plasmid pBIG4MRHrev were gifts from Yasuyuki Kubo (Kyoto Prefectural University, Japan). The strains, plasmids and primers used in this study are listed in Additional file 1: Table S1 and Table S2, respectively. The composition of the LB-Mg agar, minimal medium (MM) and inducing medium (IM), which were used for the culture, transformation and infection of *Agrobacterium tumefaciens*, respectively, have been described previously [19]. GY medium (20 g L^{-1} glucose, 10 g L^{-1}yeast extract and 20 g L^{-1}agar) was used for the transformation and screening of *Mortierella alpina*. *M. alpina* strains were grown on PDA medium (potatoes 200 g L^{-1}, glucose 20 g L^{-1}, and agar 20 g L^{-1}) slants at 25 °C. The seed culture

medium contained 30 g L^{-1} glucose, 6 g L^{-1} yeast extract, 3 g L^{-1} KH$_2$PO$_4$, 2.8 g L^{-1} NaNO$_3$, and 0.5 g L^{-1} MgSO$_4$•7H$_2$O. The medium for shake-flask culture contained 80 g L^{-1} glucose, 11 g L^{-1} yeast extract, 3.8 g L^{-1} KH$_2$PO$_4$, 3.5 g L^{-1} NaNO$_3$, and 0.5 g L^{-1} MgSO$_4$•7H$_2$O. The medium for bioreactor fermentation contained 30 g L^{-1} glucose, 20 g L^{-1} yeast extract, 4 g L^{-1} KH$_2$PO$_4$, 3.8 g L^{-1} NaNO$_3$, and 0.6 g L^{-1} MgSO$_4$•7H$_2$O. The *Escherichia coli* strains were grown in LB medium (10 g L^{-1} tryptone, 5 g L^{-1} yeast extract, and 10 g L^{-1} NaCl) at 37 °C.

Codon analysis and optimization
The codon preference table was found by analyzing the codon preference of the gene of *M. alpina* by the online software Codon Usage Database (http://www.kazusa.or.jp/codon/). The *vgb* gene (GenBank accession no.M30794.1) of *Vitreoscilla* was optimized by software Optimizer (http://genomes.urv.cat/OPTIMIZER/) based on the codon preference table. The codon-optimized nucleotide sequence of *vgb* was synthesized by GenScript (Nanjing) Co., Ltd. (Jiangsu, China).

Vector construction
The hygromycinB phosphotransferase gene (HPH) expression cassette containing the hisH4.1 promoter, the *hpt* resistance gene and the trpC terminator was amplified with primer pair P1/T1 from the vector pD4 [20]. This HPH expression cassette was inserted into pMD19-T to yield the plasmid pMD19T-HPH. The 18S sequence, which was amplified from *M. alpina* ATCC 32222 cDNA with primer pair 18S-F/18S-R, was ligated to the pMD19T-HPH vector, and the resulting construct was named pMD19T-HPH-18S. The carboxin resistance gene (CBXB) [21] was amplified with the primer pair CBXB-F / CBXB-R and used to replace the *hpt* gene, thus forming plasmid pMD19T-CBXB-18S. The CBXB expression cassette was amplified from plasmid pMD19T-CBXB-18S with primer pair P2/T2, digested with restriction endonucleases *Cla*I and *Eco*RI, and then ligated to T-DNA binary vector pBIG4MRHrev. The resulting plasmid was named pBIG-CBXB. The HisH4.1 promoter and trpC terminator were first amplified with primer sets P3-F/P3-R and T4-F/T4-R from the vector pBIG-CBXB, thus yielding fragments of 1.1 kb and 0.7 kb, respectively.

The optimized *vgb* sequence was inserted into pUC57 to create the plasmid pUC57-vgb (Additional file 1: Table S1). The primer pair VHb-F/VHb-R was used to amplify the optimized *vgb* gene from the vector pUC57-vgb, forming a 441 bp fragment. The *vgb* gene cassette containing the hisH4.1 promoter, the *vgb* gene and the trpC terminator was assembled by overlap PCR. The obtained 2.1 kb *vgb* gene cassette was digested with *Eco*RI and *Sma*I and then ligated to pBIG-CBXB, thus forming the plasmid pBIG-CBXB-VHb (Additional file 1: Figure S2).

Agrobacterium tumefaciens-mediated transformation of M. alpina

The transformation method was optimized on the basis of previously described *Agrobacterium tumefaciens*-mediated transformation (ATMT) methods [19, 22]. *M. alpina* spores were collected from cultures growing on PDA agar medium, and the suspension was filtered through three layers of lens paper. *A. tumefaciens* C58C1 was electrotransformed with the plasmid pBIG-CBXB-VHb. The positive transformants were confirmed by PCR, and then a single colony containing plasmid pBIG-CBXB-VHb was selected to grow overnight in 5 mL of liquid LB-Mg medium containing 50 µg mL^{-1} kanamycin and 50 µg mL^{-1} rifampin at 28 °C and 200 rpm. The bacterial cells were collected by centrifugation at 5000 rpm for 3 min at room temperature and washed once with fresh IM, and then fresh IM was used to adjust the OD to 0.2-0.3. After being incubated at 28 °C until an OD660 of 0.8 was reached, 100 µL of the *Agrobacterium* culture was mixed with 100 µL of *M. alpina* spore suspension (10^8 mL^{-1}) and then spread onto cellophane membranes, placed on co-cultivation medium (IM with 1.5% agar) and incubated at 23 °C for 2 days. After co-cultivation, the cellophane membranes were transferred to GY plates containing 150 µg mL^{-1} carboxin, 50 µg mL^{-1} spectinomycin and 50 µg mL^{-1} cefotaxime to inhibit the growth of *Agrobacterium*. The plates were incubated at 25 °C until CBXB-resistant colonies became visible. The transformed candidates were transferred to GY agar plates containing 150 µg mL^{-1} carboxin. To obtain stable transformants, this procedure was repeated three times, and all experiments were conducted in triplicate.

PCR verification of the transformants

PCR was used to identify whether the mycelium was successfully transformed. Genomic DNA of *M. alpina* strains was prepared by a method described previously [23]. The genomic DNAs of the transformants were used as PCR templates to confirm the integration of the plasmid pBIG-CBXB-VHb by using the *vgb* gene specific primers VHb-F and VHb-R (Additional file 1: Table S2). Wild type *M. alpina* and the plasmid pBIG-CBXB-VHb were used as negative and positive PCR controls, respectively. The PCR products were run on a 1.0% agarose gel and stained with ethidium bromide.

Analytical methods for dry mycelial weight, total lipid and arachidonic acid production

Cell dry weight determination

In total, 50 mL cell suspensions were harvested by filtration at room temperature. The cell pellet was washed twice with distilled water and dried to constant weight at 50 °C. The dry cell weight was determined by weight analysis and represented by dry cell weight (DCW).

Lipid extraction and fatty acid composition analysis

Fatty acid extraction and methylation were carried out according to the previously described methods [24]. Total lipids were extracted by using approximately 100 mg of mycelia (dry weight). Fatty acid methyl esters (FAMEs) were obtained by reacting the lipids at 85 °C for 2.5 h in the presence of 2% sulfuric acid/methanol (v/v). FAMEs were extracted in hexane and analyzed by gas chromatography (Agilent Technologies, 7890B) with an HP-INNOWAX capillary column (30 m by 0.25 mm, 0.25 µm film thickness). The oven temperature was set at 100 °C for 1 min, was raised to 250 °C at a rate of 15 °C per minute and then was held at 250 °C for 5 min. The split ratio was 1:19, and the carrier gas was nitrogen. The injection volume was 1 µL. Peak detection used a flame ionization detector (FID). The temperature of the flame ionization port and injection port was 280 °C.

Shake-flask culture and bioreactor scale fermentation

To study the effect of VHb on *M. alpina*, especially under the oxygen-restricted conditions, we compared the DCW, total lipid content, and arachidonic acid production between the transformant VHb-20 and the wild-type strain in shake flasks and a bioreactor under dO$_2$ limiting and non-limiting conditions. First, the VHb-20 transformants and wild type were inoculated into six 250 mL shake flasks with baffles. Three shake flasks contained 50 mL medium (normal, dO2 non-limiting condition), and three contained 150 mL medium (dO2 restriction). They were shaken at 25 °C and 200 rpm for 7 days. Meanwhile, bioreactor cultures were grown in 5-L Biostat B plus bioreactors (Sartorius Stedim Biotech, Germany) with 3 L of medium. The rate of ventilation and agitation during growth in the bioreactor was set to non-limiting oxygen (1.7 vvm, 200 rpm) and limiting oxygen (0.8 vvm, 100 rpm) conditions. The transformant and wild-type cultures were grown for 7 days at 25 °C, pH 6.5. The glucose concentration was maintained at approximately 20 g L^{-1} by feeding the 80% (w / v) glucose solution, which was measured with a SBA-40D Biosensor (Institute of Biology, Shandong Academy of Sciences, China). Twenty milliliter samples were taken at 24 h intervals in all experiments. The entire fermentation process used bioprocess-automated software for monitoring and control.

Quantitative real-time PCR (qRT-PCR) validation

The expression levels of some desaturase genes, such as Δ9, Δ12, Δ6, and Δ5 desaturase, were further quantified by real-time PCR to validate the effects of VHb expression on the biosynthesis of unsaturated fatty acids. The primer sequences (Additional file 1: Table S3) were designed according to the genome sequencing data (NCBI BioProjects PRJNA41211). Total RNA was extracted using a thermo Scientific GeneJET RNA Purification kit (No. K0731). After DNA degradation by DNase, 2 µg of total RNA was reverse transcribed using a TIANScript reverse transcription Kit (Tiangen) in a 20 µl reaction volume.

qRT-PCR was performed in a LightCycler R480 Real-time Detection System (Roche). A melting curve analysis of the amplification products was performed to confirm the specificity of amplification. To quantify the transcription of each gene, the copy number was determined by generating a standard curve by using a serial 10-fold dilution of the targeted PCR product inserted into the pMD™ 19-T vector (TaKaRa Bio Group). For sample normalization, 18S rRNA was used as an internal standard. All the reactions were performed in triplicate, and the data were normalized by using the average values for the internal standard.

Calculation and statistical analysis

All results are given as the mean ± standard error of the mean (SEM), and significant differences ($P < 0.05$, $P < 0.01$) between means were compared by the T-test.

Results

The stabilized expression of VHb in *M. alpina*

To express VHb efficiently in *M. alpina*, we designed and optimized the *vgb* gene according to the codon preference in *M. alpina*. The optimized *vgb* gene sequence is shown in Additional file 1: Figure S1.

We first tested the sensitivity of *M. alpina* ATCC 32222 to carboxin and another series of antibiotics (Additional file 1: Table S4). The strain did not grow in the presence of 150 µg mL^{-1} carboxin but showed high resistance to other antibiotics. Therefore, the carboxin resistance gene (*CBXB*) was used as the selection marker [21, 25] at a screening concentration of 150 µg mL^{-1}.

The plasmid pBIG-CBXB-VHb was transformed into *M. alpina* ATCC 32222 by ATMT. All clones were verified by PCR using genomic DNA as templates. To select the high-yield ARA strain, 14 correct clones verified by PCR were randomly selected to check ARA production in limiting and non-limiting oxygen conditions. After 7 d of cultivation, a positive clone exhibiting the highest ARA production (VHb-20) was identified. Thus, VHb-20 strain was chosen for all further experiments. The *vgb* gene expression cassette inserted in clone VHb-20 was confirmed by genomic PCR amplification (see Fig. 1a and b).

To confirm the functional expression of VHb in VHb-20, the carbon monoxide (CO) difference spectra were measured, and the wild-type strain was used as a control.

VHb-20 showed a typical VHb CO-binding absorption peak at 419 nm, whereas this absorption peak was not observed in the control (Fig. 1c). This result indicated that the VHb expressed in VHb-20 was biologically active. Afterwards, VHb-20 was continuously inoculated and cultivated on carboxin-free PDA solid medium, and the integrated *vgb* gene in the genome was monitored by PCR. The result indicated that the recombinant strain was genetically stable after 5 sub-cultivation cycles.

Effects of VHb expression on the growth and ARA production of *M. alpina*

Both VHb-20 and the wild-type strain were cultivated in 250 mL shake flasks under limiting and non-limiting oxygen conditions, and their growth patterns and ARA production were investigated. The results showed that the VHb-20 strain grew faster than the wild-type strain and formed more homogeneous mycelial pellets under both conditions. In addition, the hyphae of *M. alpina* ATCC 32222 were looser.

Under aerobic conditions (cultivated in a 250-mL baffled flask containing 50 mL of medium at 25 °C and 200 rpm), the glucose consumption rates and dry weight of VHb-20 and wild-type strains were not obviously different after cultivation for 7 days (Fig. 2a and b), but the total lipid yield of strain VHb-20 (7.97 g L^{-1}) was 3.15-fold higher than that of the wild-type strain (2.53 g L^{-1}; Fig. 2c). In addition, the ARA production of the strain VHb-20 was 3.62 g L^{-1}, which was 5.10 times higher than that of the wild-type strain (0.71 g L^{-1}; Fig. 2d). To obtain a microaerobic condition with a lower level of dissolved oxygen, the liquid medium in the flask was increased from 50 mL to 150 mL. Under these conditions, the VHb-20 strain had a higher glucose consumption rate and growth rate compared with that of the control strain (Fig. 2a and b), and lipid and ARA production of VHb-20 was also higher by 4.03 folds (4.64 g L^{-1} vs. 1.15 g L^{-1}) and 8.36 folds (1.84 g L^{-1} vs. 0.22 g L^{-1}), respectively (Fig. 2c, and d).

Effect of VHb expression on the composition of fatty acids in *M. alpina*

The concentration of dissolved oxygen affects not only cell growth but also the synthesis of fatty acids. Therefore, we further compared the fatty acid composition and content of VHb-20 and the wild-type strain under aerobic or microaerobic conditions. VHb-20 and the wild-type strain produced similar types of fatty acids (Fig. 3), but the proportion of various fatty acids in the lipids had changed significantly (Table 1). The results showed that the proportion of palmitic acid (C16: 0), stearic acid (C18: 0) and oleic acid (C18: 1) in the total fatty acids of wild

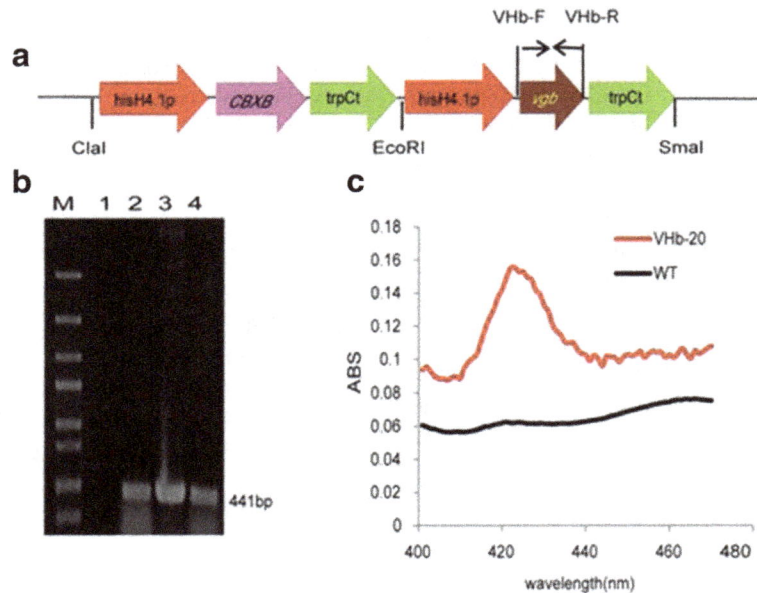

Fig. 1 Transformant verification. **a** Scheme for expression of vgb in *Mortierella alpina* ATCC 32222; **b** PCR analysis of transformants of *Mortierella alpina*. The PCR product corresponding to VHb is 441 bp (Lanes: M, 5000 bp marker; 1, negative control, wild-type strain ATCC 32222; 2, transformant VHb-20; 3, positive control, plasmid pBIG-CBXB-VHb; 4, transformant VHb-10); **c** CO binding spectra of cell extracts from original strain ATCC 32222 and transformant VHb-20

type strains was 15.82%, 17.98% and 17.58%, significantly higher than other fatty acids. Especially in the limited oxygen condition, the accumulation of oleic acid (C18: 1) is very huge, the content in the total fatty acids as high as 39.61%, resulting in ARA content was only 19.27%. However, the contents of palmitic acid (C16: 0) and oleic acid (C18: 1) in VHb-20 strain were significantly decreased, and the ARA content was increased to 45.68% and 38.51%, under the aerobic or microaerobic conditions respectively, which were nearly 1.7 folds and 2.0 folds as high as that of the control.

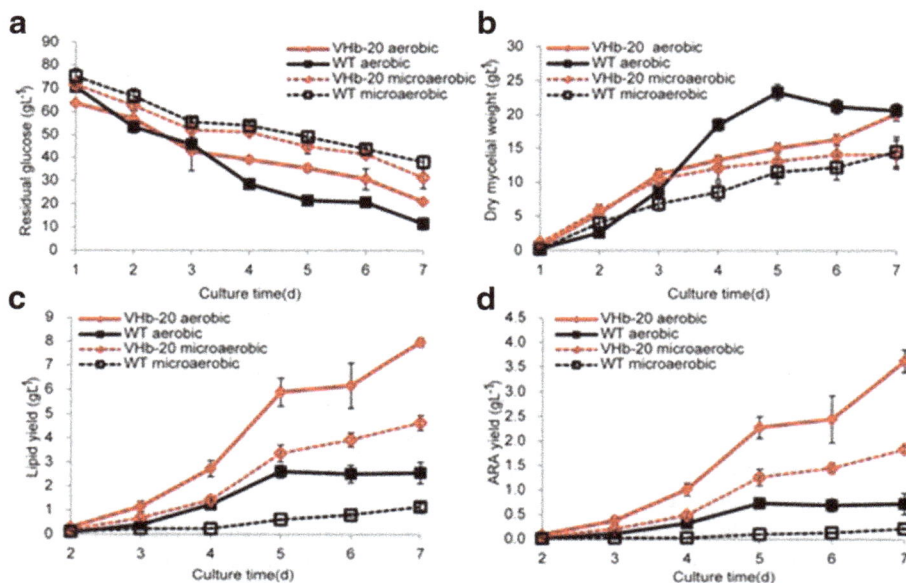

Fig. 2 Comparison of growth profiles in shake flasks under aerobic and microaerobic conditions. **a** Residual glucose; **b** Dry mycelial weight; **c** Total lipid content; (**d**) Arachidonic acid production; The data are the mean of three repeats

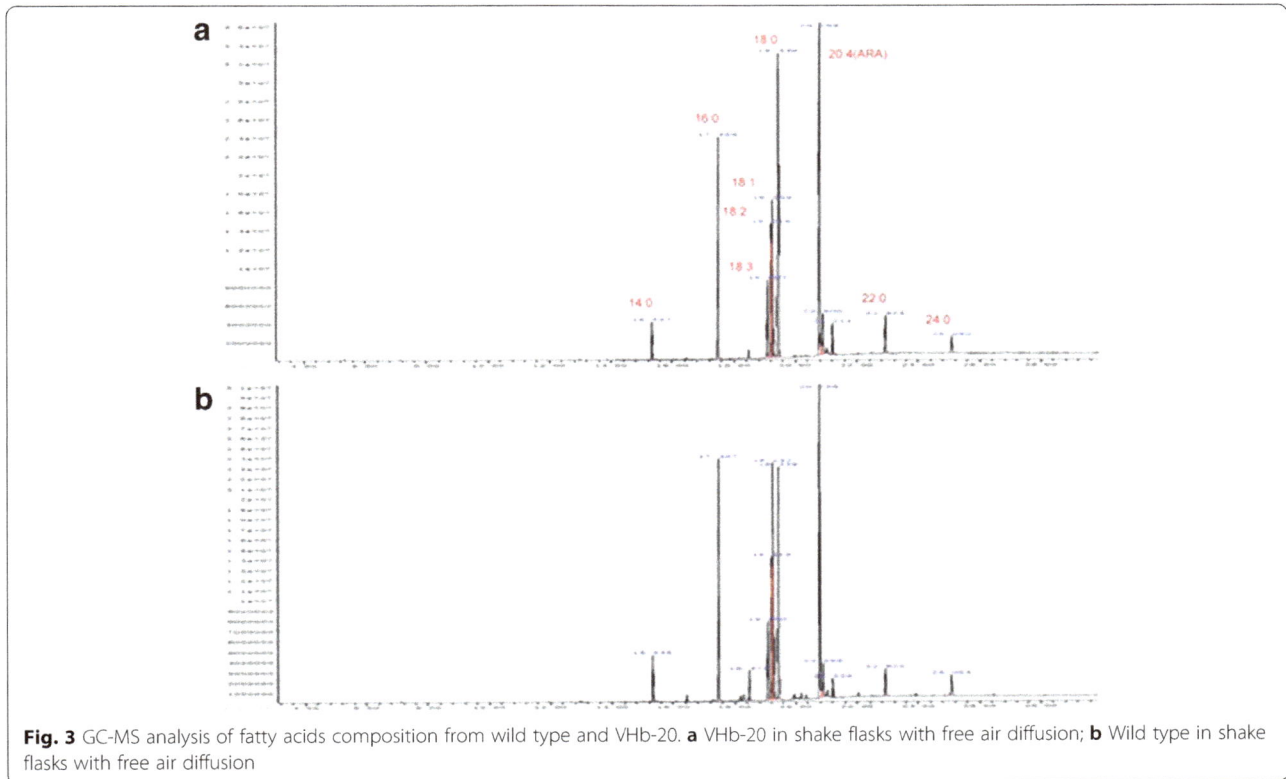

Fig. 3 GC-MS analysis of fatty acids composition from wild type and VHb-20. **a** VHb-20 in shake flasks with free air diffusion; **b** Wild type in shake flasks with free air diffusion

Effects of VHb expression on the expression of desaturase genes

The biosynthesis of unsaturated fatty acids in cell is catalyzed by desaturases, which require oxygen as a substrate [3]. Therefore, the expression of desaturase genes, including Δ9 desaturase, Δ12 desaturase, Δ6 desaturase, Δ5 desaturase, was investigated to verify the effects of VHb expression on the biosynthesis of unsaturated fatty acids at the molecular level. All the expression levels of these desaturase genes were higher in VHb-20, as compared with the wild type strain, in both aerobic and microaerobic conditions (Fig. 4a). From the results, Δ9 desaturase, which is the first enzyme in unsaturated fatty acid biosynthesis, was most sensitive to the expression of VHb, which resulted 3.2 and 4.2-fold higher expression of the Δ9 desaturase gene under aerobic and microaerobic conditions, respectively. A lesser effect was observed in the increased expression of Δ12 and Δ6 desaturase, which were only 1.4-fold higher in VHb-20, and 1.6-fold higher expression of Δ5 desaturase was observed under aerobic conditions. However, under microaerobic conditions, the expression level of Δ12 and Δ6 desaturase was increased by 2.3 and 8.7 times higher than that of the control strain. In agreement with the

Table 1 Fatty Acid Compositions of wild type and VHb-20 in shake flasks

Fatty acid	Content (% total fatty acids)			
	WT(aerobic)	VHb-20(aerobic)	WT(microaerobic)	VHb-20(microaerobic)
14:0	3.22 ± 0.14	0.73 ± 0.02**	2.84 ± 0.04	1.46 ± 0.17**
16:0	15.82 ± 0.33	9.72 ± 0.36**	13.35 ± 0.10	11.51 ± 0.24**
18:0	17.98 ± 0.49	19.63 ± 0.73**	13.34 ± 0.95	23.19 ± 0.48**
18:1	17.58 ± 0.58	6.92 ± 0.55**	39.61 ± 2.26	8.34 ± 0.24**
18:2	7.43 ± 0.44	5.82 ± 0.22**	2.87 ± 0.35	5.94 ± 0.56**
18:3	3.37 ± 0.11	3.74 ± 0.08*	2.27 ± 0.38	3.71 ± 0.22*
20:3	3.43 ± 0.06	2.84 ± 0.19*	3.71 ± 0.08	2.38 ± 0.26**
20:4(ARA)	27.30 ± 0.77	45.68 ± 1.09**	19.27 ± 1.18	38.51 ± 0.44**

**p < 0.01,*p < 0.05

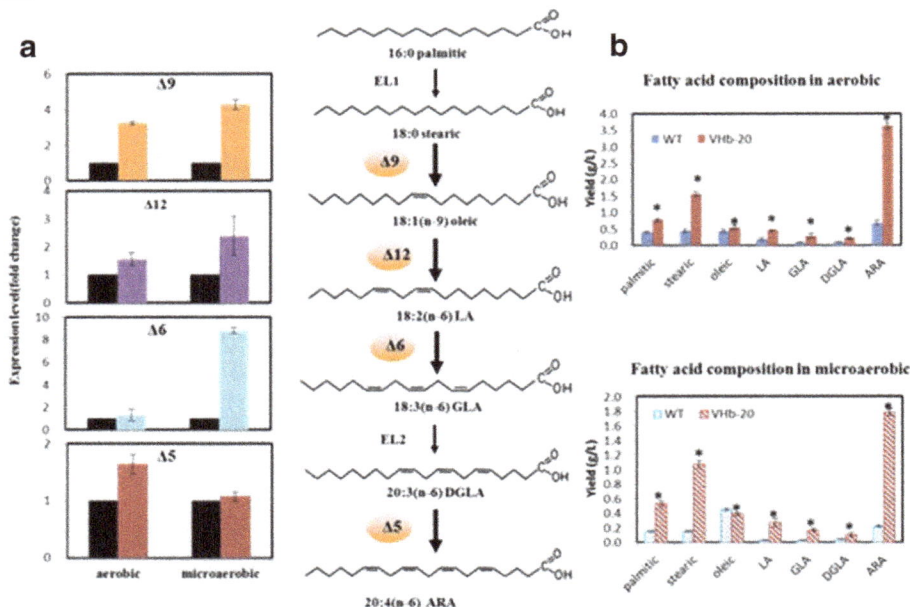

Fig. 4 The changes of the expression of desaturases (Δ9 desaturase, Δ12 desaturase, Δ6 desaturase, Δ5 desaturase) and ARA production between the WT and VHb-20 strains under aerobic and microaerobic conditions. **a** Changes in the expression of desaturases, Black: WT; Colored: VHb-20 strain; **b** Changes in unsaturated fatty acid production (*$p < 0.05$)

differences in the levels of these desaturases, the biosynthesis of unsaturated fatty acids was augmented, and the ARA production of VHb-20 was increased to 3.62 g L^{-1} and 1.84 g L^{-1} under the aerobic and microaerobic conditions (Fig. 4b), respectively.

Confirmation of the effects of VHb expression on cell growth and ARA production by using bioreactor fermentation

Bioreactor cultures were grown to confirm the influence of VHb expression on cell growth and lipid production under both non-limiting (1.7vvm, 200 rpm) and limiting (0.8vvm, 100 rpm) dO2 conditions. From a morphological viewpoint (Fig. 5), VHb-20 grew faster and formed more uniform mycelial pellets than did the wild-type strain *M. alpina* ATCC 32222. In hypoxic conditions, the VHb-20 strain grew faster than the wild-type strain and reached the maximum biomass (12.90% more than the wild type) sooner (Fig. 6a). In addition, the culture of the wild-type strain clumped heavily and became difficult to stir after 3 days of cultivation, whereas uniform mycelial pellets were observed in the VHb-20 culture, and the culture broth maintained good fluid properties through 7 days of fermentation.

In agreement with the results of the shake flask assay, VHb expression also enhanced lipid and ARA biosynthesis in the bioreactor experiments, especially under limiting oxygen conditions. There was almost no difference in the lipid content of the DCW of VHb-20 in the

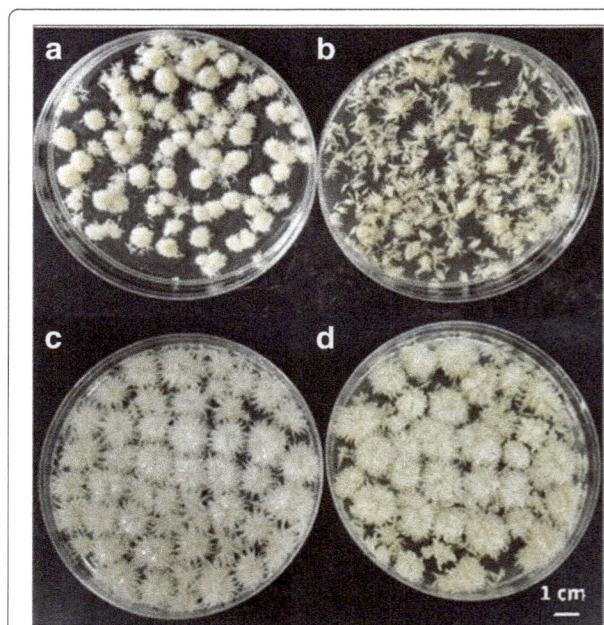

Fig. 5 Morphology of hyphae in 5 L fermenter. **a** VHb-20 transformant under aerobic conditions; **b** wild type under vaerobic conditions; **c** VHb-20 transformant under microaerobic conditions; **d** wild type under microaerobic conditions

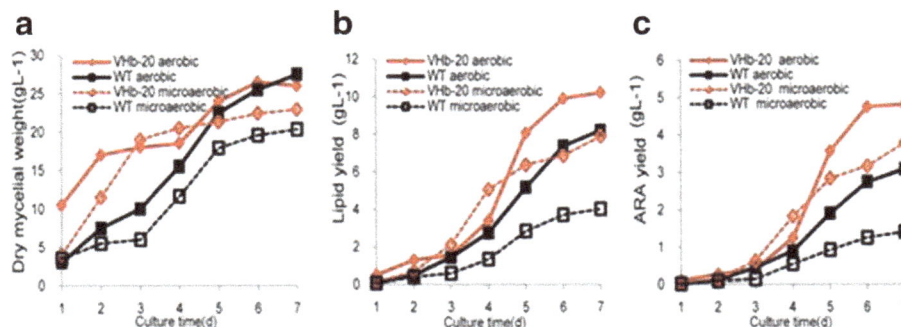

Fig. 6 Comparison of growth profiles in a 5-L fermenter under aerobic and microaerobic conditions. **a** Dry mycelial weight; **b** Total lipid content; **c** Arachidonic acid production

two conditions (39.30% vs 34.40% for the normal and limiting oxygen conditions, respectively), but in the control, the lipid content was 34% lower in limiting oxygen conditions (29.80% vs 19.80%) at the end of fermentation. The maximum lipid yield of VHb-20 strain was 10.21 g L^{-1} and 7.89 g L^{-1}, which was nearly 1.25 and 1.96 fold higher than that of the original strain under the two aerobic conditions (Fig. 6b).

Under aerobic conditions, VHb-20 had a higher ARA yield by 1.56-fold, from a value of 3.08 g L^{-1} to 4.81 g L^{-1}, compared with the original strain *M. alpina* ATCC 32222. At the end of fermentation, the ARA yield of VHb-20 was 2.70 times (3.78 g L^{-1} vs 1.40 g L^{-1}) higher than that of the wild type under microaerobic conditions (Fig. 6c).

Discussion

ARA is an essential polyunsaturated fatty acid in human nutrition that has broad applications in many industries [26]. Although *M. alpina* is considered to be a good production strain for high level of ARA production, the low percentage of ARA in total lipids results in a production bottleneck [27]. In this study, we demonstrated that expression of *Vitreoscilla* hemoglobin significantly improves ARA and total lipid production in *M. alpina*.

Heterologous expression of VHb in *M. alpina* increased cell growth as well as oxygen uptake under non-limiting and limiting dO2 conditions. Furthermore, VHb expression shortened the lag phase for the growth of *M. alpina* and the appearance of ARA. A similar observation has previously been reported in *S. occidentalis*, in which alpha-amylase appears earlier in the culture medium of VHb-expressing cells [28], and the result is also consistent with the expression of VHb in *Gluconobacter oxydans* [29] and *Phellinus igniarius* [8]. Importantly, the morphology of the VHb-20 strain changed significantly. The mutant VHb-20 stain formed small homogeneous globules, whereas the wild-type strain formed loose mycelia (Fig. 5). Previous experiments have shown that in late fermentation stages, the wild-type strain readily agglomerates, thus

resulting in decreased dissolved oxygen. In contrast, mycelial pellets increased the fermentation fluidity, which is more conducive to dissolved oxygen and accumulation of lipids and ARA [7]. Moreover, the growth pattern of *M. alpina* has a large effect on ARA production [30]. These results are also consistent with findings by Higashiyama et al. [31] indicating that high concentrations of dissolved oxygen are necessary for obtaining smooth pellets, and the number of spherical fluffy pellets and the dissolved oxygen concentration are positively correlated.

Dissolved oxygen levels affect not only cell growth but also the synthesis of fatty acids in *M. alpina*. Therefore, the VHb-20 and wild-type strains were analyzed for fatty acid and ARA content under normal culture conditions and dissolved oxygen-limiting conditions. Studies have shown that the expression of VHb increases the yield of ARA and total lipids, especially under oxygen-limiting conditions. It has been reported previously that VHb enhances ATP production and respiratory activity [32]. Evidence has shown that VHb improves the efficiency of oxidative phosphorylation of cells by regulating the activity of the relevant respiratory oxidase, thereby changing the central carbon metabolic pathway, and ultimately promoting cell growth and protein expression under hypoxic conditions [33, 34]. In addition, VHb also has peroxidase activity, and many heterologous hosts protect against oxidative invasion by expressing this gene, in agreement with the hypothesis that VHb is a carrier of oxygen in the cell [12, 35]. On the basis of these observations, VHb expression may promote fatty acid synthesis and increase total lipid accumulation, owing to the enhancement of respiratory activity and ATP supply.

In addition, another prominent finding was that VHb expression significantly changed the fatty acid composition, particularly by decreasing the ratio of C16:0 and C18:1 and increasing the ratio of C20:4 (ARA; Table 1). The biosynthesis of ARA strongly depends on oxygen accessibility, and some specific catalytic functions of various fatty acid desaturases and elongation enzymes involved in ARA synthesis in *M. alpina* have been elucidated [36, 37]. The oxygen requirement for desaturation was higher than that for cell growth and total lipid production. Therefore, the increase in ARA content in VHb-20 may be a result of the overall increase of the fatty acid desaturase caused by the presence of heterologous hemoglobin. This speculation is consistent with the expression level of desaturases that we determined by qRT-PCR. It can be seen from Fig. 4 that four kinds of desaturases in the ARA metabolic pathway are Δ9-, Δ12-, Δ6-, Δ5- desaturases respectively, and their expression levels are significantly up-regulated under both the aerobic and microaerobic conditions. Moreover, these desaturases were the key enzymes catalyzed the palmitic acid (C16: 0), stearic acid (C18: 0) and Oleic acid (C18: 1) converted to ARA. Therefore, we hypothesized that the increase of ARA production of *M. alpina* is probably due to the increase of the expression of desaturase and the improved desaturase activities in ARA metabolic pathway and the promotion of metabolic flow to ARA synthesis. More detailed mechanisms need further study.

Conclusions

In conclusion, the successful expression of VHb in *M. alpina* significantly increases ARA and lipid production as well as promotes cell growth and increase viability under both lower and normal dissolved oxygen conditions. Therefore, we suggest that expression of the *vgb* gene in *M. alpina* might have good prospects in high-density fermentation. However, the specific mechanisms underlying the effects of VHb in *M. alpina* are not clear and require further investigation.

Abbreviations

ARA: Arachidonic acid; ATMT: *Agrobacterium tumefaciens*-mediated transformation; CO: Carbon monoxide; DCW: Dry cell weight; DGLA: Dihomo-γ-linolenic acid; DO: Dissolved oxygen; FAMEs: Fatty acid methyl esters; FID: Flame ionization detector; GLA: γ-linolenic acid; HPH: hygromycinb phosphotransferase gene; IM: Inducing medium; LA: Linoleic acid; *M. alpina*: *Mortierella alpina*; MM: Minimal medium; PUFAs: Polyunsaturated fatty acids; VHb: *Vitreoscilla* hemoglobin

Acknowledgements

The authors are thankful to Prof. Yasuyuki Kubo for providing the *Agrobacterium tumefaciens* strain C58C1 and plasmid pBIG4MRHrev. This study is a contribution to the international IMBER project.

Funding

This work is financially supported by the National Key Research and Development Program (no. 2016YFA0601400), the National Natural Science Foundation of China (no. 41306132), and the National High

Technology Research and Development Program of China (no. 2014AA021701). This study is a contribution to the international IMBER project.

Authors' contributions

Huidan Zhang performed the experiments as part of her doctoral work. All this work was carried out under the supervision of Qiu Cui and Xiaojin Song. Xiaojin Song assisted in the bioreactor experiment. Qiu Cui and Yingang Feng also assisted in editing the manuscript. All the authors have read and approved the final manuscript.

Competing interest

The authors declare that they have no competing interests.

Author details

[1]Shandong Provincial Key Laboratory of Energy Genetics, Qingdao Institute of Bioenergy and Bioprocess Technology, Chinese Academy of Sciences, No.189 Songling Road, Laoshan District, Qingdao, Shandong Province 266101, China. [2]Key Laboratory of Biofuels, Qingdao Institute of Bioenergy and Bioprocess Technology, Chinese Academy of Sciences, Qingdao, Shandong 266101, China. [3]Qingdao Engineering Laboratory of Single Cell Oil, Qingdao, Shandong 266101, China. [4]University of Chinese Academy of Sciences, Beijing 100049, China.

References

1. You JY, Peng C, Liu X, Ji XJ, Lu J, Tong Q, et al. Enzymatic hydrolysis and extraction of arachidonic acid rich lipids from *Mortierella alpina*. Bioresour Technol. 2011;102:6088–94.
2. Wang HC, Yang B, Hao GF, Feng Y, Chen HQ, Feng L, et al. Biochemical characterization of the tetrahydrobiopterin synthesis pathway in the oleaginous fungus *Mortierella alpina*. Microbiology. 2011;157:3059–70.
3. Dedyukhina EG, Chistyakova TI, Vainshtein MB. Biosynthesis of arachidonic acid by micromycetes (review). Appl Biochem Microbiol. 2011;47:109–17.
4. Nie ZK, Ji XJ, Shang JS, Zhang AH, Ren LJ, Huang H. Arachidonic acid-rich oil production by *Mortierella alpina* with different gas distributors. Bioprocess Biosyst Eng. 2014;37:1127–32.
5. Wu WJ, Yan JC, Ji XJ, Zhang X, Shang JS, Sun LN, et al. Lipid characterization of an arachidonic acid-rich oil producing fungus *Mortierella alpina*. Chin J Chem Eng. 2015;23:1183–7.
6. Dedyukhina EG, Chistyakova TI, Mironov AA, Kamzolova SV, Minkevich IG, Vainshtein MB. The effect of pH, aeration, and temperature on arachidonic acid synthesis by *Mortierella alpina*. Appl Biochem Microbiol. 2015;51:242–8.
7. Higashiyama K, Fujikawa S, Park EY, Okabe M. Image analysis of morphological change during arachidonic acid production by *Mortierella alpina* 1S-4. J Biosci Bioeng. 1999;87:489–94.
8. Zhu H, Sun S, Zhang S. Enhanced production of total flavones and exopolysaccharides via *Vitreoscilla* hemoglobin biosynthesis in *Phellinus igniarius*. Bioresour Technol. 2011;102:1747–51.
9. Li M, Wu J, Lin J, Wei D. Expression of *Vitreoscilla* hemoglobin enhances cell growth and Dihydroxyacetone production in *Gluconobacter oxydans*. Curr Microbiol. 2010;61:370–5.
10. Suen YL, Tang H, Huang J, Chen F. Enhanced production of fatty acids and astaxanthin in *Aurantiochytrium* sp. by the expression of *Vitreoscilla* hemoglobin. J Agric Food Chem. 2014;62:12392–8.
11. Tao L, Sedkova N, Yao H, Ye RW, Sharpe PL, Cheng Q. Expression of bacterial hemoglobin genes to improve astaxanthin production in a methanotrophic bacterium *Methylomonas* sp. Appl Microbiol Biotechnol. 2006;74:625–33.
12. Stark BC, Pagilla KR, Dikshit KL. Recent applications of *Vitreoscilla* hemoglobin technology in bioproduct synthesis and bioremediation. Appl Microbiol Biotechnol. 2015;99:1627–36.
13. Frey AD, Kallio PT. Bacterial hemoglobins and flavohemoglobins: versatile proteins and their impact on microbiology and biotechnology. FEMS Microbiol Rev. 2003;27:525–45.

14. Zhang S, Wang J, Wei Y, Tang Q, Ali MK, He J. Heterologous expression of VHb can improve the yield and quality of biocontrol fungus *Paecilomyces lilacinus*, during submerged fermentation. J Biotechnol. 2014;187:147–53.

15. Bhave SL, Chattoo BB. Expression ofvitreoscilla hemoglobin improves growth and levels of extracellular enzyme in *Yarrowia lipolytica*. Biotechnol Bioeng. 2003;84:658–66.

16. Wang X, Sun Y, Shen X, Ke F, Zhao H, Liu Y, et al. Intracellular expression of *Vitreoscilla* hemoglobin improves production of *Yarrowia lipolytica* lipase LIP2 in a recombinant *Pichia pastoris*. Enzym Microb Technol. 2012;50:22–8.

17. Hao G, Chen H, Gu Z, Zhang H, Chen W, Chen YQ, Cullen D. Metabolic engineering of *Mortierella alpina* for enhanced Arachidonic acid production through the NADPH-supplying strategy. Appl Environ Microbiol. 2016;82(11):3280–8.

18. Takeno S, Sakuradani E, Tomi A, Inohara-Ochiai M, Kawashima H, Ashikari T, Shimizu S. Improvement of the fatty acid composition of an oil-producing filamentous fungus, *Mortierella alpina* 1S-4, through RNA interference with delta12-desaturase gene expression. Appl Environ Microbiol. 2005;71(9):5124–8.

19. Ando A, Sumida Y, Negoro H, Suroto DA, Ogawa J, Sakuradani E, et al. Establishment of *Agrobacterium* tumefaciens-mediated transformation of an oleaginous fungus, *Mortierella alpina* 1S-4, and its application for eicosapentaenoic acid producer breeding. Appl Environ Microbiol. 2009;75:5529–35.

20. Mackenzie D, Wongwathanarat P, Carte A, Archer D. Isolation and use of a homologous Histone H4 promoter and a ribosomal DNA region in a transformation vector for the oil-producing fungus *Mortierella alpina*. Appl Environ Microbiol. 2000;66:4655–61.

21. Ando A, Sakuradani E, Horinaka K, Ogawa J, Shimizu S. Transformation of an oleaginous zygomycete *Mortierella alpina* 1S-4 with the carboxin resistance gene conferred by mutation of the iron-sulfur subunit of succinate dehydrogenase. Curr Genet. 2009;55:349–56.

22. Hao G, Chen H, Wang L, Gu Z, Song Y, Zhang H, et al. Role of malic enzyme during fatty acid synthesis in the oleaginous fungus *Mortierella alpina*. Appl Environ Microbiol. 2014;80:2672–8.

23. Takeno S, Sakuradani E, Tomi A, Inohara-Ochiai M, Kawashima H, Shimizu S. Transformation of oil-producing fungus, *Mortierella alpina* 1S-4, using Zeocin, and application to arachidonic acid production. J Biosci Bioeng. 2005;100:617–22.

24. Cheng YR, Sun ZJ, Cui GZ, Song XJ, Cui Q. A new strategy for strain improvement of *Aurantiochytrium* sp. based on heavy-ions mutagenesis and synergistic effects of cold stress and inhibitors of enoyl-ACP reductase. Enzym Microb Technol. 2016;93-94:182–90.

25. Kikukawa H, Sakuradani E, Ando A, Okuda T, Ochiai M, Shimizu S, et al. Disruption of lig4 improves gene targeting efficiency in the oleaginous fungus *Mortierella alpina* 1S-4. J Biotechnol. 2015;208:63–9.

26. Ji XJ, Zhang AH, Nie ZK, Wu WJ, Ren LJ, Huang H. Efficient arachidonic acid-rich oil production by *Mortierella alpina* through a repeated fed-batch fermentation strategy. Bioresour Technol. 2014;170:356–60.

27. Li X, Lin Y, Chang M, Jin Q, Wang X. Efficient production of arachidonic acid by *Mortierella alpina* through integrating fed-batch culture with a two-stage pH control strategy. Bioresour Technol. 2015;181:275–82.

28. Suthar DH, Chattoo BB. Expression of Vitreoscilla hemoglobin enhances growth and levels of alpha-amylase in *Schwanniomyces occidentalis*. Appl Microbiol Biotechnol. 2006;72:94–102.

29. Li MH, Wu JA, Lin JP, Wei DZ. Expression of *Vitreoscilla* hemoglobin enhances cell growth and Dihydroxyacetone production in *Gluconobacter oxydans*. Curr Microbiol. 2010;61:370–5.

30. Koike Y, Cai HJ, Higashiyama K, Fujikawa S, Park EY. Effect of consumed carbon to nitrogen ratio of mycelial morphology and arachidonic acid production in cultures of *Mortierella alpina*. J Biosci Bioeng. 2001;91:382–9.

31. Higashiyama KM, K Tsujimura H. Effects of dissolved oxygen on the morphology of an Arachidonic acid production by *Mortierella alpina* 1S-4. Biotechnol Bioeng. 1998;63:442–8.

32. Kallio PT, Kim DJ, Tsai PS, Bailey JE. Intracellular expression of *Vitreoscilla* hemoglobin alters Escherichia-Coli energy-metabolism under oxygen-limited conditions. Eur J Biochem. 1994;219:201–8.

33. Ramandeep HKW, Raje M, Kim KJ, Stark BC, Dikshit KL, et al. *Vitreoscilla* hemoglobin - intracellular localization and binding to membranes. J Biol Chem. 2001;276:24781–9.

34. Zhang L, Li Y, Wang Z, Xia Y, Chen W, Tang K. Recent developments and future prospects of *Vitreoscilla* hemoglobin application in metabolic engineering. Biotechnol Adv. 2007;25:123–36.

35. Li W, Zhang Y, Xu H, Wu L, Cao Y, Zhao H, et al. pH-induced quaternary assembly of *Vitreoscilla* hemoglobin: the monomer exhibits better peroxidase activity. Biochim Biophys Acta (BBA) - Proteins and Proteomics. 2013;1834:2124–32.

36. Sakuradani E. Advances in the production of various polyunsaturated fatty acids through oleaginous fungus *Mortierella alpina* breeding. Biosci Biotechnol Biochem. 2010;74:908–17.

37. Vongsangnak W, Ruenwai R, Tang X, Hu X, Zhang H, Shen B, et al. Genome-scale analysis of the metabolic networks of oleaginous Zygomycete fungi. Gene. 2013;521:180–90.

19

Comprehensive in silico allergenicity assessment of novel protein engineered chimeric Cry proteins for safe deployment in crops

Maniraj Rathinam, Shweta Singh, Debasis Pattanayak and Rohini Sreevathsa*（ID）

Abstract

Background: Development of chimeric Cry toxins by protein engineering of known and validated proteins is imperative for enhancing the efficacy and broadening the insecticidal spectrum of these genes. Expression of novel Cry proteins in food crops has however created apprehensions with respect to the safety aspects. To clarify this, premarket evaluation consisting of an array of analyses to evaluate the unintended effects is a prerequisite to provide safety assurance to the consumers. Additionally, series of bioinformatic tools as in silico aids are being used to evaluate the likely allergenic reaction of the proteins based on sequence and epitope similarity with known allergens.

Results: In the present study, chimeric Cry toxins developed through protein engineering were evaluated for allergenic potential using various in silico algorithms. Major emphasis was on the validation of allergenic potential on three aspects of paramount significance viz., sequence-based homology between allergenic proteins, validation of conformational epitopes towards identification of food allergens and physico-chemical properties of amino acids. Additionally, in vitro analysis pertaining to heat stability of two of the eight chimeric proteins and pepsin digestibility further demonstrated the non-allergenic potential of these chimeric toxins.

Conclusions: The study revealed for the first time an all-encompassing evaluation that the recombinant Cry proteins did not show any potential similarity with any known allergens with respect to the parameters generally considered for a protein to be designated as an allergen. These novel chimeric proteins hence can be considered safe to be introgressed into plants.

Keywords: Allergenicity, *Bacillus thuringiensis*, Cry proteins, Food crops, Transgenics, Insect resistance

Background

Agricultural biotechnology has gained enormous thrust since the latter half of twentieth century. One of the major revolutions has been the successful adoption of transgenic technology both in lab as well as in land. Development of insect resistant plants was given primary impetus as one of the primary utilities of the technology. There has been a continued quest for search of novel genes and technologies to confer insect resistance to crop plants. *Bacillus thuringiensis*-mediated insect management has been in practice since 1938 [1];

alongside its usefulness as a biopesticide, the crystalline protein genes (*cry*) of the bacterium have been used as insecticidal proteins [2]. Despite the identification of a large number of *cry* genes globally, the search for novel genes is still on [3]. However, the success of *cry* gene–based transgenic insect resistant crops has demonstrated the potential of these proteins and the adoption of the technology worldwide [4].

Among the various classes of *cry* genes, the *cry1* series of genes are generally effective against Lepidoptera and are species specific [1, 5]. In spite of the superiority in the efficacy of the genes identified thus far, the need of the hour is towards development of novel toxins with broad spectrum insecticidal activity due to the growing

* Correspondence: rohinisreevathsa@gmail.com
ICAR-National Research Centre on Plant Biotechnology, LBS Centre, Pusa Campus, New Delhi 110012, India

concern about resistance development in the insects to these toxins. This has resulted in increased focus on the identification and development of novel chimeric Cry proteins using technologies like protein engineering through domain swapping. Exploitation of these hybrid or chimeric *Bt* genes is proposed to be advantageous in providing good and long term protection against a range of pests [5–9]. The concept of chimeric proteins was developed based on the hypothesis that combining domains of validated and effective Cry proteins would not only result in improved efficacy of the resultant hybrid but would also delay the onset of resistance towards these proteins in the target insects. Several studies in our laboratory have demonstrated the efficacy of these chimeric genes that were synthesised by domain swapping of proven effective genes [8, 10, 11]. Nonetheless, it is important to analyse concomitantly whether such chimeric genes are safe and non-allergenic for consumption prior to commercialization.

Allergenicity through food and feed is one of the primary concerns to mankind as food allergy due to various substances is increasing in both adults and children [12]. With respect to transgenic food crops, it is essential to have clarity that the food crops are being engineered with proteins that do not cause any allergic symptoms in the consumers. Hence, demonstration that *cry* genes are non-allergenic is one of the ways to improve acceptance of transgenic food crops for insect resistance. Bioinformatics has come in as one of the quickest means to demonstrate whether a protein is allergenic or not based on sequence and epitope similarities. There are many reliable softwares and tools that unambiguously predict whether or not a given protein is allergenic [13, 14] based on sequence similarity, presence of allergenicity related linear motifs like IgE epitopes and physico-chemical properties of amino acids. Nevertheless, it is essential to use more than one software for explicit proof due to the increased probability of occurrence of false positives if the search is narrowed [15]. There exists a plethora of information about in silico analysis of various Cry proteins being used in the development of transgenic crops [16, 17]. However, there is no information available about the allergenicity of novel chimeric Cry proteins developed through protein engineering and domain swapping. The present study is the first of its kind to demonstrate non-allergenic potential of selected chimeric Cry proteins using a comprehensive in silico and supportive in vitro analyses.

Methods

Source of genes for in silico analysis
Cry genes with proven efficacy against lepidopteran pests were selected for protein engineering through domain swapping. Sequences of thus developed novel chimeric *cry* genes (Table 1) were converted to protein sequences in FASTA format and used for bioinformatics analysis.

Assessment of homology between query and database protein using full length FASTA search
Allergen online database 17.0
Allergen online Database (AOL) is a peer reviewed open access database maintained by Food Allergy Research and Resource Program (FARPP) and was introduced by Department of Food Science and Technology at University of Nebraska, Lincoln. The database is functional from 2007 with constant updating and has a list of 2035 allergenic and putative allergenic protein sequences (http://www.allergenonline.org/AllergenOnlineV17.pdf) and 808 taxonomical protein groups. BLOSUM 50 score matrix is used in the database to predict homology between query and database proteins. All the database entries in AOL are linked to the sequences in National Centre for Biotechnology Information (NCBI) of National Institute of Health (NIH). In the present study, full length FASTA3 search was carried out for the respective chimeric proteins using the updated database (Version#17; January 2017) with default settings; percentage similarity more than 70% was considered as putative allergen and below 50% not likely to be an allergen [18]. E-score was generated based on the similarity in protein sequence or functional analogy in amino acids. Low degree of similarity between allergenic sequences present in the database reflected high E value.

Structural database of allergenic proteins (SDAP)
SDAP is an online free web server store that stocks structural and sequence information of known allergens from international union of immunological societies (IUIS) and is linked to various protein servers – Protein Data Base (PDB), SWISS-PROT, Protein Information Resource (PIR), GenBank. This database has information on 1526 allergens and isoallergens. Full length FASTA alignment was performed in our study for all the Cry proteins against allergenic proteins in SDAP and aligned by FASTA 3.45. Query with E-value below 0.1 was considered as a putative allergen.

Assessment of homology between query and database proteins using 80mer sliding window search
The sequences of the selected chimeric Cry proteins were evaluated by FASTA search for the alignment with the listed sequences in the database. For this, each of the protein sequences was aligned in an online 80mer window in 1–80 and 2–81 sliding fashion. Any protein with >35% sequence identity as a default threshold value with known allergens was presumed to produce cross reactivity with IgEs.

Table 1 Novel protein engineered chimeric *cry* genes used in the study

Sl. No	Chimeric genes	Parental genes/Domains	Target insects
1	Cry1Aabc	Cry1Aa(D-I) Cry1Ab(D-II) Cry1Ac(D-III)	Lepidoptera
2	Cry1AcF	Cry1Ac(D-I-II) Cry1F(D-III)	Lepidoptera
3	Cry1AbBaBa	Cry1Ab(D-I) Cry1Ba(D-II-III)	Lepidoptera/Diptera
4	Cry1BaBaAb	Cry1Ba(D-I-II) Cry1Ab(D-III)	Lepidoptera/Diptera
5	Cry1AbBaAb	Cry1Ab(D-I) Cry1Ba(D-II) Cry1Ab(D-III)	Lepidoptera/Diptera
6	Cry1AcJAc	Cry 1Ac(D-I) Cry1J(D-II) Cry1Ac(III)	Lepidoptera
7	Cry1Aala5la5	Cry1Aa (D-I) Cry1Ia5(D-II-III)	Lepidoptera
8	Cry1AaB	Cry1Aa(D-1) Cry1B(D-II-III)	Lepidoptera

Assessment of homology between query and database proteins using DELTA-BLAST

Domain Enhanced look-up Time Accelerated (DELTA)-BLAST [19] predicts remote homology through blast search, providing comparatively more homologous alignment. It performs long presumed homologous alignment with conserved domains to construct its Position Specific Score Matrix (PSSMS) and facilitates efficient search compared to other blast search engines [20]. As entrez query limits the search based on the keyword entered, the selected *cry* gene sequences were run on DELTA-BLAST with Entrez Query using keyword "Allergen". Best alignment was selected based on the least E-value (E-value <0.01).

Assessment of allergenic properties of *cry* genes through machine learning language

Algpred 1.0

Algpred (www.imtech.res.in/ragava/algpred/) [21] is a web server that predicts the allergenic protein and maps IgE epitopes on the protein. Four different tools are used to predict allergenic proteins – Firstly, it predicts the allergen by using Support Vector Machine (SVM) by taking input as amino acid and dipeptide composition of proteins. Secondly, it adopts MEME/MAST tool for motif-based allergenic protein prediction. Thirdly, it aligns against 2890 Allergen Representative Peptides (ARPs) and fourthly, it predicts the known IgE epitopes on the query protein sequence. This forms a hybrid approach of high accuracy for the prediction of allergenic potential of proteins. This hybrid approach to identify the allergenic potential was performed for all the selected chimeric Cry proteins to analyse for the presence of IgE epitopes.

AllerTop 1.0

AllerTop (www.pharmfac.net/allertop) [22] is another web based server that was used in the study for validation of the chimeric proteins. It is an alignment-free server that predicts allergenicity of proteins based on the physicochemical properties of amino acids present in the protein sequence. Firstly, it uses Z-descriptors (Z_1-hydrophobicity, Z_2 - molecular size and Z_3-polarity) to represent amino acids in peptide sequence and later uses Auto and Cross-Covariance (ACC) transformation for conversion of peptide sequences to uniform vectors. It eventually uses "*k* Nearest Neighbors (*k*NN)" method that predicts the route of exposure based on three nearest neighbors of known allergens to distinguish between allergenic and non allergenic proteins.

In silico assessment of pepsin digestion sites in the chimeric proteins

In silico protein digestion test was performed by ExPASy-peptide cutter server (http://web.expasy.org/peptide_cutter/) [23]. Protein sequences of two of the chimeric proteins, Cry1AcF and Cry1Aabc were subjected to in silico pepsin digestion with recommended pH conditions (pH 1.5 and 2.0). The assay pH was described by FAO/WHO [24].

In vitro assessment of the chimeric proteins for allergenic potential

Expression and purification of Cry1Aabc and Cry1AcF proteins from E. coli

The genes *cry1AcF* and *cry1Aabc* were cloned separately in the expression vector pET-29a and mobilized into *E. coli* strain BL-21. For protein expression, 6 ml of an overnight grown culture was added to a 200 ml LB medium and incubated with vigorous shaking at 37 °C. At an OD_{600} of 0.6–1.0 of the culture, protein expression was induced with 1 mM IPTG at 37 °C for 4 h. Cells were later collected by centrifugation and resuspended in 20 ml buffer A (50 mM carbonate buffer, 150 mM NaCl, 2 mM PMSF and 10 mM βME) followed by sonication at 4 °C for 30 cycles with each cycle consisting of 10 s of sonication and 20 s of resting time. The pellets were later resuspended in buffer B (50 mM carbonate buffer, 300 mM NaCl, 2 mM PMSF, pH 9.0). A 'Protein A gravity flow column' (Biorad) was washed with 20% ethanol and was loaded onto 3 ml Ni-NTA beads (Probond resin, Novex by Life technologies) and the solid matrix (50% slurry in 20% ethanol) was equilibrated upto 2 ml mark. The column was further equilibrated with buffer B and 10 mM imidazole was added prior to protein loading. After the flow through was eluted, the slurry was washed twice with 25 ml buffer B (wash buffer I) and twice with wash buffer II (buffer B + 20 mM imidazole). Later, the protein was eluted in 1.5 ml collection tubes following the addition of 4 ml elution buffer (buffer B + 300 mM imidazole). Concentration of the purified proteins was calculated and used for further experiments.

Thermal stability of the chimeric proteins

The purified Cry1Aabc and Cry1AcF proteins were dissolved in carbonate buffer (pH 9.5) at a concentration of 0.1 mg/ml in 1.5 ml microcentrifuge tubes and incubated at 100 °C for different time intervals i.e., 10, 30, and 60 min. The experiment was stopped by placing the tubes on ice after allotted time intervals and SDS sample buffer (50 mM Tris-HCl, 8% sucrose, 2% SDS, with 5% 2-mercaptoethanol, and 0.02% bromophenol blue) was added in each tube. Control sample was prepared with 0 min incubation of the proteins (kept at 4 °C) [25]. The proteins were further separated on SDS-PAGE (silver staining) and analysed by western blot.

In vitro digestibility of the chimeric proteins with pepsin

The stability of the purified Cry1Aabc and Cry1AcF proteins was evaluated with purified porcine pepsin following the standardized procedure [26] with modifications. The purified proteins (0.1 mg/ml) were taken in simulated intestinal fluid (SIF) (Sigma, USA) with pH set at 1.2 and 2.0 [24]. Pepsin was added to the sample mixture at the concentration of 10 U/μg of test protein. The sample mixtures were incubated at 37 °C and taken out at the intervals of 0 s, 30 s, 60 s, 5 min, 10 min and 30 min. Control samples consisted of test protein in SIF buffer without pepsin with 5 min incubation, test protein in pepsin without SIF buffer with 5 min incubation and undigested intact proteins with incubation of 5 min at 37 °C. The digestibility was later evaluated by subjecting the assay mixtures to SDS-PAGE.

Results and discussion

Sequence homology-based allergenicity assessment of the chimeric proteins

Novel *cry* genes developed through protein engineering tools like domain swapping between different *cry* genes are being increasingly identified and synthesized for improved resistance management strategies. The present study, a first of its kind, demonstrated the non-allergenic potential of such protein engineered novel *cry* genes using an array of bioinformatic tools (diverse set of sequence and epitope-based algorithms) and in vitro analysis.

As the prediction of allergenicity potential of any protein cannot be achieved following a single step analysis,

a comprehensive strategy has been designed by FAO/WHO and Codex Alimentarius Commission (2003) (Fig. 1). In this investigation, full length FASTA search was performed on Allergen Online Database (AOL) with default settings and it was observed that the selected genes did not show any similarity between the proteins present in the database with expected E value <1.0; however, Cry1AcF and Cry1Ba-BaAb exhibited an E value below 1.0 (Table 2). It was observed that Cry1AcF showed 23.7% similarity against Gamma-Gliadin food allergen from bread wheat (*Triticum aestivum*) with 0.49 E-value whereas Cry1BaBaAb shared similarity with the food allergen gamma-gliadin precursor (26.8%) from bread wheat (*Triticum aestivum*) and Der f Mal f 6 allergen (24%) from *Dermatophagoides farinae* with E values 0.65 and 0.96 respectively. Results of full length FASTA search reflected that the recombinant proteins did not meet the suggested threshold value of 50% similarity [18] and served as an initial proof to demonstrate lack of homology with any of the known allergens.

SDAP revealed that all the proteins encoded by the chimeric *cry* genes showed minimal (not more than 10%) similarity with the allergenic proteins and were far below the set threshold level (Table 2). This indicated that the protein engineering had not introduced any unintended effects in the chimeric proteins and were thus safe to be integrated into food crops. Further supportive evidence was obtained when the sliding 80mer window search of the chimeric proteins did not show >35% similarity against any known allergen (Table 2) demonstrating that the search did not meet the criteria of codex for cross reactivity between the allergenic proteins and were thus safe and non-allergenic.

BLAST alignment was performed for the cry proteins with ENTREZ query limits "ALLERGEN", so that the blast would align the query with known allergenic proteins. This blast is more sensitive for the protein–protein alignment and the frequency of occurrence of false positives is less compared to other blast search engines [19]. The search demonstrated that the chimeric proteins Cry1Aabc, Cry1AbBaBa, Cry1AcJAc, Cry1AaIa5Ia5 and Cry1AaB did not show any similarity with the known allergenic proteins with an E score of <0.01. Cry1AcF showed 17% similarity with a pectate lyase/Amb allergen from different sources like *Clostridium thermocellum* and *Fibrobacter succinogenes* with E-values of 9e-07 and 2e-09 respectively and 15% identity with Fibronectin type III domain protein from *Paneni bacillus sps* JDR-2 with E-value of 6e-04. Cry1Babab and Cry1AbBaAb showed 14% similarity with bacterial allergen - Pectate lyase/Amb allergen from *Ruminiclostridium thermocellum*. Further, the study demonstrated that all the query proteins did not shown more than 35% similarity with any of the allergenic proteins reiterating their safe nature

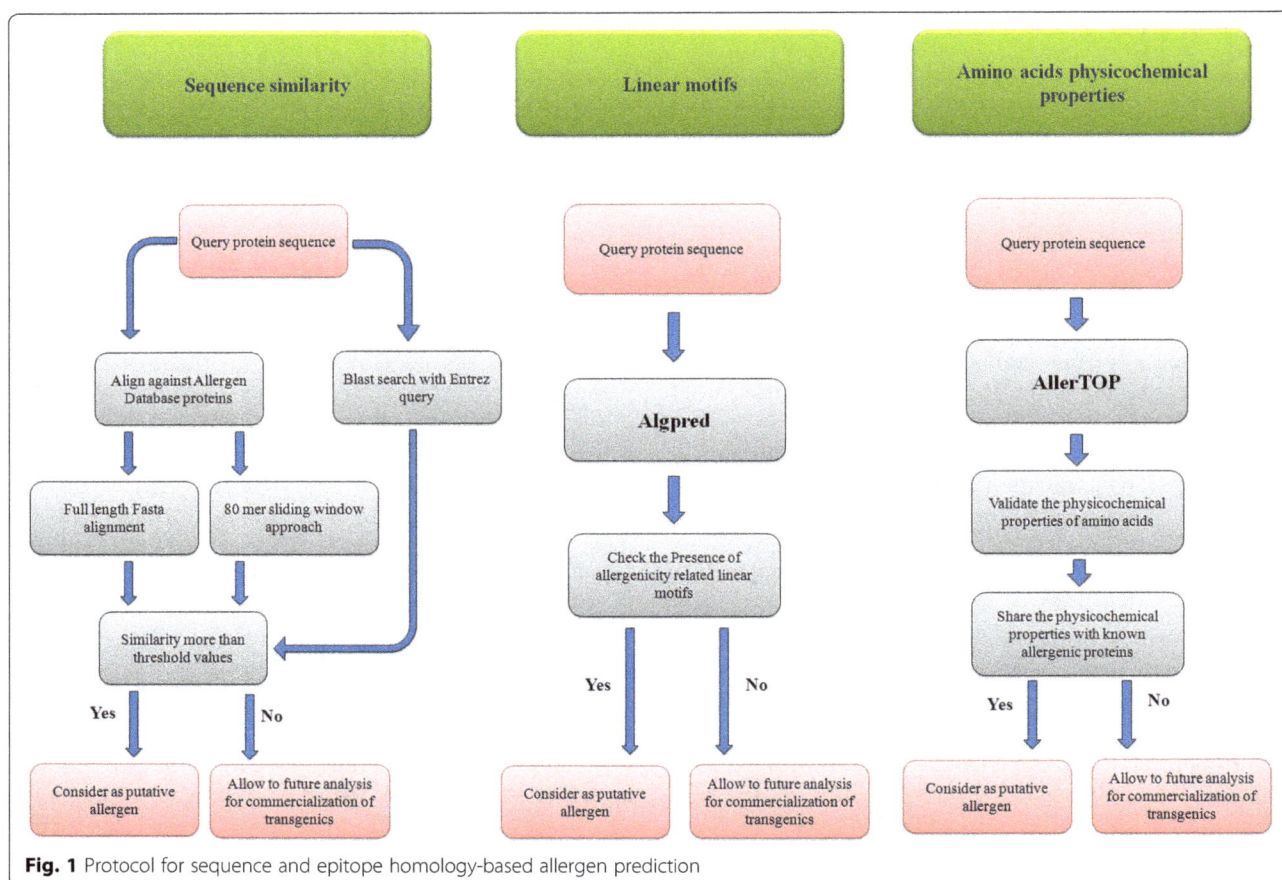

Fig. 1 Protocol for sequence and epitope homology-based allergen prediction

as per the guidelines recommended by Codex Alimentarius Commission, 2003.

The sequence homology-based identification of similarity with allergenic proteins is the generally followed pipeline for assessment of cry proteins and this was unequivocally established with respect to the chimeric proteins of the present study. However, the major concern with genetic engineering of food crops is the possibility of it being resistant to gastric digestion, which is a key characteristic nature of food allergens. It is proposed that the food matrix thus could be protecting the allergens from digestion which will allow them to retain their native structure, making the validation of conformational epitopes very important. This aspect assumes significance and needs to be given a serious thought with respect to Genetically Modified (GM) foods. Therefore, assessment of allergenicity using in silico tools exploiting the conformational epitopes is very important.

Allergenicity assessment of the chimeric proteins based on linear motifs and physicochemical properties of amino acids

In our study, analysis of the chimeric proteins based on the conformational epitopes was further authenticated

by using two specific algorithms, Algpred and AllerTop for better clarity about the non-allergenic potential of the chimeric proteins. Algpred utilizes a hybrid approach and combines four steps of motif search involving SVM, MEME/MAST, IgE epitopes and Allergen-Representative peptides (ARPs) [27] as allergen-specific protein structures and motifs have been reported in a few families of proteins [28]. Clear evidence was obtained in the present study that the eight chimeric cry proteins did not share sequence similarity with IgE epitopes of known allergens (Table 2). In the same way, it was also demonstrated in the study that there was no sequence similarity of the chimeric protein genes with the collection of ARPs [27]. Identification of motifs occurring commonly in allergens but rarely in ordinary proteins is a very important aspect before designating a protein as a non-allergen. The absence of similarity with the ARPs, which are specific to allergens showed that the genes were safe for deployment in food crops.

Algpred and other local alignment tools recommended by codex commission predict the allergenic potential of any protein based on the sequence homology by considering linear epitopes with immunogenic properties. In contrast, it is an interesting fact that IgE binding B cell epitopes are not only linear but also conformational

Table 2 Summary of in silico analyses for assessment of allergenic potential of the chimeric cry proteins

Cry Proteins	Full length FASTA search Allergen online 17.0 (% identity& E-value)	SDAP (% identity)	80mer sliding window search Allergen online 17.0 (% identity)	SDAP (% identity)	NCBI DELTA-BLAST (% identity& E-value)	Algpred 1.0	AllerTop 1.0
Cry1Aabc	No matches with E score < 1.0	9.02% Gly m Bd2 8 K (BAB21619) 5.80% Ana c 2 (BAA21849) 5.80% Tri a gliadin (AAA34285)	No matches with >35%	28.75% Asp f 13 (P28296) 26.25% Der f 1.0102 (2428875) 27.50% Alt a 2 (AAD00097)	No matches with E score < 0.01	Non –Allergen	Probably Non - Allergen
Cry1AcF	23.7% Gamma-Gliadin (170738) E - 0.49	8.99% Gly m Bd2 8 K (BAB21619) 5.78% Tri a gliadin (AAA34285) 5.78% Eur m 3 (O97370)	No matches with >35%	27.50% Hev b 1 (P15252) 27.50% Alt a 2 (AAD00097) 26.25% Lig v 1 (O82015)	17% Pectate lyase/Amb allergen (EEU00825.1) E – 9e-07 17% Pectate lyase/Amb allergen (ACX76150.1) E – 2e-06 15% Fibronectin type III domain protein (ACT03889.1) E – 6e-04	Non –Allergen	Probably Non - Allergen
Cry1AbBaBa	No matches with E score < 1.0	6.28% Ani s11.0101(BAJ78220) 4.71% Lig v 1(O82015)	No matches with >35%	28.75% Pis v 3.0101 (EF116865) 26.25% Pet c PR10 (CAA67246) 26.25% Lig v 1 (O82015)	No matches with E score < 0.01	Non –Allergen	Probably Non - Allergen
Cry1BaBaAb	26.8% Gamma-Gliadin precursor (1063270) E – 0.65 24.0% Der f Mal f 6 allergen (37958141) E - 0.96	5.58% Mala s 1 (Q01940) 4.50% Asp f 17 (CAA12162) 4.19% Pis v 4.0101 (EF470980)	No matches with >35%	32.50% Mala s 1 (Q01940) 27.50% Asp f 17 (CAA12162) 26.25% Asp f 10 (CAA59419)	14% Pectate lyase/Amb allergen (ABN54148.1) E - 8e-08 14% Pectate lyase/Amb allergen (EEU00825.1) E - 7e-08	Non –Allergen	Probably Non - Allergen
Cry1AbBaAb	No matches with E score < 1.0	5.77% Mala s 1 (Q01940) 4.81% Lig v 1 (O82015) 4.81% Asp f 17 (CAA12162)	No matches with >35%	32.50% Mala s 1 (Q01940) 27.50% Asp f 17 (CAA12162) 26.25% Asp f 10 (CAA59419)	14% Pectate lyase/Amb allergen (EEU00825.1) E - 9e-08 14% Pectate lyase/Amb allergen (ABN54148.1) E - 1e-07	Non –Allergen	Probably Non - Allergen
Cry1AcJAc	No matches with E score < 1.0	7.62% Gly m 6.0501 (Q7GC77)	No matches with >35%	32.50% Api m 3.0101 (NP_00101337)	No matches with E score < 0.01	Non –Allergen	Probably Non - Allergen

Table 2 Summary of in silico analyses for assessment of allergenic potential of the chimeric cry proteins (*Continued*)

Protein	FASTA full-length (E score)	80-mer window (>35%)	8-mer / E score < 0.01		
Cry1Aala5la5	No matches with E score < 1.0 7.28% Api m 3.0101 (NP_00101337) 5.79% Eur m 3 (O97370)	No matches with >35% 8.41% Per a 3.0101 (Q25641) 6.23% Sol i 1.0101 (AAT95008) 5.51% Asp f 4 (O60024)	No matches with E score < 0.01 30.0% Dol m 1.0101 (Q06478) 28.75% Gly m 1 (AAB09252) 26.5% Lig v 1 (O82015) 26.5% Pol f 5 (P35780) 25% Ana c 1 (AAK54835)	Non-Allergen	Probably Non-Allergen
Cry1AaB	No matches with E score < 1.0 6.21% Ani s 11.0101 (BAJ78220) 5. Car p papain (AAB02650) 4.94% Alt a 2 (AAD00097)	No matches with >35%	No matches with E score < 0.01 27.50% Alta2 (AAD00097) 26.50% Ligv1 (O82015) 25% Pen c 24 (AAR17475)	Non-Allergen	Probably Non-Allergen

epitopes, and therefore share less homology with proteins [29]. It has been earlier reported that allergen specific patches in proteins are dominated by hydrophobic amino acids on the surface [30], leading to the inability of the alignment-based approaches to detect allergenicity in an unambiguous manner. This defines the need for prediction of allergenicity based on the physicochemical properties of amino acids like hydrophobicity, molecular size and polarity. Our study presents the utility of Allertop, a server for allergen prediction which is an alignment-free bioinformatics tool based on the ACC transformation of protein sequences into uniform equal length vectors followed by evaluation based on similarity in physico-chemical properties between the test protein and nearest neighbors. When analyzed individually by AllerTop server, all the eight chimeric Cry proteins showed the absence of allergenicity reiterating the earlier analyses of absence of allergenic potential in them.

Most of the dietary proteins that are ingested in the human gut are immediately exposed to hydrolytic digestion and/or degradation. However, an important character of allergenic proteins is the stability to the activity of digestive enzymes like pepsin etc. In silico as well as in vitro analysis can be used to delineate them based on the ability of pepsin to digest the proteins. In our study, two of the eight chimeric proteins, Cry1AcF and Cry1Aabc were further assessed using ExPASy-peptide cutter server to demonstrate the presence of pepsin digestibility sites [Additional file 1]. It was seen that the number of pepsin sites in the chimeric toxins were similar to that of the already analysed Cry1C and Cry1Ac toxins (igmoris.nic.in/files/Biosafety_data/Biosafety/metahelix) further confirming the non allergenicity of the chimeric cry proteins. This study is the first of its kind to use this tool and explicitly demonstrate the non-allergenic potential of the selected protein engineered toxins.

In vitro assessment of the chimeric proteins for allergenic potential

An integral part of the safety assessment of GM food lies in the fool proof knowledge and demonstration that the transgene deployed in the crop is not a food allergen. Concomitant to the in silico analyses to establish this, various in vitro analyses are carried out as a part of risk assessment before judging it to be safe [31]. This involves a number of biochemical and toxicological studies as outlined by the biosafety regulators. Processing of the GM plant material prior to or after ingestion in the human gut is a vital aspect to be considered for

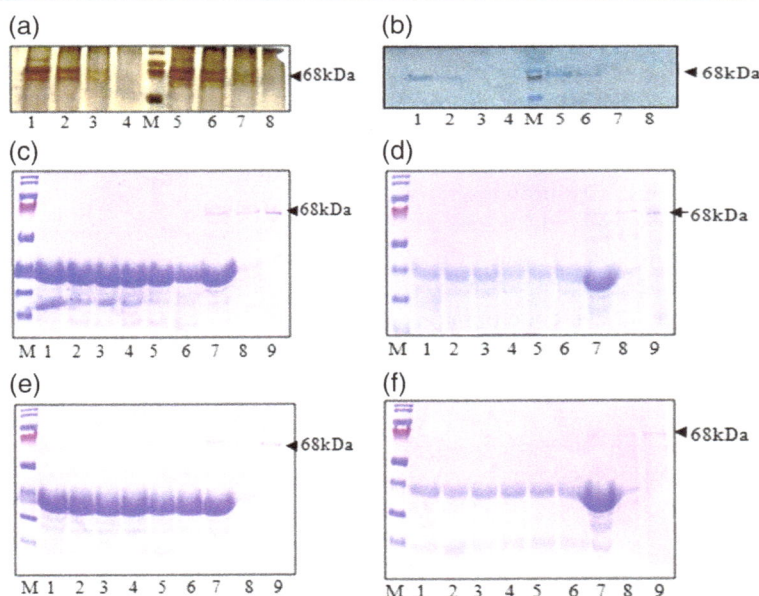

Fig. 2 In vitro assessment of the chimeric proteins for allergenic potential. Analysis of the chimeric proteins for thermal stability: a and b SDS PAGE and Western blot analysis respectively showing stability of Cry1AcF and Cry1Aabc proteins after incubation at 100 °C for different time intervals. Lane 1- Cry1AcF without incubation, Lane 2–10 min incubation, Lane 3–30 min incubation, Lane 4–60 min incubation, M-Marker, Lane 5-Cry1Aabc without incubation, Lane 6–10 min incubation, Lane 7–30 min incubation, Lane 8–60 min incubation. Analysis of the chimeric proteins for pepsin digestibility: c and d SDS PAGE for in vitro digestibility of Cry1AcF protein in SIF (simulated intestinal fluid) with pepsin at pH 1.2 and 2.0 respectively and incubated at 37 °C for different time periods. e and f SDS PAGE for in vitro digestibility of Cry1Aabc protein in SIF with pepsin at pH 1.2 and 2.0 respectively and incubated at 37 °C for different time periods. (M- Marker, Lane 1–0 s incubation, Lane 2–30s incubation, Lane 3–60s incubation, Lane 4-5 min incubation, Lane 5–10 min incubation, Lane 6–30 min incubation, Lane 7- test protein with pepsin and without SIF and 30 min incubation, Lane 8- test protein with SIF and without pepsin and 30 min incubation, Lane 9- Intact test protein with 30 min incubation)

toxicological safety assessment. Availability of the intact protein during absorption in the gut not only shows that it is heat stable but also depicts its resistance to the action of digestive enzymes. These parameters provide strong evidence to demonstrate the allergenic potential of proteins, as allergens are known to be both heat stable as well as resistant to digestion by the enzymes in the gut.

Thermal stability of the chimeric proteins
In the present study, two chimeric proteins, Cry1AcF and Cry1Aabc were chosen for the assessment of two important parameters, resistance to heat and the digestive enzyme pepsin. Incubation of the chimeric proteins at 100 °C for different time periods ranging from 0 to 60 mins demonstrated that both the chimeric proteins degraded completely after 10 mins (Fig. 2 a and b). Silver staining and western blot analysis of the assay mixtures corroborated the time of degradation indicating that the cry proteins could be degraded into simpler structures and therefore were safe for consumption.

In vitro digestibility of the chimeric proteins with pepsin
Evaluation of the allergenic potential of proteins based on the digestibility by purified porcine pepsin is one of the widely used methodologies worldwide [31]. Porcine pepsin is an aspartic endopeptidase with broad substrate specificity and optimal activity between pH 1.2 and 2.0. A validated in vitro assay [26] that uses a fixed porcine pepsin: protein ratio and simulated intestinal fluid (SIF) under both pH 1.2 and 2.0 (as per the recommendations of FAO/WHO) was used in the present study. The assay demonstrated that both the chimeric Cry proteins were completely digested as soon as they were added into the assay mix (0 s) in both pH 1.2 as well as pH 2.0 because of the presence of active pepsin (Fig. 2 c–f). However, intact cry proteins were observed under conditions either lacking SIF or porcine pepsin indicating the susceptibility of the chimeric proteins to pepsin activity. This is in coherence with the observation that, following pepsin digestion, the non-allergenic food proteins were digested within approximately 30 s while major food allergens exhibited pepsin-stable fragments that were detectable even after 8–60 min.

Conclusion
This study therefore reconfirms that the selected chimeric proteins encoded by the selected chimeric genes are of significance both towards scientific and safety perspective. Based on the comprehensive bioinformatic and supportive in vitro analyses, we demonstrated that the selected chimeric Cry proteins complied with parameters used to identify a protein as non-allergenic. Consumption of these proteins following deployment in transgenic crops will

therefore not result in triggering of any allergenic response. The study also demonstrates the usefulness of protein engineering as an additional alternative to engineer insect resistance.

Abbreviations
ACC: Auto and cross-covariance; ARPs: Allergen representative peptides; *Bt*: *Bacillus thuringiensis*; DELTA-BLAST: Domain enhanced look-up time accelerated–BLAST; GM: Genetically modified; KNN: *k* nearest neighbors; PSSMS: Position specific score matrix; SVM: Support vector machine

Acknowledgements
The authors wish to acknowledge Dr. P. Ananda Kumar, Dr. K. Sushmitha, Dr. Paul Anderson K, Dr. Shabana Khan for providing the sequences of the chimeric Cry proteins.

Funding
The authors acknowledge financial support from ICAR-National Agricultural Science Fund (NFBSFARA/PB-2010).

Authors' contributions
MR and RS developed the idea for the study. The in vitro experiments were conducted by SS. The manuscript was drafted by MR and RS and all authors contributed in analyzing the data and writing the draft. DP critically edited the manuscript and overall data presentation. All authors read and approved the final manuscript.

Competing interests
The authors declare that they have no competing interests.

References
1. Kumar PA, Bambawale OM. Advances in Microbial toxin Research: Insecticidal proteins of *Bacillus thuringiensis* and their applications in Agriculture. In: Upadhyay, editor. New York: Kluwer Academic/Plenum Publishers; 2002. p. 259–280.
2. Bates SL, Zhao JZ, Roush RT, Shelton AM. Insect resistance management in GM crops: past, present and future. Nat Biotechnol. 2005;23:57–62.
3. Mahalakshmi A, Shenbagarathai R, Sujatha K. Identification of novel indigenous *Bacillus thuringiensis* isolates. Indian J Exp Biol. 2005;43:867–72.
4. James C. Global status of commercialized biotech/GM crops. The International Service for the Acquisition of Agri-biotech Applications; 2014. http://www.isaaa.org/resources/publications/briefs/49/. Accessed 5 Jan 2017.
5. Chakrabarti SK, Mandaokar AD, Kumar PA, Sharma RP. Synergistic effect of cry1Ac and cry1F delta endotoxins of *Bacillus thuringiensis* on cotton bollworm, *Helicoverpa armigera*. Curr Sci. 1998;75:663–4.
6. Ramu SV, Rohini S, Keshavareddy G, Neelima MG, Shanmugam NB, Kumar ARV, Sarangi SK, Kumar PA, Udayakumar M. Expression of a synthetic *cry1AcF* gene in transgenic pigeon pea confers resistance to *Helicoverpa armigera*. J Appl Entomol. 2012;136:675–87.
7. Frutos R, Rang C, Royer M. Managing insect resistance in plants producing *Bacillus thuringiensis* toxins. Crit Rev Biotechnol. 1999;19:227–76.
8. Sushmita K. Gene pyramiding in Bt transgenic tomato and development of novel *Bt* toxin by protein engineering. PhD thesis, Ch. Charan Singh University, Meerut, Uttar Pradesh; 2012.
9. Ho NH, Baisakh N, Oliva N, Datta K, Frutos R, Datta SK. Translational fusion hybrid *Bt* genes confer resistance against yellow stem borer in transgenic elite vietnamese rice (*Oryza sativa* L.) cultivars. Crop Sci. 2006;46:781–9.

10. Reddy KG, Rohini S, Ramu SV, Sundaresha S, Kumar ARV, Ananda Kumar P, Udayakumar M. Transgenics in groundnut (*Arachis hypogaea* L.) expressing *cry1AcF* gene for resistance to *Spodoptera litura* (F.). Physiol Mol Biol Plants. 2013;19:343–52.

11. Anderson PK. Production and characterization of transgenic *Brassica oleracea* carrying delta-endotoxin genes of *Bacillus thuringiensis*. Ph.D thesis, Jamia Milia Islamia, New Delhi; 2002.

12. Sicherer SH, Sampson HA. Food allergy: epidemiology, pathogenesis, diagnosis and treatment. J Allergy Clin Immunol. 2014;133:291–307.

13. Thomas K, Bannon G, Hefle S, Herouet C, Holsapple M, Ladics G, MacIntosh S, Privalle L. *In silico* methods for evaluating human allergenicity to novel proteins: international bioinformatics workshop meeting report, 23–24 February. Toxicol Sci. 2005;88:307–10.

14. Sekhar CB, Sachin CH, Raman BV, Bondili JS. Molecular characterization and *in silico* analysis of sorghum panallergens: profilin and polcalin. Ind J Exp Biol. 2015;53:726–31.

15. Herman RA, Song P, Kumpatla S. Percent amino acid identity thresholds are not necessarily conservative for predicting allergenic cross-reactivity. Food Chem Toxicol. 2015;8:141–2.

16. Randhawa GJ, Singh M, Grover M. Bioinformatic analysis for allergenicity assessment of *Bacillus thuringiensis* cry proteins expressed in insect-resistant food crops. Food Chem Toxicol. 2011;49:356–62.

17. Mathur C, Kathuria PC, Dahiya P, Singh AB. Lack of detectable allergenicity in genetically modified maize containing "cry" proteins as compared to native maize based on *in silico* & in vitro analysis. PLoS One. 2015. doi:10.1371/journal.pone.0117340.

18. Aalberse RC. Structural biology of allergens. J Allergy Clin Immunol. 2000;106:228–38.

19. Boratyn GM, Schäffer AA, Agarwala R, Altschul SF, Lipmanand DJ, Madden TL. Domain enhanced lookup time accelerated BLAST. Biol Direct. 2012;7:12.

20. Marchler BA, Lu S, Anderson B, Chitsaz F. CDD: a conserved domain database for the functional annotation of proteins. Nucleic Acids Res. 2011;39:225.

21. Saha S, Raghava GPS. Algpred: prediction of allergenic proteins and mapping of IgE epitopes. Nucleic Acids Res. 2006;34:202–9.

22. Dimitro I, Flower RD, Docytchinova I. AllerTOP – a server for *in silico* prediction of allergens. BMC Bioinfo. 2013;14:1471–2105.

23. Gasteiger E, Hoogland C, Gattiker A, Duvaud S, Wilkins MR, Appel RD, Bairoch A. The proteomics protocols handbook: protein identification and analysis tools on the ExPASy server. New York: Humana Press; 2005.

24. FAO/WHO. Evaluation of the allergenicity of genetically modified foods: report of a joint FAO/WHO expert consultation on the allergenicity of foods derived from biotechnology. Rome; 2001.

25. Xu W, Cao S, He X, Luo Y, Guo X, Yuan Y, Huang K. Safety assessment of Cry1Ab/ac fusion protein. Food Chem Toxicol. 2009;47:1459–65.

26. Astwood JD, Leach JN, Fuchs RL. Stability of food allergens to digestion in vitro. Nature Biotech. 1996;14:1269–73.

27. Bjorklind AK, Soeria-Atmadja D, Zorzet A, Hammerling U, Gustafsson MG. Supervised identification of allergen representative peptides for *in silico* detection of potentially allergenic proteins. Bioinfo. 2005;21:39–50.

28. Radauer C, Bublin M, Wangner S, Mari A, Breiteneder H. Allergens are distributed into few protein families and possess a restricted number of biochemical functions. J Allergy Clin Immunol. 2008;121:847–52.

29. Chruszcz M, Pomés A, Glesner J, Vailes DL, Osinski T, Porebski JP, Majorek A. K.Heymann WP, Platts-Mills TAE, minor W, Chapman DM. Molecular determinants for antibody binding on group 1 house dust mite allergens. J Biol Chem. 2012;287:7388–98.

30. Furmonaviciene R, Sutton JB, Glaser F, Laughton AC, Jones N, Sewell FH, Shakib F. An attempt to define allergen-specific molecular surface features: a bioinformatic approach. Bioinfo. 2005;21:4201–4.

31. Koch MS, Ward JM, Levine SL, Baum JA, Vicini JL, Hammond BG. The food and environmental safety of *Bt* crops. Frontiers Plant Sci. 2015. doi:10.3389/fpls.2015.00283.

20

Effect of acetic acid on ethanol production by *Zymomonas mobilis* mutant strains through continuous adaptation

Yu-Fan Liu[1,2], Chia-Wen Hsieh[3*], Yao-Sheng Chang[3] and Being-Sun Wung[3]

Abstract

Background: Acetic acid is a predominant by-product of lignocellulosic biofuel process, which inhibits microbial biocatalysts. Development of bacterial strains that are tolerant to acetic acid is challenging due to poor understanding of the underlying molecular mechanisms.

Results: In this study, we generated and characterized two acetic acid-tolerant strains of *Zymomonas mobilis* using N-methyl-N'-nitro-N-nitrosoguanidine (NTG)-acetate adaptive breeding. Two mutants, ZMA-142 and ZMA-167, were obtained, showing a significant growth rate at a concentration of 244 mM sodium acetate, while the growth of *Z. mobilis* ATCC 31823 were completely inhibited in presence of 195 mM sodium acetate. Our data showed that acetate-tolerance of ZMA-167 was attributed to a co-transcription of *nha*A from ZMO0117, whereas the co-transcription was absent in ATCC 31823 and ZMA-142. Moreover, ZMA-142 and ZMA-167 exhibited a converstion rate (practical ethanol yield to theorical ethanol yield) of 90.16% and 86% at 195 mM acetate-pH 5 stress condition, respectively. We showed that acid adaptation of ZMA-142 and ZMA-167 to 146 mM acetate increased ZMA-142 and ZMA-167 resulted in an increase in ethanol yield by 32.21% and 21.16% under 195 mM acetate-pH 5 stress condition, respectively.

Conclusion: The results indicate the acetate-adaptive seed culture of acetate-tolerant strains, ZMA-142 and ZMA-167, could enhance the ethanol production during fermentation.

Keywords: Lignocellulosic hydrolysates, Acetic acid, Bioethanol, *Zymomonas mobilis*, Acid adaptation

Background

Lignocellulosic biomass is attractive as a feedstock because of its renewability and remarkable availability in industrial bioconversion. Use of lignocellulosic biomass for ethanol genesis through biochemical processes presents technical difficulties [1]. The main challenge is suggested to be pretreatment and enzymatic hydrolysis steps with which fermentable sugars are released from the biomass. These steps have been suggested to generate a variety of toxic compounds and form a stressful environment which inhibits microbial fermentations [2–4]. Inhibitors include furans, phenols, organic acids, aldehydes and alcohols [5, 6] of these inhibitors, acetic acid is known at the predominant inhibitor formed from lignocellulosic hydrolysates during pretreatment [7].

Toxicity of acetic acid to ethanologens is suggested to be pH based growth inhibition and disruption of pH and anion pool of cytoplasm. Strategies have been developed to overcome such issues for high ethanol production including removal of fermentation inhibitors and use of inhibitor-tolerant strains. However, there is currently no ideal bacterial ethanologen available for use in an industrial setting.

Zymomonas mobilis, a Gram-negative facultative bacterium, is considered as an ideal organism for large scale ethanol production. It has several favorable industrial characteristics including high ethanol production, excellent ethanol tolerance, use of the Entner-Doudoroff pathway and a broad range of working pH. Wild type *Z. mobilis* strains are known for a lack of pentose metabolism pathway and sensitivity to inhibitors formed during pretreatment [7–9]. Several mutant strains of *Z. mobilis* have been engineered to overcome substrate limitation and growth inhibition including ZM4/AcR [10], AX101

* Correspondence: cwhsieh@mail.ncyu.edu.tw
[3]Department of Microbiology, Immunology and Biopharmaceuticals, National Chiayi University, Chiayi, Taiwan
Full list of author information is available at the end of the article

[11] and ZM4(pZB5) [12]. Techniques employed to obtain mutant *Z. mobilis* including chemical mutagenesis, genetic recombination, metabolic engineering and evolutionary adaptation [13–15]. It is of interest to develop an appropriate approach that facilitate industrial strain improvement.

In this study, two sodium acetate-tolerant mutant of *Z. mobilis* ZMA-142 and ZMA-167 were generated by chemical mutagenesis and adaptive evolution from parental strain *Z. mobilis* ATCC 31823 (ZM481). Full characterization of the cell growth behavior, ethanol fermentation characteristics, and metabolic profiles were conducted.

Methods
Bacterial stains and culture conditions
Zymomonas mobilis ZM481 (ATCC31823) was chosen and used as a starting strain for adaption. ZM481 was maintained in Rich medium (RM; 20.0 g/L glucose, 10.0 g/L yeast Extract, 2.0 g/L KH_2PO_4, pH 5.0) at 30 °C without shaking. Adapted strains were cultured with RM medium containing 100 g/L glucose for investigation on its profiling of cell growth, glucose utilization, and ethanol yield under normal or stress conditions. Bacterial strains were grown overnight, mixed with equal volume of 60%(w/v) glycerol solution and stored at −80 °C.

Fermentations
Bacterial strains were revived from glycerol stock in RM broth for 6-8 h, followed by streaking the culture on RM agar plate. Colonies taken from plates were grown for 18 h without shaking at 30 °C in RM broth. Resulting pre-seed cultures were inoculated at a ratio of 2% (v/v) into seed medium (100 g/L glucose, 5 g/L yeast extract (Oxoid), 1 g/L $MgSO_4.7H_2O$, 1 g/L $(NH_4)_2SO_4$ and 2 g/L KH_2PO_4). Seed cultures were grown at 30 °C without shaking to mid growth phase and subsequently inoculated at 10%(v/v) into fermentation medium (FM), which was identical in composition to the seed medium except that final concentrations of up to 196 mM sodium acetate (Ajax Chemicals, NSW, Australia, AR grade). Batch fermentation was conducted in a 5-L jar fermentor (New Brunswick Scientific, Enfield, CT, USA) with 3 L working volume fermenters at controlled pH 5 and temperature 30 °C. Fermentation medium was inoculated (10% v/v) with seed culture and incubated with agitation at 200 rpm. The culture medium was FM contained with/out 195 mM sodium acetate and the pH in 5.0. The initial number of cells (CFU) for all tests was approximately equal 2.0×10^7 mL-1.

N-methyl-N-nitro-N-nitrosoguanidine (NTG) treatment
The acetate- tolerant mutant of *Z. mobilis* was selected and mutated with NTG. 1 mL of NTG solutions (0.4 mg/mL) were added to 9 mL of the cell suspension

of ZM481. The mixture was incubated at 37 °C and agitated at 200 rpm for 30 and 45 min, followed by a centrifugation at 8000×g for 10 min. The pellet was washed thrice with 3 mL of 0.85% NaCl and then resuspended in 1 mL of the seed medium. The culture was incubated for 24 h at 30 °C and subsequently repetitively cultivated in the adaptive evolution medium.

Adaptive evolution experiments
The concentrations of acetic acid in lignocellulosic hydrolysates are known in the range from 16.7 to 258 mM (1, 23). In this study, initial concentration of acetate for adaption was 195 mM and the concentrations increased up to 270 mM at pH 5. The procedures of adaptive mutation were described in brief as follows. Firstly, NTG treated ZM481 was grown in RM broth in the absence of sodium acetate acetic acid at pH 7.0 and 30 °C without shaking overnight. The culture was inoculated into a new 10-ml Falcon tubes containing 5 ml of RM broth supplemented with 195 mM sodium acetate, followed by transferring into the same medium using the method of serial batch transfer for 3 times. The resulting culture was plated onto RM agar plate supplemented with 195 mM sodium acetate and incubated at 30 °C. 400 individual colonies were selected and subject to assessment for cell growth under stress conditions as 220, 250 and 270 mM sodium acetate pH 5, respectively. Two strains, namely ZMA-142 and ZMA-167 were obtained and selected for further analyses.

Analytical methods
The concentration of glucose was determined using YSI 2300 STAT plus analyser (Yellow Springs Instrument Co., USA). Ethanol concentration was estimated using gas chromatography with a glass 4 mm ID 32 m Porapak Q 100/120 mesh column operated isothermally with N_2 as carrier and a flame ionisation detector. Peak areas were determined with an integrator.

RNA preparation and analysis
Total RNA was prepared as previously described (24). Briefly, cells were harvested by centrifugation, followed by a total RNA extraction using TRIzol reagent (Invitrogen). Purification of total RNA was perfoemd using NucleoSpin® RNA clean-up (MACHEREY-NAGEL, Germany) according to the manufacturer's instructions. The RNA quality and quantity were assessed by formaldehyde agarose gel electrophoresis and measured by calculating absorbance ratio at both 260/280.

Reverse transcription PCR
Reverse transcription was carried out using four primers, ZMO0117F (TGTGATGGTATC AAAAGCGGTC), ZM O0117R (CCAAATCGGTGACACGGAA), ZMO0119F

(CTGCTCTTATCCGCCCTTC), ZMO0119R (GGAAA GAAGCCAGATGTCCC) in combinations. Reverse transcription PCR was performed using Eppendorf® Mastercycler Personal within Super 2 RT-PCR Mix (Protech). The reverse transcription PCR conditions were set as following: 30 min at 45 °C, 10 min at 95 °C, 25 cycles at 95 °C for 30 s, followed by 55 °C for 30 s and 72 °C for 70 s, and final extension at 72 °C for 10 min. The reverse transcription PCR product was subject to electrophoresis in 1.5%

agarose gel incorporating with Omics 100 bp Plus DNA RTU Ladder.

Genomic DNA capture and sequencing

Genomic DNA samples measured with O.D. 260/280 in the range 1.8 ~ 2.0, and quantity ratio by Qubit versus NanoDrop over 0.7 were considered as acceptable for further processing. The sequencing library construction and sequencing were performed by Welgene Biotech Co., Ltd., Taiwan. Isolated DNA was sonicated by Misonix 3000 sonicator to the size ranging from 400 to 500 bp. DNA sizing was checked by bioanalyzer DNA 1000 chip (Agilent TechnologiesTM). All subsequent steps were performed on the instrument, including end-repaired, A-tailed and adaptor-ligated. After library construction, samples were mixed with MiSeq Reagent Kit (600-cycle) and loaded onto Illumina Solexa Miseq platform, which was performed by Welgene Biotech Co., Ltd.

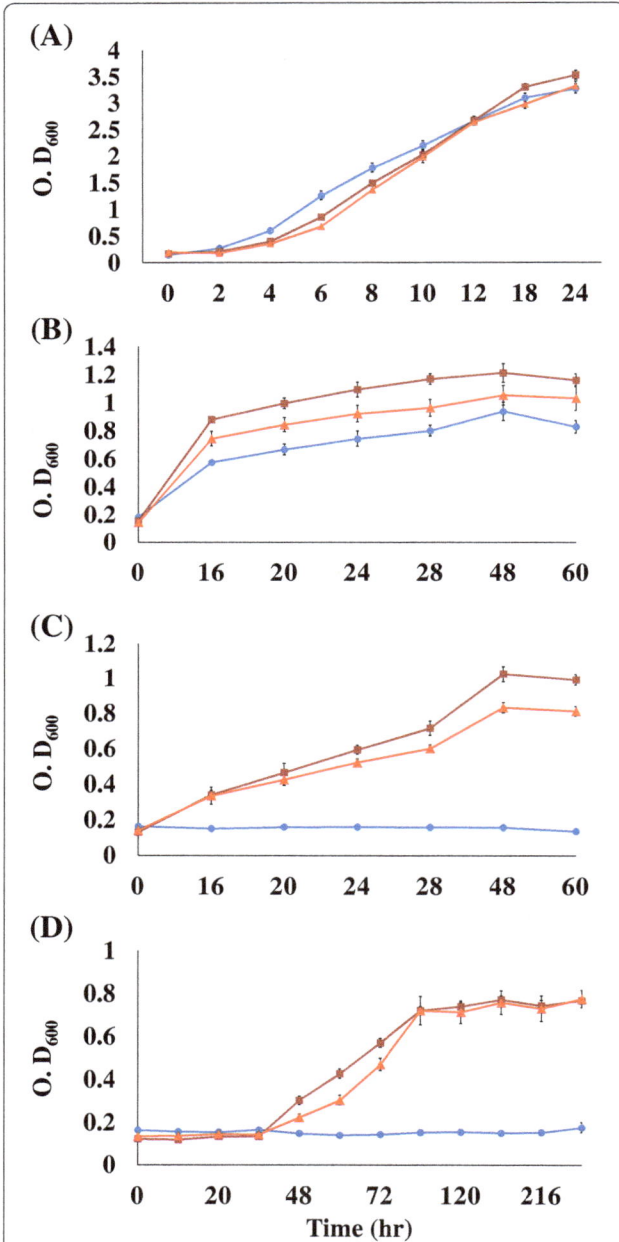

Fig. 1 Comparison of growth by the acetate-tolerant mutant ZMA-167(■), ZMA-142(▼) and ZM481(●) with 0 mM (a), 146 mM (b), and 195 mM (c) and 244 mM (d) sodium acetate (500-mL flask, pH 5.0, 30oC). The data were presented in mean values of triplicate experiments

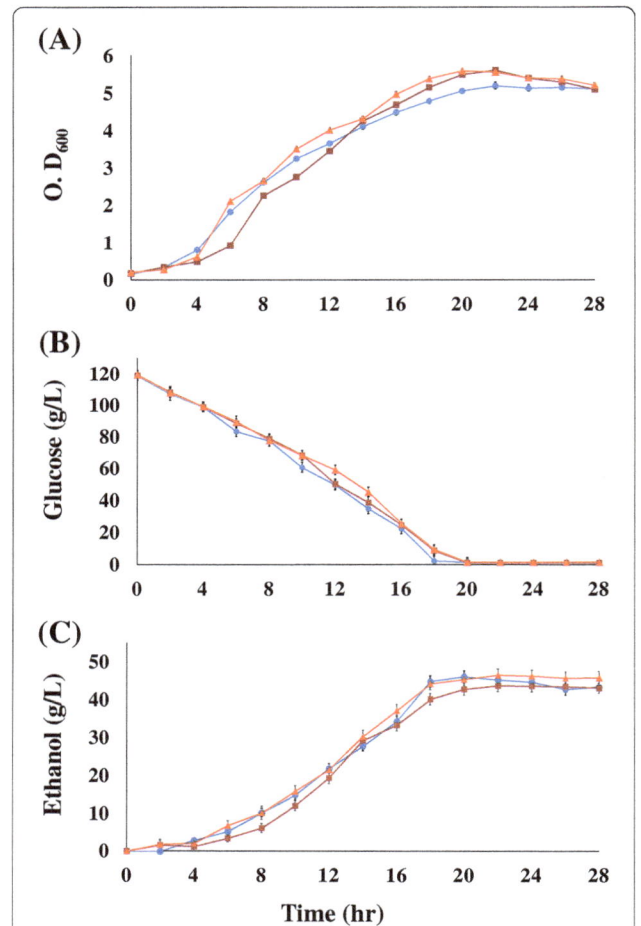

Fig. 2 Profiling of cell growth (a), glucose utilization (b), and ethanol yield (c) of the acetate-tolerant mutant ZMA-167(■), ZMA-142(▲) and ZM481(●) grown in 5 L fermenter(pH 5.0, 30oC) glucose utilization (b), and ethanol yield (c) of the acetate-tolerantwere presented in mean values of triplicate experiments

(Taipei, Taiwan) [16]. Automated cluster generation and a 2 × 300 bp paired-end sequencing run were performed.

Sequence analysis and alignment

The Miseq reads were trimmed and filtered by Illumina protocols to ensure the high quality data for all following analyses. To evaluate the raw reads quality by CLC-relative standard NGS packages was used to trim reads that did not achieve sufficient quality PHRED score > 20), excluding short sequences (<30 bp). Sequences were aligned to the ZM4 ATCC31821 (accession number, NC006526) reference genome using BWA-SW read alignment programs [17, 18]. To visualize reads coverage and insertion/deletion junctions, estimate transcripts structure and evaluate different expression patterns in the different mutants by Integrative genomics viewer (IGV) [19].

Results

Improvement of acetate tolerance through NTG mutagenesis

ZM481 is known for its limited growth in presence of high concentration of acetate. To obtain a library of mutants of ZM 481, chemical mutagenesis using NTG was performed. Stepwise adaptive evolution was employed for screening acetate-tolerant mutants. NTG-treated ZM481 was repetitively cultured in the adaptive evolution medium with stepwise increasing concentrations of sodium acetate up to 270 mM. In total, nine rounds of mutagenesis and selection were performed for the improvement of acetate tolerance. A total of 400 mutants

were obtained from NTG-treated ZM 481 grown in present of 195 mM sodium acetate. Adapting ZM481 to serial concentrations of sodium acetate, there were 20 mutants identified with abilities to grow in a condition of 270 mM sodium acetate. Of the 20 mutants, two mutants, namely ZMA-142 and ZMA-167, were genetically stable for over 20 generations and therefore chosen for further characterization.

Characterization of acetate-tolerant ZMA 142 and ZMA 167

We next characterized the growth features of ZM481 and two acetate-tolerant mutants, ZMA142 and ZMA-167, in culture supplemented with sodium acetate at serial concentrations of 0, 146, 195 and 243 mM. As shown in Fig. 1A, the differences in cell growth among three strains after 12-h under normal FM culture condition (pH 5.0) were insignificant, showing that ZM481, ZMA-142 and ZMA-167 reached maximum optical densities at OD600 of 3.5, 3.3 and 3.5, respectively. The data showed that increasing concentrations of sodium acetate led to a reduction in cell growth rate in three strains. ZMA-167 exhibited higher growth rate at concentrations of 146 and 195 mM after an incubation of 60 h compared with those of ZMA-142 (Fig. 1b and c). Interestingly, the growth rates of two mutants at high concentration of acetate (243 mM) were identical after 240 h incubation, whereas ZMA-167 exhibited better growth than ZMA-142 in first 60 h incubation. The growth of ZM481 was completely inhibited by sodium acetate at a concentration of 195 mM.

Table 1 Effect of acetate-adaptive seed culture on ZMA-142 and ZMA-167 in a 5 L fermenter using medium containing 120 g glucose/L, 195 mM sodium acetate and controlled at pH 5.0

Strains	Fermentation time (hr)	Glucose utilization (g/L)	Ethanol Production (g/L)	$q_s$max[a]	$q_p$max[b]	Yp/s[c]	Conversion Rate (%)[d]
Seed culture with RM; Fermentation in FM							
ZM 481	20	107.2	46.1	5.4	2.3	0.43	84.0
ZMA-142	22	99.2	46.5	4.5	2.1	0.47	91.8
ZMA-167	22	100.0	43.7	3.9	1.7	0.44	86.0
Seed culture with RM; Fermentation in FM, 195 mM NaOAc, pH 5							
ZM 481	64	11.40	2.3	0.2	0.04	0.20	39.1
ZMA-142	56	103.9	48.0	1.86	0.9	0.46	90.2
ZMA-167	64	100.7	44.3	1.57	0.7	0.44	86
Seed culture with RM, 146 mM NaOAc; Fermentation in FM, 195 mM NaOAc, pH 5							
ZM 481	48	12.4	2.9	0.3	0.1	0.23	45.0
ZMA-142	36	97.2	60.8	2.7	1.7	0.63	122.4
ZMA-167	48	86.6	47.4	1.8	1.0	0.55	107.2

[a]qs max: Total sugar utilization rate (g l-1 h-1)
[b]qp max: Ethanol production rate (g l-1 h-1)
[c]Yp/s: Ethanol yield on total sugars (g g-1)
[d]Conversion Rate(%) = (Yp/s) x 100%/(Theorical Yp/s), Theorical Yp/s = (46*2)/180.16 = 0.51143

Batch fermentations with ZM481 and selected acetate-tolerant mutants

To evaluate the performances of the ZM481 and the selected acetate-tolerant mutants in ethanol production, batch fermentations with these strains were performed. Sodium acetate concentrations were chosen according to the highest concentrations tolerated by ZM481 and two selected mutants. Glucose consumption and ethanol production were measured using HPLC. As shown in Fig. 2, in normal culture conditions, the fermentative performances of ZM481 and two mutants were comparable as in sugar utilization rate and ethanol yield rate. Ethanol concentration of ZM481, ZMA-142 and ZMA-167 were 46.1, 46.5, and 43.7 g ethanol/L, respectively. The ratio of ethanol produced to moles of glucose consumed in ZM481, ZMA-142 and ZMA167 was 0.43, 0.47 and 0.44, respectively. The results revealed that ZMA-142 or ZMA-167, in 5-L batch fermentation with sodium acetate-adaptive seed

culture, exhibited an accelerated fermentation process compared with that of normal seed culture (Table 1).

We next examined fermentative pattern of ZM481 and two selected mutants in response to high acetate concentrations of 195 mM in a 5-L fermenter with 120 g glucose/L. The results revealed that ZMA-142 exhibited higher growth rate than that of ZMA-167, whereas the growth of ZM481 was significantly inhibited (Fig. 3a). In addition, the glucose consumption rates of ZMA-142 and ZMA-167 were declined to constant levels after 44 and 56 h incubation, respectively (Fig. 3b). The data showed that ZMA-142 reached the highest ethanol concentration (48 g/L) in 44 h, whereas highest ethanol production was observed in mutant ZMA-167 after an incubation of 64 h.

We determined the effects of seed culture on ethanol production of ZMA-142 and ZMA-167 in response to high sodium acetate concentration. Seed culture adapted

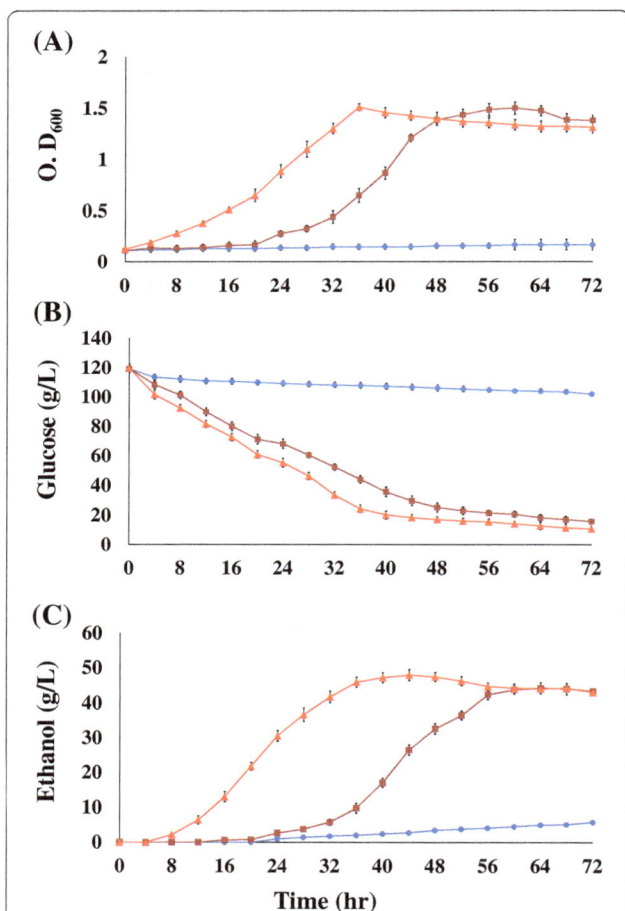

Fig. 3 Profiling of cell growth (**a**), glucose consumption (**b**), and ethanol yield (**c**) of the acetate-tolerant mutant ZMA-167(■), ZMA-142(▲) and ZM481(●) grown in 5 L fermenter(pH 5.0, 30 °C, 200 rpm) with 120 g/L glucose under 195 mM sodium acetate stress condition. The data come from mean values of triplicate experiments

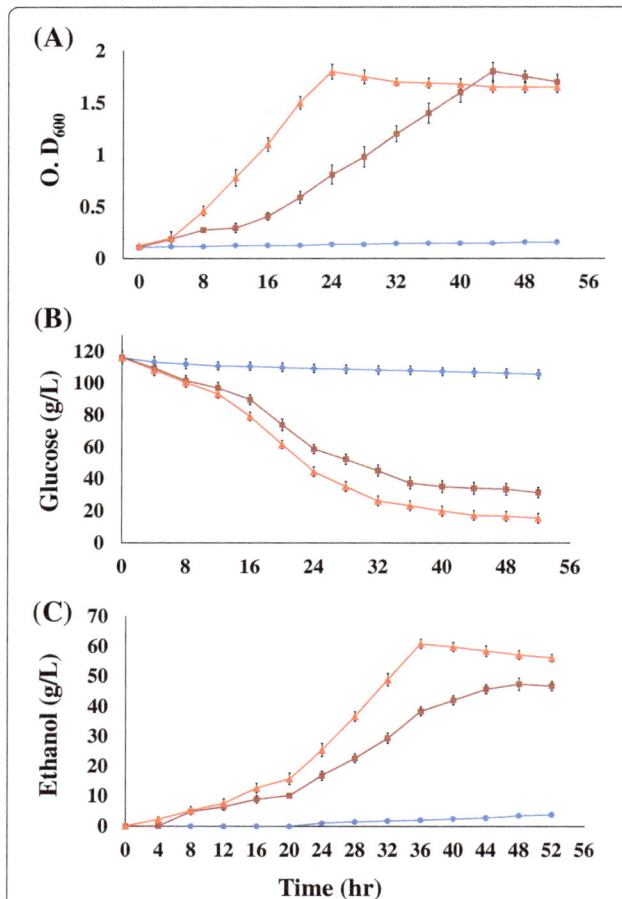

Fig. 4 Profiling of cell growth (**a**), glucose utilization (**b**), and ethanol yield (**c**) of the acetate-tolerant mutant ZMA-167(■), ZMA-142(▲) and ZM481(●) grown in 5 L fermenter(pH 5.0, 30 °C, 200 rpm) 120 g/L glucose under 195 mM sodium acetate stress condition. Inoculation with 146 mM sodium acetate-adaptive seed culture. The data come from mean values of triplicate experiments

to 146 mM sodium acetate with an inoculum size of 10% (*v*/v) was added to a medium containing 195 mM sodium acetate. We found that ZMA-142 and ZMA-167 reached the stationary phase of cell growth in 24 and 44 h, respectively (Fig. 4a). The data showed that glucose consumption of ZMA-142 was 96.5 g/L and maximum ethanol concentration was 60.8 g/L in 36 h, whereas ZMA-167 reached its maximum glucose consumption in 48 h and produced ethanol of 47.4 g/L (Fig. 4b and c). In addition, ethanol yield and ethanol conversion rate of ZMA-142 and ZMA-167 after 36 h and 48 h fermentation were 0.6 g/g and 122.4%, 0.55 g/g and 107.2%, respectively (Table 1). Our data showed that ZMA-142 and ZMA-167 exhibited relatively higher the glucose uptake rate and ethanol fermentation performance in acetate-adaptive seed culture used fermentation.

Genomic variants for improvement of acetate tolerance

To determine the genomic variants in acetate-tolerant mutants, next-generation sequence (NGS) by reversible dye terminators sequencing, Illumina solexa Miseq, was employed following combine with genome-wide bio-informatics approaches. Based on the all genomic DNA libraries were aligned on the same strand as the targeted feature genes of reference genome using SAMtools package v.1.19, ZMA-167 had a deletion of 303-bp, which was confirmed using PCR. The deletion truncated 3′ part of ZMO0117 by 255-bp and upstream of ZMO0119 (Fig. 5). TransTerm HP search showed that there was a putative ρ-dependent terminator 66 bp downstream of ZMO0117 in ZM481 (HP score is −3.141 and 100% confidence), whereas ZMA-167 had truncated ρ-dependent terminator by 28 bp. Using the Neutral Network

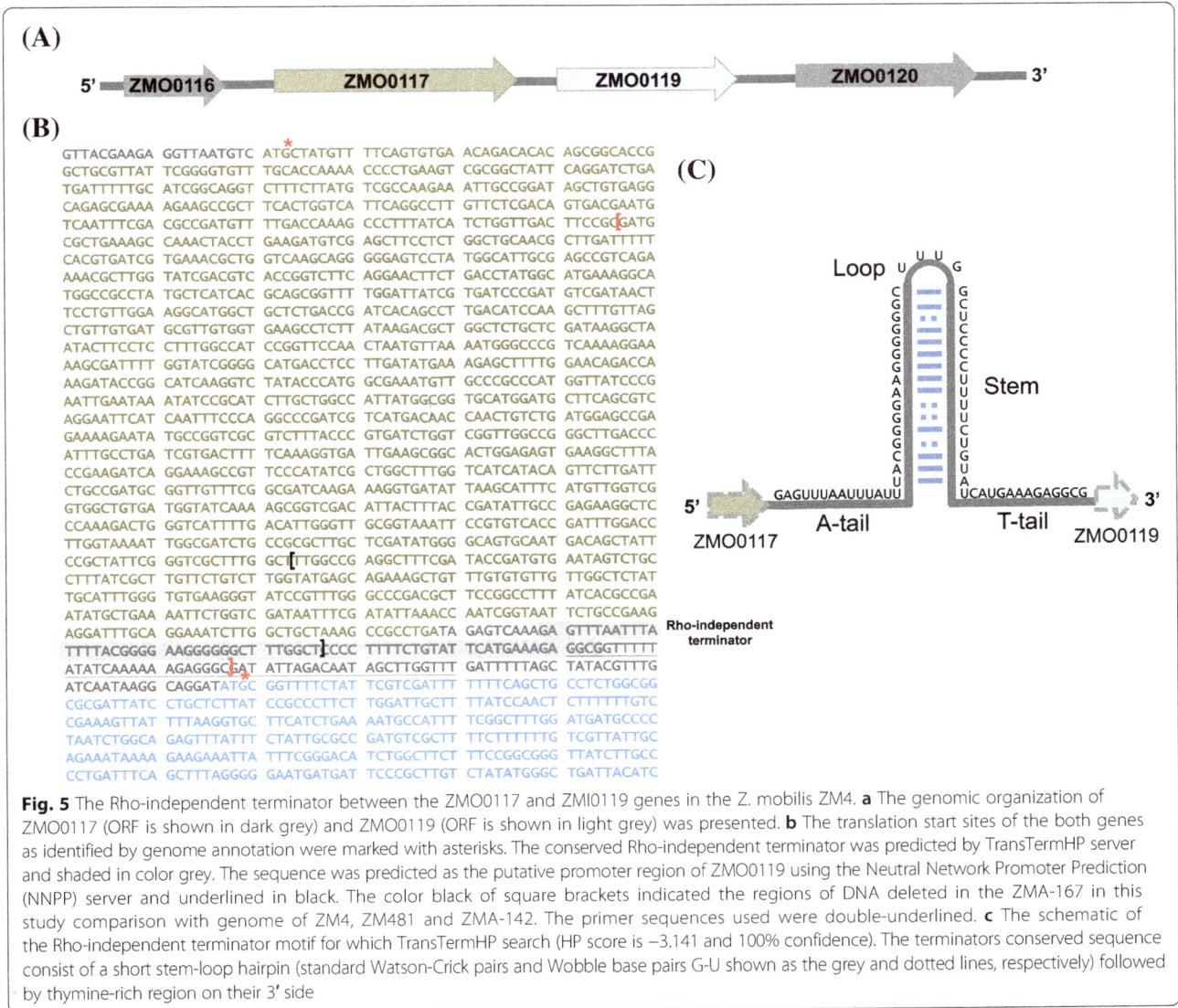

Fig. 5 The Rho-independent terminator between the ZMO0117 and ZMI0119 genes in the Z. mobilis ZM4. **a** The genomic organization of ZMO0117 (ORF is shown in dark grey) and ZMO0119 (ORF is shown in light grey) was presented. **b** The translation start sites of the both genes as identified by genome annotation were marked with asterisks. The conserved Rho-independent terminator was predicted by TransTermHP server and shaded in color grey. The sequence was predicted as the putative promoter region of ZMO0119 using the Neutral Network Promoter Prediction (NNPP) server and underlined in black. The color black of square brackets indicated the regions of DNA deleted in the ZMA-167 in this study comparison with genome of ZM4, ZM481 and ZMA-142. The primer sequences used were double-underlined. **c** The schematic of the Rho-independent terminator motif for which TransTermHP search (HP score is −3.141 and 100% confidence). The terminators conserved sequence consist of a short stem-loop hairpin (standard Watson-Crick pairs and Wobble base pairs G-U shown as the grey and dotted lines, respectively) followed by thymine-rich region on their 3′ side

Promoter Prediction (NNPP), server to predict the sequence of putative promoter region of ZMO0119 in ZMA-167. The putative promoter region of ATCC 31823 ZMO0117 was 60-bp, which was the same with those in ZMA-167. We next analyzed the 0.3-Kbp deletion in ZMA-167 using reverse-transcription PCR. Using the PCR primer set, ZMO0117F and ZMO0117R, a 0.3-Kbp RT-PCR product from ZMA-167 and ZM481 were obtained respectively (Fig. 6a, lane 1, 2, 10 and 11). Use of primer set, ZMO0119F and ZMO0119R, resulted in 0.2-Kbp RT-PCR products from ZMA-167 and ZM481 respectively (Fig. 6a, lane 4, 5, 10 and 11). Interestingly, a 1.1-Kbp RT-PCR product was obtained from ZMA-167 with a primer set of ZMO0117F and ZMO0119R, whereas there was no product generated from ZM481 (Fig. 6a, lane 7, 8, 10 and 11). Our RT-PCR data showed

that a ZMO0117-ZMO0119 co-transcription product of the 0.3-Kbp deletion ZMA-167 was detected 195 mM sodium acetate culture condition (Fig. 6). In addition to 0.3-Kbp deletion in ZMA-167, there was no genome change in this region in ZMA-142 compared with ZM481 parental strain.

Discussion

In the present study, we reported the inhibitory effects of acetate and low pH on an ethanol-tolerant strain. Using combination of NTG mutagenesis and adaptive evolution, we obtained two acetate-tolerant mutants exhibiting favorable ethanol production under high-acetate condition. We found that the acetate-tolerant property of mutant ZMA-167 was attributed to a deletion of sequence of terminator of nhaA encoding hydroxylamine reductase.

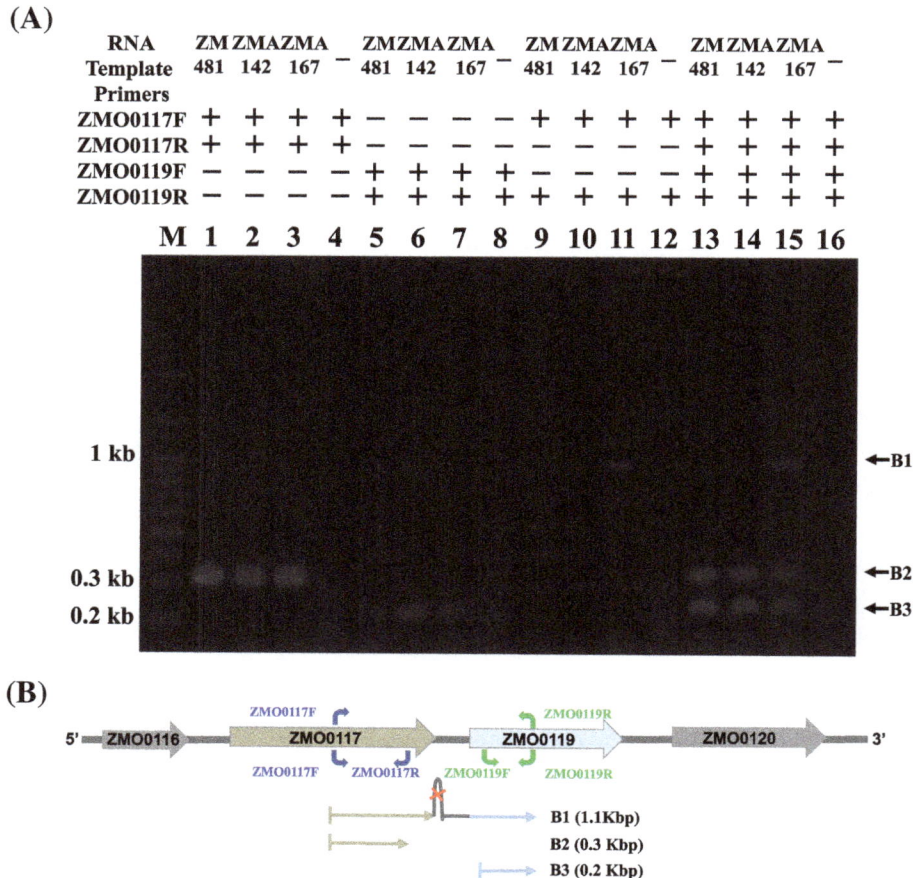

Fig. 6 RT-PCR detection of ZMO0117 and ZMO0119 mRNA from the acetate-tolerant mutant ZMA-142, ZMA-167 and ZM481. **a** Total RNA from the acetate-tolerant mutant ZMA-142(treated with 196 mM sodium acetate)(lane 2, 6, 10 and 14), ZMA-167 (treated with 196 mM sodium acetate)(lane 3, 7, 11 and 15) and ZM481 (without treated with sodium acetate) (lane 1, 5, 9 and 13) cells in the middle exponential phase were RT-PCR performed with various combinations of primers (ZMO0117F, ZMO0117R, ZMO0119F and ZMO0119R, respectively), in three independent reactions. Ten micrograms of each RNA was electrophoresed through 1% agarose–2.2 M formaldehyde gels and stained with ethidium bromide. The negative (lane 4, 8, 12, 16) controls sterile water. Lane M contained 0.1 ~ 3 kb markers. **b** Schematic representation of the ZM481, ZMA-142 and ZMA-167 chromosome containing the ZMO0117 and ZMI0119 genes. Arrows indicate positions of the predicted transcripts, B1 (transcipted from ZMO0117 to ZMO0119), B2 (ZMO0117), and B3(ZMO0119)

Z. mobilis is known for its features of rapid glucose uptake and favorable alcohol production. Several *Z. mobilis* strains have been developed to have high ethanol tolerance and to use a wide range of sugars [9, 15, 20–22]. With improved strains of *Z. mobilis*, the main challenges have become generating an ethanologen tolerant to inhibitors formed and released during the pretreatment and hydrolysis process. Many inhibitors have been reported including furans, phenols, organic acids and alcohols [7, 23–25]. Acetic acid has been shown to be a ubiquitous and strong inhibitor on ethanologen growth during ethanol fermentation [26–28]. Our results that the growth of ZM481 was decreased corresponding to increasing concentration of acetate were in consistent with previous studies. pH value has been shown to have an impact on inhibitory effect of acetic acid on the growth of ethanologen [29–31]. In the present experimental setting, the percent dissociation of acetic acid in the growth medium at pH 5 was 64%, resulting in 36% uncharged, undisscoated acetic acid that across cell membrane freely and compromise the biological balance. In addition to acetic acid, sodium ion has been suggested to synergistically contribute to decreased cell growth of *Z. mobilis* [7, 31]. Many genetic approaches and strategies have been employed to improve tolerance to variety of inhibitors in *Z. mobilis* [32]. Rogers et al. reported an acetate-tolerant mutant of *Z. mobilis* (AcR) generated using chemical mutagenesis and adaptation [10]. Several strains of *Z. mobilis* have been generated for tolerance to inhibitors by transposon mutagenesis [33–36]. Evolutionary adaptation has been employed to improve fermentation capability and resistance to inhibitory stress [29]. In this study, we obtained two mutants, ZMA-142 and ZMA-167 generated from ZM481, which exhibited improved acetate tolerance with favorable ethanol production at low pH. Our results suggest that chemical mutagenesis incorporating with adaptive evolution represents a practical technique to develop strains adapted to various inhibitory substrates and stress. However, molecular responses of two mutants to acetic acid and glucose are dynamic and complicated, which require further studies to elucidate.

Acetic acid toxicity is suggested to hinder the ethanologenesis by *Z. mobilis*. Acetate-resistant mutant (AcR) of *Z. mobilis* has been generated from ZM4, which has high ethanol production in the presence of 20 g/L sodium acetate [10]. The genome of ZM4 has been sequenced and its annotation has been improved [17, 37]. A recent study has identified the mutant loci that contribute to sodium acetate tolerance in the AcR strain [31]. Yang et al. has reported that a regulator Hfq protein acting as an RNA chaperone was involved in the acetate-tolerance in *Z. mobilis* [38]. In the present study, two mutants generated were examined for genome changes in comparison with their parental strain ZM481. We identified a 303-bp of deletion within the ρ-

dependent terminator putative sequences of ZMO0117 and upstream of ZMO0119 in ZMA-167 mutant. AcR mutant has reported to have a 1.5 kbp of deletion in ZMO0117- ZMO0119 resulting in acetate-tolerant phenotype [31]. Our results showing a deletion in terminator sequence of ZMO0117 and upstream of ZMO0119 in ZMA-167 are in agreement with previous study. It is suggested that elimination of a potential ZMO0117 transcriptional terminator lead to an over-expression of ZMO0119 encoding Na+/H+ antiporter by promotor of ZMO0117 in ZMA-167. In addition to ZMO0119, deletion of 303 bp in ZMA-167 resulted in removal of terminator putative sequences of ZMO0117 encoding hydroxylamine reductase. Hydroxylamine reductase is suggested to function as scavenger of other intermediates of nitrate ammonification in bacteria [39, 40]. It is implied that removal of terminator of ZMO0117 putatively results in high expression of hydroxylamine reductase, leading to a resistance to stress. However, further studies are required to elucidate the role of hydroxylamine reductase in acetate tolerance.

Conclusion
In conclusion, a combination of NTG mutagenesis and adaptive evolution represents an ideal strategy to generate mutants in response to environmental stress such as acetic acid. Taken the approach, two acetate-tolerant mutant *Z. mobilis* ZMA-142 and ZMA-167 exhibited enhanced tolerance to acetate and significantly improved fermentation performances. The mutants are suggested to have potential for industrial bioconversion process.

Abbreviations
AcR: Acetate-resistant mutant; CGS: Comparative Genome Sequencing; FM: Fermentation medium; HPLC: High performance liquid chromatography; Kbp: Kilo base pairs; NNPP: Neutral Network Promoter Prediction; NTG: N-methyl-N'-nitro-N-nitrosoguanidine; OD: Optical density; PCR: Polymerase chain reaction; RM: Rich medium; RT-PCR: Reverse transcription PCR

Acknowledgements
Authors thank Yo-Chuan Yang, Der-Sheng Chiou, Yi-Shan Yang and Min-Shuan Liu for their laboratory assistance.

Funding
The financial support by Ministry of Science and Technology, Republic of China (103-2221-E-415-026-MY3).

Authors' contributions
CWH conceived of the study, and participated in its design, acquisition of data, interpretation of data and have been involved in drafting the manuscript. YFL carried out the molecular genetic studies, participated in the sequence alignment and drafted the manuscript. YSC participated in the bacteria breeding and fermentation performed the molecular genetic studies and the statistical analysis. BSW participated in study's design and helped to draft the manuscript. All authors read and approved the final manuscript.

Competing interests

The authors declare that they have no competing interests.

Author details

[1]Division of Allergy, Department of Pediatrics, Chung-Shan Medical University Hospital, Taichung, Taiwan. [2]Department of Biomedical Sciences, College of Medicine Sciences and Technology, Chung Shan Medical University, Taichung, Taiwan. [3]Department of Microbiology, Immunology and Biopharmaceuticals, National Chiayi University, Chiayi, Taiwan.

References

1. Rabemanolontsoa H, Saka S. Various pretreatments of lignocellulosics. Bioresour Technol. 2016;199:83–91.

2. Clark TA, Mackie KL. Fermentation inhibitors in wood hydrolysates derived from the softwood Pinus Radiata. J Chem Technol Biotechnol. 1984;34(2): 101–10.

3. Palmqvist E, Grage H, Meinander NQ, Hahn-Hagerdal B. Main and interaction effects of acetic acid, furfural, and p-hydroxybenzoic acid on growth and ethanol productivity of yeasts. Biotechnol Bioeng. 1999;63(1):46–55.

4. Palmqvist E, Almeida JS, Hahn-Hagerdal B. Influence of furfural on anaerobic glycolytic kinetics of Saccharomyces Cerevisiae in batch culture. Biotechnol Bioeng. 1999;62(4):447–54.

5. Olsson L, Hahn-Hägerdal B. Fermentation of lignocellulosic hydrolysates for ethanol production. Enzym Microb Technol. 1996;18(5):312–31.

6. Klinke HB, Thomsen AB, Ahring BK. Inhibition of ethanol-producing yeast and bacteria by degradation products produced during pre-treatment of biomass. Appl Microbiol Biotechnol. 2004;66(1):10–26.

7. Franden MA, Pilath HM, Mohagheghi A, Pienkos PT, Zhang M. Inhibition of growth of Zymomonas mobilis by model compounds found in lignocellulosic hydrolysates. Biotechnol Biofuels. 2013;6(1):99.

8. Franden MA, Pienkos PT, Zhang M. Development of a high-throughput method to evaluate the impact of inhibitory compounds from lignocellulosic hydrolysates on the growth of Zymomonas mobilis. J Biotechnol. 2009;144(4): 259–67.

9. Zhang M, Eddy C, Deanda K, Finkelstein M, Picataggio S. Metabolic engineering of a pentose metabolism pathway in Ethanologenic Zymomonas mobilis. Science. 1995;267(5195):240–3.

10. Joachimsthal E, Haggett KD, Jang J-H, Rogers PL. A mutant of Zymomonas mobilis ZM4 capable of ethanol production from glucose in the presence of high acetate concentrations. Biotechnol Lett. 1998;20(2):137–42.

11. Mohagheghi A, Evans K, Chou YC, Zhang M. Cofermentation of glucose, xylose, and arabinose by genomic DNA-integrated xylose/arabinose fermenting strain of Zymomonas mobilis AX101. Appl Biochem Biotechnol. 2002;98-100:885–98.

12. Kim IS, Barrow KD, Rogers PL. Kinetic and nuclear magnetic resonance studies of xylose metabolism by recombinant Zymomonas mobilis ZM4(pZB5). Appl Environ Microbiol. 2000;66(1):186–93.

13. Shui ZX, Qin H, Wu B, Ruan ZY, Wang LS, Tan FR, Wang JL, Tang XY, Dai LC, Hu GQ, et al. Adaptive laboratory evolution of ethanologenic Zymomonas mobilis strain tolerant to furfural and acetic acid inhibitors. Appl Microbiol Biotechnol. 2015;99(13):5739–48.

14. Zhang X, Wang T, Zhou W, Jia X, Wang H. Use of a Tn5-based transposon system to create a cost-effective Zymomonas mobilis for ethanol production from lignocelluloses. Microb Cell Factories. 2013;12:41.

15. Yanase H, Miyawaki H, Sakurai M, Kawakami A, Matsumoto M, Haga K, Kojima M, Okamoto K. Ethanol production from wood hydrolysate using genetically engineered Zymomonas mobilis. Appl Microbiol Biotechnol. 2012;94(6):1667–78.

16. Caporaso JG, Lauber CL, Walters WA, Berg-Lyons D, Huntley J, Fierer N, Owens SM, Betley J, Fraser L, Bauer M, et al. Ultra-high-throughput microbial community analysis on the Illumina HiSeq and MiSeq platforms. ISME J. 2012;6(8):1621–4.

17. Yang S, Pappas KM, Hauser LJ, Land ML, Chen GL, Hurst GB, Pan C, Kouvelis VN, Typas MA, Pelletier DA, et al. Improved genome annotation for Zymomonas mobilis. Nat Biotechnol. 2009;27(10):893–4.

18. Li H, Durbin R. Fast and accurate short read alignment with burrows-wheeler transform. Bioinformatics. 2009;25(14):1754–60.

19. Thorvaldsdottir H, Robinson JT, Mesirov JP. Integrative genomics viewer (IGV): high-performance genomics data visualization and exploration. Brief Bioinform. 2013;14(2):178–92.

20. Lawford HG, Rousseau JD, Mohagheghi A, McMillan JD. Continuous culture studies of xylose-fermenting Zymomonas mobilis. Appl Biochem Biotechnol. 1998;70-72:353–67.

21. Yanase H, Sato D, Yamamoto K, Matsuda S, Yamamoto S, Okamoto K. Genetic engineering of Zymobacter palmae for production of ethanol from xylose. Appl Environ Microbiol. 2007;73(8):2592–9.

22. Mohagheghi A, Evans K, Finkelstein M, Zhang M. Cofermentation of glucose, xylose, and arabinose by mixed cultures of two genetically engineered Zymomonas mobilis strains. Appl Biochem Biotechnol. 1998;70-72:285–99.

23. Jobses IM, Roels JA. The inhibition of the maximum specific growth and fermentation rate of Zymomonas mobilis by ethanol. Biotechnol Bioeng. 1986;28(4):554–63.

24. Tan FR, Dai LC, Wu B, Qin H, Shui ZX, Wang JL, Zhu QL, Hu QC, Ruan ZY, He MX. Improving furfural tolerance of Zymomonas mobilis by rewiring a sigma factor RpoD protein. Appl Microbiol Biotechnol. 2015;99(12):5363–71.

25. Gu H, Zhang J, Bao J. High tolerance and physiological mechanism of Zymomonas mobilis to phenolic inhibitors in ethanol fermentation of corncob residue. Biotechnol Bioeng. 2015;112(9):1770–82.

26. Kumari R, Pramanik K. Improvement of multiple stress tolerance in yeast strain by sequential mutagenesis for enhanced bioethanol production. J Biosci Bioeng. 2012;114(6):622–9.

27. Guadalupe Medina V, Almering MJ, van Maris AJ, Pronk JT. Elimination of glycerol production in anaerobic cultures of a Saccharomyces Cerevisiae strain engineered to use acetic acid as an electron acceptor. Appl Environ Microbiol. 2010;76(1):190–5.

28. Takahashi CM, Takahashi DF, Carvalhal ML, Alterthum F. Effects of acetate on the growth and fermentation performance of Escherichia Coli KO11. Appl Biochem Biotechnol. 1999;81(3):193–203.

29. Lawford HG, Rousseau JD. Improving fermentation performance of recombinant Zymomonas in acetic acid-containing media. Appl Biochem Biotechnol. 1998;70-72:161–72.

30. Casey E, Sedlak M, Ho NW, Mosier NS. Effect of acetic acid and pH on the cofermentation of glucose and xylose to ethanol by a genetically engineered strain of Saccharomyces Cerevisiae. FEMS Yeast Res. 2010;10(4):385–93.

31. Yang S, Land ML, Klingeman DM, Pelletier DA, Lu TY, Martin SL, Guo HB, Smith JC, Brown SD. Paradigm for industrial strain improvement identifies sodium acetate tolerance loci in Zymomonas mobilis and Saccharomyces Cerevisiae. Proc Natl Acad Sci U S A. 2010;107(23):10395–400.

32. He MX, Wu B, Qin H, Ruan ZY, Tan FR, Wang JL, Shui ZX, Dai LC, Zhu QL, Pan K, et al. Zymomonas mobilis: a novel platform for future biorefineries. Biotechnol Biofuels. 2014;7:101.

33. Jia X, Wei N, Wang T, Wang H. Use of an EZ-Tn5-based random mutagenesis system to create a Zymomonas mobilis with significant tolerance to heat stress and malnutrition. J Ind Microbiol Biotechnol. 2013;40(8):811–22.

34. Nakatsu CH, Providenti M, Wyndham RC. The cis-diol dehydrogenase cbaC gene of Tn5271 is required for growth on 3-chlorobenzoate but not 3,4-dichlorobenzoate. Gene. 1997;196(1-2):209–18.

35. Pappas KM, Galani I, Typas MA. Transposon mutagenesis and strain construction in Zymomonas mobilis. J Appl Microbiol. 1997;82(3):379–88.

36. Wang JL, Wu B, Qin H, You Y, Liu S, Shui ZX, Tan FR, Wang YW, Zhu QL, Li YB, et al. Engineered Zymomonas mobilis for salt tolerance using EZ-Tn5-based transposon insertion mutagenesis system. Microb Cell Factories. 2016; 15(1):101.

37. Seo JS, Chong H, Park HS, Yoon KO, Jung C, Kim JJ, Hong JH, Kim H, Kim JH, Kil JI, et al. The genome sequence of the ethanologenic bacterium Zymomonas mobilis ZM4. Nat Biotechnol. 2005;23(1):63–8.

38. Yang S, Pelletier DA, Lu TY, Brown SD. The Zymomonas mobilis regulator hfq contributes to tolerance against multiple lignocellulosic pretreatment inhibitors. BMC Microbiol. 2010;10:135.

39. Boutrin MC, Wang C, Aruni W, Li X, Fletcher HM. Nitric oxide stress resistance in Porphyromonas gingivalis is mediated by a putative hydroxylamine reductase. J Bacteriol. 2012;194(6):1582–92.

40. Yurkiw MA, Voordouw J, Voordouw G. Contribution of rubredoxin:oxygen oxidoreductases and hybrid cluster proteins of Desulfovibrio Vulgaris Hildenborough to survival under oxygen and nitrite stress. Environ Microbiol. 2012;14(10):2711–25.

21

Simultaneous detection of α-Lactoalbumin, β-Lactoglobulin and Lactoferrin in milk by Visualized Microarray

Zhoumin Li[1,2], Fang Wen[3], Zhonghui Li[1], Nan Zheng[3*], Jindou Jiang[4] and Danke Xu[1*]

Abstract

Background: α-Lactalbumin (a-LA), β-lactoglobulin (β-LG) and lactoferrin (LF) are of high nutritional value which have made ingredients of choice in the formulation of modern foods and beverages. There remains an urgent need to develop novel biosensing methods for quantification featuring reduced cost, improved sensitivity, selectivity and more rapid response, especially for simultaneous detection of multiple whey proteins.

Results: A novel visualized microarray method was developed for the determination of a-LA, β-LG and LF in milk samples without the need for complex or time-consuming pre-treatment steps. The measurement principle was based on the competitive immunological reaction and silver enhancement technique. In this case, a visible array dots as the detectable signals were further amplified and developed by the silver enhancement reagents. The microarray could be assayed by the microarray scanner. The detection limits ($S/N = 3$) were estimated to be 40 ng/mL (α-LA), 50 ng/mL (β-LG), 30 ng/mL (LF) ($n = 6$).

Conclusions: The method could be used to simultaneously analyze the whey protein contents of various raw milk samples and ultra-high temperature treated (UHT) milk samples including skimmed milk and high calcium milk. The analytical results were in good agreement with that of the high performance liquid chromatography. The presented visualized microarray has showed its advantages such as high-throughput, specificity, sensitivity and cost-effective for analysis of various milk samples.

Keywords: Visualized microarray, α-Lactoalbumin, β-Lactoglobulin, Lactoferrin

Background

Milk whey protein represents a rich and mixture proteins with wide ranging nutritional, biological and food functional attributes. The main constituents are α-lactalbumin (α-LA), β-lactoglobulin (β-LG) and lactoferrin (LF), which account for approximately 70–80% of total whey protein. α-LA, β-LG and LF are of high nutritional value which have made ingredients of choice in the formulation of modern foods and beverages. They may also have physiological activity through moderating gut microflora, mineral absorption and immune function [1, 2].

Although several methods have been reported for α-LA, β-LG and LF, either alone or concomitant with other whey proteins, including chromatographic analysis (High performance liquid chromatography (HPLC) [3–11], Ultra high performance liquid chromatography (UHPLC) [12], High performance liquid chromatography -mass spectra (HPLC-MS) [13–21], Ultra high performance liquid chromatography - mass spectra (UHPLC-MS) [22–27], Immunoaffinity chromatography (IAC) [26, 27]), Radial Immunodiffusion (RID) [28], sodium dodecyl sulfate polyacrylamide gel electropheresis (SDS-PAGE) [29, 30], Capillary Electrophoresis(CE) [10, 31–34], Enzyme-llinked Immunosorbent Assay (ELISA) [17, 35–42], Fluorescent Immunosorbent Assay(FIA) [43, 44], Surface Plasmon Resonance (SPR) [45–49] and Sensors [50–52]. In general, chromatographic analysis requires pre-treated samples, high initial sample volumes and long analysis times, which lead to high cost. In addition, analytical

* Correspondence: zhengnan_1980@126.com; xudanke@nju.edu.cn
[3]Ministry of Agriculture-Key Laboratory of Quality & Safety Control for Milk and Dairy Products, Institute of Animal Science, Chinese Academy of Agricultural Sciences, Beijing 100193, People's Republic of China
[1]State Key Laboratory of Analytical Chemistry for Life Science, School of Chemistry and Chemical Engineering, Nanjing University, Nanjing 210093, China
Full list of author information is available at the end of the article

chromatographic technologies are unable to identify protein denaturation or modification that may occur during processing and storage. This is an important factor for public health and food commodities marketing. Some of these drawbacks can be overcome using traditional immunological methods, such as ELISA. It also offers the advantages of working directly with complex fluids, such as whole milk and other dairy fluids, but only one whey protein can be detected. However, there remains an urgent need to develop alternative methods for quantification featuring reduced cost, improved sensitivity, selectivity and more rapid response, especially for simultaneous detection of multiple whey proteins.

Development of new tools, minimizing limitations imposed by these methodologies and leveraging the high specificity of traditional immunological methods, is of great interest. In this sense, visualized microarray are envisaged as a valid alternative to classical methods for analysis of protein, because they are amenable to direct readout by eyes and well suited to rapid detection with high sensitivity and selectivity using low-cost instrumentation that is adaptable to portable, field-deployable embodiments, which is ideal for routine determination in the dairy industry [53–56].

In this paper, we described the development of visualized microarray method for simultaneous, high-throughput quantitative immune-detection of three commercially important whey proteins (α-LA, β-LG, and LF) in samples at a time, from various milk sources. To the best of our knowledge, no visualized microarray has been described thus far for the determination of a-LA, β-LG, and LF simultaneously. Visualized microarray method allowed the analysis of milk without the need for sample preparation, including pre-enrichment or purification steps, "extraction" of target analytes from the complex matrix, and measurement of signal in a "clean" environment. The assay was then used to simultaneously analyze the whey protein contents of various raw milk samples and UHT milk samples including skimmed milk and high calcium milk and the analytical results were in good agreement with that of the HPLC.

Methods
Materials and instruments
α-LA, β-LG, LF and silver enhancement solution including solution A (AgNO$_3$) and solution B (Hydroquinone) were all purchased from Sigma-Aldrich. NaCl, KCl, Na$_2$HPO$_4$·12H$_2$O, KH$_2$PO$_4$, Tween-20, Ethylenediaminetetraacetic acid (EDTA) was from Nanjing Chemical Reagent Co., Ltd. (Nanjing, China). Pure water of 18.2 MΩcm-1 was generated in-lab from a Milli-Q water system. Bovine serum albumin (BSA) was purchased from Merck. Goat polyclonal to α-lactalbumin (α-LA), goat polyclonal to β-lactoglobulin (β-LG), goat polyclonal to lactoferrin (LF) and AgNPs labeled donkey

anti-goat IgG were kindly supplied by Nanjing Xiangzhong Biotechnology Co. Ltd. (Nanjing, China).

All solutions were made by triply deionized water (Milli-Q water purification system, Millipore, Billerica, MA, USA). A 10 mM phosphate buffered saline (PBS) at pH 7.2 was used as the assay buffer which was prepared as following: 137 mmol/L NaCl, 2.7 mmol/L KCl, 10 mmol/L Na$_2$HPO$_4$·12H$_2$O and 2 mmol/L KH$_2$PO$_4$. A 10 mM PBS containing 0.01% Tween-20 and 1 mM EDTA (PBST- EDTA) at pH 7.2 was used for milk sample preparation and dilution. The wash buffer was a PBS containing 0.05% Tween 20 (PBST). The blocking solution was 1% BSA in 10 mM PBS. All buffers were filtered through 0.22 µm pore size filter before use.

The microarrays were prepared by TMAR microarray spotter (Tsinghua University, Beijing, China). Automated plate washer (BioTek Instruments, Inc. America) was used as washing platform. LXJ-II centrifuge (Shanghai Anting Instrument Co., Shanghai, China) were used for the centrifugation. Clear flat-bottom 96-well plate, thermo-shaker and microarrays scanner (QARRAY 2000) were from Nanjing Xiangzhong Biotechnology Co. Ltd. (Nanjing, China).

Microarray preparation
The obtained of α-LA, β-LG and LF were spotted on clear flat-bottom 96-well plate. A volume of 10 µL of each coating antigens diluted by spotting buffer were arrayed with a 500 µm spot-to-spot pitch using a microarray spotter, each antigen solutions was in triplicate. After spotted, microarray was incubated for 2 h at 37 °C. In this step the coating antigens were immobilized on the microplate wells by absorption over the surface of the support of polystyrene. After immobilization, microarray surface was treated with 200 µL 1% BSA for 1 h at 37 °C in order to minimize further unspecific bindings. After incubation, the microarray plate was washed with 1 × PBST buffer using an automated plate washer and then sealed in foil packets for storage at 2–8 °C.

Indirect competitive microarray immunoassay protocol
The indirect competitive microarray immunoassay principle was presented in Scheme 1. In a microarray immunoassay analysis the following experimental procedure was performed. The competition is established by the addition of a mixture of 25 µL the standard (or the sample), a known amount of 25 µL mixed antibodies, The reaction is incubated at 25 °C for 45 min on a thermoshaker (shaking at 600 rpm). After the corresponding washing step, AgNPs labeled donkey anti-goat IgG in a total volume of 50 µL/well. The reaction is incubated at 37 °C for 30 min on a thermoshaker (shaking at 600 rpm). After the corresponding washing step, 50 µL silver enhancement solution including solution A (AgNO$_3$) and solution B (Hydroquinone) was then added to each well, and incubated for 12 min at 37 °C

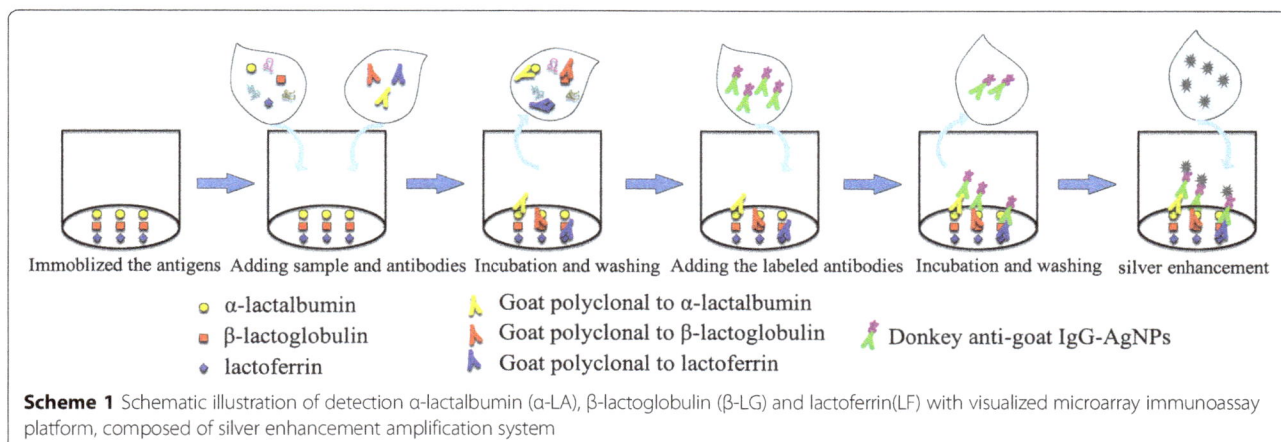

Scheme 1 Schematic illustration of detection α-lactalbumin (α-LA), β-lactoglobulin (β-LG) and lactoferrin(LF) with visualized microarray immunoassay platform, composed of silver enhancement amplification system

in dark. At the end of colorimetric reaction, each well was washed 3 times with 250 µL pure water.

Microarray imaging and data processing

The microarray was imaged with microarray scanner (QARRAY 2000) and performed using the corresponding software to quantify the signal over the sample spot area and expressed as relative light units (RLUs). The calibration curve was represented by a linear relationship.

Cross reactivity calculation

Cross reactivity (CR) is generally defined as the necessary amount of mass or concentration of interference able to produce an equal signal as when the analyte is assayed to provoke a signal inhibition of 50%. Therefore, in this work CR rates, in terms of percentage (%), were calculated according to the expression (eq. 1).

$$CR = [(IC_{50}(analyte)/IC_{50}(interference))] \times 100\%$$

(1)

IC_{50} is the necessary concentration of analyte or interference to induce a signal inhibition of 50%.

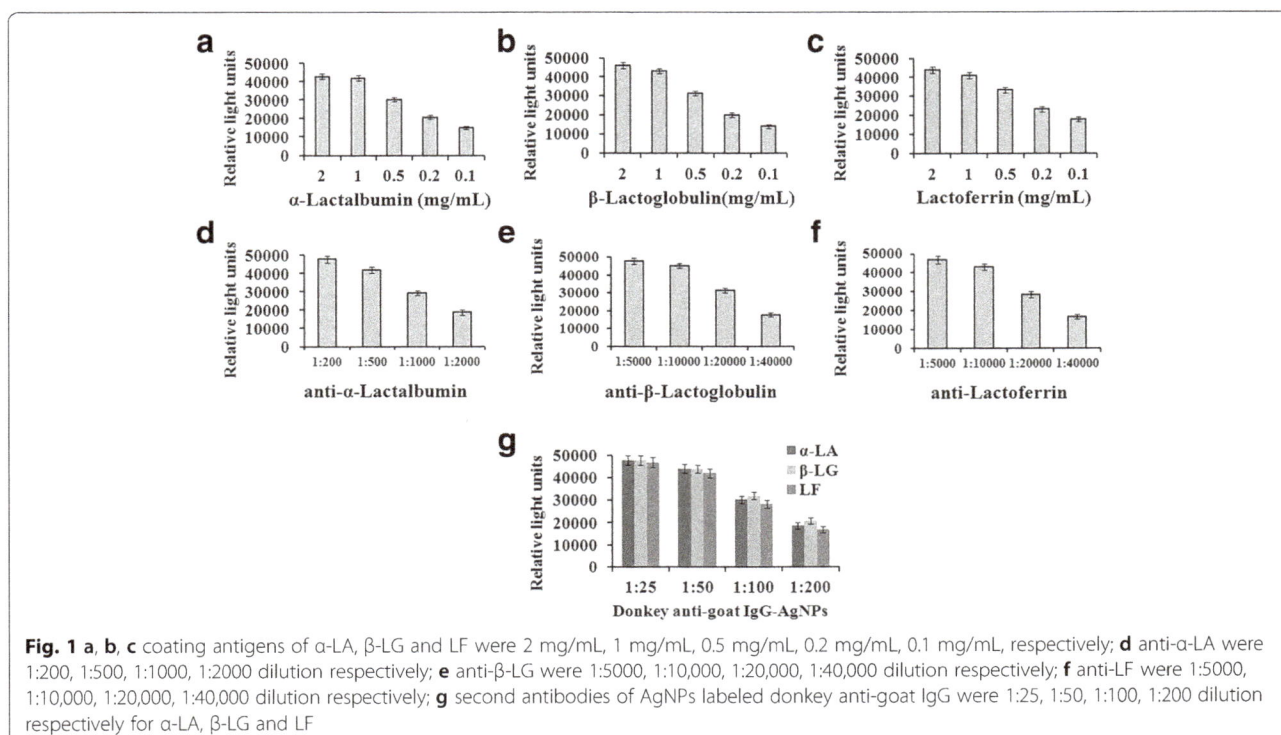

Fig. 1 a, b, c coating antigens of α-LA, β-LG and LF were 2 mg/mL, 1 mg/mL, 0.5 mg/mL, 0.2 mg/mL, 0.1 mg/mL, respectively; **d** anti-α-LA were 1:200, 1:500, 1:1000, 1:2000 dilution respectively; **e** anti-β-LG were 1:5000, 1:10,000, 1:20,000, 1:40,000 dilution respectively; **f** anti-LF were 1:5000, 1:10,000, 1:20,000, 1:40,000 dilution respectively; **g** second antibodies of AgNPs labeled donkey anti-goat IgG were 1:25, 1:50, 1:100, 1:200 dilution respectively for α-LA, β-LG and LF

Milk samples

Milk consists of metal ions such as calcium, iron, magnesium and zinc. For actual sample analysis, it should be considered that α-LA, β-LG and LF had a high possibility of forming chelation complex with these metal ions. Thus, prior to actual sample analysis, milk was diluted 200-fold with PBST-EDTA at pH 7.2. Milk was purchased from local supermarket.

HPLC method
Solutions

Binding buffer (BB): 1.211 g Tris was dissolved with 800 mL 6 mol/L HCl, adjusted to pH 7.4 and then volumed to final volume to 1000 mL.

Elution buffer (EB): 0.15 mol/L sodium phosphate, pH 12.

Buffer for adjusting pH of EB (AB): 1 mol/L sodium dihydrogen phosphate.

Treatment of milk sample for analysis α- Lactalbumin and β- lactoglobulin

5 mL milk sample was mixed with 14 mL water and adjusted to pH 4.6. Next water was added to the mixture making final volume to 20 mL. Then the above mixture was centrifuged under 10,000 rpm and 4 °C for 10 min. Finally, supernatant was filter with 0.22 μm filter and injected to HPLC system.

Treatment of milk sample for analysis lactoferrin

Milk samples were centrifuged under 8000 g and 4 °C for 10 min to remove fat. Then 15 mL skim milk was loaded onto lactoferrin immune-affinity column that was pre-equilibrated with 10 mL BB. After washing with 20 mL BB, lactoferrin was eluted with 3.6 mL EB. Then the 3.6 mL elution was mixed with 0.4 mL EB. Finally the mixture was filtered with 0.22 μm filter and injected to HPLC system. The lactoferrin immune-affinity column was washed with 10 mL BB and stored at 4 °C for further use.

HPLC system

The chromatographic analysis of lactoferrin was carried out on a HPLC system (2695 Separations Module, Waters; Milford, MA, USA) coupled with a photodiode array detector (PDA 2996 detector, Waters; Milford, MA, USA). Separation was performed using a Symmetry C4 Column (300 Å, 5 μm, 4.6 mm × 250 mm, Waters). Acetonitrile (eluent A) and 0.1% trifluoroacetic acid in water (eluent B) were used as mobile phase. The flow rate was set at 1.0 mL/min and the LC elution gradient was as follows: initial 30% A, 5 min 55% A, 10 min 60% A, 12 min 30% A and hold on for a further 4 min for re-equilibration, giving a total run time 16 min. The column temperature was kept at 25 °C and the injection volume was 50 μL for

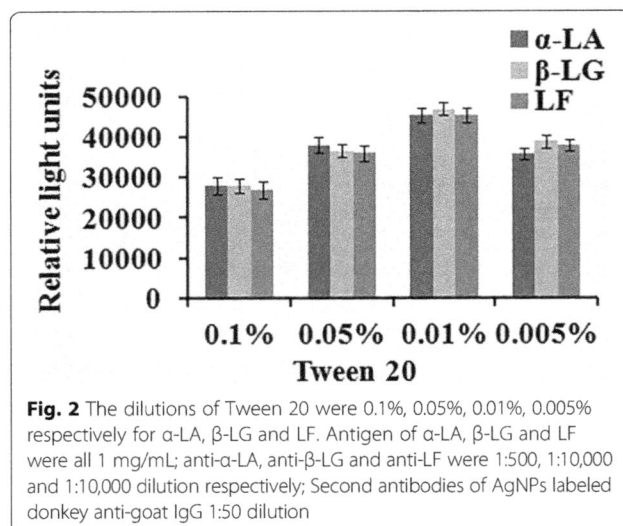

Fig. 2 The dilutions of Tween 20 were 0.1%, 0.05%, 0.01%, 0.005% respectively for α-LA, β-LG and LF. Antigen of α-LA, β-LG and LF were all 1 mg/mL; anti-α-LA, anti-β-LG and anti-LF were 1:500, 1:10,000 and 1:10,000 dilution respectively; Second antibodies of AgNPs labeled donkey anti-goat IgG 1:50 dilution

standards and sample solutions. The wavelengths was set at 280 nm for detection. Waters Empower 2.0 chromatography software package was used for HPLC system control, data acquisition and management.

Results and discussion
Optimization

To develop a highly sensitive and specific indirect competitive immunoassay, the conditions including the concentrations of coating antigens and antibodies, should be carefully optimized by a checkboard titration of antigen and antibody simultaneously. In addition, it was necessary to evaluate the effect of presence or absence of EDTA and Tween 20 in assay buffer.

Concentrations of coating antigens and antibodies

To develop highly sensitive competitive immunoassay, the conditions including the concentrations of coating antigens and dilutions of antibodies should be carefully optimized. In this study, coating antigens of α-LA, β-LG and LF all were 2 mg/mL, 1 mg/mL, 0.5 mg/mL, 0.2 mg/mL, 0.1 mg/mL, respectively; anti-α-LA were 1:200, 1:500,

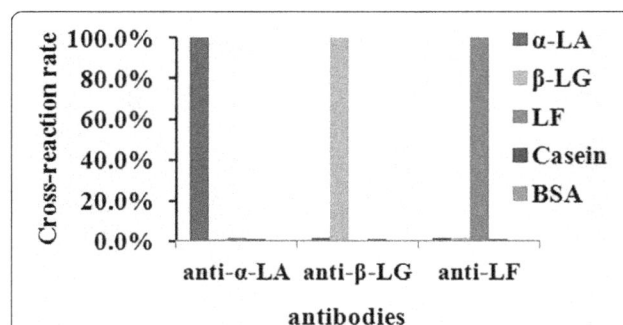

Fig. 3 Anti-α-LA, anti-β-LG and anti-LF were cross-reactivity with α-LA, β-LG, LF, Casein and BSA

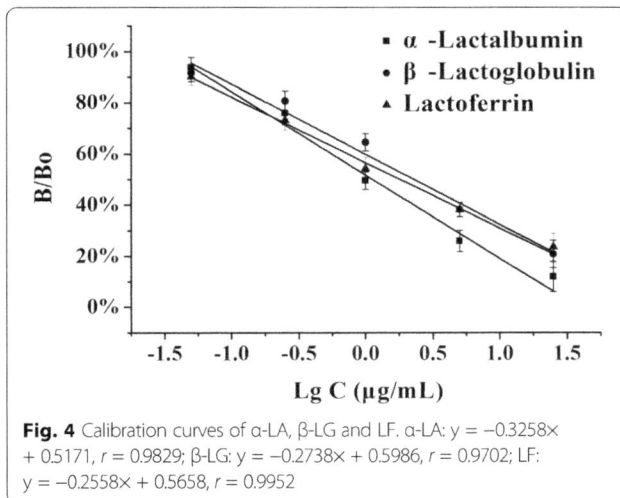

Fig. 4 Calibration curves of α-LA, β-LG and LF. α-LA: $y = -0.3258x + 0.5171$, $r = 0.9829$; β-LG: $y = -0.2738x + 0.5986$, $r = 0.9702$; LF: $y = -0.2558x + 0.5658$, $r = 0.9952$

1:1000, 1:2000 dilution respectively; anti-β-LG were 1:5000, 1:10,000, 1:20,000, 1:40,000 dilution respectively; anti-LF were 1:5000, 1:10,000, 1:20,000, 1:40,000 dilution respectively. In addition, second antibodies of AgNPs labeled donkey anti-goat IgG were 1:25, 1:50, 1:100, 1:200 dilution respectively. The results can be seen in Fig. 1.

Lower the concentration of antigen and antibody can increase detection sensitivity, but the signal value will be lower. So the optimal assay conditions were as follows: Appropriate concentrations of α-LA, β-LG and LF were all 1 mg/mL; The appropriate concentrations of anti-α-LA, anti-β-LG and anti-LF were 1:500, 1:10,000 and 1:10,000 dilution respectively; Appropriate concentrations of second antibodies of AgNPs labeled donkey anti-goat IgG were 1:50 dilution.

Effect of EDTA and Tween 20

However, milk has metal ions such as calcium, iron, magnesium, potassium, sodium and zinc ion, so it is

necessary to consider the high potential for forming chelating complexes between α-LA, β-LG and LF with these metal ions. To prevent these interferences, EDTA was incorporated in the assay buffer. EDTA has a greater affinity for the calciums than α-LA, β-LG and LF, thus it can block the interaction of α-LA, β-LG, LF with calciums.

Tween 20 is a non-ionic surfactant, which has emulsification, diffusion, solubilization, stabilizing effect with samples. Moreover, it provides a protective of antigen-antibody in buffers, and reduces the nonspecific binding of antibodies to antigens and interfering proteins. Thus it can reduce the background and improve the sensitivity. However, an excessive concentration could inhibit binding of antibody and antigen. Finally, Tween 20 concentration was selected to be 0.01%. The results can be seen in Fig. 2.

Method development

Assay specificity indicates the ability of antibody to generate a measurable response only for the target molecule. The cross-reactivity of antibodies was evaluated under indirect competitive immunoassay conditions in order to confirm specificity. Here, a study was performed using five main proteins in milk, such as α-LA, β-LG, LF, Casein and BSA. The cross-reactivity studies were carried out by adding various free cross reactants at different concentrations to compete with antigen coated on the surface, to bind with the antibody. The cross-reactivity for each compound was calculated according to the expression (eq. 1) and given in Fig. 3.

The anti-α-LA, anti-β-LG, anti-LF were determined to be highly specific for α-LA, β-LG, LF respectively, although there was a minor dose–response relationship for Casein and BSA (cross-reactivity <1.0%), the binding responses for these proteins were analytically insignificant at

Table 1 The recoveries of different concentrations of α-Lactalbumin, β-Lactoglobulin, Lactoferrin

Proteins	Spiked concentration (μg/mL)	Average (μg/mL)	SD (μg/mL)	RSD	recovery
α-Lactalbumin	20	18.23	2.14	11.74%	91.15%
	100	108.19	12.27	11.34%	108.19%
	400	409.81	40.57	9.90%	102.45%
	2000	2032.93	252.61	12.43%	101.65%
β-Lactoglobulin	20	19.13	2.56	13.38%	95.65%
	100	105.77	11.76	11.12%	105.77%
	400	419.8	48.36	11.52%	104.95%
	2000	2131.46	206.85	9.70%	106.57%
Lactoferrin	20	19.13	2.76	14.43%	95.65%
	100	95.77	12.08	12.61%	95.77%
	400	402.81	40.36	10.02%	100.70%
	2000	1931.46	263.44	13.64%	96.57%

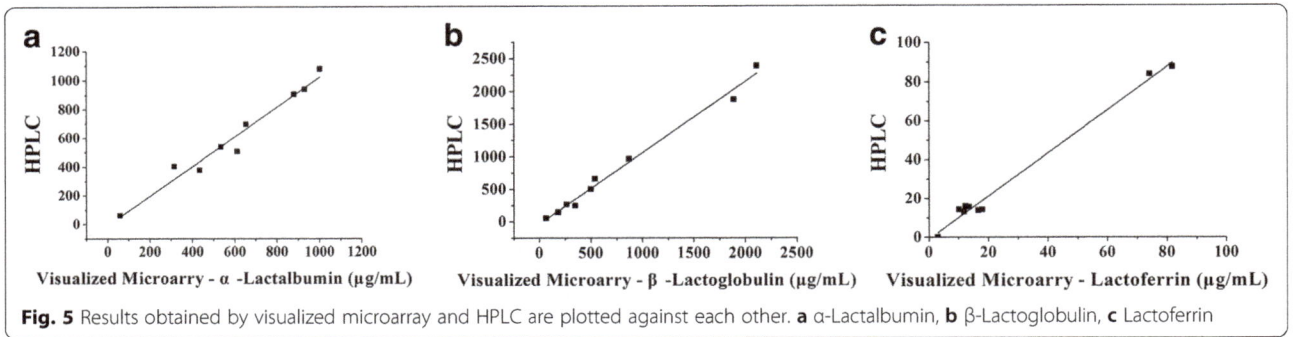

Fig. 5 Results obtained by visualized microarray and HPLC are plotted against each other. **a** α-Lactalbumin, **b** β-Lactoglobulin, **c** Lactoferrin

concentrations equivalent to those of diluted milk samples.

Method performance

In order to be able to determine multiplex format concentrations of α-LA, β-LG and LF, the assay was calibrated independently using a cocktail of the α-LA, β-LG, LF antibodies and different concentrations of α-LA, β-LG and LF. As a matter of fact, the competition occurs for all target molecules and the specific signal obtained on each probe decreases with the analyte concentration, as expected in a competitive immunoassay.

Over the optimized working calibration range (α-LA, β-LG and LF were all 0.05, 0.25, 1, 5, 25 μg/mL), a semi-log curve fit adequately described the dose-response relationship which can be seen in Fig. 4. Their calibration curves were calculated as follows, α-LA: y = −0.3258×

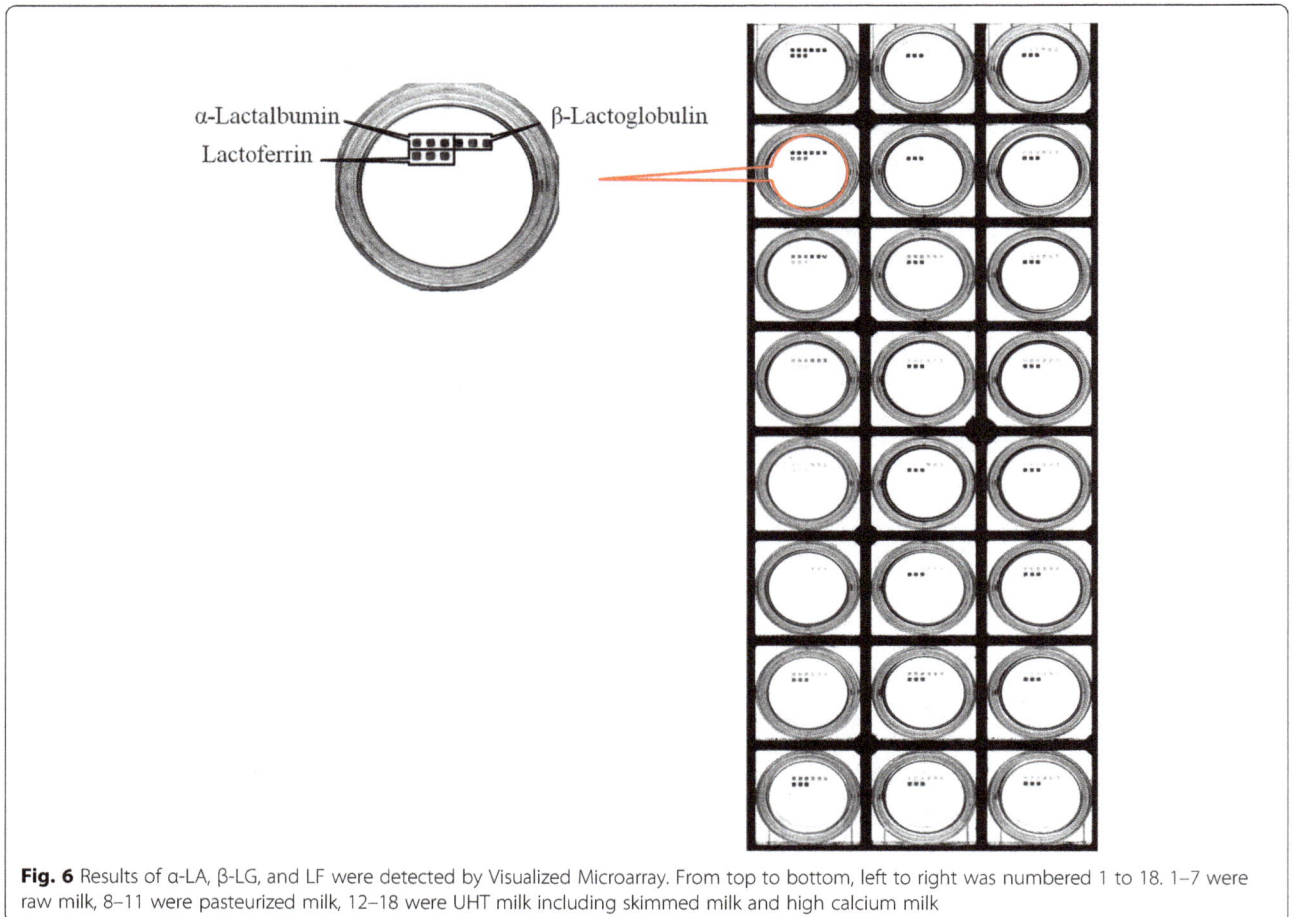

Fig. 6 Results of α-LA, β-LG, and LF were detected by Visualized Microarray. From top to bottom, left to right was numbered 1 to 18. 1–7 were raw milk, 8–11 were pasteurized milk, 12–18 were UHT milk including skimmed milk and high calcium milk

+ 0.5171, r = 0.9829; β-LG: y = −0.2738× + 0.5986, r = 0.9702; LF: y = −0.2558× + 0.5658, r = 0.9952. In the calculation formula, y: B/B0%, x: lg C. (B/B0 is the ratio of response B to the maximum response when no analyte is present B0.)

The method detection limits (response 3 standard deviations of blank over several independent runs) were estimated to be 0.04 μg/mL (α-LA), 0.05 μg/mL (β-LG), 0.03 μg/mL (LF) (n = 6). Method precision was estimated from the aggregate of a single-level control α-LA (1 μg/mL), β-LG (1 μg/mL), LF (1 μg/mL) over multiple independent runs, and the measured RSD were 6.71%, 7.82%, 5.13%, respectively(n = 6). Between-run precision may be further assessed with RSD 12.31%, 13.52%, 14.15%, respectively (n = 6).

After a simple dilution of commercial milk (200-fold in PBST-EDTA, pH 7.2), use this calibration curve to calculate the concentration of milks. The recovery study was performed samples of milk purchased from local supermarkets. Free α-LA, β-LG and LF (20 μg L^{-1}, 100 μg L^{-1}, 400 μg L^{-1}and 2000 μg L^{-1}) were spiked in milk solution. The recovery study was performed in three replicates and the results were quite satisfactory as seen in Table 1.

Recovery = (C1-C2)/C3 × 100%

C1: Sample concentration after adding standard.
C2: Sample concentration before adding standard.
C3: concentration of adding standard.

Comparison with a reference method-HPLC

To verify the reliance and accuracy of visualized microarray system, the results of 9 milk samples were compared with an HPLC method. The results obtained by visualized microarray and HPLC are plotted against each other in Fig. 5. The correlation index r was very good with a linear regression curve of y = 1.031×-9.30, r = 0.9604 (α-Lactalbumin); y = 1.094×-35.33, r = 0.9872 (β-Lactoglobulin); y = 1.1096×-1.054, r = 0.9889 (Lactoferrin); These results confirm those of the validation experiments. The findings indicate that reliable results can be obtained over the whole concentration range.

Method applications

The developed procedure was then applied to quantify the concentration of native α-LA, β-LG and LF in three different kinds of milks. The results were shown in Additional file 1, Fig. 6 and Table 2. The precision of the results were well (RSD < 15%). As samples, two bovine milks with different processing treatments have been analyzed. Taking into account the calibration curve, it has been determined that raw milk which numbered 1–7 presented highest concentration of α-LA, β-LG and LF,then pasteurized milk (72–85 °C for 15 s) which numbered 8–11, UHT milk (135–150 °C for 4–15 s) including skimmed milk and high calcium milk which numbered 12–18. As compared to other references that mention the concentration of α-LA, β-

Table 2 Results of α-LA, β-LG, and LF detected by Visualized Microarray

Proteins	number	α-Lactalbumin			β-Lactoglobulin			Lactoferrin		
		Average (μg/mL)	SD (μg/mL)	RSD (n = 3)	Average (μg/mL)	SD (μg/mL)	RSD (n = 3)	Average (μg/mL)	SD (μg/mL)	RSD (n = 3)
Raw Milk	1	230.53	21.79	9.45%	490.20	60.93	12.43%	124.2	13.23	10.65%
	2	80.86	8.36	10.34%	70.70	8.29	11.73%	11.8	1.44	12.21%
	3	1315.13	130.20	9.90%	983.09	104.01	10.58%	6.1	0.69	11.23%
	4	653.01	79.99	12.25%	964.84	139.42	14.45%	7.9	1.01	12.78%
	5	150.21	16.88	11.24%	211.13	27.68	13.11%	28.5	3.62	12.69%
	6	454.64	38.83	8.54%	440.27	41.17	9.35%	10.3	1.12	10.88%
	7	811.91	107.82	13.28%	348.66	44.84	12.86%	21.0	3.10	14.78%
pasteurized milk	8	534.7	75.45	14.11%	499.84	67.23	13.45%	18.7	2.58	13.80%
	9	199.45	19.89	9.97%	166.57	17.79	10.68%	23.6	2.34	9.93%
	10	461.9	54.23	11.74%	308.7	46.12	14.94%	15.6	2.00	12.81%
	11	409.4	58.83	14.37%	243.6	32.57	13.37%	0.0	-	-
UHT Milk	12	263.6	34.00	12.90%	229.5	27.88	12.15%	0.0	-	-
	13	348.2	46.87	13.46%	205.5	29.65	14.43%	0.0	-	-
	14	264.3	33.51	12.68%	307.6	31.87	10.36%	10.2	1.26	12.39%
	15	503.0	66.30	13.18%	327.4	39.68	12.12%	6.7	0.82	12.17%
	16	251.3	35.73	14.22%	169.6	23.10	13.62%	8.1	1.01	12.51%
	17	579.6	60.57	10.45%	737.0	97.21	13.19%	12.4	1.43	11.56%
	18	312.8	34.10	10.90%	255.5	30.02	11.75%	0.0	-	-

LG and LF in milk [12, 34, 50]. Now, it is well known that α-LA, β-LG and LF were highly sensitive to temperature.

Conclusions

In this work, visualized microarray for the high-throughput, specific and sensitive determination of a-LA, β-LG and LF in milk samples was developed for the first time, without the need for complex or time-consuming pre-treatment steps, following dilution with an appropriate working buffer. The applicability of the visualized microarray as-developed was underlined by the implementation and analysis of different milk samples, and the results were validated successfully against a HPLC. The visualized microarray performance is in accordance with such an ELISA kit in terms of rapidity, sensitivity, simplicity and inexpensive, However, ELISA detect α-lactoalbumin, β-lactoglobulin and lactoferrin in milk, it needs at least three times of experiments. Therefore, it has potential as an alternative analytical tool to screen for the presence of a-LA, β-LG and LF in the dairy industry and pediatric foods. Moreover, the implementation of disposable conjunction with the simplicity, automation and miniaturization of the instrumentation constitute important advantages leading towards the integration of the method in portable (in-field), reliable and user-friendly analytical systems for milk and infant formula quality control.

Additional file

> **Additional file 1: Figure S1.** The microarray of α-LA, β-LG, and LF on clear flat-bottom 96-well plate after silver enhancement was imaged with microarray scanner (QARRAY 2000). From top to bottom, left to right was numbered 1 to 18. 1–7 were raw milk, 8–11 were pasteurized milk, 12–18 were UHT milk including skimmed milk and high calcium milk.

Abbreviations
AB: Buffer for adjusting pH of EB; a-LA: α-Lactalbumin; BB: Binding buffer; BSA: Bovine serum albumin; CE: Capillary Electrophoresis; CR: Cross reactivity; EB: Elution buffer; EDTA: Ethylenediaminetetraacetic acid; ELISA: Enzyme-linked immunosorbent assay; FIA: Fluorescent Immunosorbent Assay; HPLC: High performance liquid chromatography; HPLC-MS: High performance liquid chromatography -mass spectra; IAC: Immunoaffinity chromatography; LF: Lactoferrin; PBS: Phosphate buffered saline; RID: Radial Immunodiffusion; RLUs: Relative light units; SDS-PAGE: sodium dodecyl sulfate polyacrylamide gel electrophoresis; SPR: Surface Plasmon Resonance; UHPLC: Ultra high performance liquid chromatography; UHPLC-MS: ultra high performance liquid chromatography - mass spectra; UHT: Ultra-high temperature treated; β-LG: β-lactoglobulin

Acknowledgements
The authors would like to thank and acknowledge the help of Nanjing Xiangzhong Biotechnology Co. Ltd.

Funding
We acknowledge financial support of the National Natural Science Foundation of China (21,405,077, 21,227,009, 21,475,060), Natural Science Foundation of Jiangsu Province (BK20140591). Research Foundation of Jiangsu Province Environmental Monitoring (1116), Special Fund for Agro-scientific research in the Public interest (201403071) and the National Science Fund for Creative Research Groups (21121091).

Authors' contributions
ZL wrote the manuscript and carried out visualized microarray experiments, including optimization of experimental conditions, determination of cross reaction rate, calculation calibration curves and recovery. FW and NZ performed HPLC measurements, including treatment of milk samples and determination the concentration of α-LA, β-LG, LF. ZL and JJ performed dairy determination, including evaluation the concentration of α-LA, β-LG, LF. DX designed the study and assisted in manuscript revision. All authors read and approved the final manuscript.

Competing interests
The authors declare that they have no competing interests.

Author details
[1]State Key Laboratory of Analytical Chemistry for Life Science, School of Chemistry and Chemical Engineering, Nanjing University, Nanjing 210093, China. [2]School of Chemistry and Biological Science, Nanjing University Jingling College, Nanjing 210089, China. [3]Ministry of Agriculture-Key Laboratory of Quality & Safety Control for Milk and Dairy Products, Institute of Animal Science, Chinese Academy of Agricultural Sciences, Beijing 100193, People's Republic of China. [4]Ministry of Agriculture Dairy Quality Supervision and Testing Center, Harbin 150090, China.

References
1. Chatterton DEW, Smithers G, Roupas P, Brodkorb A. Bioactivity of β-lactoglobulin and α-lactalbumin—technological implications for processing. Int Dairy J. 2006;16(11):1229–40.
2. Wakabayashi H, Yamauchi K, Takase M. Lactoferrin research, technology and applications. Int Dairy J. 2006;16(11):1241–51.
3. Ferraro V, Madureira AR, Sarmento B, Gomes A, Pintado ME. Study of the interactions between rosmarinic acid and bovine milk whey protein α-Lactalbumin, β-Lactoglobulin and Lactoferrin. Food Res Int. 2015;77:450–9.
4. Mayer HK, Raba B, Meier J, Schmid A. RP-HPLC analysis of furosine and acid-soluble β-lactoglobulin to assess the heat load of extended shelf life milk samples in Austria. Dairy Sci Technol. 2010;90(4):413–28.
5. Anandharamakrishnan C, Rielly CD, Stapley AGF. Loss of solubility of α-lactalbumin and β-lactoglobulin during the spray drying of whey proteins. LWT Food Sci Technol. 2008;41(2):270–7.
6. Mudgal P, Daubert CR, Foegeding EA. Kinetic study of β-lactoglobulin thermal aggregation at low pH. J Food Eng. 2011;106(2):159–65.
7. Yao X, Bunt C, Cornish J, Quek S, Wen J. Improved RP-HPLC method for determination of bovine lactoferrin and its proteolytic degradation in simulated gastrointestinal fluids. Biomed Chromatogr. 2013;27(2):197–202.
8. Sostmann K, Guichard E. Immobilized β-lactoglobulin on a HPLC-column: a rapid way to determine protein—flavour interactions. Food Chem. 1998;62(4):509–13.
9. Palmano KP, Elgar DF. Detection and quantitation of lactoferrin in bovine whey samples by reversed-phase high-performance liquid chromatography on polystyrene–divinylbenzene. J Chromatogr A. 2002;947(2):307–11.
10. Ding X, Yang Y, Zhao S, Li Y, Wang Z. Analysis of α-lactalbumin, β-lactoglobulin A and B in whey protein powder, colostrum, raw milk, and infant formula by CE and LC. Dairy Sci Technol. 2011;91(2):213–25.
11. Jackson JG, Janszen DB, Lonnerdal B, Lien EL, Pramuk KP, Kuhlman CF. A multinational study of α-lactalbumin concentrations in human milk. J Nutr Biochem. 2004;15(9):517–21.
12. Boitz LI, Fiechter G, Seifried RK, Mayer HK. A novel ultra-high performance liquid chromatography method for the rapid determination of β-lactoglobulin as heat load indicator in commercial milk samples. J Chromatogr A. 2015;1386:98–102.
13. Muhammad G, Saïd B, Thomas C. Structural consequences of dry heating on Beta-Lactoglobulin under controlled pH. Procedia Food Sci. 2011;1:391–8.
14. Gulzar M, Bouhallab S, Jardin J, Briard-Bion V, Croguennec T. Structural consequences of dry heating on alpha-lactalbumin and beta-lactoglobulin at pH 6.5. Food Res Int. 2013;51(2):899–906.

15. Corzo-Martínez M, Moreno FJ, Olano A, Villamiel M. Structural characterization of bovine β-Lactoglobulin–Galactose/Tagatose Maillard complexes by Electrophoretic, chromatographic, and spectroscopic methods. J Agric Food Chem. 2008;56(11):4244–52.

16. Yan R, Qu L, Luo N, Liu Y, Liu Y, Li L, Chen L. Quantitation ofα -Lactalbumin by liquid chromatography tandem mass spectrometry in medicinal adjuvant lactose. Int J Anal Chem. 2014;2014:1–4.

17. Stojadinovic M, Burazer L, Ercili-Cura D, Sancho A, Buchert J, Velickovic TC, Stanic-Vucinic D. One-step method for isolation and purification of native β-lactoglobulin from bovine whey. J Sci Food Agr. 2012;92(7):1432–40.

18. Silveira ST, Martínez-Maqueda D, Recio I, Hernández-Ledesma B. Dipeptidyl peptidase-IV inhibitory peptides generated by tryptic hydrolysis of a whey protein concentrate rich in β-lactoglobulin. Food Chem. 2013;141(2):1072–7.

19. Yang W, Liqing W, Fei D, Bin Y, Yi Y, Jing W. Development of an SI-traceable HPLC–isotope dilution mass spectrometry method to quantify β-Lactoglobulin in milk powders. J Agric Food Chem. 2014;62(14):3073–80.

20. Cunsolo V, Costa A, Saletti R, Muccilli V, Foti S. Detection and sequence determination of a new variantβ-lactoglobulin II from donkey. Rapid Commun Mass Sp. 2007;21(8):1438–46.

21. Czerwenka C, Maier I, Potocnik N, Pittner F, Lindner W. Absolute Quantitation of β-Lactoglobulin by protein liquid chromatography–mass spectrometry and its application to different milk products. Anal Chem. 2007;79(14):5165–72.

22. Chen Q, Zhang J, Ke X, Lai S, Li D, Yang J, Mo W, Ren Y. Simultaneous quantification of α-lactalbumin and β-casein in human milk using ultra-performance liquid chromatography with tandem mass spectrometry based on their signature peptides and winged isotope internal standards. Biochim Biophys Acta. 2016;1864(9):1122–7.

23. Xing K, Chen Q, Pan X. Quantification of lactoferrin in breast milk by ultra-high performance liquid chromatography-tandem mass spectrometry with isotopic dilution. RSC Adv. 2016;6(15):12280–5.

24. Ren Y, Han Z, Chu X, Zhang J, Cai Z, Wu Y. Simultaneous determination of bovine α-lactalbumin and β-lactoglobulin in infant formulae by ultra-high-performance liquid chromatography–mass spectrometry. Anal Chim Acta. 2010;667(1–2):96–102.

25. Zhang J, Lai S, Cai Z, Chen Q, Huang B, Ren Y. Determination of bovine lactoferrin in dairy products by ultra-high performance liquid chromatography-tandem mass spectrometry based on tryptic signature peptides employing an isotope-labeled winged peptide as internal standard. Anal Chim Acta. 2014;829:33–9.

26. Puerta A, Diez-Masa JC, de Frutos M. Immunochromatographic determination of β-lactoglobulin and its antigenic peptides in hypoallergenic formulas. Int Dairy J. 2006;16(5):406–14.

27. Puerta A, Diez-Masa JC, de Frutos M. Development of an immunochromatographic method to determine β-lactoglobulin at trace levels. Anal Chim Acta. 2005;537(1–2):69–80.

28. Mazri C, Sánchez L, Ramos SJ, Calvo M, Pérez MD. Effect of high-pressure treatment on denaturation of bovine β-lactoglobulin and α-lactalbumin. Eur Food Res Technol. 2012;234(5):813–9.

29. Alomirah HF, Alli I. Separation and characterization of β-lactoglobulin and α-lactalbumin from whey and whey protein preparations. Int Dairy J. 2004;14(5):411–9.

30. Giacinti G, Basiricò L, Ronchi B, Bernabucci U. Lactoferrin concentration in buffalo milk. Ital J Anim Sci. 2013;12(1):e23.

31. Cheang B, Zydney AL. Separation of -Lactalbumin and -Lactoglobulin using membrane Ultrafiltration. Biotechnol Bioeng. 2003;83(2):201–9.

32. Li J, Ding X, Chen Y, Song B, Zhao S, Wang Z. Determination of bovine lactoferrin in infant formula by capillary electrophoresis with ultraviolet detection. J Chromatogr A. 2012;1244:178–83.

33. Gutierrez JEN, Jakobovits L. Capillary electrophoresis of α-Lactalbumin in milk powders. J Agric Food Chem. 2003;51(11):3280–6.

34. Chen H, Busnel J, Gassner A, Peltre G, Zhang X, Girault HH. Capillary electrophoresis immunoassay using magnetic beads. Electrophoresis. 2008;29(16):3414–21.

35. Liu L, Kong D, Xing C, Zhang X, Kuang H, Xu C. Sandwich immunoassay for lactoferrin detection in milk powder. Anal Methods UK. 2014;6(13):4742.

36. Wroblewska B, Karamac M, Amarowicz R, Szymkiewicz A, Troszynska A, Kubicka E. Immunoreactive properties of peptide fractions of cow whey milk proteins after enzymatic hydrolysis. Int J Food Sci Technol. 2004;39(8):839–50.

37. de Luis R, Lavilla M, Sánchez L, Calvo M, Pérez MD. Development and evaluation of two ELISA formats for the detection of β-lactoglobulin in model processed and commercial foods. Food Control. 2009;20(7):643–7.

38. Huang YQ, Morimoto K, Hosoda K, Yoshimura Y, Isobe N. Differential immunolocalization between lingual antimicrobial peptide and lactoferrin in mammary gland of dairy cows. Vet Immunol Immunopathol. 2012;145(1–2):499–504.

39. Pelaez-Lorenzo C, Diez-Masa JC, Vasallo I, de Frutos M. Development of an optimized ELISA and a sample preparation method for the detection of β-Lactoglobulin traces in baby foods. J Agric Food Chem. 2010;58(3):1664–71.

40. Manzo C, Pizzano R, Addeo F. Detection of pH 4.6 insoluble β-Lactoglobulin in heat-treated milk and mozzarella cheese. J Agric Food Chem. 2008;56(17):7929–33.

41. Mehta R, Petrova A. Biologically active breast milk proteins in association with very preterm delivery and stage of lactation. J Perinatol. 2011;31(1):58–62.

42. Kleber N, Maier S, Hinrichs J. Antigenic response of bovine β-lactoglobulin influenced by ultra-high pressure treatment and temperature. Innov Food Sci Emerg. 2007;8(1):39–45.

43. Finetti C, Plavisch L, Chiari M. Use of quantum dots as mass and fluorescence labels in microarray biosensing. Talanta. 2016;147:397–401.

44. Yang A, Zheng Y, Long C, Chen H, Liu B, Li X, Yuan J, Cheng F. Fluorescent immunosorbent assay for the detection of alpha lactalbumin in dairy products with monoclonal antibody bioconjugated with CdSe/ZnS quantum dots. Food Chem. 2014;150:73–9.

45. Billakanti JM, Fee CJ, Lane FR, Kash AS, Fredericks R. Simultaneous, quantitative detection of five whey proteins in multiple samples by surface plasmon resonance. Int Dairy J. 2010;20(2):96–105.

46. Tomassetti M, Martini E, Campanella L, Favero G, Sanzò G, Mazzei F. Lactoferrin determination using flow or batch immunosensor surface plasmon resonance: comparison with amperometric and screen-printed immunosensor methods. Sensors Actuators B Chem. 2013;179:215–25.

47. Indyk HE, McGrail IJ, Watene GA, Filonzi EL. Optical biosensor analysis of the heat denaturation of bovine lactoferrin. Food Chem. 2007;101(2):838–44.

48. Indyk HE. Development and application of an optical biosensor immunoassay for α-lactalbumin in bovine milk. Int Dairy J. 2009;19(1):36–42.

49. Indyk HE, Filonzi EL. Determination of lactoferrin in bovine milk, colostrum and infant formulas by optical biosensor analysis. Int Dairy J. 2005;15(5):429–38.

50. Ruiz-Valdepeñas Montiel V, Campuzano S, Torrente-Rodríguez RM, Reviejo AJ, Pingarrón JM. Electrochemical magnetic beads-based immunosensing platform for the determination of α-lactalbumin in milk. Food Chem. 2016;213:595–601.

51. Eissa S, Tlili C, L'Hocine L, Zourob M. Electrochemical immunosensor for the milk allergen β-lactoglobulin based on electrografting of organic film on graphene modified screen-printed carbon electrodes. Biosens Bioelectron. 2012;38(1):308–13.

52. Hohensinner V, Maier I, Pittner F. A 'gold cluster-linked immunosorbent assay': optical near-field biosensor chip for the detection of allergenic β-lactoglobulin in processed milk matrices. J Biotechnol. 2007;130(4):385–8.

53. Li Z, Li Z, Zhao D. Smartphone-based visualized microarray detection for multiplexed harmful substances in milk. Biosens Bioelectron. 2017;87:874–80.

54. Li Z, Li Z, Niu Q. Visual microarray detection for human IgE based on silvernanoparticles. Sensors Actuators B Chem. 2017;239:45–51.

55. Li Z, Li Z, Jiang J. Simultaneous detection of various contaminants in milk based on visualized microarray. Food Control. 2017;73:994–1001.

56. Li Z, Li Z, Xu D. Simultaneous detection of four nitrofuran metabolites in honey simultaneous detection of four nitrofuran metabolites in honey by using a visualized microarray screen assay. Food Chem. 2017;221:1813–21.

22

An *att* site-based recombination reporter system for genome engineering and synthetic DNA assembly

Michael J. Bland[1,2], Magaly Ducos-Galand[1,2], Marie-Eve Val[1,2] and Didier Mazel[1,2]* (ID)

Abstract

Background: Direct manipulation of the genome is a widespread technique for genetic studies and synthetic biology applications. The tyrosine and serine site-specific recombination systems of bacteriophages HK022 and ΦC31 are widely used for stable directional exchange and relocation of DNA sequences, making them valuable tools in these contexts. We have developed site-specific recombination tools that allow the direct selection of recombination events by embedding the *attB* site from each system within the β-lactamase resistance coding sequence (*bla*).

Results: The HK and ΦC31 tools were developed by placing the *attB* sites from each system into the signal peptide cleavage site coding sequence of *bla*. All possible open reading frames (ORFs) were inserted and tested for recombination efficiency and *bla* activity. Efficient recombination was observed for all tested ORFs (3 for HK, 6 for ΦC31) as shown through a cointegrate formation assay. The *bla* gene with the embedded *attB* site was functional for eight of the nine constructs tested.

Conclusions: The HK/ΦC31 *att-bla* system offers a simple way to directly select recombination events, thus enhancing the use of site-specific recombination systems for carrying out precise, large-scale DNA manipulation, and adding useful tools to the genetics toolbox. We further show the power and flexibility of *bla* to be used as a reporter for recombination.

Keywords: Site-specific recombination, Tyrosine recombinase, Serine recombinase, Genetic engineering

Background

The ability to precisely and directly manipulate DNA is important for functional studies and the synthetic assembly of large genetic constructs. Site-specific recombinase (SSR) systems are widely used as tools to rearrange, insert, remove, and join DNA with virtually no upper limit in size. For biotechnology purposes, this can include the insertion of exogenous DNA into chromosomes, the fusing of DNA molecules, or the construction of synthetic gene networks [1]. The tyrosine (Y-rec) and serine (S-rec) recombination families are named for the catalytic residue of their respective integrase (Int) protein. Important members of the Y-rec family include the λ-like phage recombination systems, which include λ and the closely related phage HK022 (hereafter referred to as HK). The ΦC31 recombinase system is an important member of the S-rec family [2]. Both HK and ΦC31 systems comprise *attB*/*attP* attachment sites that serve as points of recombination, and the recombinases that catalyze recombination. In each family, DNA exchange requires host-encoded proteins for recombination that differ between systems. These systems are attractive due to their directionality and stability, and both systems are functional in prokaryotic and eukaryotic organisms [3–5].

Mechanistically, *attB* and *attP* integrative recombination forms *attL* and *attR* sites. The reverse *attL* x *attR* excisive reaction also requires Int as well as a recombination directionality factor (RDF), named Xis in the HK system and gp3 in the ΦC31 system [6], typically supplied *in trans* from a helper plasmid, a non-replicating DNA molecule, or as mRNA [7]. Structurally, HK and

* Correspondence: mazel@pasteur.fr
[1]Unité Plasticité du Génome Bactérien, Département Génomes et Génétique, Institut Pasteur, 75015 Paris, France
[2]UMR3525, Centre National de la Recherche Scientifique, 75015 Paris, France

ΦC31 *att* sites differ in size, with the HK *attB* sites being generally shorter than the HK *attP* sites, 21 base pairs (bp) vs 234 bp [8, 9]; in addition, *attP* contains binding sites for Int and Xis along with host-encoded proteins Fis and IHF [8–11]. ΦC31 *attB* and *attP* sites are similar in size (~50 bp) and do not require additional proteins to carry out recombination [12].

The use of SSRs generally involves selecting the recombination event through the use of a marker gene within the inserted sequence whose presence or absence would indicate successful integration [1]. Genes can be activated following recombination through either removal of blocking DNA sequences or by bringing together physically separated congruous sequences, with the recombination site embedded within the gene or between the promoter and coding sequence. This approach has long been used with the popular *CRE/loxP* [13] and Flp/FRT [14] systems. The β-lactamase (*bla*) gene is an attractive marker, as it is a useful reporter gene for both pro- and eukaryotic applications [15]. Protein chimeras of β-lactamase demonstrate tolerance to exogenous peptide insertions [16], even for domains of unknown function [17]. A split gene reassembly approach using *bla* has also been developed to discover directed evolution-modified SSR enzymes capable of recombining designer sequences [18]. The *bla* signal peptide is an attractive region for peptide insertion [19], as insertions between the signal peptide sequence and the rest of the coding gene have minimal interference with protein function [20]. As we wished to expand the available molecular toolbox, we created a set of recombination reporters consisting of the *attB* of HK and ΦC31 inserted in frame with *bla*, allowing expression of the gene and enabling the direct selection of recombination events. The selective agent is not expressed when the *att* sites are in *attL* and *attR* form, as the reporter gene fragments are physically separated (Fig. 1a).

This approach has been used to explore the physical structure of the *E. coli* genome [21, 22]. Genome engineering of the two *Vibrio cholerae* chromosomes used this tool to understand the evolutionary and genetic implications of multi-chromosomal bacteria [23]. We have used HK recombination in tandem with the λ-*lacZ* system from [21] to exchange DNA between the two *V. cholerae* chromosomes in a recombination-mediated cassette exchange (RMCE), resulting in large-scale chromosomal rearrangements [23]. Because the *lacZ* reporter allows the observation of recombination events but not to select for them, we developed a reporter system for HK recombination based on antibiotic selection. We have used an HK *attB* site placed in-frame within the β-lactamase (*bla*) gene to carry out relocation of the *S10-spec-α* ribosomal locus in *V. cholerae* in order to study the consequences of essential gene positioning as it relates to

dosage [24]. We further used HK-*bla* to carry out large-scale genome inversions around the origin region (*ori*) of *V. cholerae* chromosome one (Chr1) to shift the timing of the initiation of chromosome two (Chr2) replication relative to Chr1 in order to study the mechanisms involved in bacterial chromosome replication timing [25].

Here, we describe the construction and validation of HK-*bla* and a similar tool using the serine ΦC31 *att* system (ΦC31-*bla*). We placed *attB* sites from each system immediately downstream of the *bla* signal peptide coding sequence, which directs transport of β-lactamase to the periplasm and is removed in the mature protein. β-lactamase is generally tolerant of insertions into this region. When each system is present as *attL* and *attR* sites, they are associated with fragment sequences *bla'* (the 5′ region upstream of the cleavage site including the promoter and signal sequence) and *'bla* (the 3′ region comprising the mature protein sequence), respectively (Fig. 1a). In addition, the cognate *att* site partners show high recombination frequencies without the presence of *bla*-resistant background from the fragmented *bla* gene. These systems are extremely useful due to their ability to directly select for recombination through resistance to β-lactam antibiotics. They also have the potential to be used within synthetic biology frameworks for constructing and precisely inserting large genetic assemblies, making them useful additions to the molecular biology toolbox for both synthetic and molecular applications.

Results

In-frame insertion of *attB*$_{HK}$ sites within the ß-lactamase gene

The β-lactamase gene has a 23-amino acid (aa) signal peptide sequence for protein transmembrane transport that is cleaved during protein maturation [26]. We inserted the *attB* sequences in frame into the junction between the encoded signal sequence and the mature protein (Fig. 1a), as this region is tolerant to sequence insertions [19]. To avoid interfering with the β-lactamase coding sequence we took into account *attB* length and the amino acid sequence of the translated *att* sequence, so as to avoid frameshift or stop codon insertion.

Recombination frequency in *attB*$_{HK}$ sites decreases with size

The *attB*$_{HK}$ site comprises a 7 bp core, or overlap, (O) region where strand exchange occurs, and flanking B and B′ arm regions of 7 bp each that are recognized by Int monomers to form a synaptic complex, although sites shorter than this 21 bp have been shown to be functional but with low efficiency [27]. To allow recombination, the O region between *attB* and *attP* must

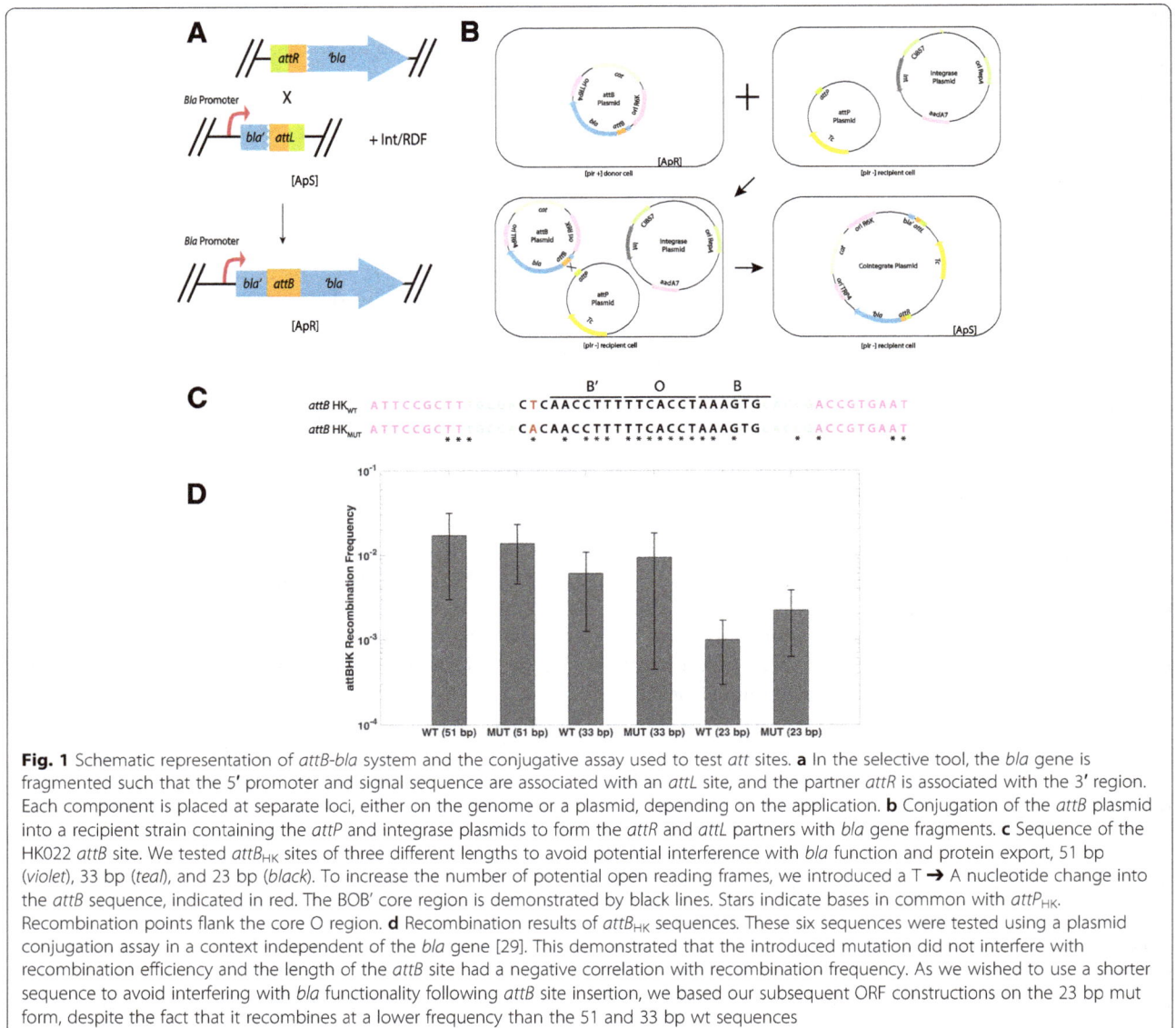

Fig. 1 Schematic representation of *attB-bla* system and the conjugative assay used to test *att* sites. **a** In the selective tool, the *bla* gene is fragmented such that the 5′ promoter and signal sequence are associated with an *attL* site, and the partner *attR* is associated with the 3′ region. Each component is placed at separate loci, either on the genome or a plasmid, depending on the application. **b** Conjugation of the *attB* plasmid into a recipient strain containing the *attP* and integrase plasmids to form the *attR* and *attL* partners with *bla* gene fragments. **c** Sequence of the HK022 *attB* site. We tested *attB*$_{HK}$ sites of three different lengths to avoid potential interference with *bla* function and protein export, 51 bp (*violet*), 33 bp (*teal*), and 23 bp (*black*). To increase the number of potential open reading frames, we introduced a T ➜ A nucleotide change into the *attB* sequence, indicated in red. The BOB′ core region is demonstrated by black lines. Stars indicate bases in common with *attP*$_{HK}$. Recombination points flank the core O region. **d** Recombination results of *attB*$_{HK}$ sequences. These six sequences were tested using a plasmid conjugation assay in a context independent of the *bla* gene [29]. This demonstrated that the introduced mutation did not interfere with recombination efficiency and the length of the *attB* site had a negative correlation with recombination frequency. As we wished to use a shorter sequence to avoid interfering with *bla* functionality following *attB* site insertion, we based our subsequent ORF constructions on the 23 bp mut form, despite the fact that it recombines at a lower frequency than the 51 and 33 bp wt sequences

perfectly overlap, and the arm regions must share similarity. Flanking the core minimal region, there are homologous nucleotides that may play an additional role in recombination efficiency [10, 28]. Insertion of *attB* into *bla* extends the gene and could affect either transport through the membrane or mature enzyme function. It is therefore necessary to test different open reading frames encoded by the *attB*$_{HK}$ sequence to avoid unwanted interference with *bla*. The native *attB*$_{HK}$ sequence encodes two open reading frames (ORFs) that do not have stop codons. As we wished to increase the potential sequences we could test within *bla*, we added a third potential ORF by mutating one bp just outside of the B′ region (Fig. 1c; Fig. 2a) [8, 27]. We compared these "mutant" *attB* sites to the "wild-type" sites to ensure there was no loss of recombination frequency (Fig. 1d).

The 23, 33, and 51 bp "wild type" and "mutant" *attB*$_{HK}$ sequences were tested by placing them on the conditionally replicating conjugative plasmid pSW23T containing an *oriT*$_{RP4}$ for plasmid conjugation and *oriV*$_{R6K\gamma}$ for π protein replication dependence (Fig. 1b); [29]. As these plasmids do not replicate in bacterial strains not expressing the π protein, conjugation into non-π expressing DH5α leads to plasmid loss unless *att* recombination occurs. The DH5α recipient strain houses plasmid pHK11Δamp, which has the *attP*$_{HK}$ partner site, and pHK-Int, which expresses the HK integrase under control of the temperature-dependent CI857 promoter [30]. Following conjugation, recombination frequency was calculated by measuring the ratio of recovered colonies (representing co-integrates) over the number of recipient colonies [31]. Recombination frequencies were

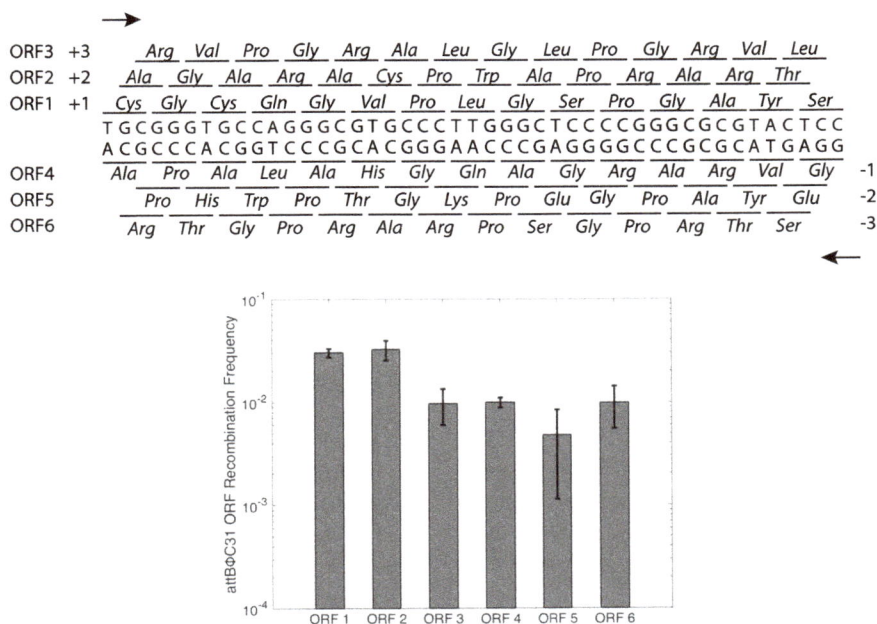

Fig. 2 Sequences and recombination frequencies of HK and ΦC31 *attB* sites. The three ORFs for HK and the six ORFs for ΦC31 were inserted into *bla* and tested using the conjugation assay as in Fig. 1a. The six open reading frames of the 23 bp *attB* HK site are shown. As in Fig. 1, black nucleotides represent the 23 bp HK sequence, with the corresponding amino acids also in *black*. The *red* nucleotide shows the base changed from the original *attB*, with the resulting amino acid changes also shown in *red*. Nucleotides and amino acids in *teal* represent sequences flanking the 23 bp site. *Horizontal arrows* indicate the direction of transcription, and asterisks indicate a stop codon. For both **a** and **b**, the sequence of recombination exchange is indicated by a *horizontal red line*. As described in the text, three open reading frames did not have a stop codon and were able to be tested for *bla* insertion. The recombination frequencies of these open reading frames compared to the 23 bp HK *attB* site are shown in the bar graph. The open reading frames are also shown in context of the *bla* sequence flanking the insertion site. Note that to keep the *attB* site in frame with *bla*, nucleotides were added to either the 5′ or 3′ end of the site, which changed the expected amino acid residue for ORFs 1 and 3 compared to the original *attB*. The background colors highlighting the sequence correspond to Fig. 1a. The recombination frequencies of the different ORFs were compared using 1-way ANOVA followed by a Tukey-Kramer test. Each of the HK ORF recombination frequencies are significantly different ($p < 0.05$). **b**. None of the six ΦC31 ORFs encode a stop codon. ORFs 1 and 2 recombine at a higher rate than ORFs 3–6 ($p < 0.001$)

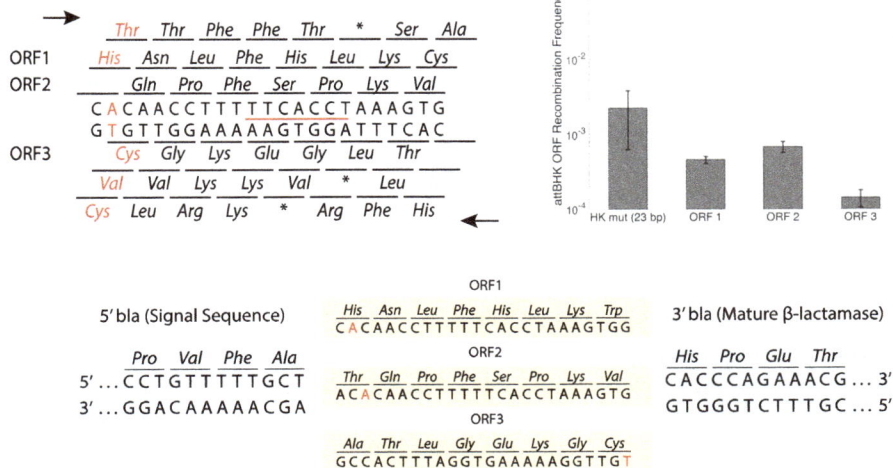

similar between the different sites, with only a 10-fold reduction in recombination observed for the 23 bp sites compared to the larger *attB* sites (Fig. 1c). As we wished to use a shorter sequence to avoid interfering with *bla* functionality following *attB* site insertion, we based our subsequent tests on the 23 bp *attB*$_{HK}$ mutant form.

Placing a single nucleotide mutation in the 23 bp $attB_{HK}$ site enables the use of three ORFs that would potentially allow *bla* function following their insertion into the gene (Fig. 2a). These ORFs were inserted separately into *bla* downstream of the signal sequence and cloned into pSW23T in a π + host. Following construction of these plasmids, we measured the ampicillin minimum inhibitory concentration (MIC) of each to test and measure *bla* function. All ORFs provided resistance to ampicillin at an MIC >256 µg/ml (Table 1). Recombination frequencies were then tested using the conjugation assay as above. The three HK ORF constructions demonstrated a wide range of recombination efficiencies, with the ORF 2 construct recombining at the highest level, and the ORF 3 construct recombining at the lowest (Fig. 2a). Thus, we used ORF 2 for the final construction of this tool.

ΦC31 *attB* x *attP* recombination is functional in all six ORFs

We designed $attB_{ΦC31}$ sites for all six possible ORFs maintaining at least the minimal sequence necessary for recombination [32] and inserted them into *bla*. Ampicillin resistance and recombination frequency were determined as with the HK system. Five of six ORFs were found to provide MICs greater than 256 µg/ml, with the ORF 5 construction being the only sequence to interfere with β-lactamase function (MIC = 6 µg/ml - Table 1). ΦC31 pSW23T-bla plasmids were conjugated into a DH5α strain harboring plasmids pΦC31-Int and pΦC31-attP. All six ORFs were able to recombine successfully, with ORF constructions 1 and 2 recombining at a higher rate, on the order of 10^{-2}, than ORFs 3–6, which recombined at an average rate of 10^{-3} (Fig. 2b). We found this difference to be significant using a 1-way ANOVA ($p < 0.001$) followed by a post-hoc Tukey-Kramer test ($p < 0.001$). Additionally, all six ΦC31 ORF constructions recombined at a higher rate than HK ORFs 1–3 (Fig. 2).

Table 1 Minimum inhibitory concentration (MIC) of $attB_{HK}$ and $attB_{ΦC31}$ ORFs inserted into β-lactamase

Ampicillin Resistance of *bla-attB* ORFs	MIC (µg/ml)
HK022 ORF1	> 256
HK022 ORF2	> 256
HK022 ORF3	> 256
ΦC31 ORF1	> 256
ΦC31 ORF2	> 256
ΦC31 ORF3	> 256
ΦC31 ORF4	> 256
ΦC31 ORF5	6
ΦC31 ORF6	> 256

Discussion

In this study, we describe the construction of two site-specific recombination tools useful for DNA manipulation applications. The utility of this *attB-bla* tool is based on its incorporation of the widely used HK and ΦC31 recombination systems. In the case of HK, the removal of sequences flanking the BOB' core region reduced *attB* x *attP* recombination. This reduction could be due to the removal of bases outside of the *attB* core that have homology with the *attP* sequence, which may act to stabilize the *attB/attP* complex. However, obtaining the highest possible recombination frequency was not critical for the design of this system, as our main concern was β-lactamase function following insertion of the *att* sites into the *bla* coding frame.

In directly comparing the two systems, the ΦC31 site appears to recombine at a similar frequency to the 51 bp HK sites and the 23 bp HK ORFs incorporated into *bla* have a lower recombination frequency (Fig. 2). This decrease is likely due to the reduction of size of the $attB_{HK}$ site, as the recombination frequencies for the smaller HK site tested independently of *bla* insertion are not different from the frequencies obtained when they are embedded in *bla* (Fig. 1). Reported differences between recombination systems in the literature may result from differences in protocols and practices. A recent review of ΦC31 found a wide range of reported recombination frequencies for this recombinase [33]. To our knowledge, the only information comparing HK and ΦC31 recombination frequencies reports HK recombining at a higher frequency than ΦC31 [34]. However, this study used a clonetegration technique where constructs were recombined into native *att* sites on either the *E. coli* chromosome for $attP_{HK}$ or *Salmonella typhimurium* for $attP_{ΦC31}$.

While testing *bla* expression with inserted ORFs, we observed that ΦC31 ORF 5 interfered with *bla* expression, while ΦC31 recombination was not affected (Table 1, Fig. 2b). The *bla* gene used for our system originates from pBR322 and belongs to the TEM-1 class of β-lactamases. The signal sequence is recognized by the Sec export pathway that transports unfolded proteins across the cytoplasmic membrane [26, 35]. DNA secondary structures could be a source of transcription interference, as ORF 5 forms a 30 bp hairpin (ΔG at 37 °C = −9.09 kcal/mol). However, hairpins are formed in all 6 ORFs at similar ΔG, making it unlikely that this factor alone prevents *bla* expression. At the translation level, the overall charge of the first 5 amino acids following the signal sequence can influence cleavage and cross-membrane transport, as they generally have an overall negative charge [36]. For ORF 5, the overall negative charge of this region is +2. Again, however, this is unlikely to explain the loss of *bla* expression, as only ORF 1 has an overall negative charge, at −1. The amino acids in the 1 and 2 position after the cleavage site can also influence protein function [37, 38]. For ORF 5, the first two amino acids are glycine and serine. Analysis of 307 proteins from the SPdb database [39] found that in Gram-negative bacteria, glycine occurs in the 1st position in 6.19% of proteins, and serine appears in the 2nd position in 5.54% of proteins. [40]. Additionally, two of the 307 Gram-negative proteins analyzed in this study begin with glycine-serine. Thus it is unlikely that the first two residues of the ORF 5 sequence alone interfere with protein transport. More experimental and analytical work is needed to determine the source of *bla* expression interference.

The high tolerance of *bla* to in-frame DNA sequence insertion downstream of the *bla* promoter and leader peptide sequence allows for further modifications of this system through insertion of potentially large ORFs. This approach has already been proposed as an "ORF-trap" to capture DNA encoding protein fragments [41]. Indeed, large ORFs in frame with *bla* may not greatly reduce β-lactamase function, although export to the periplasm can be inhibited [42]. Additionally, as *attB* and *attP* site reactivity can be modified through mutations to their respective core sequences, variable non-reacting "synthetic" *att* sites can be designed for sequential introduction into the bacterial chromosome [43].

Integration of exogenous DNA sequences into genomes by SSRs generally involves the recombination of an *attP* site on the inserted sequence with an endogenous chromosomal *attB* or pseudo-*attB* site [1]. The use of genome editing technologies allows the insertion of recombination sites that differ from native sites in location and sequence. Native *att* sites may be located in undesirable regions of the genome, for example, in an active gene locus, or a locus subject to silencing. Additionally, dosing effects can be observed in bacterial species dependent on a gene's location in the chromosome [24]. Engineering *att* site recognition by Int proteins allows the creation of semi-synthetic partner sites [27, 43]. This would avoid recombination with other native *att* sites, and could allow rapid construction of synthetic gene networks. The addition of FRT sites flanking the *bla-attB* cassette would further allow for removal of the resistance selection

marker gene. Similarly, gene-editing technologies could allow the targeted insertion of *att* sites to serve as landing pads for insertion. In this way, the *bla'-attL* sequence from our system can be inserted into a genome, into which a sequence containing the partner *attR-'bla* can be inserted through *attL* x *attR* recombination. This framework has already been proposed for the construction and insertion of metabolic networks into eukaryotic cell lines [44]. Our system adds the advantage of avoiding marker expression until recombination, making it versatile for synthetic applications as well as genome-scale engineering.

Conclusions

We describe here the construction of new tools based on two different site-specific recombination systems, the tyrosine recombinase HK, and the serine recombinase ΦC31. Recombination for each system is reported based on the reconstitution of the *bla* ampicillin resistance gene, providing resistance to β-lactam antibiotics as a selective agent. Both HK-*bla* and ΦC31-*bla* are useful for selecting recombination events in a genomic context due to a high rate of recombination frequency, directionality based on the recombination proteins supplied *in trans*, and the ability to carry out in vivo genomic rearrangements. We have previously used this tool in our lab to carry out large-scale reorganization of the *V. cholerae* chromosomes to study the importance of chromosome size in multichromosomal bacteria [23], the relevance of genome position and chromosome location for gene dosage and its evolutionary importance [24], and the timing of *V. cholerae* chromosome replication [25]. The importance of these tools lie in their capacity to exist simultaneously in the cell at two separate loci without expression of the marker gene until expression of the recombination proteins is induced.

Methods

Bacterial strains and media

Bacterial strains used in this study are described in Table 2. All strains were grown in lysogeny broth (LB) medium at 30 °C, 37 °C, or 42 °C depending on plasmid temperature-sensitivity. Antibiotic and nutritional supplement concentrations were as follows: ampicillin (Ap):

Table 2 Bacterial strains used in this study

E. coli		
Name	Genotype	Reference/Source
β2163	(F⁻) RP4–2-Tc::Mu ΔdapA::(erm-pir) [Km^R Em^R]	[29]
π1	DH5αΔthyA::(erm-pir116) [Em^R]	[29]
MFDpir	MG1655 RP4–2-TC::[Mu1::aac(3)IV-ΔaphA-Δnic35-ΔMu2::zeo] ΔdapA::(erm-pir)ΔrecA[Apra^R Zeo^{R}Erm^R]	[47]
PGB-8557	DH5α strain containing plasmids pHKΔ-Amp and pHK-Int [Tc^R Sp^R]	this study
PGB-E274	DH5α strain containing plasmids pΦC31-attP and pΦC31-Int [Tc^R Sp^R]	this study
One Shot ® Top10	F- mcrA Δ(mrr-hsdRMS-mcrBC) Φ80lacZΔM15 Δ lacX74 recA1 araD139 Δ(araleu)7697 galU galK rpsL (StrR) endA1 nupG	ThermoFisher Scientific

100 µg/ml, carbenicillin (Carb): 100 µg/ml, kanamycin (Km): 25 µg/ml, chloramphenicol, (Cm): 25 µg/ml, tetracyclin (Tc): 15 µg/ml, spectinomycin (Sp): 100 µg/ml, erythromycin (Em): 20 µg/ml, with nutritional supplements diaminopimelic acid (DAP): 300 µM, and thymine (dT): 300 µM.

Cloning

Basic cloning steps were performed using the following tools and appropriate protocols: for DNA purification, a QIAquick PCR purification kit (QIAGEN) was used. Plasmid minipreps were performed using the GeneJET Plasmid Miniprep kit (Life Technologies). All PCR reactions for plasmid construction were performed using the Phusion High-Fidelity PCR Master Mix (Life Technologies), and all diagnostic PCR reactions were performed using DreamTaq DNA Polymerase (Life Technologies). Oligonucleotides were synthesized by Sigma-Aldrich and Eurofins Genomics. Oligonucleotides were phosphorylated by T4 polynucleotide kinase (NEB). DNA was sequenced by GATC Biotech and Eurofins Genomics.

Construction of plasmids

Insertion of *attB* sequences into pSW23T was performed by annealing phosphorylated oligos containing the respective *att* sequence with overhangs overlapping with *Bam*HI and *Eco*RI restriction sites, followed by cloning of these sequences into the pSW23T fragment. Insertion of *attP* sequences into pHK11-Amp was similarly performed. The various *attB* ORFs for both HK and ΦC31 were inserted into the β-lactamase (*bla*) by overlapping PCR, in which the 5′ region of *bla* was amplified from pMP58 using oligos MV26 and the appropriate reverse *attB* oligo, and the 3′ *bla* region amplified using a forward *attB* oligo and JB13. These products were gel purified and co-amplified using oligos MV26 – JB13 to form a DNA fragment containing *bla* with the inserted *attB*. This product was digested with EagI and EcoRI and cloned into pSW23T and transformed into MFDpir. The pMP58 *bla* gene comes from pUC19.

To make plasmid pPhiC31-Int, we first deleted the XbaI site in pZJ7 (a kind gift of Jia Zhao and Sean Colloms) by digestion with SpeI-XbaI followed by religation to make plasmid pZJ7ΔXbaI. The ΦC31

Table 3 Plasmids used in this study

Name	Description	Reference/Source
pSW23T	pSW23::oriTRP4; [Cm^R]; oriVR6K	[29]
pSU38Δ	orip15A [Km^R]	[48]
pHK-Int	pGB2ts::cI857-λ-P_R-HKInt, [Sp^R]	[30]
pHK11-Amp	pLDR11::attP_HK, [Ap^R,Tc^R]	[30]
pSC101	pSC101ts, repA [Tc^R]	[49]
pUC19	oriColE1, lacZα [Ap^R]	[50]
pBAD43	oripSC101, PBAD::MCS,[Sp^R]	[51]
pHK11Δamp	pHK11-Amp::attP_HK,ΔAmp, [Tc^R]	this study
pMP96	pSC101ts::cI857-λ-P_R-(HK_Xis-HK_Int λ_Xis-λ_Int), [Sp^R]	[23]
pMP58	pSC101ts::oriTRP4;repA, [Cm^R,Ap^R]	this study
pMDG1	pMP58;bla::attB_HK,[Ap^R,Cm^R]	this study
pMDG2	pSW23T::bla::attB_HK from pMDG1	this study
pMDG3	α/pSU38::attR_HK, [Ap^R]	this study
pMDG4	pSW23T::attL_HK, [Cm^R]	this study
pMJM1	pSW23T::attB_HKwt, [Cm^R]	this study
pMJM2	pSW23T::attL_HKmut, [Cm^R]	this study
pMJM3	pSW23T::attL_HK40, [Cm^R]	this study
pMJM4	pSW23T::attL_HK30, [Cm^R]	this study
pJB6	pSU38Δ::attR_HK-attL_λ, [Ap^R]	this study
pJB7	pSW23T::attR_HK-attL_λ, [Cm^R]	this study
pJB8	pBAD43::HK_Xis-HK_Int λ_Xis-λ_Int, [Sp^R]	this study
pZJ7	pBAD33::ΦC31Int, [Cm^R]	J. Zhao and S. Colloms
pZJ7ΔXbaI	pZJ7 with SpeI – XbaI fragment deleted	this study
pPhiC31-Int	pGB2ts::cI857-λ-PR-ΦC31Int, [Sp^R]	this study
pPhiC31-attP	pHK11Δamp::attP_ΦC31, [Tc^R]	this study

integrase gene was amplified using oligos PhiC31 IntF and PhiC31 IntR. The pHK-Int backbone was amplified using oligos JB485 and JB486. These oligos produce DNA fragments with overlapping ends, which were then joined by Gibson assembly [45]. Plasmids used in this study are listed in Table 3 and oligonucleotides in Table 4.

Recombination assay

Recombination frequencies were tested by performing a conjugation assay in which the plasmid pSW23T containing the $oriT_{RP4}$ transfer region and $oriV_{R6K}$ π-controlled replication origin were transferred from the π+/DAP- donor strain MFDpir to a recipient strain

Table 4 Oligonucleotides used in this study

Oligonucleotide	Sequence 5' – 3'
PhiC31 Int F	ATGTACTAATCTAGAGAAGAGGATCAGAAATGGACACGTACGCGGGTGC
PhiC31 Int R	CAAGCTTGCATGCCTGCAGG
JB13	AGCGGGTGTTCCTTCTTCACTG
JB485	TCTTCTCTAGATTAGTACATGCAACCA
JB486	CGACTAGAGTCGACCTGCAGCCAAGCTTAGTAAAGCCCTC
MV26	ACGGCTGACATGGGAATTGC
MV143	CCTCTTACGTGCCGATCAACGTCTC
MV145	GCTGGTGATTCCGCTTTGCGACTCAACCTTTTTCACCTAAAGTGCACCGACCGTGA
MV146	ACATCAGCGATCACCTGGCAGAC
attBHKwtERI	AATTCCGCTTTGCGACTCAACCTTTTTCACCTAAAGTGCACCGACCGTGAATG
attBHKwtREV	GATCCATTCACGGTCGGTGCACTTTAGGTGAAAAAGGTTGAGTCGCAAAGCGG
attBHKmutERI	AATTCCGCTTTGCGACACAACCTTTTTCACCTAAAGTGCACCGACCGTGAATG
attBHKmutREV	GATCCATTCACGGTCGGTGCACTTTAGGTGAAAAAGGTTGTGTCGCAAAGCGG
40wtERI	AATTCTGCGACTCAACCTTTTTCACCTAAAGTGCACCG
40wtREV	GATCCCGGTGCACTTTAGGTGAAAAAGGTTGAGTCGCAG
40attBHKmutERI	AATTCTGCGACACAACCTTTTTCACCTAAAGTGCACCG
40attBHKmutREV	GATCCCGGTGCACTTTAGGTGAAAAAGGTTGTGTCGCAG
30wtERI	AATTCTCAACCTTTTTCACCTAAAGTG
30wtREV	GATCCACTTTAGGTGAAAAAGGTTGAG
30attBHKmutERI	AATTCACAACCTTTTTCACCTAAAGTG
30attBHKmutREV	GATCCACTTTAGGTGAAAAAGGTTGTG
30attBHKamp2ORF1min	TTTGCTCACAACCTTTTTCACCTAAAGTGGCACCCAGAAACGCTGGTGAAAGTAAAAGATGCTGAAGATCAGTT
30attBHKamp1ORF1min	CTTTCACCAGCGTTTCTGGGTGCCACTTTAGGTGAAAAAGGTTGTGAGCAAAAACAGGAAGGCAAAATGCCGC
30attBHKamp2ORF2min	TTTGCTACACAACCTTTTTCACCTAAAGTGCACCCAGAAACGCTGGTGAAAGTAAAAGATGCTGAAGATCAGTT
30attBHKamp1ORF2min	CTTTCACCAGCGTTTCTGGGTGCACTTTAGGTGAAAAAGGTTGTGTAGCAAAAACAGGAAGGCAAAATGCCGC
30attBHKamp2ORF3min	TTTGCTGCCACTTTAGGTGAAAAAGGTTGTCACCCAGAAACGCTGGTGAAAGTAAAAGATGCTGAAGATCAGTT
30attBHKamp1ORF3min	CTTTCACCAGCGTTTCTGGGTGACAACCTTTTTCACCTAAAGTGGCAGCAAAAACAGGAAGGCAAAATGCCGC
phiC31 ORF1 F	TTTGCTTGCGGGTGCCAGGGCGTGCCCTTGGGCTCCCCGGGCGCGTACTCCCACCCAGAAACGCTGGTGAAAG
phiC31 ORF2 F	TTTGCTGCGGGTGCCAGGGCGTGCCCTTGGGCTCCCCGGGCGCGTACTCCCCACCCAGAAACGCTGGTGAAAG
phiC31 ORF3 F	TTTGCTCGGGTGCCAGGGCGTGCCCTTGGGCTCCCCGGGCGCGTACTCCCCCACCCAGAAACGCTGGTGAAAG
phiC31 ORF4 F	TTTGCTGGAGTACGCGCCCGGGGAGCCCAAGGGCACGCCCTGGCACCCGCACACCCAGAAACGCTGGTGAAAG
phiC31 ORF5 F	TTTGCTGGGAGTACGCGCCCGGGGAGCCCAAGGGCACGCCCTGGCACCCGCCACCCAGAAACGCTGGTGAAAG
phiC31 ORF6 F	TTTGCTGGGGAGTACGCGCCCGGGGAGCCCAAGGGCACGCCCTGGCACCCGCACCCAGAAACGCTGGTGAAAG
phiC31 ORF1 R	CTTTCACCAGCGTTTCTGGGTGGGAGTACGCGCCCGGGGAGCCCAAGGGCACGCCCTGGCACCCGCAAGCAAAAACAGGAAGGCAAAATG
phiC31 ORF2 R	CTTTCACCAGCGTTTCTGGGTGGGGAGTACGCGCCCGGGGAGCCCAAGGGCACGCCCTGGCACCCGCAGCAAAAACAGGAAGGCAAAATG
phiC31 ORF3 R	CTTTCACCAGCGTTTCTGGGTGGGGGAGTACGCGCCCGGGGAGCCCAAGGGCACGCCCTGGCACCCGAGCAAAAACAGGAAGGCAAAATG
phiC31 ORF4 R	CTTTCACCAGCGTTTCTGGGTGTGCGGGTGCCAGGGCGTGCCCTTGGGCTCCCCGGGCGCGTACTCCAGCAAAAACAGGAAGGCAAAATG
phiC31 ORF5 R	CTTTCACCAGCGTTTCTGGGTGGCGGGTGCCAGGGCGTGCCCTTGGGCTCCCCGGGCGCGTACTCCCAGCAAAAACAGGAAGGCAAAATG
phiC31 ORF6 R	CTTTCACCAGCGTTTCTGGGTGCGGGTGCCAGGGCGTGCCCTTGGGCTCCCCGGGCGCGTACTCCCCAGCAAAAACAGGAAGGCAAAATG

containing an *attP* plasmid and a helper plasmid expressing the appropriate integrase gene under control of the temperature-sensitive CI857 promoter. Prior to conjugation, strains were diluted 1/100 from an overnight starter culture and grown to OD_{600} = 0.3. Conjugations were performed by two techniques: for the *attB* $HK_{WT/MUT}$ strains, 0.5 ml of donor was mixed with 4.5 ml of recipient and applied to a 0.45 μm filter (Millipore) by vacuum-filtration through a glass column. The *attB* ORF insertions into *bla* were performed by mixing 0.2 ml of donor with 1.8 ml of recipient, and following centrifugation at 6000 RPM for 5 min, ~1.8 ml of supernatant was removed, the pellet resuspended in the remaining liquid media, and similarly placed onto a 0.45 μm filter. For both techniques, the filters were then incubated on an LB-DAP plate for approx. 16 h prior to resuspension and plating. Recombinants were recovered by selecting for Cm resistance in DAP-free media, and recombination frequencies were measured as the ratio of recovered recombinants over donor CFUs. Each *att* site was tested three times.

Minimum inhibitory concentration (MIC)

The MICs of *E. coli* strains containing plasmids with either *attB* inserted into *bla*, or *bla* fragments associated with *attL* and *attR* were performed by plating and aspirating 2 ml of a 1/100 dilution of an overnight culture onto an LB/DAP agar petri dish. An Etest (bioMérieux) ampicillin antibiotic strip was placed onto the plate and incubated overnight at 37 °C, and the level of antibiotic resistance was scored the following day.

Data analysis

Recombination frequencies were analyzed for statistical significance using MATLAB software (The MathWorks, Inc., Natick, MA). 1 and 2-way analysis of variance (ANOVA) tests were performed using the anova1 and anova2 functions. Tukey-Kramer post-hoc tests were performed using the multcompare function.

DNA folding and protein structure analysis

Secondary DNA structures were analyzed using the mfold software [46]. Protein residue charges were calculated by counting negatively charged residues Asp and Glu as −1, and positively charged His, Lys, and Arg as +1.

Abbreviations

Bla: β-lactamase; HK022: HK; Int: Integrase; MIC: Minimum inhibitory concentration; O region: Overlap region; ORF: Open reading frame; RDF: Recombination directionality factor; S-rec: Serine recombinase; SSR: Site-specific recombinase; Y-rec: Tyrosine recombinase

Acknowledgments

The authors thank Sean Colloms (Institute of Molecular Cell and Systems Biology, University of Glasgow, Glasgow, Scotland, UK) for providing ΦC31 plasmids. The authors thank Aleksandra Nivina and Jessica Bryant for critical reading of the manuscript.

Funding

Work in the Mazel laboratory is funded by the Institut Pasteur, the Institut National de la Santé et de la Recherche Médicale (INSERM), the Centre National de la Recherche Scientifique (CNRS-UMR 3525), the French National Research Agency (ANR-14-CE10–0007), the French Government's Investissement d'Avenir program, Laboratoire d'Excellence "Integrative Biology of Emerging Infectious Diseases" (grant n°ANR-10-LABX-62-IBEID) and the European Union Seventh Framework Programme (FP7-HEALTH-2011-single-stage) "Evolution and Transfer of Antibiotic Resistance" (EvoTAR). MJB was supported by the Pasteur-Paris University (PPU) International PhD program.

Authors' contributions

MJB, MEV, and DM conceived and designed the experiments; MJB and MDG carried out the experiments; MJB, MDG, MEV, and DM analyzed the data; MJB and DM wrote the paper. All authors read and approved the final version of the manuscript.

Competing interests

The authors declare that they have no competing interests.

References

1. Olorunniji FJ, Rosser SJ, Stark WM. Site-specific recombinases: molecular machines for the genetic revolution. Biochem J. 2016;473(6):673–84.
2. Grindley ND, Whiteson KL, Rice PA. Mechanisms of site-specific recombination. Annu Rev Biochem. 2006;75:567–605.
3. Thyagarajan B, Olivares EC, Hollis RP, Ginsburg DS, Calos MP. Site-specific genomic integration in mammalian cells mediated by phage phiC31 integrase. Mol Cell Biol. 2001;21(12):3926–34.
4. Bischof J, Maeda RK, Hediger M, Karch F, Basler K. An optimized transgenesis system for drosophila using germ-line-specific phiC31 integrases. Proc Natl Acad Sci U S A. 2007;104(9):3312–7.
5. Hirano N, Muroi T, Takahashi H, Haruki M. Site-specific recombinases as tools for heterologous gene integration. Appl Microbiol Biotechnol. 2011;92(2):227–39.
6. Khaleel T, Younger E, McEwan AR, Varghese AS, Smith MC. A phage protein that binds phiC31 integrase to switch its directionality. Mol Microbiol. 2011; 80(6):1450–63.
7. Petersen KV, Martinussen J, Jensen PR, Solem C. Repetitive, marker-free, site-specific integration as a novel tool for multiple chromosomal integration of DNA. Appl Environ Microbiol. 2013;79(12):3563–9.
8. Yagil E, Dolev S, Oberto J, Kislev N, Ramaiah N, Weisberg RA. Determinants of site-specific recombination in the lambdoid coliphage HK022. An evolutionary change in specificity. J Mol Biol. 1989;207(4):695–717.
9. Azaro MA, Landy A. In: Craig RC NL, Gellert M, Lambowitz AM, editors. Mobile DNA II. Washington, DC: ASM Press; 2002. p. 118–48.
10. Weisberg RA, Gottesmann ME, Hendrix RW, Little JW. Family values in the age of genomics: comparative analyses of temperate bacteriophage HK022. Annu Rev Genet. 1999;33:565–602.
11. Groth AC, Calos MP. Phage integrases: biology and applications. J Mol Biol. 2004;335(3):667–78.
12. Smith MC, Brown WR, McEwan AR, Rowley PA. Site-specific recombination by phiC31 integrase and other large serine recombinases. Biochem Soc Trans. 2010;38(2):388–94.
13. Tungsuchat T, Kuroda H, Narangajavana J, Maliga P. Gene activation in plastids by the CRE site-specific recombinase. Plant Mol Biol. 2006;61(4–5):711–8.
14. Nakano M, Odaka K, Ishimura M, Kondo S, Tachikawa N, Chiba J, Kanegae Y, Saito I. Efficient gene activation in cultured mammalian cells mediated by FLP recombinase-expressing recombinant adenovirus. Nucleic Acids Res. 2001;29(7):E40.
15. Qureshi SA. Beta-lactamase: an ideal reporter system for monitoring gene expression in live eukaryotic cells. BioTechniques. 2007;42(1):91–6.
16. Collinet B, Herve M, Pecorari F, Minard P, Eder O, Desmadril M. Functionally accepted insertions of proteins within protein domains. J Biol Chem. 2000; 275(23):17428–33.

17. Vandevenne M, Filee P, Scarafone N, Cloes B, Gaspard G, Yilmaz N, Dumoulin M, Francois JM, Frere JM, Galleni M. The *Bacillus licheniformis* BlaP beta-lactamase as a model protein scaffold to study the insertion of protein fragments. Protein Sci. 2007;16(10):2260–71.

18. Gersbach CA, Gaj T, Gordley RM, Barbas CF, 3rd. Directed evolution of recombinase specificity by split gene reassembly. Nucleic Acids Res 2010; 38(12):4198-206.

19. Kadonaga JT, Gautier AE, Straus DR, Charles AD, Edge MD, Knowles JR. The role of the beta-lactamase signal sequence in the secretion of proteins by *Escherichia coli*. J Biol Chem. 1984;259(4):2149–54.

20. Itoh Y, Kanoh K, Nakamura K, Takase K, Yamane K. Artificial insertion of peptides between signal peptide and mature protein: effect on secretion and processing of hybrid thermostable alpha-amylases in *Bacillus Subtilis* and *Escherichia coli* cells. J Gen Microbiol. 1990;136(8):1551–8.

21. Valens M, Penaud S, Rossignol M, Cornet F, Boccard F. Macrodomain organization of the *Escherichia coli* chromosome. EMBO J. 2004;23(21):4330–41.

22. Thiel A, Valens M, Vallet-Gely I, Espeli O, Boccard F. Long-range chromosome organization in *E. coli*: a site-specific system isolates the Ter macrodomain. PLoS Genet. 2012;8(4):e1002672.

23. Val ME, Skovgaard O, Ducos-Galand M, Bland MJ, Mazel D. Genome engineering in *Vibrio cholerae*: a feasible approach to address biological issues. PLoS Genet. 2012;8(1):e1002472.

24. Soler-Bistue A, Mondotte JA, Bland MJ, Val ME, Saleh MC, Mazel D. Genomic location of the major ribosomal protein gene locus determines *Vibrio cholerae* global growth and infectivity. PLoS Genet. 2015;11(4):e1005156.

25. Val ME, Marbouty M, de Lemos MF, Kennedy SP, Kemble H, Bland MJ, Possoz C, Koszul R, Skovgaard O, Mazel D. A checkpoint control orchestrates the replication of the two chromosomes of *Vibrio cholerae*. Sci Adv. 2016; 2(4):e1501914.

26. Sutcliffe JG. Nucleotide sequence of the ampicillin resistance gene of *Escherichia coli* plasmid pBR322. Proc Natl Acad Sci U S A. 1978;75(8):3737–41.

27. Kolot M, Malchin N, Elias A, Gritsenko N, Yagil E. Site promiscuity of coliphage HK022 integrase as tool for gene therapy. Gene Ther. 2015;22(7):602.

28. Nagaraja R, Weisberg RA. Specificity determinants in the attachment sites of bacteriophages HK022 and lambda. J Bacteriol. 1990;172(11):6540–50.

29. Demarre G, Guerout AM, Matsumoto-Mashimo C, Rowe-Magnus DA, Marliere P, Mazel D. A new family of mobilizable suicide plasmids based on broad host range R388 plasmid (IncW) and RP4 plasmid (IncPalpha) conjugative machineries and their cognate *Escherichia coli* host strains. Res Microbiol. 2005;156(2):245–55.

30. Rossignol M, Moulin L, Boccard F. Phage HK022-based integrative vectors for the insertion of genes in the chromosome of multiply marked *Escherichia coli* strains. FEMS Microbiol Lett. 2002;213(1):45–9.

31. Herrero M, de Lorenzo V, Timmis KN. Transposon vectors containing non-antibiotic resistance selection markers for cloning and stable chromosomal insertion of foreign genes in gram-negative bacteria. J Bacteriol. 1990; 172(11):6557–67.

32. Groth AC, Olivares EC, Thyagarajan B, Calos MP. A phage integrase directs efficient site-specific integration in human cells. Proc Natl Acad Sci U S A. 2000;97(11):5995–6000.

33. Brown WR, Lee NC, Xu Z, Smith MC. Serine recombinases as tools for genome engineering. Methods. 2011;53(4):372–9.

34. St-Pierre F, Cui L, Priest DG, Endy D, Dodd IB, Shearwin KE. One-step cloning and chromosomal integration of DNA. ACS Synth Biol. 2013;2(9):537–41.

35. Pradel N, Delmas J, Wu LF, Santini CL, Bonnet R. Sec- and tat-dependent translocation of beta-lactamases across the *Escherichia coli* inner membrane. Antimicrob Agents Chemother. 2009;53(1):242–8.

36. Li P, Beckwith J, Inouye H. Alteration of the amino terminus of the mature sequence of a periplasmic protein can severely affect protein export in *Escherichia coli*. Proc Natl Acad Sci U S A. 1988;85(20):7685–9.

37. Pluckthun A, Knowles JR. The consequences of stepwise deletions from the signal-processing site of beta-lactamase. J Biol Chem. 1987;262(9):3951–7.

38. Barkocy-Gallagher GA, Bassford PJ Jr. Synthesis of precursor maltose-binding protein with proline in the +1 position of the cleavage site interferes with the activity of *Escherichia coli* signal peptidase I in vivo. J Biol Chem. 1992; 267(2):1231–8.

39. Choo KH, Tan TW, Ranganathan S. SPdb–a signal peptide database. BMC Bioinf. 2005;6:249.

40. Choo KH, Ranganathan S. Flanking signal and mature peptide residues influence signal peptide cleavage. BMC Bioinf. 2008;(9, Suppl 12):S15.

41. Zacchi P, Sblattero D, Florian F, Marzari R, Bradbury AR. Selecting open reading frames from DNA. Genome Res. 2003;13(5):980–90.

42. Seehaus T, Breitling F, Dubel S, Klewinghaus I, Little M. A vector for the removal of deletion mutants from antibody libraries. Gene. 1992;114(2):235–7.

43. Colloms SD, Merrick CA, Olorunniji FJ, Stark WM, Smith MC, Osbourn A, Keasling JD, Rosser SJ. Rapid metabolic pathway assembly and modification using serine integrase site-specific recombination. Nucleic Acids Res. 2014; 42(4):e23.

44. Duportet X, Wroblewska L, Guye P, Li Y, Eyquem J, Rieders J, Rimchala T, Batt G, Weiss R. A platform for rapid prototyping of synthetic gene networks in mammalian cells. Nucleic Acids Res. 2014;42(21):13440–51.

45. Gibson DG, Young L, Chuang RY, Venter JC, Hutchison CA 3rd, Smith HO. Enzymatic assembly of DNA molecules up to several hundred kilobases. Nat Methods. 2009;6(5):343–5.

46. Zuker M. Mfold web server for nucleic acid folding and hybridization prediction. Nucleic Acids Res. 2003;31(13):3406–15.

47. Ferrieres L, Hemery G, Nham T, Guerout AM, Mazel D, Beloin C, Ghigo JM. Silent mischief: bacteriophage mu insertions contaminate products of *Escherichia coli* random mutagenesis performed using suicidal transposon delivery plasmids mobilized by broad-host-range RP4 conjugative machinery. J Bacteriol. 2010;192(24):6418–27.

48. Biskri L, Bouvier M, Guerout AM, Boisnard S, Mazel D. Comparative study of class 1 integron and *Vibrio cholerae* superintegron integrase activities. J Bacteriol. 2005;187(5):1740–50.

49. Cohen SN, Chang AC. Revised interpretation of the origin of the pSC101 plasmid. J Bacteriol. 1977;132(2):734–7.

50. Yanisch-Perron C, Vieira J, Messing J. Improved M13 phage cloning vectors and host strains: nucleotide sequences of the M13mp18 and pUC19 vectors. Gene. 1985;33(1):103–19.

51. Guzman LM, Belin D, Carson MJ, Beckwith J. Tight regulation, modulation, and high-level expression by vectors containing the arabinose PBAD promoter. J Bacteriol. 1995;177(14):4121–30.

PERMISSIONS

The contributors of this book come from diverse backgrounds, making this book a truly international effort. This book will bring forth new frontiers with its revolutionizing research information and detailed analysis of the nascent developments around the world.

We would like to thank all the contributing authors for lending their expertise to make the book truly unique. They have played a crucial role in the development of this book. Without their invaluable contributions this book wouldn't have been possible. They have made vital efforts to compile up to date information on the varied aspects of this subject to make this book a valuable addition to the collection of many professionals and students.

This book was conceptualized with the vision of imparting up-to-date information and advanced data in this field. To ensure the same, a matchless editorial board was set up. Every individual on the board went through rigorous rounds of assessment to prove their worth. After which they invested a large part of their time researching and compiling the most relevant data for our readers.

The editorial board has been involved in producing this book since its inception. They have spent rigorous hours researching and exploring the diverse topics which have resulted in the successful publishing of this book. They have passed on their knowledge of decades through this book. To expedite this challenging task, the publisher supported the team at every step. A small team of assistant editors was also appointed to further simplify the editing procedure and attain best results for the readers.

Apart from the editorial board, the designing team has also invested a significant amount of their time in understanding the subject and creating the most relevant covers. They scrutinized every image to scout for the most suitable representation of the subject and create an appropriate cover for the book.

The publishing team has been an ardent support to the editorial, designing and production team. Their endless efforts to recruit the best for this project, has resulted in the accomplishment of this book. They are a veteran in the field of academics and their pool of knowledge is as vast as their experience in printing. Their expertise and guidance has proved useful at every step. Their uncompromising quality standards have made this book an exceptional effort. Their encouragement from time to time has been an inspiration for everyone.

The publisher and the editorial board hope that this book will prove to be a valuable piece of knowledge for researchers, students, practitioners and scholars across the globe.

LIST OF CONTRIBUTORS

Yujin Cao, Tao Cheng, Guang Zhao, Mo Xian and Huizhou Liu
CAS Key Laboratory of Biobased Materials, Qingdao Institute of Bioenergy and Bioprocess Technology, Chinese Academy of Sciences, Qingdao 266101, China

Wei Niu and Jiantao Guo
Department of Chemistry, University of Nebraska-Lincoln, Lincoln, NE 68588, USA

Lin Ge, Anna Chen, Jianjun Pei, Linguo Zhao and Xianying Fang
Co-Innovation Center for Sustainable Forestry in Southern China, Nanjing Forestry University, 159 Long Pan Road, Nanjing 210037, China
College of Chemical Engineering, Nanjing Forestry University, 159 Long Pan Road, Nanjing 210037, China

Gang Ding, Zhenzhong Wang and Wei Xiao
Jiangsu Kanion Pharmaceutical Co., Ltd, 58 Haichang South Road, Lianyungang 222001, Jiangsu Province, China

Feng Tang
International centre for bamboo and rattan, 8 FuTong East Street, Beijing 100714, China

Latifur Rehman, Xiaofeng Su, Huiming Guo, Xiliang Qi and Hongmei Cheng
Biotechnology Research Institute, Chinese Academy of Agricultural Sciences, Beijing 100081, China

Wanwitoo Wanmolee
The Joint Graduate School for Energy and Environment (JGSEE), King Mongkut's University of Technology Thonburi, Prachauthit Road, Bangmod, Bangkok 10140, Thailand

Navadol Laosiripojana
The Joint Graduate School for Energy and Environment (JGSEE), King Mongkut's University of Technology Thonburi, Prachauthit Road, Bangmod, Bangkok 10140, Thailand
BIOTEC-JGSEE Integrative Biorefinery Laboratory, Innovation Cluster 2 Building, 113 Thailand Science Park, Phahonyothin Road, Khlong Luang, Pathumthani 12120, Thailand

Warasirin Sornlake and Surisa Suwannarangsee
Enzyme Technology Laboratory, National Center for Genetic Engineering and Biotechnology (BIOTEC), 113 Thailand Science Park, Phahonyothin Road, Khlong Luang, Pathumthani 12120, Thailand

Verawat Champreda
Enzyme Technology Laboratory, National Center for Genetic Engineering and Biotechnology (BIOTEC), 113 Thailand Science Park, Phahonyothin Road, Khlong Luang, Pathumthani 12120, Thailand
BIOTEC-JGSEE Integrative Biorefinery Laboratory, Innovation Cluster 2 Building, 113 Thailand Science Park, Phahonyothin Road, Khlong Luang, Pathumthani 12120, Thailand

Nakul Rattanaphan
Bioprocess Laboratory, National Center for Genetic Engineering and Biotechnology (BIOTEC), 113 Thailand Science Park, Phahonyothin Road, Khlong Luang, Pathumthani 12120, Thailand

Aiswarya Chenthamarakshan, Nayana Parambayil, Nafeesathul Miziriya, P. S. Soumya, M. S. Kiran Lakshmi, Anala Ramgopal, Anuja Dileep and Padma Nambisan
Department of Biotechnology, Cochin University of Science and Technology, Cochin-22, Kerala, India

Zhengyu Shu, Hong Lin, Shaolei Shi, Xiangduo Mu, Yanru Liu and Jianzhong Huang
National & Local United Engineering Research Center of Industrial Microbiology and Fermentation Technology, Ministry of Education, Fujian Normal University, Fuzhou 350117, China
Engineering Research Center of Industrial Microbiology, Ministry of Education, Fujian Normal University, Fuzhou 350117, China
College of Life Sciences, Fujian Normal University (Qishan campus), Fuzhou 350117, China

Yihua Zhan, Xiangyu Sun, Yingying Huang, Dean Jiang and Xiaoyan Weng
College of Life Science, Zhejiang University, Hangzhou 310058, China

Guozeng Rong
CixiAgricultural Technology Promotion Center, Cixi 315300, China

Chunxiao Hou
The Institute of Rural Development and Information Institute, Zhejiang Academy of Agricultural Sciences, Hangzhou 310021, China

Caie Wu and Gongjian Fan
Co-Innovation Centre for Sustainable Forestry in Southern China, Nanjing Forestry University, Nanjing 210037, China
College of Light Industry Science and Engineering, Nanjing Forestry University, Nanjing 210037, China

Zhebin Hua and Fuliang Cao
Co-Innovation Centre for Sustainable Forestry in Southern China, Nanjing Forestry University, Nanjing 210037, China
College of Forestry, Nanjing Forestry University, Nanjing 210037, China

Zhenxing Tang
College of Light Industry Science and Engineering, Nanjing Forestry University, Nanjing 210037, China

Sunita V. S. Bandewar
McLaughlin-Rotman Centre for Global Health, Toronto, Canada

Florence Wambugu
Africa Harvest Biotech Foundation International Inc., Nairobi, Kenya

Emma Richardson
Centre for Ethical, Social & Cultural Risk, Li Ka Shing Knowledge Institute of St. Michael's Hospital, Toronto, Canada
Clinical Epidemiology & Biostatistics Department, Faculty of Health Sciences, McMaster University, Hamilton, Canada

James V. Lavery
Centre for Ethical, Social & Cultural Risk, Li Ka Shing Knowledge Institute of St. Michael's Hospital, Toronto, Canada
Dalla Lana School of Public Health and Joint Centre for Bioethics, University of Toronto, Toronto, Canada

Song Liu
Key Laboratory of Industrial Biotechnology, Ministry of Education, School of Biotechnology, Jiangnan University, Wuxi, China

Jian Chen
Key Laboratory of Industrial Biotechnology, Ministry of Education, School of Biotechnology, Jiangnan University, Wuxi, China
National Engineering Laboratory for Cereal Fermentation Technology, Jiangnan University, Wuxi, China

Guocheng Du
Key Laboratory of Industrial Biotechnology, Ministry of Education, School of Biotechnology, Jiangnan University, Wuxi, China
Key Laboratory of Carbohydrate Chemistry and Biotechnology, Ministry of Education, School of Biotechnology, Jiangnan University, Wuxi, China

Miao Wang
School of Food Science and Technology, Jiangnan University, Wuxi 214122, China

Asuka Mukai, Aya Ichiraku and Kazuki Horikawa
Division of Bioimaging, Institute of Biomedical Sciences, Tokushima University Graduate School, 3 18-15 Kuramoto-cho, Tokushima City, Tokushima 770-8503, Japan

Xuefeng Lu
Key Laboratory of Biofuels, Shandong Provincial Key Laboratory of Energy Genetics, Qingdao Institute of Bioenergy and Bioprocess Technology, Chinese Academy of Sciences, Qingdao 266101, China

Qing Wang and Luyao Bao
Key Laboratory of Biofuels, Shandong Provincial Key Laboratory of Energy Genetics, Qingdao Institute of Bioenergy and Bioprocess Technology, Chinese Academy of Sciences, Qingdao 266101, China
University of Chinese Academy of Sciences, Beijing 100049, China

Jian-Jun Li
Key Laboratory of Biofuels, Shandong Provincial Key Laboratory of Energy Genetics, Qingdao Institute of Bioenergy and Bioprocess Technology, Chinese Academy of Sciences, Qingdao 266101, China
National Key Laboratory of Biochemical Engineering, Institute of Process Engineering, Chinese Academy of Sciences, Beijing 100190, China

Chenjun Jia
University of Chinese Academy of Sciences, Beijing 100049, China
National Laboratory of Biomacromolecules, Institute of Biophysics, Chinese Academy of Sciences, Beijing 100101, China

Mei Li
National Laboratory of Biomacromolecules, Institute of Biophysics, Chinese Academy of Sciences, Beijing 100101, China

André Pick, Barbara Beer, Jochen Schmid, and Volker Sieber
Technical University of Munich, Straubing Center of Science, Chair of Chemistry of Biogenic Resources, Schulgasse 16, 94315 Straubing, Germany

Risa Hemmi, Rena Momma and Kenji Miyamoto
Department of Biosciences and Informatics, Keio University, 3-14-1 Hiyoshi, 2238522 Yokohama, Japan

Justin E. Anderson, Jean-Michel Michno, Thomas J. Y. Kono, Adrian O. Stec, Benjamin W. Campbell, Shaun J. Curtin and Robert M. Stupar
Department of Agronomy & Plant Genetics, University of Minnesota, 1991 Upper Buford Circle, 411 Borlaug Hall, St. Paul MN 55108, USA

Dede Abdulrachman and Antonius Suwanto
Faculty of Biotechnology, Atmajaya Catholic University, Jl. Jend. Sudirman 51, Jakarta 12930, Indonesia

Paweena Thongkred, Kanokarn Kocharin, Verawat Champreda, Lily Eurwilaichitr, Thidarat Nimchua and Duriya Chantasingh
Microbial Biotechnology and Biochemicals Research Unit, National Center for Genetic Engineering and Biotechnology, 113 Thailand Science Park, Pahonyothin Rd, Khlong Luang, Patumthani 12120, Thailand

Monthon Nakpathom, Buppha Somboon and Nootsara Narumol
Textile Laboratory, Polymers Research Unit, National Metal and Materials Technology Center, 114 Thailand Science Park, Pahonyothin Rd, Khlong Luang, Patumthani 12120, Thailand

Nick van Biezen, Linda Peters, Eric van de Zilver, Nicole Jacobs-van Dreumel and Christien Lokman
HAN BioCentre, University of Applied Sciences, P.O. Box 6960, 6503 GL Nijmegen, The Netherlands

Dennis Lamers
HAN BioCentre, University of Applied Sciences, P.O. Box 6960, 6503 GL Nijmegen, The Netherlands
Bioprocess Engineering, Wageningen University and Research Centre, P.O. Box 8129, 6700 EV Wageningen, The Netherlands

Dirk Martens
Bioprocess Engineering, Wageningen University and Research Centre, P.O. Box 8129, 6700 EV Wageningen, The Netherlands

René H. Wijffels
Bioprocess Engineering, Wageningen University and Research Centre, P.O. Box 8129, 6700 EV Wageningen, The Netherlands
University of Nordland, Faculty of Biosciences and Aquaculture, N-8049 Bodø, Norway

Xinzhe Wang, Huihua Ge, Dandan Zhang, Shuyu Wu and Guangya Zhang
Fujian Provincial Key Laboratory of Biochemical Technology, Huaqiao University, Xiamen, Fujian 361021, China

Qiu Cui
Shandong Provincial Key Laboratory of Energy Genetics, Qingdao Institute of Bioenergy and Bioprocess Technology, Chinese Academy of Sciences, No.189 Songling Road, Laoshan District, Qingdao, Shandong Province 266101, China
Key Laboratory of Biofuels, Qingdao Institute of Bioenergy and Bioprocess Technology, Chinese Academy of Sciences, Qingdao, Shandong 266101, China
Qingdao Engineering Laboratory of Single Cell Oil, Qingdao, Shandong 266101, China

Yingang Feng and Xiaojin Song
Shandong Provincial Key Laboratory of Energy Genetics, Qingdao Institute of Bioenergy and Bioprocess Technology, Chinese Academy of Sciences, No.189 Songling Road, Laoshan District, Qingdao, Shandong Province 266101, China
Qingdao Engineering Laboratory of Single Cell Oil, Qingdao, Shandong 266101, China

Huidan Zhang
Shandong Provincial Key Laboratory of Energy Genetics, Qingdao Institute of Bioenergy and Bioprocess Technology, Chinese Academy of Sciences, No.189 Songling Road, Laoshan District, Qingdao, Shandong Province 266101, China
Qingdao Engineering Laboratory of Single Cell Oil, Qingdao, Shandong 266101, China
University of Chinese Academy of Sciences, Beijing 100049, China

Maniraj Rathinam, Shweta Singh, Debasis Pattanayak and Rohini Sreevathsa
ICAR-National Research Centre on Plant Biotechnology, LBS Centre, Pusa Campus, New Delhi 110012, India

Yu-Fan Liu
Division of Allergy, Department of Pediatrics, Chung-Shan Medical University Hospital, Taichung, Taiwan
Department of Biomedical Sciences, College of Medicine Sciences and Technology, Chung Shan Medical University, Taichung, Taiwan

Chia-Wen Hsieh, Yao-Sheng Chang and Being-Sun Wung
Department of Microbiology, Immunology and Biopharmaceuticals, National Chiayi University, Chiayi, Taiwan

Zhonghui Li and Danke Xu
State Key Laboratory of Analytical Chemistry for Life Science, School of Chemistry and Chemical Engineering, Nanjing University, Nanjing 210093, China

Zhoumin Li
State Key Laboratory of Analytical Chemistry for Life Science, School of Chemistry and Chemical Engineering, Nanjing University, Nanjing 210093, China
School of Chemistry and Biological Science, Nanjing University Jingling College, Nanjing 210089, China

Fang Wen and Nan Zheng
Ministry of Agriculture-Key Laboratory of Quality & Safety Control for Milk and Dairy Products, Institute of Animal Science, Chinese Academy of Agricultural Sciences, Beijing 100193, People's Republic of China

Jindou Jiang
Ministry of Agriculture Dairy Quality Supervision and Testing Center, Harbin 150090, China

Michael J. Bland, Magaly Ducos-Galand, Marie-Eve Val and Didier Mazel
Unité Plasticité du Génome Bactérien, Département Génomes et Génétique, Institut Pasteur, 75015 Paris, France
UMR3525, Centre National de la Recherche Scientifique, 75015 Paris, France

Index